Vincent Laude

Phononic Crystals

De Gruyter Studies in Mathematical Physics

Volume 26

Vincent Laude

Phononic Crystals

Artificial Crystals for Sonic, Acoustic, and Elastic Waves

DE GRUYTER

Physics and Astronomy Classification Scheme 2010
02.70.Dh, 43.35.+d, 43.40.+s, 46.15.Cc, 46.40.Cd, 62.30.+d, 63.20.-e, 63.20.D, 63.20.Pw, 77.65.-j

Author
Dr. Vincent Laude
Centre National de la Recherche Scientifique
Institut FEMTO-ST
15B Avenue des Montboucons
25030 Besançon Cedex
France
vincent.laude@femto-st.fr

ISBN 978-3-11-030265-3
e-ISBN (PDF) 978-3-11-030266-0
e-ISBN (EPUB) 978-3-11-038791-9
Set-ISBN 978-3-11-030267-7
ISSN 2194-3532

Library of Congress Cataloging-in-Publication Data
A CIP catalog record for this book has been applied for at the Library of Congress.

Bibliographic information published by the Deutsche Nationalbibliothek
The Deutsche Nationalbibliothek lists this publication in the Deutsche Nationalbibliografie; detailed bibliographic data are available on the Internet at http://dnb.dnb.de.

Typesetting: PTP-Berlin, Protago-TEX-Production GmbH
Printing and binding: CPI books GmbH, Leck
♾ Printed on acid-free paper
Printed in Germany

www.degruyter.com

Preface

Phononic crystals are artificial periodic structures that can efficiently alter the flow of sound, acoustic waves, or elastic waves in solids. They were introduced about twenty years ago and have gained increasing interest since then, both because of their amazing physical properties and because of their potential applications. The topic of phononic crystals stands at the crossroad of physics (condensed matter physics, wave propagation in inhomogeneous and periodic media) and engineering (acoustics, ultrasonics, mechanical engineering, or electrical engineering). Phononic crystals cover a wide range of scales, from meter-size periodic structures for sound in air to nanometer-size structures for information processing or thermal phonon control in integrated circuits. I have tried to reflect this variety in the present book.

Phononic crystals have a definite relation with the topic of photonic crystals in optics. The marriage of phononic and photonic crystals also provides a promising structural basis for enhanced sound and light interaction in the context of optomechanics or of phoxonic crystals. Such interactions are not treated in the present book because the field is still new and rapidly evolving. Depending on how this book on phononic crystals is received, it may be a subsequent book project for the author.

Since the topic of phononic crystal is getting popular, it is nowadays presented and discussed at various international conferences. After the first ten years (1993–2004, approximately) during which the topic remained mainly theoretical with a few proof-of-concept demonstrations appearing in the literature, the latest evolution has been towards applications (including technology development), instrumentation (e.g. "seeing" phonons with optical systems), and novel designs (negative refraction, spatial dispersion control). The physical explanations for various effects are now quite well understood and efficient numerical methods and analysis tools have been developed. These are the reasons why the field has become mature for books. Other books that have appeared recently are general contributions, and although they are collections written by authoritative authors who are also current researchers in the field, they somehow lack the homogeneity of a monograph. The present book was written single handedly, which is both its strength and its weakness. Though it probably contains many mistakes and misses certain developments and contributions, it can but only reflect the sincere knowledge of its author. To all distinguished colleagues, collaborators, and often friends, I wish to present my apologies for any omissions in my text.

As a researcher, I first entered the field of phononic crystals sometime around 2002. Before that, I had been a researcher in optical information processing and wave propagation for a few years. The topic of elastic waves in solids is one I am particularly fond of, because of its formal perfection and simplicity when written in the mathematical language of tensors. The reader will probably identify many traces of this inclination toward applied mathematics in the text. This is especially true of the choice I have made of illustrating most numerical examples by results obtained using the finite element method. Part of the book may be used as an introduction to the Galerkin finite element method as applied to the solution of problems in the physics of wave propagation.

The level at which this monograph aims is rather advanced, at the master level or above. The author is a full-time researcher and has extensive experience in training post-graduate students. It is doubtful that lectures will be given focusing only on the topic of phononic crystals. The subject, however, is of interest to many scientists nowadays and I thought a global reference was missing.

This book is dedicated to those individuals who have had most influence on my scientific life so far. There have been many and I do not wish to enter into the embarrassing process of citing all their names. As an exception to this rule, I wish to express special thanks to Dr Abdelkrim Khelif and to Dr Sarah Benchabane with whom I have been researching phononic crystals for more than ten years. Finally, I am wholeheartedly grateful to my wife and family for their patience and support during all the writing of this book.

Besançon, March 31, 2015 Vincent Laude

To Jean-Pierre Laude, for whom this book represents most after myself

Contents

1 Introduction

Phononic crystals were officially born in 1993 after a famous paper by Kushwaha and co-authors [74] drawing the analogy to acoustic waves of the photonic crystal concept for light, which itself had been inspired by the theory of electrons in crystal lattices. Of course this statement is a simplification, and deciding that phononic crystals were born at a certain time is just putting a name on an idea that was ready to start living its own life as it detached from the tree of the physics of waves in periodic media. Over the last twenty years, the phononic concept has developed gracefully. It has been researched by many authors and has reached maturity in the sense that it became ready to be the subject of a monograph. The purpose of this book is to propose a self-contained presentation of the topic, going from the basics of the physics of waves upto the description of potential applications.

Preliminary comments. The topic of phononic crystals is clearly a branch of physical acoustics, and even more so a branch of the physics of waves. For completeness, in Chapter 2 we have gathered a synthetic presentation of many concepts of wave propagation in periodic media. The material in that chapter is elementary and independent from the concepts and the specific equations of physical acoustics. The presentation is limited to scalar waves, but introduces some important and fruitful tools such as Bloch's theorem, the band structure, evanescent waves, basic mechanisms for band gap formation, the classification of artificial crystals according to their Bravais lattices, and primitive cells and the first Brillouin zone. All these are used abundantly in the subsequent chapters.

The materials inside which waves will be considered to propagate can be broadly classified as either fluids (liquids, gases) or solids (natural crystals, amorphous solids, metals, piezoelectric solids, etc.). A note on terminology is needed here. Although the physical concepts are globally shared, the wave equations in fluids and solids differ mathematically. We will speak of acoustic waves as the pressure waves propagating in fluids; they can be described by a time-dependent scalar field in three-dimensional space, which we denote $p(t, \mathbf{x})$. In Chapter 3 we have gathered a self-consistent presentation of linear acoustic waves in fluids, including water and air. In the case of solids, elastic waves are described most conveniently by a vector field of particle displacements, which we denote $\mathbf{u}(t, \mathbf{x})$. The implication is that elastic waves have a polarization and that even homogeneous solids can be naturally anisotropic, including the important case of natural crystals. We shall try to reserve the adjective "acoustic" for pressure waves, but the practice in the literature is very often to use it for elastic waves in solids as well. Chapter 5 summarizes the topic of elastic waves in homogeneous solids, introducing such important concepts as anisotropy, piezoelectricity, surface elastic waves, and plate waves.

As the material in Chapters 2, 3, and 5 is not specific to phononic crystals, those chapters may be skipped by the knowledgeable reader. Please note, however, that frequent or implicit reference is made to their contents in the other chapters.

Artificial crystals. By artificial crystals we mean periodic material systems that were made by technology or craft. The adjective artificial is used to differentiate them from natural crystals, which are the topic of solid state physics and are the support of phonons – the particles of sound in matter. Artificial crystals usually have a periodicity measured by a length, the lattice constant, that is much larger than the interatomic distances in natural crystals. For this reason they can be treated macroscopically using the equations of continuum mechanics. Still elastic waves in solids can be considered as the asymptotic limit of acoustic phonons, which is a reason for the choice of the name "phononic crystal".

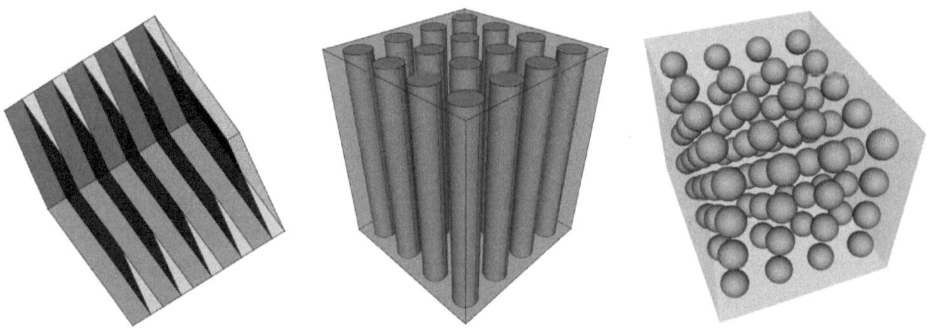

Fig. 1.1: Artificial crystals for waves in 1D, 2D, and 3D. The crystals depicted are composed of a periodic repetition of a certain distribution of solid or fluid matter, or possibly voids. The smallest volume that can be used to pave all space by periodicity is a primitive cell of the crystal. Material distribution is implicitly assumed to be continuous within the different regions but discontinuous at their interfaces.

The phononic crystal appeared around 1993, as we said earlier, and is an artificial crystal for elastic waves in solids. It is typically composed of a periodic alternation of at least two different materials, or a periodic heterostructure, as depicted in Figure 1.1. Depending on the number of periodicities, we shall classify them as one-dimensional (1D), two-dimensional (2D), or three-dimensional (3D) phononic crystals. Of course, it is perfectly possible to consider in theory phononic crystals defined from continuous variations of elastic parameters (mass density and elastic constants), but practical realizations almost always involve discontinuous regions of different materials. It is also perfectly possible to replace one material with holes, as the surface boundary conditions at the surface of the holes will provide the necessary scattering properties. As a note for readers interested in photonic crystals, a vacuum does not support elastic wave propagation, in contrast to optical waves in free space; this is a major difference between phononic and photonic crystals. Another difference is the polarization of elastic waves, which admit three independent modes of propagation in general, whereas there are two independent optical modes of propagation in dielectric materials. Phononic crystals are first introduced in Chapter 6.

Sonic crystals are artificial crystals for pressure waves in fluids. In the literature, the terms phononic crystal and sonic crystal are often used indifferently to encompass all cases of artificial crystals for mechanical waves. For clarity, we shall generally imply more restrictive and separate meanings: sonic crystals for acoustic or pressure waves, and phononic crystals for elastic waves. Sonic crystals are first introduced in Chapter 4 and support only one mode of propagation. They can be composed of a periodic distribution of different fluids, like air bubbles in water, but true voids or holes will not be permitted in practice. By extension, the very common case of solid inclusions in a fluid matrix will also be referred to as a sonic crystal. Indeed, although all three propagation modes of elastic waves exist in the solid inclusions, measurements of the properties of fluid-solid crystals are made via pressure waves. For exactly the inverse reason, fluid inclusions in a solid matrix will be termed a phononic crystal. Coupling of elastic and acoustic waves, and some associated effects, are discussed in Chapter 8.

As artificial crystals, sonic and phononic crystals owe a lot to condensed matter physics and borrow concepts that have been used to describe electronic and photonic band structures, as well as phonons in crystal lattices. The concept of Bloch waves – the natural modes of periodic media – is ubiquitous. Their dispersion properties are presented in diagrams called band structures, plotted with reference to first Brillouin zones, as Chapter 2 explains. Maybe the most salient property of artificial crystals is the existence of band gaps, or frequency ranges inside which propagating Bloch waves are absent whatever the wavenumber and thus propagation is not allowed. A partial band gap is valid for at least one direction in k-space, but not necessarily for others. A complete band gap is valid for all values of the wavevector $\mathbf{k}$, that is for any direction of propagation, but possibly for only one polarization mode. A full band gap is a complete band gap that is valid for all polarization modes and thus for all possible waves in the crystal. The distinction between full and complete band gaps is only meaningful for phononic crystals, both definitions being equivalent for sonic crystals. As a note, the distinction between full and complete band gap is not generally made in the literature, but we shall try to stick to it in this book for clarity. As an example, Figure 1.2 displays the band structure of a 2D square-lattice sonic crystal of steel rods in water, together with the corresponding first Brillouin zone. The diameter of the rods has been chosen such that a complete – or full in this case – band gap appears.

Mechanisms for band gap formation. It is perfectly possible to go a long way with sonic and phononic crystals with just numerical simulations, solving the acoustic or elastodynamic equations to obtain Bloch waves and to plot phononic band structures,

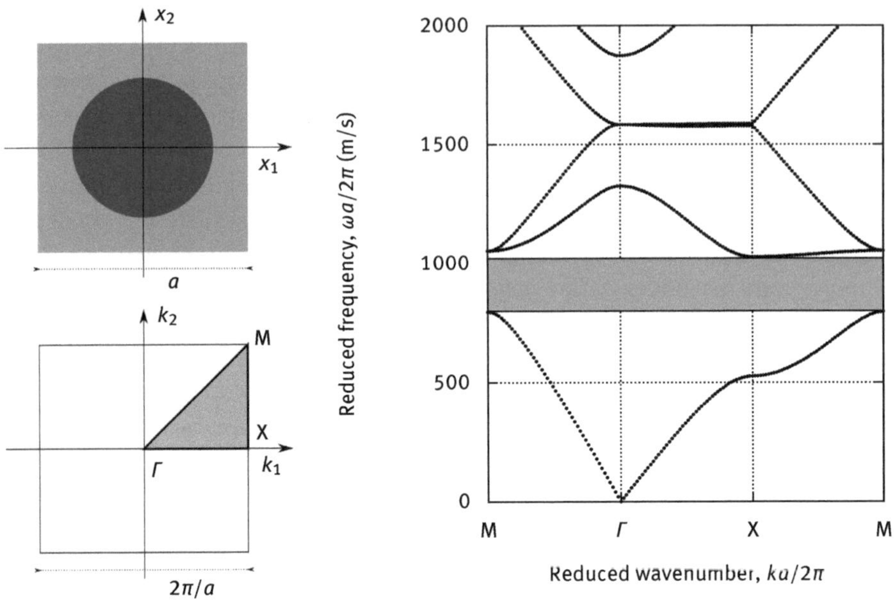

Fig. 1.2: Band structure example for a 2D square-lattice sonic crystal. The sonic crystal is composed of cylindrical steel rods in water. The ratio of the rod diameter to the lattice constant a is 0.83. The first Brillouin zone is sketched at the bottom left. For the square lattice, the first Brillouin zone is also a square whose high symmetry points are noted Γ, X, and M. The irreducible first Brillouin zone is further defined with respect to all symmetries of the unit cell and shaded in gray. The band structure on the right shows the dispersion of propagating Bloch waves as the relation between wavenumber k and angular frequency ω. It is plotted along the boundary of the irreducible first Brillouin zone as path [Γ–X–M–Γ]. The area shaded in gray in the band structure indicates a complete – or full, in the sonic crystal case, – band gap.

or computing the transmission through a finite crystal to locate frequency band gaps. Beyond numerical computation, however, an important question is the physical origin for the opening of a band gap. The most common and the historically first invoked physical mechanism is Bragg interference. Bragg scattering occurs at every periodic plane of scatterers inside a crystal. Bragg band gaps can be created anytime a forward and a backward propagating Bloch wave are phase matched, i.e. when they would have degenerated in the band structure space $(\mathbf{k}, \omega)$ if they did not couple. In the limit of a vanishingly small material contrast, such band crossings occur in the empty lattice model at every high symmetry point of the first Brillouin zone. This is the phenomenological reason why Bragg band gaps appear at symmetry points in reciprocal space, and hence along the irreducible Brillouin zone. In Figure 1.2, band foldings occur at points Γ, X, and M of the first Brillouin zone. The author however humbly recognizes he does not know of a mathematical demonstration of this property in the general case of medium or strong material contrast.

A second possible physical mechanism for band gap formation is the existence of a local resonance in the internals of the unit cell of the crystal [89]. In this case, there is possibility for part of the incoming wave energy to be trapped for some time in the resonator before it is radiated away to all permitted propagation modes. The local resonance principle is discussed in Chapter 10 and can be viewed as a classical analog of the Fano resonance, i.e. as the coupling of a continuum of propagating waves with a localized resonant state.

There are quite different types of phononic crystals that have been considered in the literature. On the one hand, when inspired by photonic crystals, or because of technological constraints when going to the microscale, the inclusions in the unit cell are quite often limited to simple shapes such as cylinders or spheres. In such conditions the possible geometries are rather limited and the engineering efforts to design artificial crystals are mostly directed to material constant contrast. On the other hand, when the lattice constant is not too small, there may be enough room for structural design with small-scale details inside the unit cell. Both types of phononic crystals, based on material contrast or structural design, can sustain the two mechanisms of band gap formation we discussed above.

Whenever a band gap is present, it is traditional to say that no wave can propagate through the crystal. Though this statement is perfectly correct, it says nothing about the other waves that can exist at frequencies within a band gap range. Band gaps are actually the realm of evanescent Bloch waves. Evanescent waves are plane waves whose wavenumber is complex valued. The imaginary part of the wavenumber is especially significant as it describes the exponential decay or amplification of the evanescent wave. The transmission through a finite length crystal at frequencies within a band gap range is governed by exponentially decaying evanescent waves. Evanescent Bloch waves and the complex band structure are introduced in Chapter 9.

Applications of phononic crystals. The potential applications of phononic and sonic crystals cover the full range of microsonics, ultrasonics, and audible frequencies. At the small scale, when the lattice constant is of the order of one micron, phononic crystals can be combined with surface acoustic wave technology (SAW), thin-film bulk acoustic resonators, and with microelectromechanical systems (MEMS). Chapter 7 introduces the fundamentals of surface and Lamb wave phononic crystals.

Full band gaps, when they exist, in principle offer a very convenient platform for designing functionalities such as mirrors, cavities, and waveguides. Indeed, adding a defect in an otherwise perfectly periodic crystal is a powerful way to localize the flow of acoustic or elastic energy. This idea is exemplified in many different instances in Chapter 11.

Sonic and phononic crystals are strongly dispersive, meaning that the conditions for propagation of waves can change drastically with only small changes in the frequency. The effects of spatial and temporal dispersion, the exploitation of crystal anisotropy to achieve positive or negative refraction, and the variations of the group velocity are discussed in Chapter 12.

This book is also about numerical simulation. One very strong motivation during the writing of the book was to offer the reader a rather complete knowledge to perform numerical simulations in most situations that occur with sonic and phononic crystal. Overall, the book can also be viewed as a practical guide to numerical simulation of sonic and phononic crystals, especially with the finite element method (FEM) and in view of obtaining Bloch waves and band structures. There is also a presentation of using FEM methods to solve problems in acoustic wave propagation and scattering, and in elastic wave propagation in bulk anisotropic and piezoelectric media, at surfaces, and in plates. The choice was made to insist on some simple variational formulations that describe the physics of wave propagation while leaving the technical details of mathematical solving to an FEM code and the technical details of domain meshing to specialized software. The examples in the book have been obtained with the open source FEM software FreeFem++ [48]. Meshing of 3D unit cells was performed with the open source software gmsh [41]. A full explanation of the plane wave expansion (PWE) method is also provided, with full consideration of anisotropic materials and piezoelectricity.

This book is not about... some possibly important topics! The extent of every subject has to be bounded by some arbitrary choices. Some topics that have a close connection to phononic crystals have not been included in our presentation, such as acoustic metamaterials, wave propagation in periodic structures of applied mechanics, structural optimization of phononic crystals, thermal phonon control, phoxonic crystals and optomechanics. A short introduction to some of these topics is given in the general conclusion.

Part I: **Acoustic waves in sonic crystals**

Tout estoit sans beauté, sans reglement, sans flamme.
Tout estoit sans façon, sans mouvement, sans ame :
Le feu n'estoit point feu, la mer n'estoit point mer,
La terre n'estoit terre, et l'air n'estoit point air :
Ou si ja se pouvoit trouver en un tel monde,
Le corps de l'air, du feu, de la terre, et de l'onde :
L'air estoit sans clarté, la flamme sans ardeur,
Sans fermeté la terre, et l'onde sans froideur.

Guillaume de Salluste du Bartas

2 Scalar waves in periodic media

This chapter is devoted to presenting the physics of waves in periodic media that are devoid of further complications such as polarization, anisotropy, tensors, loss, etc. These various physical concepts and their influence will be introduced in subsequent chapters. This chapter is thus a self-contained presentation of periodic media for scalar waves. Nevertheless, it is probably the most dense and mathematically difficult chapter of this book, introducing many concepts in a synthetic presentation. We encourage the reader who finds some derivations difficult to seek further examples and illustrations in subsequent chapters, or to consult more introductory or comprehensive texts.

2.1 Scalar waves in homogeneous media

We start with an introduction to scalar wave propagation in homogeneous media. The scalar wave equation and the scalar Helmholtz equation are introduced.

2.1.1 One-dimensional wave propagation

A wave is generally speaking a perturbation of the equilibrium state of some medium that propagates both in space and time. Let us consider functions $u(t, x)$ of time t and of a single spatial coordinate x. A scalar wave equation is a partial differential equation of the form

$$\frac{\partial^2 u}{\partial t^2} - c^2 \frac{\partial^2 u}{\partial x^2} = 0. \tag{2.1}$$

The celerity c has the units of a velocity, m/s. The constant c^2 in the equation is independent of position x, as expected for a homogeneous medium: material properties do not depend on the particular position of space where the equation is evaluated. For mechanical waves, the wave equation results from the balance of kinetic and potential energy: the first variation of the Lagrangian is zero, which is known as Hamilton's principle (the Lagrangian is the difference of kinetic minus potential energy, $\mathcal{L} = E_k - E_p$; its first variation is written $\delta\mathcal{L} = 0$). For optical or electromagnetic waves, the wave equation can be derived from Maxwell's equations.

It is peculiar to the one-dimensional case that a general solution to the wave equation is known analytically. Indeed, it can be easily verified that the general solution is given by

$$u(t, x) = F(t - x/c) + G(t + x/c), \tag{2.2}$$

where F and G are arbitrary functions of one variable that are at least twice differentiable. The functions F and G are independent and represent a wave traveling to

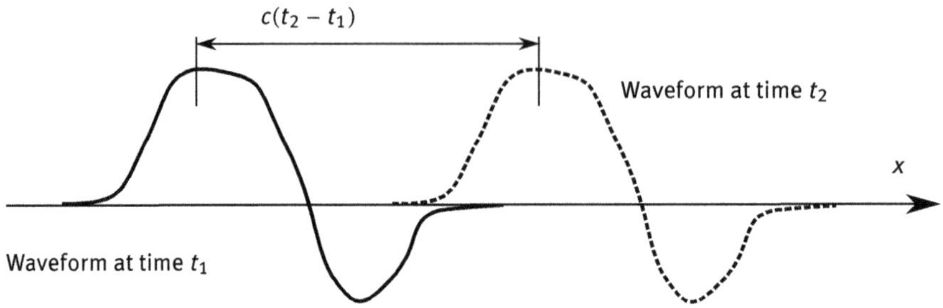

Fig. 2.1: Schematic representation of a general solution of the one-dimensional wave equation. Only the right-propagating part of the solution, or waveform $F(t - x/c)$ in (2.2), is shown. The time and ordinate axes are essentially the same in one-dimensional space.

the right and to the left, respectively. As an example, let us consider the vibration $F(t) = \sin(\omega t)$, where $\omega = 2\pi f$ is the angular frequency and f is the frequency (units: Hz). Then the part of the wave traveling to the right is $u(t, x) = \sin(\omega t - kx)$, with $k = \omega/c$ the wavenumber (units: rad/m).

Figure 2.1 depicts one possible waveform, a solution of the one-dimensional scalar wave equation. Only the right-propagating part is shown, for simplicity. The arbitrarily shaped waveform propagates without any change, it is simply translating to the right along the x axis as time increases, with velocity c.

Plane wave spectrum. Every sufficiently regular function of one variable $F(t)$ has a Fourier transform $\bar{F}(\omega)$ and reciprocally

$$F(t) = \frac{1}{2\pi} \int_{-\infty}^{+\infty} \mathrm{d}\omega \bar{F}(\omega) \exp(\imath \omega t) \; ; \quad \bar{F}(\omega) = \int_{-\infty}^{+\infty} \mathrm{d}t F(t) \exp(-\imath \omega t). \tag{2.3}$$

We refer the reader to specialized books on Fourier transforms for their detailed definition, properties, and extension to distributions. Please note the multiplicative factor $(2\pi)^{-1}$ in the inverse Fourier transform formula which is not present in the direct formula. Because the general solution to the one-dimensional wave equation has the very simple form given by (2.2), with time and space variables playing essentially a similar role, we are led to the following property.

Property 2.1 (*Plane wave spectrum*). The plane wave spectrum of a solution of the one-dimensional wave equation is

$$u(t, x) = \frac{1}{2\pi} \int_{-\infty}^{+\infty} \mathrm{d}\omega \bar{F}(\omega) \exp(\imath(\omega t - k(\omega)x)) \tag{2.4}$$

with $k(\omega) = \omega/c$. The left-propagating part of the solution admits a similar plane wave spectrum with $\bar{G}$ instead of $\bar{F}$ and with $k(\omega) = -\omega/c$. $k^2(\omega) = \omega^2/c^2$ is a dispersion relation gathering both the right- and the left-propagating solutions.

Proof. Let us consider only the right-propagating part of a general solution to the 1D wave equation, so that $G = 0$ in (2.2). $F(t)$ admits a Fourier transform $\bar{F}(\omega)$. Changing variables from t to $t - x/c$ in (2.3) leads immediately to (2.4). □

The significance of the plane wave spectrum in this case is that the spectral contents, represented by function $\bar{F}(\omega)$, are not changed upon propagation. Furthermore, for a constant celerity c, all frequency components of the wave travel with the same speed. This situation is typical of nondispersive media.

Dispersion. Whenever the celerity is dispersive, i.e. when it depends on frequency, the dispersion relation $k(\omega) = \pm\omega/c(\omega)$ is no longer linear. In the dispersion diagram obtained by plotting ω as a function of k, the dispersion relation no longer defines straight lines (see Figure 2.2). Relation (2.2) is no longer valid, but the plane wave spectrum in (2.4) remains valid, though we need to refine the wave propagation problem a bit by introducing the Helmholtz equation.

Proof. To prove that the plane wave spectrum property still holds in the presence of dispersion, we first consider the Fourier transform of (2.1), with c first considered a constant, and write

$$-\frac{\partial^2 \bar{u}(\omega, x)}{\partial x^2} = \left(\frac{\omega}{c}\right)^2 \bar{u}(\omega, x).$$

This is a Helmholtz equation for $\bar{u}(\omega, x)$ where ω is now only a parameter. Hence, we can let the celerity depend explicitly on frequency and write

$$-\frac{\partial^2 \bar{u}(\omega, x)}{\partial x^2} = \left(\frac{\omega}{c(\omega)}\right)^2 \bar{u}(\omega, x).$$

Thus, every frequency component of the wave travels at a different speed. At this point, we remark that, for the sake of considering dispersion, we have implicitly changed the definition of the wave equation, since an inverse Fourier transform of the last equation would not lead back to (2.1).

Next we define the temporal and spatial Fourier transform as

$$\hat{u}(\omega, k) = \int_{-\infty}^{+\infty} dt \int_{-\infty}^{+\infty} dx\, u(t, x) \exp(-\imath\omega t) \exp(+\imath kx). \tag{2.5}$$

The choice of sign for the spatial Fourier transform is of pure convenience. $\hat{u}$ satisfies

$$(k^2 - \omega^2/c(\omega)^2)\, \hat{u}(\omega, k) = 0.$$

Thus it has nonvanishing values only when the dispersion relation applies and we can write

$$\hat{u}(\omega, k) = \delta(k - \omega/c(\omega))\bar{F}(\omega) + \delta(k + \omega/c(\omega))\bar{G}(\omega)$$

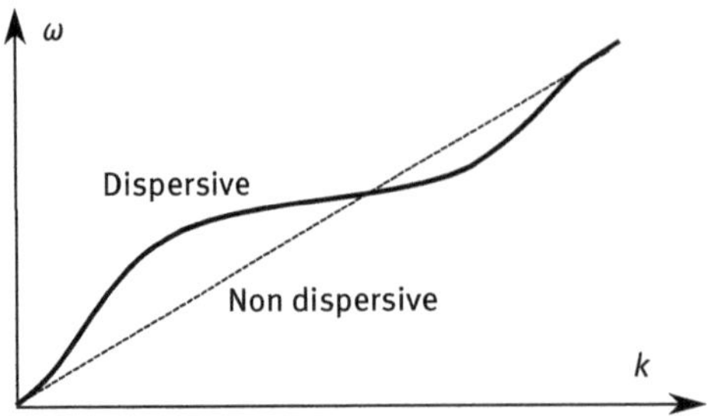

Fig. 2.2: Dispersion relation for dispersive media. The dispersion relation is conventionally presented as angular frequency ω as a function of wavenumber k. Propagation is dispersive whenever the functional relation between angular frequency and wavenumber is nonlinear. Note that we could equally well have presented k as a function of ω.

with two arbitrary functions $\bar{F}$ and $\bar{G}$. Considering the inverse temporal and spatial Fourier transform

$$u(t,x) = \frac{1}{(2\pi)^2} \int_{-\infty}^{+\infty} \mathrm{d}\omega \int_{-\infty}^{+\infty} \mathrm{d}k\, \bar{u}(\omega,k) \exp(\imath\omega t) \exp(-\imath kx) \tag{2.6}$$

and using the properties of delta functions, we are readily led back to (2.4). □

Given a general explicit dispersion relation $k(\omega)$, the phase velocity is defined by

$$v(\omega) = \omega/k(\omega). \tag{2.7}$$

The slowness is the inverse of the phase velocity, $s(\omega) = 1/v(\omega)$ (units: s/m). The group velocity is by definition

$$v_g(\omega) = \frac{\mathrm{d}\omega}{\mathrm{d}k} = \left(\frac{\mathrm{d}k(\omega)}{\mathrm{d}\omega}\right)^{-1}. \tag{2.8}$$

Both the phase velocity and the group velocity have units of m/s; they differ in dispersive media but are equal in nondispersive media. Note that the group velocity defined here is of temporal origin and is a measure of dispersion; later we will introduce a spatial group velocity that is a measure of anisotropy.

Deformation of a pulse subject to dispersion. The spectral phase acquired along a given propagation length L is $\phi(\omega) = k(\omega)L$, as (2.4) implies. The time it takes for the wave to travel this length, in an energetic sense, is known to be given by the group delay

$$t_g(\omega) = \frac{\mathrm{d}\phi}{\mathrm{d}\omega} = \frac{L}{v_g(\omega)}. \tag{2.9}$$

Let us specify the meaning of such a statement by expressing the following property.

Property 2.2 (*Mean arrival time*). The mean arrival time of a wavepacket is the propagation distance divided by the group velocity averaged over its spectral contents, or the relation

$$\int_{-\infty}^{+\infty} \mathrm{d}t\; t|u(t,x)|^2 = \frac{1}{2\pi}\int_{-\infty}^{+\infty} \mathrm{d}\omega\; \frac{x}{v_g(\omega)}|\bar{F}(\omega)|^2, \tag{2.10}$$

with the hypothesis that the wavepacket is centered on $t = 0$ at $x = 0$. The energy is assumed normalized according to

$$\int_{-\infty}^{+\infty} \mathrm{d}t\; |u(t,x)|^2 = \frac{1}{2\pi}\int_{-\infty}^{+\infty} \mathrm{d}\omega\; |\bar{F}(\omega)|^2 = 1, \tag{2.11}$$

where the first equality is Parseval's theorem.

Proof. Consider the plane wave spectrum in (2.4). $\bar{F}(\omega)\exp(-\imath k(\omega)x)$ is the Fourier transform of $u(t,x)$, or

$$\bar{F}(\omega)\exp(-\imath k(\omega)x) = \int_{-\infty}^{+\infty} \mathrm{d}t\, u(t,x)\exp(-\imath\omega t).$$

We separate the spectral amplitude (which is a complex-valued function of frequency) into its modulus and its phase (i.e. two real-valued functions of frequency), according to

$$\bar{F}(\omega)\exp(-\imath k(\omega)x) = A(\omega)\exp(-\imath\phi(\omega)),$$

where $\phi(\omega) = \psi(\omega) + k(\omega)x$ is the sum of an initial phase ψ, accounting for the possible inherent dispersion of the wavepacket at $x = 0$, and of the dispersion arising from propagation over distance x, $k(\omega)x$.

Differentiating with respect to ω yields

$$(A'(\omega) - \imath\phi'(\omega)A(\omega))\exp(-\imath\phi(\omega)) = -\imath\int_{-\infty}^{+\infty} \mathrm{d}t\, tu(t,x)\exp(-\imath\omega t).$$

Multiplying by $A(\omega)\exp(\imath\phi(\omega))$ then leads to

$$A'(\omega)A(\omega) - \imath\phi'(\omega)A(\omega)^2 = -\imath\int_{-\infty}^{+\infty} \mathrm{d}t\, tu(t,x)\int_{-\infty}^{+\infty} \mathrm{d}t' u^*(t',x)\exp(-\imath\omega(t-t')).$$

The final step is the spectral integration $\frac{1}{2\pi}\int_{-\infty}^{+\infty}\mathrm{d}\omega(\ldots)$. Integration of the left-hand side leads to

$$\frac{1}{2\pi}\left[\frac{1}{2}A(\omega)^2\right]_{-\infty}^{+\infty} - \imath\frac{1}{2\pi}\int_{-\infty}^{+\infty} \mathrm{d}\omega\; \phi'(\omega)A(\omega)^2.$$

For a well-behaved wave, the spectrum vanishes at infinity, so that the first term is zero. Integration of the right-hand side leads to

$$-\imath \int_{-\infty}^{+\infty} \mathrm{d}t\, t u(t,x) \int_{-\infty}^{+\infty} \mathrm{d}t' u^*(t',x) \frac{1}{2\pi} \int_{-\infty}^{+\infty} \mathrm{d}\omega \exp(-\imath\omega(t-t')).$$

The last integral is the delta function

$$\delta(t-t') = \frac{1}{2\pi} \int_{-\infty}^{+\infty} \mathrm{d}\omega \exp(-\imath\omega(t-t')),$$

so that the triple integral reduces to

$$-\imath \int_{-\infty}^{+\infty} \mathrm{d}t\, t|u(t,x)|^2.$$

Gathering the results we obtain the result that

$$\frac{1}{2\pi} \int_{-\infty}^{+\infty} \mathrm{d}\omega\, \phi'(\omega) A(\omega)^2 = \int_{-\infty}^{+\infty} \mathrm{d}t\, t|u(t,x)|^2.$$

We have the group delay time relation $\phi'(\omega) = \psi'(\omega) + k'(\omega)x$. The hypothesis that the wavepacket is centered on $t = 0$ at $x = 0$ implies that

$$\frac{1}{2\pi} \int_{-\infty}^{+\infty} \mathrm{d}\omega\, \psi'(\omega) A(\omega)^2 = 0,$$

which concludes the derivation, since the group delay time on propagation is $k'(\omega)x = v_g(\omega)^{-1}x$. □

The previous theorem only gives an indication about the central position of the waveform, as measured by its first temporal moment. It is possible to continue a similar derivation for higher order moments. We introduce the following notation for the temporal and the spectral means of a function:

$$\langle f \rangle_t = \int_{-\infty}^{+\infty} \mathrm{d}t\, f(t), \tag{2.12}$$

$$\langle \bar{f} \rangle_\omega = \frac{1}{2\pi} \int_{-\infty}^{+\infty} \mathrm{d}\omega\, \bar{f}(\omega). \tag{2.13}$$

Parseval's theorem has a more general form that previously used in the derivation of Property 2.2

$$\langle fg \rangle_t = \langle \bar{f}\bar{g} \rangle_\omega \tag{2.14}$$

for two arbitrary functions f and g.

Property 2.3 (*Pulse width*). The temporal variance of a wavepacket satisfies the following identity

$$\langle (t - t_0)^2 |u(t, x)|^2 \rangle_t = \langle A'^2 \rangle_\omega + \left(\langle \phi'^2 A^2 \rangle_\omega - \langle \phi' A^2 \rangle_\omega^2 \right). \tag{2.15}$$

The first spectral term is related to the natural temporal width of the wavepacket (at $x = 0$), while the second spectral term is the group delay variance. In this expression, $t_0 = \langle t |u(t, x)|^2 \rangle_t = \langle \phi' A^2 \rangle_\omega$ is the mean arrival time introduced in Property 2.2 and the energy normalization $\langle |u(t, x)|^2 \rangle_t = 1$ is assumed.

As a comment, the expression $\sqrt{\langle (t - t_0)^2 |u(t, x)|^2 \rangle_t}$ is an estimate for the temporal width of the wavepacket at position x. The property shows that this temporal width, obtained as a standard deviation, results from the combination of a minimum temporal width and of the added effect of dispersion. In particular, the minimum temporal width is $\sqrt{\langle A'^2 \rangle_\omega}$, which represents the width of the pulse at constant phase, i.e. when all frequency components in the spectrum have the same phase.

Proof. The extended form of Parseval's theorem (2.14) is the basis of the derivation of (2.15). We first note that $\langle (t - t_0)^2 |u(t, x)|^2 \rangle_t = \langle t^2 |u(t, x)|^2 \rangle_t - t_0^2$. Then we choose $f = tu$ and $g = tu^*$. The Fourier transform of f is

$$(A'(\omega) - \imath \phi'(\omega) A(\omega)) \exp(-\imath \phi(\omega)),$$

as we obtained in the proof of Property 2.2. Similarly, the Fourier transform of g is

$$(A'(\omega) + \imath \phi'(\omega) A(\omega)) \exp(+\imath \phi(\omega)).$$

Parseval's theorem (2.13) tells us that

$$\langle t^2 |u(t, x)|^2 \rangle_t = \langle fg \rangle_t = \langle (A' - \imath \phi' A)(A' + \imath \phi' A) \rangle_\omega,$$

which simplifies to

$$\langle t^2 |u(t, x)|^2 \rangle_t = \langle A'^2 \rangle_\omega + \langle \phi'^2 A^2 \rangle_\omega,$$

from which it is easy to conclude. □

In practice, a polynomial model is often considered for the spectral phase, according to a Taylor expansion around some angular frequency ω_0

$$\phi(\omega) = \phi(\omega_0) + (\omega - \omega_0) t_g(\omega_0) + \cdots + \frac{1}{n!} (\omega - \omega_0)^n \frac{\mathrm{d}^n \phi}{\mathrm{d}\omega^n}(\omega_0) + \cdots$$

The different derivatives have a special meaning and their effect on the wavepacket can be clearly identified. The constant phase $\phi(\omega_0)$ has no particular physical significance since it depends essentially on the chosen origin for space and time. $t_g(\omega_0)$ is a group delay (units: s). $\frac{\mathrm{d}^2 \phi}{\mathrm{d}\omega^2}(\omega_0)$ is called the group delay dispersion (GDD, units: s^2).

The group delay $t_g(\omega_0)$ gives a first order approximation of the arrival time of the wavepacket, valid only if the group delay does not vary too much in the considered spectral region; it is much cruder than the definition $t_0 = \langle \phi' A^2 \rangle_\omega$ we introduced earlier. To see this, we can evaluate

$$\langle \phi' A^2 \rangle_\omega \approx \left\langle (t_g(\omega_0) + (\omega - \omega_0) \frac{d^2\phi}{d\omega^2}(\omega_0)) A^2 \right\rangle_\omega = t_g(\omega_0).$$

For the last equality, we have assumed that the spectrum is centered around ω_0, so that $\langle (\omega - \omega_0) A^2 \rangle_\omega = 0$. Similarly, the group delay dispersion gives an indication of the temporal width that is due to dispersion in a given bandwidth. Actually, we can approximate the variance

$$\begin{aligned} \langle \phi'^2 A^2 \rangle_\omega - \langle \phi' A^2 \rangle_\omega^2 &\approx \langle (t_g(\omega_0) + (\omega - \omega_0) \frac{d^2\phi}{d\omega^2}(\omega_0))^2 A^2 \rangle_\omega - t_g(\omega_0)^2 \\ &= \frac{d^2\phi}{d\omega^2}(\omega_0)^2 \langle (\omega - \omega_0)^2 A^2 \rangle_\omega . \end{aligned}$$

Defining the spectral width by the standard deviation

$$B^2 = \langle (\omega - \omega_0)^2 A^2 \rangle_\omega ,$$

an estimate for the temporal width that is due to dispersion is then

$$\sqrt{\langle \phi'^2 A^2 \rangle_\omega - \langle \phi' A^2 \rangle_\omega^2} \approx \frac{d^2\phi}{d\omega^2}(\omega_0) \times B.$$

As an illustration, Figure 2.3 displays the case of a short pulse undergoing second- and third-order spectral dispersion. To be specific, we consider an optical spectrum extending around the frequency 375 THz (corresponding to a wavelength for light in a vacuum of 800 nm). The spectrum of the pulse displayed in Figure 2.3 (a) is a Gaussian function with a full width at half maximum (FWHM) of 44 THz ($B/(2\pi) = 18.7$ THz). The real part of the inverse Fourier transform is shown in Figure 2.3 (b). It is also a Gaussian function, with a temporal FWHM of 10 fs (standard deviation 4.2 fs). A second-order spectral phase is then added to the wavepacket, with GDD=100 fs^2. The FWHM of the pulse is enlarged by approximately 30 fs (standard deviation 12.5 fs), as shown in Figure 2.3 (c). As a comparison, the estimate GDD $\times B \approx 11.7$ fs. When a third-order dispersion (TOD) is added to the spectral phase of the wavepacket, with TOD = 2000 fs^3, the pulse is dispersed in an asymmetrical way. As Figure 2.3 (d) illustrates, the trailing edge of the pulse is more affected than its leading edge.

Green's functions for 1D wave propagation. Let us come back to nondispersive propagation and consider now that the wave equation has a right-hand side

$$\frac{\partial^2 u}{\partial t^2} - c^2 \frac{\partial^2 u}{\partial x^2} = h(t, x). \tag{2.16}$$

The right-hand side term $h(t, x)$ is driving the response of the medium and hence the generation of waves inside it. In other words, it is a source term for the wave equation.

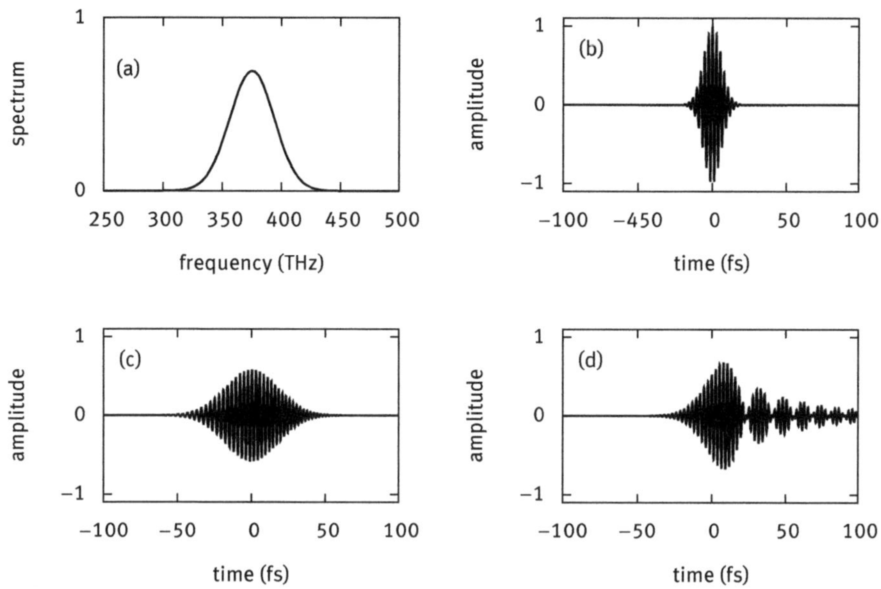

Fig. 2.3: Influence of dispersion on the temporal shape of a wavepacket. (a) A Gaussian spectrum distribution around the optical frequency 375 THz is considered. (b) The shortest pulse in the time domain corresponding to the spectrum in (a) has a Gaussian envelope and is obtained for a constant or linearly varying spectral phase. (c) When group delay dispersion (GDD) is added to the spectral phase, the pulse is enlarged symmetrically. (d) When third-order dispersion (TOD) is added instead, the pulse is deformed asymmetrically in time.

We have so far considered “free” solutions of the wave equation without this term acting (or with $h(t, x) = 0$). “Forced” solutions can be obtained by a general procedure involving particular solutions called Green’s functions.

The Green’s function of (2.16) is a particular solution obtained for $h(t, x) = \delta(t)\delta(x)$, i.e. it is the response to an infinitely sharp excitation in both time and space. Once the Green’s function $g(t, x)$ is known, the solution of (2.16) with an arbitrary excitation can be obtained as the convolution product of the excitation with the Green’s function

$$u(t, x) = [h \star g](t, x) = \int_{-\infty}^{+\infty} \mathrm{d}t' \int_{-\infty}^{+\infty} \mathrm{d}x' h(t', x') g(t - t', x - x'). \tag{2.17}$$

Property 2.4 (*1D Green’s function for the wave equation*). The 1D Green’s function for the wave equation assumes the following forms depending on the choice of representation space

$$\hat{g}(\omega, k) = \frac{1}{c^2 k^2 - \omega^2}, \tag{2.18}$$

$$g(t,x) = \frac{1}{2c} H(t - |x|/c), \tag{2.19}$$

$$\bar{g}(\omega,x) = \frac{1}{2\imath c\omega} \exp(-\imath\omega|x|/c), \tag{2.20}$$

with H the Heaviside distribution defined by

$$H(x) = \begin{cases} 1 & \text{if } x > 0, \\ 0 & \text{if } x < 0. \end{cases} \tag{2.21}$$

The different forms of the one-dimensional Green's function are depicted in Figure 2.4.

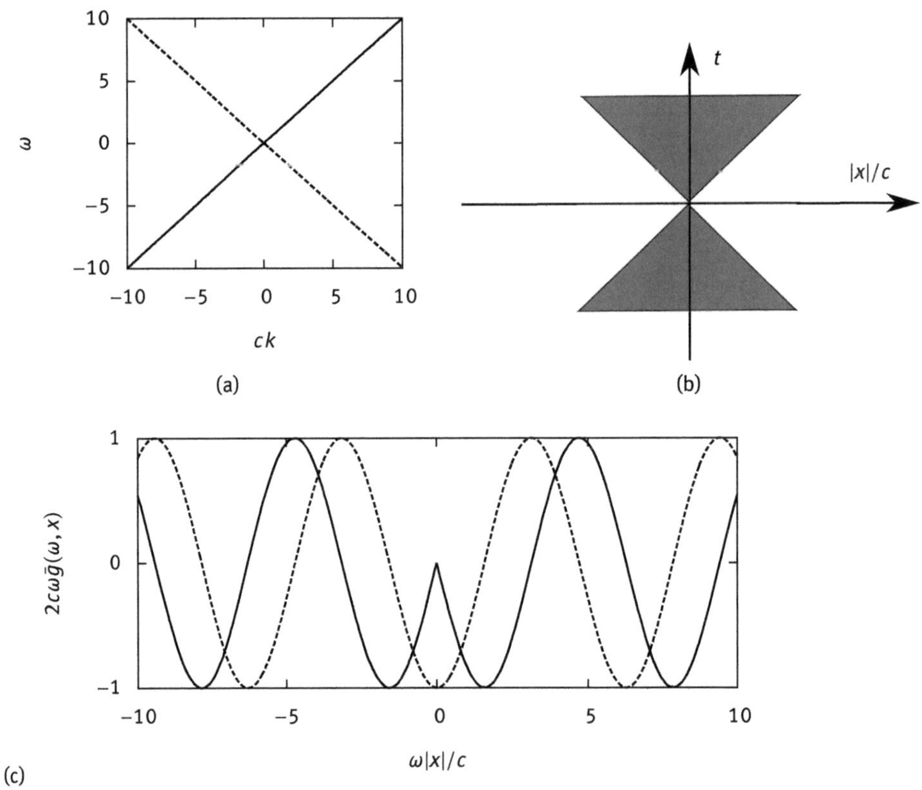

Fig. 2.4: Schematic representation of the three expressions for the 1D Green's function. (a) In the reciprocal (ω, k) space, (2.18) is singular on the dispersion relation for the right- and the left-propagating plane waves. (b) In the direct (t, x) space of waves, (2.19) has the shape of a double cone respecting causality. (c) In the mixed (ω, x) space of monochromatic waves, (2.20) is composed of two oscillating parts, one traveling to the right ($x > 0$) and one traveling to the left ($x < 0$). The real (imaginary, respectively) part is shown with a solid (dashed, resp.) line.

Proof. Starting with the driven wave equation

$$\frac{\partial^2 g}{\partial t^2} - c^2 \frac{\partial^2 g}{\partial x^2} = \delta(t)\delta(x),$$

we take the time and space Fourier transform and immediately obtain

$$(-\omega^2 + c^2 k^2)\hat{g}(\omega, k) = 1,$$

from which (2.18) follows.

We next verify that (2.20) is indeed a solution to the transformed (Helmholtz) equation

$$-\omega^2 \bar{g} - c^2 \frac{\partial^2 \bar{g}}{\partial x^2} = \delta(x).$$

Taking the first spatial derivative

$$\frac{\partial \bar{g}}{\partial x} = \begin{cases} \frac{-1}{2c^2} \exp(-\imath\omega x/c) & \text{if } x > 0, \\ \frac{1}{2c^2} \exp(\imath\omega x/c) & \text{if } x < 0. \end{cases}$$

It can be seen that this function is discontinuous at $x = 0$ with a jump from $\frac{1}{c^2}$ at $x = 0^-$ down to $\frac{-1}{c^2}$ at $x = 0^+$. Hence we write

$$\frac{\partial \bar{g}}{\partial x} = \frac{-\operatorname{sgn}(x)}{2c^2} \exp(-\imath\omega|x|/c),$$

where $\operatorname{sgn}(x)$ is the *signum* or sign distribution. The second derivative is

$$\frac{\partial^2 \bar{g}}{\partial x^2} = \frac{\imath\omega}{2c^3} \exp(-\imath\omega|x|/c) - \frac{1}{c^2}\delta(x),$$

where the last term results from the sign distribution derivative $\mathrm{d}\operatorname{sgn}(x)/\,\mathrm{d}x = 2\delta(x)$. Replacement in the Helmholtz equation above shows that (2.20) is indeed a solution.

Finally, we derive (2.19) from the inverse Fourier transform of (2.20). This inverse Fourier transform is

$$g(t, x) = \frac{1}{2c}\frac{1}{2\pi} \int_{-\infty}^{+\infty} \mathrm{d}\omega \frac{\exp(\imath\omega(t - |x|/c))}{\imath\omega}.$$

Then

$$\frac{\partial g(t, x)}{\partial t} = \frac{1}{2c}\frac{1}{2\pi} \int_{-\infty}^{+\infty} \mathrm{d}\omega \exp(\imath\omega(t - |x|/c)) = \frac{1}{2c}\delta(t - |x|/c).$$

Since δ is the derivative of the Heaviside function, (2.19) follows. □

2.1.2 Three-dimensional wave propagation

The three-dimensional isotropic and homogeneous wave equation has the form

$$\frac{\partial^2 u}{\partial t^2} - c^2 \Delta u = 0 \tag{2.22}$$

with $u(t, \mathbf{x})$ a function of time and space. $\mathbf{x}$ is a vector with Cartesian coordinates (x_1, x_2, x_3) in Euclidean space, as depicted in Figure 2.5. $\Delta = \frac{\partial^2}{\partial x_1^2} + \frac{\partial^2}{\partial x_2^2} + \frac{\partial^2}{\partial x_3^2}$ is the Laplacian operator. Isotropy implies that c is a constant independent of the direction of propagation. Homogeneity still means that c does not depend on the position in three-dimensional space.

In this chapter and the rest of this book we shall use the following conventions to simplify mathematical expressions with vectors and tensors.

- y_i is the i-th component of vector $\mathbf{y}$.
- $u_{,i} = \frac{\partial u}{\partial x_i}$, for $i = 1, 2, 3$, is a partial derivative of the scalar function $u(\mathbf{x})$. The gradient ∇u is the vector with components $u_{,i}$:

$$\nabla u = \begin{pmatrix} u_{,1} \\ u_{,2} \\ u_{,3} \end{pmatrix}.$$

- Let $\mathbf{y}(\mathbf{x})$ be a vector function of position. Then $y_{i,j} = \frac{\partial y_i}{\partial x_j}$. The gradient $\nabla \mathbf{y}$ of the vector function is the matrix (or rank-2 tensor) with components $(\nabla \mathbf{y})_{ij} = y_{i,j}$.
- Summation is implied whenever an index is repeated. For instance, the norm of vector $\mathbf{y}$ is the scalar $|\mathbf{y}| = \sqrt{y_i y_i}$. $\Delta u = \nabla \cdot (\nabla u) = u_{,ii}$ is the Laplacian of the scalar field u (it is a scalar).

By isotropic wave equation we mean that the properties of the propagation medium are invariant by any rotation of the coordinate system. As a consequence, the propagation velocity for plane waves is a constant in an isotropic medium. When we consider phononic crystals for elastic waves, we will encounter anisotropic media. We defer the full discussion of this case until then, but we note for completeness that the anisotropic scalar wave equation in a homogeneous medium is of the form

$$\frac{\partial^2 u}{\partial t^2} - c_{ij}^2 u_{,ij} = 0. \tag{2.23}$$

The rank-2 tensor c_{ij} is introduced here to stress that the velocity depends on the direction of propagation.

Plane wave. A plane wave in 3D space is of the form

$$u(t, \mathbf{x}) = F(t - \mathbf{s} \cdot \mathbf{x}) = F(t - s\, n_i x_i), \tag{2.24}$$

with F some arbitrary function and with $\mathbf{n}$ a unit vector giving the propagation direction, as depicted in Figure 2.6. $\mathbf{s}$ is the slowness vector and $s = |\mathbf{s}|$ is its magnitude.

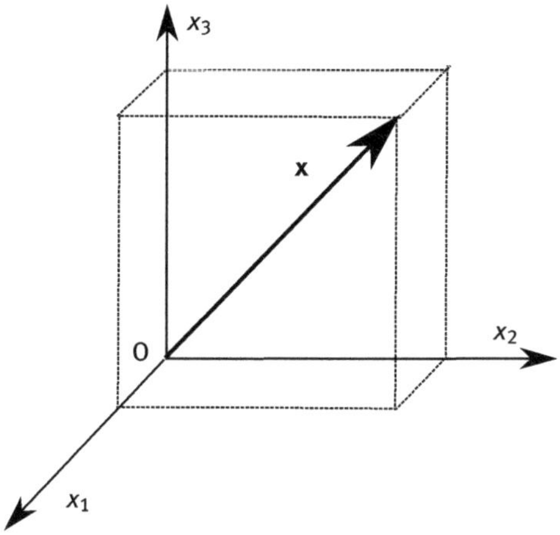

Fig. 2.5: Cartesian coordinate system for three-dimensional wave propagation in Euclidean space. The notation (x_1, x_2, x_3) for coordinates is chosen to conform with tensorial notations.

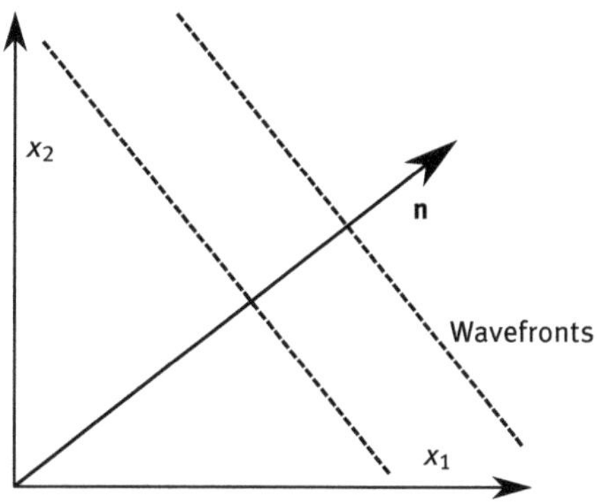

Fig. 2.6: Representation of a plane wave of the form $F(t - s\mathbf{n} \cdot \mathbf{x})$ at some time t. s is the slowness (the inverse of the phase velocity) and **n** is a unit vector in the direction of propagation, orthogonal to the wavefronts.

Obviously, this expression generalizes (2.2). The general solution to the 3D wave equation, however, is no longer given by (2.2), i.e. there are other solutions that are *not* of this particular form.

Inserting (2.24) in the 3D isotropic wave equation results in

$$(1 - c^2 s^2 n_i n_i) F''(t - s n_i x_i) = 0.$$

Discarding the case that $F'' = 0$ identically, and since $n_i n_i = 1$ for a unit vector, we have that $1/s = c$. The propagation velocity is thus simply c for any direction, as we anticipated above. Considering the 3D anisotropic wave equation results instead in the relation $(1/s)^2 = n_i c_{ij}^2 n_j$. Propagation is hence anisotropic with a direction dependent velocity given by $\sqrt{n_i c_{ij}^2 n_j}$.

A particular type of plane wave is the harmonic plane wave

$$u(t,\mathbf{x}) = \exp(\imath(\omega t - \mathbf{k}\cdot\mathbf{x})) = \exp(\imath(\omega t - k n_i x_i)). \tag{2.25}$$

The angular frequency is ω and the wavevector is $\mathbf{k} = k\,\mathbf{n} = \omega\,\mathbf{s}$.

Plane wave spectrum. Can we generalize the 1D plane wave spectrum to solutions of the 3D wave equation? The inverse temporal and spatial Fourier transform is written

$$u(t,\mathbf{x}) = \frac{1}{(2\pi)^4}\int_{-\infty}^{+\infty} \mathrm{d}\omega \int_{-\infty}^{+\infty} \mathrm{d}\mathbf{k}\,\bar{u}(\omega,\mathbf{k})\exp(\imath\omega t)\exp(-\imath\mathbf{k}\cdot\mathbf{x}). \tag{2.26}$$

The integral is over harmonic plane waves. If u is a solution to the 3D wave equation, then every harmonic plane wave must satisfy the dispersion relation. We thus choose to express, say, k_3 as a function of ω, k_1, and k_2. Then we have

$$u(t,\mathbf{x}) = \frac{1}{(2\pi)^3}\int_{-\infty}^{+\infty} \mathrm{d}\omega \int_{-\infty}^{+\infty} \mathrm{d}k_1\mathrm{d}k_2\,\bar{u}(\omega,\mathbf{k})\exp(\imath\omega t) \times \exp(-\imath(k_1x_1 + k_2x_2 + k_3(\omega,k_1,k_2)x_3)). \tag{2.27}$$

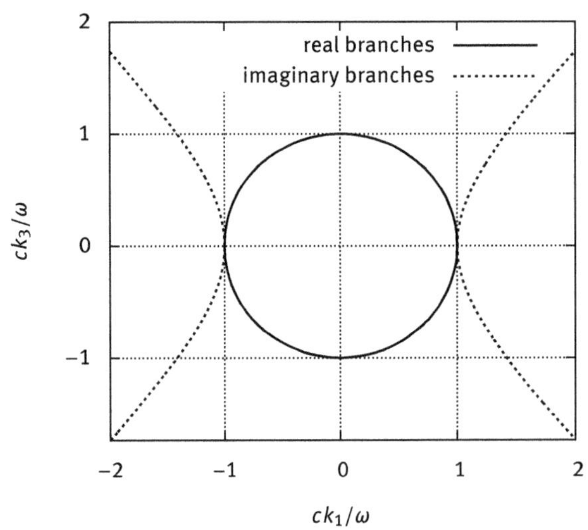

Fig. 2.7: Cross section of the equifrequency surface for isotropic wave propagation. The equifrequency surface given by (2.28) is cut by the plane $k_2 = 0$. The real part forms a circle (solid line), while the imaginary branches are hyperbolic (dotted line).

As an example, for isotropic propagation we have $k^2 = \omega^2/c^2$ and then

$$k_3 = \begin{cases} \pm\sqrt{\omega^2/c^2 - k_1^2 - k_2^2} & \text{if } k_1^2 + k_2^2 < \omega^2/c^2, \\ \pm\imath\sqrt{|\omega^2/c^2 - k_1^2 - k_2^2|} & \text{if } k_1^2 + k_2^2 > \omega^2/c^2. \end{cases} \tag{2.28}$$

This relation defines a complex surface in the ω, k_1, k_2 space. A cross section at constant frequency is called an equifrequency surface. Considering in addition a cross section by plane $k_2 = 0$ leads to the equifrequency contour plotted in Figure 2.7. The integration over k_3 is then seen to be replaced by two complex branches, corresponding physically to a right- and a left-propagating plane harmonic wave. One difficulty with the plane wave spectrum just introduced is that it is not so easy to extend to anisotropic wave propagation. If one is only interested in the far field, away from any source of waves, it is possible to neglect evanescent waves corresponding to the imaginary branches of the dispersion relation and to consider alternatively the angular wave spectrum along the following lines.

Angular wave spectrum. Suppose we know the dispersion relation in the form $k(\omega, \mathbf{n})$, i.e. the wavenumber is a function of frequency and of the direction of propagation. This formalization will be useful to describe the anisotropy of elastic waves in Chapter 6 but also to study refraction in and reflection on sonic and phononic crystals in Chapter 12. The following representation can be used instead of (2.26) and is typical of the far field

$$u(t, \mathbf{x}) = \frac{1}{2\pi} \int_{-\infty}^{+\infty} \mathrm{d}\omega \int \mathrm{d}\mathbf{n}\bar{u}(\omega, \mathbf{n}) \exp(\imath\omega t) \exp(-\imath k(\omega, \mathbf{n})\mathbf{n} \cdot \mathbf{x}). \tag{2.29}$$

The integral over $\mathbf{n}$ must be understood as over all directions of space.

We introduce the following definitions for wave propagation in dispersive and anisotropic media.

- $v(\omega, \mathbf{n}) = \omega/k(\omega, \mathbf{n})$ is the phase velocity and its inverse $s(\omega, \mathbf{n}) = k(\omega, \mathbf{n})/\omega$ is the slowness. If propagation is not dispersive, the slowness is solely a function of the direction of propagation. If propagation is dispersive, then equifrequency contours inform about spatial propagation.
- $v_g(\omega, \mathbf{n}) = (\partial k/\partial\omega)^{-1}$ is the temporal group velocity which is a measure of the propagation velocity of a signal.
- $\mathbf{v}_g(\omega, \mathbf{n}) = \nabla_{\mathbf{k}}\omega$ is the group velocity vector giving information on the direction of propagation and velocity of the energy carried by the wave. The relation between the group velocity vector and the Poynting vector will be specified later in Chapter 5 when we discuss elastic waves in solids.

It may seem weird that we have to define both a temporal and a spatial group velocity, as actually they must be related physically as they are obtained from the same dispersion relation. It can actually be noted that the two particular forms of the dispersion relation that we have assumed, first $k_3(\omega, k_1, k_2)$ for the plane wave spectrum and second $k(\omega, \mathbf{n})$, are different explicit forms of the same physical reality. A more general approach is to consider the dispersion relation in implicit form, for instance as

$$D(\omega, k_1, k_2, k_3) = 0. \tag{2.30}$$

This implicit equation actually defines a hypersurface, or a 3-dimensional surface in the 4-dimensional space $(\omega, \mathbf{k})$. The partial derivatives that define the temporal and spatial group velocities can then be given a much less mathematically ambiguous meaning than those just given. This type of notation will be used in Chapter 12 when we discuss anisotropy in sonic and phononic crystals. On this occasion, we will also demonstrate that the spatial group velocity is orthogonal to the equifrequency surface, defined for a given frequency ω as the locus of $k(\omega, \mathbf{n})$ as a function of the direction $\mathbf{n}$. For nondispersive media, the spatial group velocity is also orthogonal to the slowness surface defined as the locus of $s(\omega, \mathbf{n})$ as a function of the direction $\mathbf{n}$. Reciprocally, $\mathbf{n}$ is also orthogonal to the wave surface, defined as the locus of $\mathbf{v}_g(\omega, \mathbf{n})$ for all values of the direction $\mathbf{n}$. These surfaces are especially useful to describe the propagation of harmonic plane waves in anisotropic media but also the far field of waves radiated from point sources.

Property 2.5 (*Far field approximation*). Energy concentrates along trajectories such that the phase is stationary in time and space, or $t = (\mathbf{n} \cdot \mathbf{x})/v_g$ and $\mathbf{x} = \mathbf{v}_g t$. The second relation implies that energy propagates in the direction of the group velocity vector. The first relation implies that the time it takes is the distance measured normally to the phase fronts divided by the temporal group velocity. Overall, both group velocities are related by $v_g = \mathbf{n} \cdot \mathbf{v}_g$.

Proof. As a note, the following arguments are heuristic and we make no attempt at true mathematical rigor. The integral (2.26) is an oscillatory integral. We will assume that the variations of $\bar{u}(\omega, \mathbf{k})$ in (2.26) are rather slow or even constant, the limiting case being the harmonic plane wave. Within the method of the *stationary phase approximation*, the contributions to the integral of most values of ω and $\mathbf{k}$ are negligible because they result in destructive interferences: the phase in the exponential term varies very quickly over a wide range of angles and the resulting mean of the exponential term tends to zero. Only those values that make the phase stationary can contribute effectively to the far field of the wave. Then the trajectory for rays must concentrate around the stationary values of the phase $\phi(\omega, \mathbf{k}) = \omega t - \mathbf{k} \cdot \mathbf{x}$. For the phase to be stationary, its first derivatives with respect to ω and $\mathbf{k}$ must vanish, or

$$\frac{\partial \phi}{\partial \omega} = t - \frac{\partial k}{\partial \omega} \mathbf{n} \cdot \mathbf{x} = 0,$$
$$\frac{\partial \phi}{\partial k_i} = \frac{\partial \omega}{\partial k_i} t - x_i = 0.$$

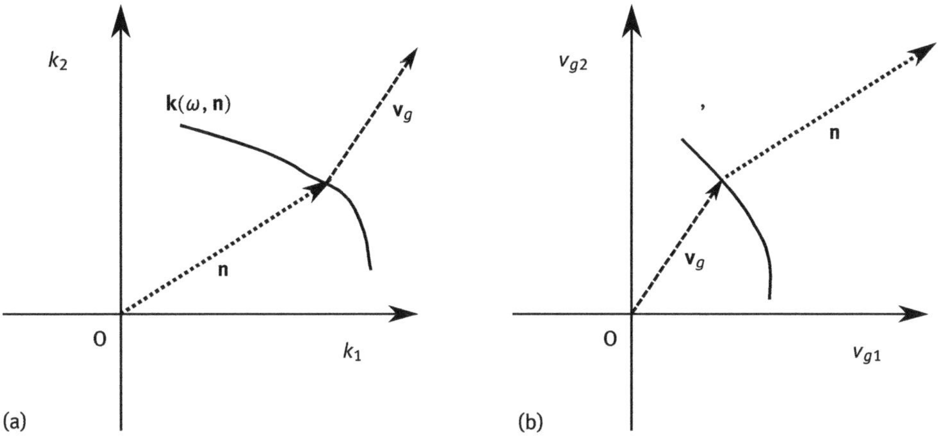

Fig. 2.8: Equifrequency surface (or slowness surface) and wave surface. (a) For a given angular frequency ω, the equifrequency surface $k(\omega, \mathbf{n})$ indicates the slowness and the phase velocity of a harmonic plane wave propagating in direction $\mathbf{n}$. It must be understood as a plot in the Fourier space of spatial frequencies. The spatial group velocity is orthogonal to the equifrequency surface and indicates the direction of energy flow of each particular plane wave component. (b) The wave surface is constructed from all values of the spatial group velocity. In physical space it represents the far field radiated by a point source located at the origin.

The two relations $t = (\mathbf{n} \cdot \mathbf{x})/v_g$ and $\mathbf{x} = \mathbf{v}_g t$ follow immediately from the definitions above. Further projecting the second relation on $\mathbf{n}$, $v_g = \mathbf{n} \cdot \mathbf{v}_g$ follows. For nondispersive media, the last relation simplifies to $v = \mathbf{n} \cdot \mathbf{v}_g$, with v the phase velocity. □

As a result, propagation in anisotropic media follows the principles depicted in Figure 2.8. The energy of a harmonic plane wave propagating in the phase direction $\mathbf{n}$ is directed along the vector $\mathbf{v}_g$. The direction of the latter vector is easily obtained graphically from the slowness surface, while its length can be obtained from $v_g = \mathbf{n} \cdot \mathbf{v}_g$. Reciprocally, the waves radiated from a point source are emitted in any direction and concentrate in time and in the far field along concentric surfaces that are homothetic to the wave surface. The wavefronts are locally orthogonal to the wave surface and can thus be reconstructed from it. In particular, it is possible to easily recover the direction $\mathbf{n}$ that corresponds to a particular point of the wave surface by drawing the perpendicular at that point.

Mirrors, cavities, and waveguides. Let us consider the reflection of a plane wave traveling to the right along axis x_1 in an isotropic medium and incident normally on a plane mirror orthogonal to this axis. As the problem is one-dimensional, this wave can be written as $F(t - sx_1)$. It is sensible to consider that the reflected wave is also a plane wave and that it travels to the left, so we can write it as $G(t + sx_1)$. We made the implicit assumption that the slowness is the same, because the two waves travel in the same medium. The total wave is thus the superposition $u(t, x_1) = F(t-sx_1)+G(t+sx_1)$.

Suppose now that the boundary condition is that the amplitude vanishes at the mirror surface, i.e. $u = 0$ at $x_1 = 0$. Such a boundary condition (BC) is known as a Dirichlet BC, or essential BC. We must have $F(t) + G(t) = 0$ at all times and consequently

$$u(t, x_1) = F(t - sx_1) - F(t + sx_1).$$

The reflected wave thus appears in the incident medium as if it had been reflected at the mirror and its sign had been changed, as depicted in Figure 2.9 (a).

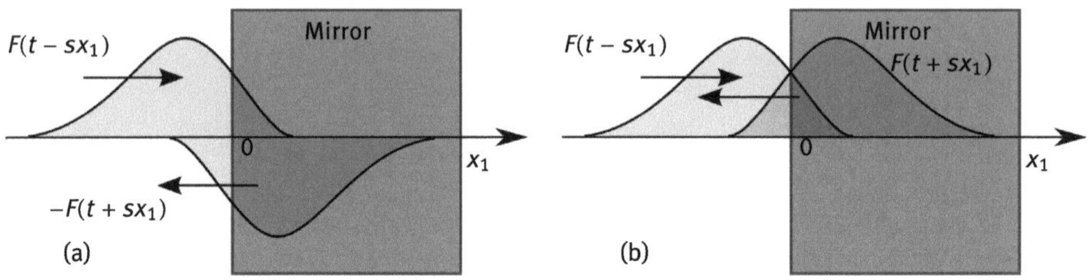

Fig. 2.9: Total reflection for a plane wave reflecting on a perfect mirror. (a) When the boundary condition at the mirror surface is of the Dirichlet type (vanishing wave amplitude), the reflected wave bounces on the mirror with a sign change. (b) When the boundary condition is of the Neumann type (vanishing normal derivative of the wavefield), the reflected wave bounces on the mirror without a sign change.

Suppose now that the boundary condition is that the normal derivative of the amplitude vanishes at the mirror surface, i.e. $\frac{\partial u}{\partial x_1} = 0$ at $x_1 = 0$. This is known as a Neumann BC, or natural BC. This time we have $F'(t) - G'(t) = 0$ at all times and then

$$u(t, x_1) = F(t - sx_1) + F(t + sx_1).$$

The reflected wave appears in the incident medium as if it had been reflected at the mirror but with its sign unchanged, as depicted in Figure 2.9 (b). As a consequence, the value of the wavefield exactly at the mirror surface is exactly doubled.

Let us consider the particular case of harmonic plane waves. For the Dirichlet BC, the superposition is

$$u(t, x_1) = -2\imath \exp(\imath\omega t) \sin(kx_1).$$

This particular solution, the interference of two harmonic plane waves traveling in opposite directions, is a standing wave. The amplitude of the standing wave vanishes at $x_1 = 0$, but also at a distance $L = \pi/k$ away, with $k = s\omega$. If we add a second mirror at a distance L, we define a one-dimensional cavity that can store waves. Resonances of the cavity are obtained whenever the wavenumber equals $k_n = n\pi/L$ with n a positive integer larger than 1, as obtained from the condition $u(t, L) = 0$, or $\sin(kL) = 0$. This condition defines the normal modes of the cavity, with angular frequencies $\omega_n = nc\pi/L$, $n \geq 1$. The first three normal modes are plotted in Figure 2.10 (a).

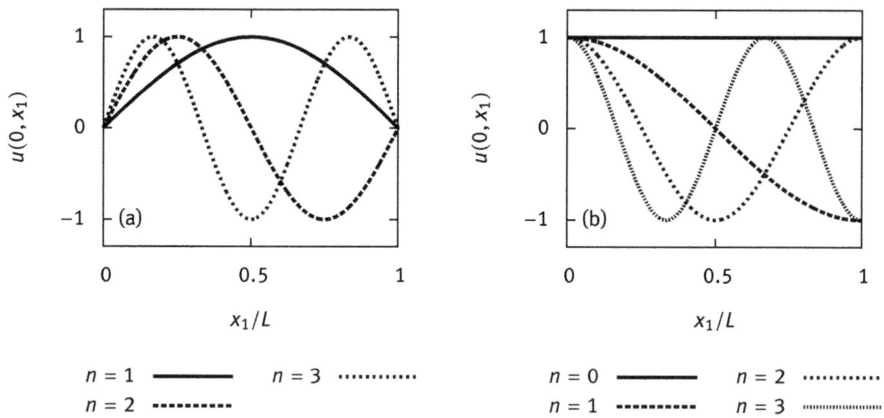

Fig. 2.10: Field distribution inside a cavity of width L closed by two perfect mirrors, for the first normal modes. (a) Case of Dirichlet boundary condition at the mirror surface. (b) Case of Neumann boundary condition at the mirror surface.

Now assuming instead that the Neumann BC applies, the superposition becomes

$$u(t, x_1) = 2\exp(\imath\omega t)\cos(kx_1).$$

Again considering a cavity of width L, the normal frequencies are unchanged since they are defined by the resonance condition $\cos(kL) = \pm 1$, though the case $n = 0$ is now allowed in addition. The normal mode distribution for $n = 0$ is a constant and has zero resonant frequency. The first four normal modes are plotted in Figure 2.10 (b).

We now consider in addition an arbitrary incidence angle on the mirrors, which are still assumed parallel and separated by a distance L. This situation is depicted in Figure 2.11 (a). The part of the wave propagating to the right is written as the harmonic plane wave $\exp(\imath(\omega t - k_1x_1 - k_2x_2))$. Reflection on the right mirror will induce a harmonic plane wave propagating to the left. We assume Dirichlet boundary conditions at the surfaces of the mirrors. In order for the superposition of two harmonic plane waves to satisfy the boundary conditions, phase matching should be achieved. Writing the reflected wave as $A\exp(\imath(\omega' t - k_1'x_1 - k_2'x_2))$, with A some unknown coefficient, the superposition is

$$u(t, x_1, x_2) = \exp(\imath(\omega t - k_1x_1 - k_2x_2)) + A\exp(\imath(\omega' t - k_1'x_1 - k_2'x_2)).$$

Since we must have

$$u(t, x_1 = 0, x_2) = \exp(\imath(\omega t - k_2x_2)) + A\exp(\imath(\omega' t - k_2'x_2)) = 0$$

whatever t and x_2, the only possibility is $A = -1$, and the phase matching conditions $\omega' = \omega$ and $k_2' = k_2$. In terms of waves, the phase matching conditions express that

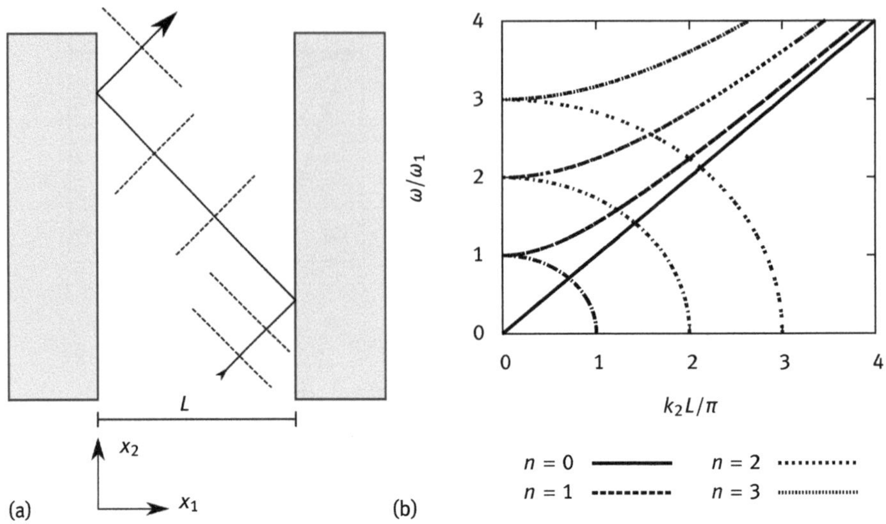

Fig. 2.11: A simple waveguide. (a) Two perfect plane mirrors are separated by a distance L. Propagation of guided modes is along axis x_2, while the wavefield is confined along axis x_1. (b) Dispersion relation of the first guided modes, $\omega(k_2)$. Both real (propagating) and imaginary (evanescent) branches of the dispersion relation are shown. The cut-off frequency of the first guided mode is $\omega_1 = \pi c/L$. All subsequent cut-off frequencies are integer multiples of ω_1. The evanescent branches of the dispersion relation appear as circle arcs with the chosen normalization for frequency and wavenumber.

both frequency and the tangential part of the wavevector are conserved upon reflection on a plane interface. In terms of particles (e.g. photons or phonons), we would say that energy and momentum are conserved. Further application of the dispersion relation and knowledge that the reflected wave travels to the left then imply $k_1' = -k_1$. The value of coefficient A is determined from the boundary conditions. For the Dirichlet boundary condition, $A = -1$; for the Neumann boundary condition, $A = 1$. We select the latter case for deriving the dispersion of the waveguide as shown in Figure 2.11. The superposition simplifies to $u(t, x_1, x_2) = 2\exp(\imath(\omega t - k_2 x_2))\cos(k_1 x_1)$. This is a guided wave that is standing along x_1 but progressive along x_2. As in the case of the cavity, k_1 can only assume discrete values given by $k_{1n} = n\pi/L$, with $n \geq 0$. Each of the guided modes then follows a dispersion relation given by

$$\omega(k_2) = \sqrt{c^2 k_2^2 + \omega_n^2},$$

with the cut-off frequencies $\omega_n = nc\pi/L$, or reciprocally

$$k_{2n}(\omega) = \frac{1}{c}\sqrt{\omega^2 - \omega_n^2}.$$

It is clear that each particular guided wave with $n > 0$ is evanescent below its cut-off frequency. Indeed, for $\omega < \omega_n$,

$$k_{2n}(\omega) = \frac{\pm\imath}{c}\sqrt{|\omega^2 - \omega_n^2|}.$$

We will meet mirrors, cavities, and waveguides again in the context of sonic and phononic crystals in Chapter 11. Evanescent waves in this context are specifically discussed in Chapter 9.

2.2 Bloch's theorem

Bloch's theorem is a classic of solid state physics. It is usually introduced in the context of crystals, which are composed of atoms organized periodically according to a lattice. It was originally presented by Felix Bloch to describe the motion of electrons in crystals, a problem which is represented in quantum physics by a Hamiltonian with a periodic potential. In the context of phononic crystals, the periodic potential is replaced by a periodic modulation of the material constants of a Helmholtz equation, and we are interested in the propagation of classical waves. In this section we propose a demonstration of Bloch's theorem that applies to the Helmholtz equation with periodic coefficients

$$-\nabla\cdot(c(\mathbf{r})\nabla u(\mathbf{r})) = \omega^2 u(\mathbf{r}). \tag{2.31}$$

An essential property of a crystal is that it is invariant under any translation that is an integer multiple of the lattice constants. Let us accordingly introduce the primitive lattice vectors $\mathbf{a}_1$, $\mathbf{a}_2$, and $\mathbf{a}_3$ (in 2D only $\mathbf{a}_1$ and $\mathbf{a}_2$ are required). They are not required to be orthogonal, but they must not be parallel. As represented in Figure 2.12, they define a parallelepiped (or rhomboid) in 3D, with volume $V = |\mathbf{a}_1\cdot(\mathbf{a}_2\times\mathbf{a}_3)|$, or a parallelogram in 2D, with surface $S = |\mathbf{a}_1\cdot\mathbf{a}_2|$. The parallelepiped or the parallelogram thus defined is a unit cell of the crystal. In Section 2.4 we specify the concepts of a primitive unit cell and a Wigner–Seitz cell, but they are not required at this stage.

Theorem 2.1 (Bloch's theorem for classical waves). *The eigenmodes of the periodic Helmholtz equation are Bloch waves of the form*

$$u(\mathbf{r}) = \exp(-\imath\mathbf{k}\cdot\mathbf{r})\tilde{u}(\mathbf{r}) \tag{2.32}$$

where $\tilde{u}(\mathbf{r})$ is a periodic function with the same periodicity as the crystal and $\mathbf{k}$ is the Bloch wavevector.

We propose a simplified proof of the theorem in the following. By simplified, we mean that we do not attempt a rigorous mathematical proof, which would be outside our

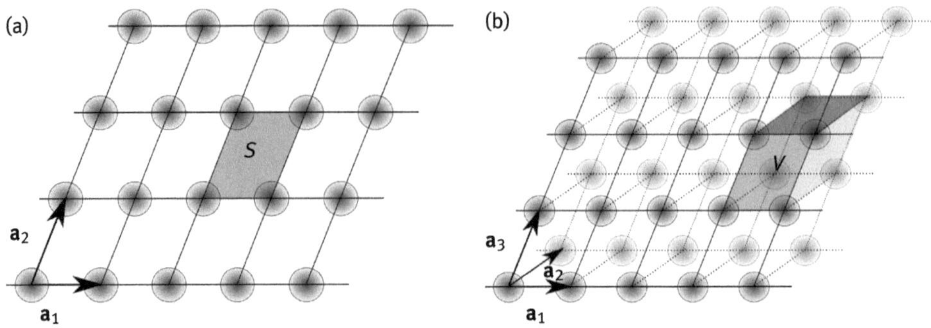

Fig. 2.12: Primitive lattice vectors of a crystal. (a) The primitive lattice vectors in 2D are two nonparallel vectors $\mathbf{a}_1$ and $\mathbf{a}_2$. They define a natural unit cell in the form of a parallelogram with surface $S = |\mathbf{a}_1 \cdot \mathbf{a}_2|$. The parallelogram can be used to pave all 2D space periodically. (b) The primitive lattice vectors in 3D are three nonparallel vectors $\mathbf{a}_1$, $\mathbf{a}_2$, and $\mathbf{a}_3$. They define a natural unit cell in the form of a parallelepiped (or rhomboid) with volume $V = |\mathbf{a}_1 \cdot (\mathbf{a}_2 \times \mathbf{a}_3)|$. The parallelepiped can be used to pave all 3D space periodically.

scope and capabilities, but that we want to outline instead the important concepts that lead to the theorem itself.

Proof. Let us first define the translation operator $T(\mathbf{R})$ that shifts every wavefield by the vector $\mathbf{R} = n_1\mathbf{a}_1 + n_2\mathbf{a}_2 + n_3\mathbf{a}_3$, where n_1, n_2, n_3 are integers. Its action is defined by $T(\mathbf{R})u(\mathbf{r}) = u(\mathbf{r}+\mathbf{R})$. Discrete periodicity in a crystal can be described as invariance in any such translation. Defining the Helmholtz operator by $\mathcal{L}u(\mathbf{r}) = -\nabla \cdot (c(\mathbf{r})\nabla u(\mathbf{r})) - \omega^2 u(\mathbf{r})$, where $c(\mathbf{r})$ is a periodic function, it is clear that $\mathcal{L}(\mathbf{r}+\mathbf{R}) = \mathcal{L}(\mathbf{r})$. Furthermore, $\mathcal{L}$ commutes with all translations, i.e.

$$T(\mathbf{R})\mathcal{L}u(\mathbf{r}) = \mathcal{L}(\mathbf{r}+\mathbf{R})u(\mathbf{r}+\mathbf{R}) = \mathcal{L}(\mathbf{r})T(\mathbf{R})u(\mathbf{r}).$$

It is also obvious that translations commute, or

$$T(\mathbf{R}')T(\mathbf{R})u(\mathbf{r}) = u(\mathbf{r}+\mathbf{R}+\mathbf{R}') = T(\mathbf{R})T(\mathbf{R}')u(\mathbf{r}).$$

We will assume that eigenvectors of all the translation operators and of the Helmholtz operator do exist. Then from the commutation property of operators, we know that there exists a common eigenvector basis to the Helmholtz operator and *all* the translation operators. In linear algebra, this is known as the fact that commuting matrices are simultaneously diagonalizable. Note that the different operators actually have different eigenvalues for every particular common eigenvector. The common eigenvectors are called – by definition – Bloch waves.

Let us consider one such Bloch wave $u(\mathbf{r})$. We denote as α_i its eigenvalues for the elementary discrete translations $T(\mathbf{a}_i)$, $i = 1, 2, 3$. Since $T(\mathbf{a}_i)u = \alpha_i u$, it follows that $T(\mathbf{R})u = \alpha_1^{n_1}\alpha_2^{n_2}\alpha_3^{n_3}u$. Hence, we have a simple expression for the eigenvalues of all discrete translation operators of the lattice. We have two possibilities from now on to conclude the demonstration.

(i) In solid state physics, it is customary to consider that the crystal is periodic but finite in size, and that it can be viewed as composed of $N_1 \times N_2 \times N_3$ unit cells. If the crystal is large enough, we can consider Born–Von Karman periodic boundary conditions for the crystal as $u(\mathbf{r} + N_i\mathbf{a}_i) = u(\mathbf{r})$. Then we have $\alpha_1^{N_1}\alpha_2^{N_2}\alpha_3^{N_3} = 1$, which implies that the α_i are N_i-th roots of unity. In particular they are complex numbers of modulus one.

(ii) If we want to avoid the Born–Von Karman periodic boundary conditions, we can remark that a translation of the wavefield preserves its energy (is a unitary transformation). Then its eigenvalues should be of unit modulus. Of course this implies that the energy of the wave can be defined. For instance, the energy of the plane wave, obtained by integrating its square modulus over the whole space, is infinite! This example shows that the requirement is not always fulfilled. One way around this is to consider the energy density integrated over any compact (closed and bounded) subset of Euclidean space.

Anyway, because they are complex numbers of unit modulus we can now pose $\alpha_i = \exp(-2\imath\pi\beta_i)$, with the β_i some real numbers. We further form a vector $\mathbf{k} = \beta_1\mathbf{b}_1 + \beta_2\mathbf{b}_2 + \beta_3\mathbf{b}_3$, where the $\mathbf{b}_i$ are basis vectors, until now arbitrary. We are in a position to prove that $\tilde{u}(\mathbf{r}) \triangleq \exp(\imath\mathbf{k}\cdot\mathbf{r})u(\mathbf{r})$ is periodic. Actually

$$\begin{aligned}\tilde{u}(\mathbf{r}+\mathbf{a}_i) &= \exp(\imath\mathbf{k}\cdot(\mathbf{r}+\mathbf{a}_i))u(\mathbf{r}+\mathbf{a}_i)\\ &= \exp(\imath\mathbf{k}\cdot\mathbf{r})\exp(\imath\mathbf{k}\cdot\mathbf{a}_i)\exp(-2\imath\pi\beta_i)u(\mathbf{r})\\ &= \exp(\imath\beta_j\mathbf{b}_j\cdot\mathbf{a}_i)\exp(-2\imath\pi\beta_i)\tilde{u}(\mathbf{r}).\end{aligned}$$

Hence it suffices to choose the $\mathbf{b}_i$ such that $\mathbf{b}_j\cdot\mathbf{a}_i = 2\pi\delta_{ij}$ to obtain $\tilde{u}(\mathbf{r}+\mathbf{a}_i) = \tilde{u}(\mathbf{r})$. This natural choice defines the reciprocal lattice basis vectors. The reciprocal lattice is a periodic pavement of wavevector space. □

2.3 Physical origin of band gaps

We will frequently meet Bloch's theorem again in the remaining chapters. It will specifically be used to define the band structure of sonic crystals in Chapter 4 and of phononic crystals in Chapter 6. But before doing so, it is instructive to directly derive some properties of wave propagation in periodic media for some simple but illustrative cases, especially in one dimension. This is particularly useful in order to understand the concepts of Bragg band gaps and of local resonance, independently of further complications.

2.3.1 1D periodic media

Let us start with the very common example of a linear chain of masses coupled by springs. This is the basic model that is often used to introduce the concept of phonons in solid state physics.

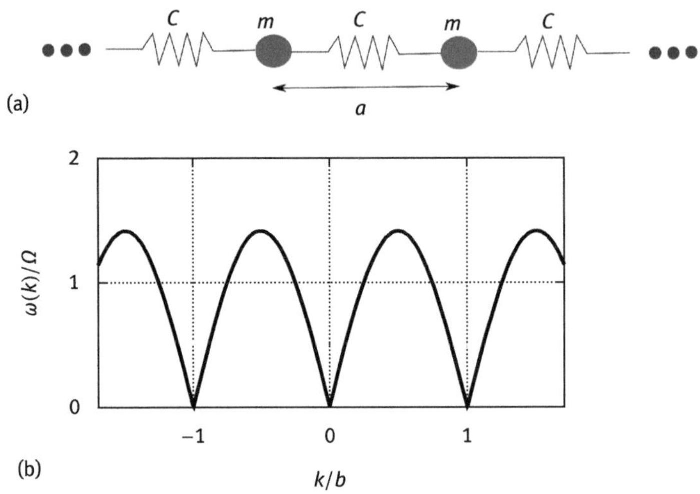

Fig. 2.13: A linear chain of masses coupled by springs. (a) The masses m are assumed to be punctual and to be connected by springs with stiffness C. The one-dimensional chain is periodic and infinite, with a lattice constant a. (b) The dispersion relation (2.36) gives the frequency of propagating Bloch waves as a function of wavenumber. $b = 2\pi/a$ is the reciprocal lattice constant and $\Omega^2 = 2C/m$.

Linear chain. Let us consider a linear chain of punctual masses m connected by springs with stiffness C (a very simple model for atoms with nearest neighbor only bonds, as depicted in Figure 2.13 (a)). The dynamical equation of motion of the masses is given by Newton's first law

$$m\frac{\partial^2 u_n}{\partial t^2} = C(u_{n+1} - u_n) + C(u_{n-1} - u_n). \tag{2.33}$$

This is an infinite system of linear coupled equations. We consider the discrete Fourier transform (DFT) of the sequence u_n as

$$\check{u}(q) = \sum_{n=-\infty}^{\infty} u_n \exp(2\imath\pi q n), \quad q \in [-1/2, 1/2], \tag{2.34}$$

with the inverse formula

$$u_n = \int_{-1/2}^{1/2} dq\, \check{u}(q) \exp(-2\imath\pi qn). \tag{2.35}$$

Taking the DFT of the linear chain equation, every Fourier component is decoupled and we have

$$m\frac{\partial^2 \check{u}(q)}{\partial t^2} = C(\exp(2\imath\pi q) + \exp(-2\imath\pi q) - 2)\check{u}(q) = 2C(\cos(2\pi q) - 1)\check{u}(q).$$

This is a harmonic equation: the solutions are time harmonic (monochromatic) with an angular frequency given by the dispersion relation

$$\omega^2 = \Omega^2(1 - \cos(2\pi q))$$

with $\Omega^2 = 2C/m$. Considering that the masses are separated by distance a, the lattice constant, we choose to write $2\pi q = ka$, with k the wavenumber.[1] Hence we cast the dispersion relation in the form

$$\omega(k)^2 = \Omega^2(1 - \cos(ka)). \tag{2.36}$$

It is not difficult to verify that every function $\exp(\imath(\omega t - kna))$ is a Bloch wave of the original equation (2.33). To show this, let us apply Bloch's theorem by defining the solution over the unit cell in the form $u(r) = u_n\delta(r - na)$. The Dirac delta function is introduced to respect the discrete physical nature of the problem. Every Bloch wave then has the dependence $u_n = u_0\delta(r - na)\exp(\imath(\omega t - kna))$, with only one degree of freedom (u_0) per Bloch wave. Inserting this expression in (2.33) directly yields the dispersion relation (2.36). With this example, we want to stress that it is not always necessary to use Bloch's theorem to obtain the eigenfunctions of a periodic wave propagation problem.

The dispersion relation of the linear chain of masses is plotted in Figure 2.13 (b). It can be observed that propagating Bloch waves belong to only one band and that there is a maximum frequency $\sqrt{2}\Omega$ that is attained at the edge of the first Brillouin zone. Dispersion relation (2.36) can also easily be expressed in the form $k(\omega)$, in which case solutions of the dispersion relation with complex imaginary values of k are readily found. We will highlight these evanescent solutions in the following example of the bilinear chain.

Bilinear chain. Let us enrich the previous model and consider a linear chain with two types of punctual masses, or two types of atoms, as depicted in Figure 2.14. The dy-

1 Note that the wavenumber k was defined previously through a continuous Fourier transform. Since we are using a discrete Fourier transform here, k is arbitrarily defined from the variable q with respect to the lattice constant.

namical equations of motion are

$$m_1 \frac{\partial^2 u_n}{\partial t^2} = C(v_n - u_n) + C(v_{n-1} - u_n), \tag{2.37}$$

$$m_2 \frac{\partial^2 v_n}{\partial t^2} = C(u_{n+1} - v_n) + C(u_n - v_n). \tag{2.38}$$

Taking the DFT, and with the notations $\Omega_1^2 = 2C/m_1$ and $\Omega_2^2 = 2C/m_2$,

$$\frac{\partial^2 \check{u}(q)}{\partial t^2} = \frac{\Omega_1^2}{2}(1 + \exp(-2\imath\pi q))\check{v}(q) - \Omega_1^2 \check{u}(q),$$

$$\frac{\partial^2 \check{v}(q)}{\partial t^2} = \frac{\Omega_2^2}{2}(1 + \exp(2\imath\pi q))\check{u}(q) - \Omega_2^2 \check{v}(q).$$

This is a pair of coupled harmonic equations, with time harmonic solutions such that

$$(\omega^2 - \Omega_1^2)\check{u}(q) + \frac{\Omega_1^2}{2}(1 + \exp(-2\imath\pi q))\check{v}(q) = 0,$$

$$(\omega^2 - \Omega_2^2)\check{v}(q) + \frac{\Omega_2^2}{2}(1 + \exp(2\imath\pi q))\check{u}(q) = 0.$$

These equations are compatible only if the determinant vanishes, leading to the dispersion relation

$$(\omega^2 - \Omega_1^2)(\omega^2 - \Omega_2^2) = \frac{1}{2}\Omega_1^2\Omega_2^2(1 + \cos(ka)) \tag{2.39}$$

where we have again used $2\pi q = ka$. As previously, exactly the same dispersion relation can be obtained directly from Bloch's theorem by considering a unit cell with two degrees of freedom, for instance

$$u(r) = u_n\delta(r + a/4 - na) + v_n\delta(r - a/4 - na).$$

Solving (2.39) for ω as a function of k we obtain

$$\omega^2 = \frac{1}{2}(\Omega_1^2 + \Omega_2^2) \pm \frac{1}{2}\sqrt{\Omega_1^4 + \Omega_2^4 + 2\Omega_1^2\Omega_2^2\cos(ka)}. \tag{2.40}$$

The dispersion relation (2.40) defines two bands, as shown in Figure 2.14 (b). Looking at frequency ω as a function of k, there are two solutions for the frequency ranges $[0, \Omega_1]$ and $[\Omega_2, \sqrt{\Omega_1^2 + \Omega_2^2}]$. We can thus notice the appearance of a band gap for frequencies between Ω_1 and Ω_2: there are no propagating Bloch waves in this frequency range. The same observation holds for all frequencies above $\sqrt{\Omega_1^2 + \Omega_2^2}$.

Suppose we place some mechanism at the origin to excite waves inside the bilinear chain of masses, for instance some system vibrating at frequency ω. If the frequency is within one of the two pass-band ranges, we can expect to see waves outgoing to the left and the right from the origin, with a phase velocity given by the ratio ω/k that can be read from Figure 2.14 (b). In order for this phase velocity to be defined uniquely, the range of wavenumbers must be restricted to the first Brillouin zone, shown inside the

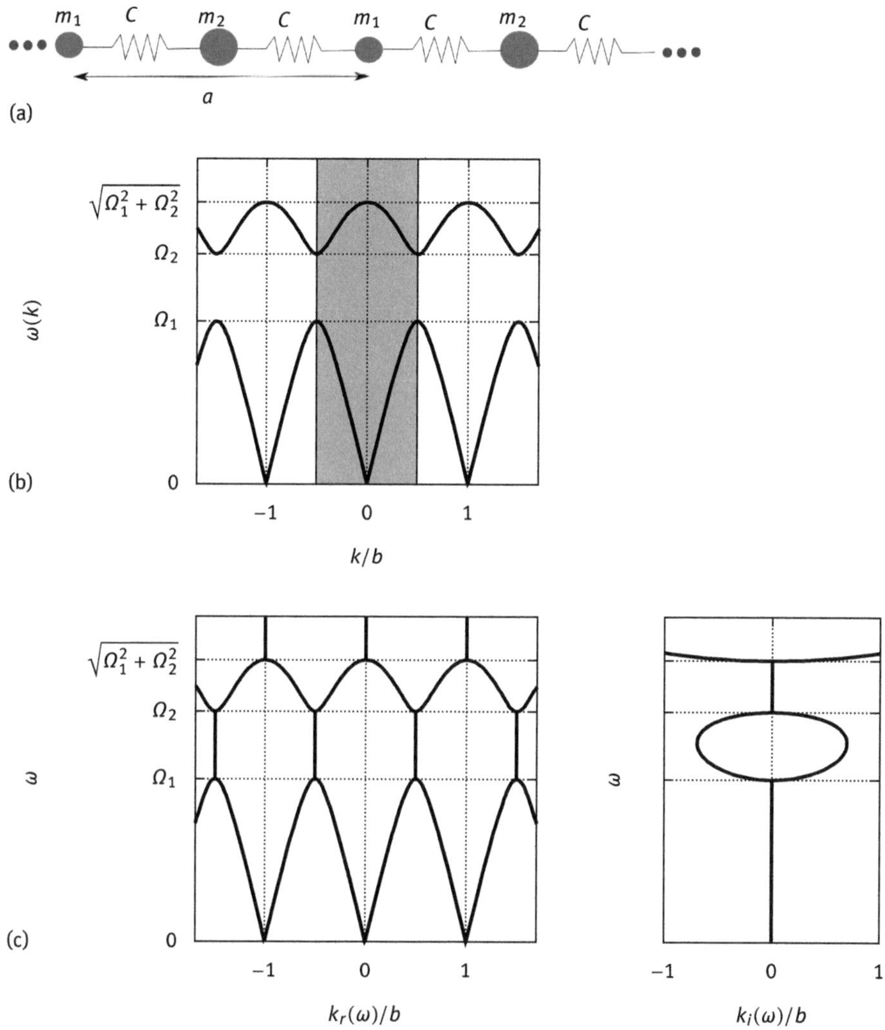

Fig. 2.14: A bilinear chain of masses coupled by springs. (a) The masses m_1 and m_2 are assumed to be punctual and connected by springs with stiffness C. The one-dimensional chain is periodic and infinite, with a lattice constant a, but there are two "atoms" per unit cell. (b) The dispersion relation (2.40) gives the frequency of propagating Bloch waves as a function of wavenumber. $b = 2\pi/a$ is the reciprocal lattice constant. $\Omega_1^2 = 2C/m_1$ and $\Omega_2^2 = 2C/m_2$. The first Brillouin zone for the one-dimensional lattice is outlined in gray. (c) The complex dispersion relation (2.41) and (2.42) give the real part (left panel) and the imaginary part (right panel) of k as a function of ω. Note that the periodicity of the real part of k is not shared by its imaginary part. Bloch waves with complex values of the wavenumber are evanescent.

gray rectangle in the figure. As we will argue in Section 2.4, this restriction is a direct consequence of periodicity. But what happens if the frequency is within one of the two band gap ranges where no propagating Bloch wave exists?

In order to answer that question, let us solve for the wavenumber as a function of frequency in (2.39). For any frequency for which a Bloch wave exists, we will find the same solutions as before, obviously. Frequencies inside the first band gap are such that the left-hand side of (2.39) is negative; frequencies in the second band gap are such that this quantity exceeds $\Omega_1^2\Omega_2^2$; in both cases no real-valued solution can be found for k. We thus let the wavenumber become complex so that the dispersion relation can be satisfied. Writing $k = k_r + \imath k_i$, and using the identity

$$\cos(x + \imath y) = \cos x \cosh y + \imath \sin x \sinh y,$$

we can solve the dispersion relation as follows.

- Inside the first band gap, $k_r a = \pi\,(2\pi)$, so that $\cos(ka) = -\cosh(k_i a)$. The notation $\pi\,(2\pi)$ means π modulo 2π. We find two complex bands as

$$k_i a = \pm\cosh^{1}\left(1 - \left(\frac{\omega^2}{\Omega_1^2} - 1\right)\left(\frac{\omega^2}{\Omega_2^2} - 1\right)\right). \tag{2.41}$$

- Above the second band, $k_r a = 0\,(2\pi)$, so that $\cos(ka) = \cosh(k_i a)$. We find again two complex bands as

$$k_i a = \pm\cosh^{-1}\left(\left(\frac{\omega^2}{\Omega_1^2} - 1\right)\left(\frac{\omega^2}{\Omega_2^2} - 1\right) - 1\right). \tag{2.42}$$

Figure 2.14 (c) displays the real part and the imaginary part of $k(\omega)$ as obtained from (2.41) and (2.42). This way of plotting the dispersion relation is a complex band structure and will be discussed specifically in Chapter 9. It is of course identical to the classical band structure $\omega(k)$ plotted in Figure 2.14 (b), providing it is limited to the frequency range of propagating Bloch waves. Inside the band gaps, a nonzero imaginary part of k indicates an evanescent wave. Actually, the Bloch wave expression (2.32) is explicitly in this one-dimensional case

$$u(x_1) = \exp(k_i x_1)\exp(-\imath k_r x_1)\tilde{u}(\mathbf{r}). \tag{2.43}$$

The first exponential term indicates either an exponentially increasing wave if $k_i x_1 > 0$, or an exponentially decreasing wave otherwise. If the boundary conditions of the problem command to select only decreasing solutions, as in the case of radiation toward infinity, then $k_i < 0$ must be chosen for the half space $x_1 > 0$ and conversely $k_i > 0$ must be chosen for the half space $x_1 < 0$. As Figure 2.14 (c) illustrates, both signs of k_i are always present, so this choice can always be made. We can now answer the question we asked previously: if the frequency is within one of the two band gap ranges where no propagating Bloch wave exists, then the response of the bilinear chain of masses will be in the form of a localized (nonpropagating) wave centered at the excitation point and decaying evanescently on both sides so that no energy can be transported toward infinity. The spatial extent of this localized wave is directly measured by the magnitude $|k_i|$, a larger value meaning a faster decay.

We could also have obtained complex values for $\omega(k)$ generalizing (2.40). We did not do so because we find it a bit less easy to attach a physical meaning to a complex frequency ω for a given real wavenumber k. The excitation would have a spatial dependence of the type $\exp(-\imath k x_1)$, but there are always a number of propagating Bloch waves with different real frequencies whatever the wavenumber, so it would be difficult to isolate complex frequencies in the response of the system. Anyway, coming back to the discussion of explicit and implicit dispersion relations, the $k(\omega)$ and the $\omega(k)$ dispersion relations are explicit, while the dispersion relation (2.39) is implicit. The latter contains the former, obviously, and has an even richer span of potential (ω, k) solutions than those we plotted in Figure 2.14.

One-dimensional sinusoidal grating. Let us consider a sinusoidal modulation of the celerity in the 1D wave equation

$$\frac{\partial^2 u}{\partial t^2} - c^2(x)\frac{\partial^2 u}{\partial x^2} = 0, \quad c^2(x) = c_0^2 + c_1^2 \sin(2\pi x/a) \tag{2.44}$$

as depicted in Figure 2.15 (a).

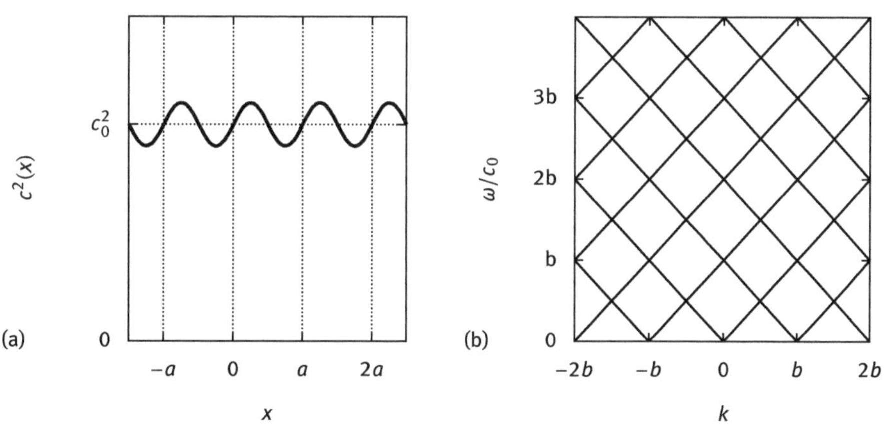

Fig. 2.15: A one-dimensional sinusoidal grating. (a) The sinusoidal modulation of the squared velocity extends to plus and minus infinity along axis x. (b) The dispersion relation in the limit that the amplitude of the sinusoidal modulation goes to zero is called the empty lattice model. In the one-dimensional case it is simply composed of two sets of parallels lines indicating left- and right-propagating Bloch waves.

Let us look for the Bloch waves of this equation in the form $u(t,x) = \tilde{u}(x)\exp(\imath(\omega t - kx))$. $\tilde{u}(x)$ is a periodic function with period a and thus admits the Fourier series representation

$$\tilde{u}(x) = \sum_{p=-\infty}^{\infty} \tilde{u}_p \exp(-2\imath\pi px/a). \tag{2.45}$$

According to the definition of the reciprocal lattice, here also one-dimensional, we put $b = 2\pi/a$. As a result, we can write

$$\frac{\partial^2 u(t,x)}{\partial t^2} = -\omega^2 \exp(\imath\omega t)\sum_p \tilde{u}_p \exp(-\imath(k+pb)x)$$

and

$$\frac{\partial^2 u(t,x)}{\partial x^2} = -\exp(\imath\omega t)\sum_p (k+pb)^2\tilde{u}_p \exp(-\imath(k+pb)x).$$

Inserting these expressions in the wave equation and noting that $\sin(bx) = \frac{1}{2\imath}(\exp(\imath bx) - \exp(-\imath bx))$ results in the infinite set of coupled equations

$$(c_0^2(k+pb)^2 - \omega^2)\tilde{u}_p + \frac{c_1^2}{2\imath}(k+(p+1)b)^2\tilde{u}_{p+1} - \frac{c_1^2}{2\imath}(k+(p-1)b)^2\tilde{u}_{p-1} = 0.$$

The matrix of this linear system of equations is Hermitian tridiagonal

$$\begin{pmatrix} \ddots & \ddots & 0 & & \\ -\frac{c_1^2}{2\imath}(k-2b)^2 & c_0^2(k-b)^2-\omega^2 & \frac{c_1^2}{2\imath}k^2 & 0 & \\ 0 & -\frac{c_1^2}{2\imath}(k-b)^2 & c_0^2k^2-\omega^2 & \frac{c_1^2}{2\imath}(k+b)^2 & \\ & 0 & -\frac{c_1^2}{2\imath}k^2 & c_0^2(k+b)^2-\omega^2 & \\ & & 0 & \ddots & \end{pmatrix}\begin{pmatrix} \vdots \\ \tilde{u}_{-1} \\ \tilde{u}_0 \\ \tilde{u}_1 \\ \vdots \end{pmatrix} = 0.$$

The system of equations is compatible only if the determinant vanishes, which will provide us with a secular dispersion relation in the end. Because the matrix is infinite, however, it is impossible the find the roots analytically. With a computer program, it would be straightforward to truncate the matrix to some finite size and to obtain the roots. Let us, however, outline a simple approximate solution. If the modulation is small, or $c_1^2 \ll c_0^2$, the determinant will be dominated by its diagonal term (which is just the product of all diagonal terms). If $c_1 = 0$, the determinant vanishes exactly for $c_0(k+pb) = \pm\omega$, for all p. This defines the empty lattice dispersion relation of the 1D periodic structure and is depicted in Figure 2.15 (b).

If c_1 is small, the determinant will vanish in the vicinity of the empty lattice dispersion relation, by continuity. We have in addition the exceptional case when two diagonal terms with different indices p and p' vanish simultaneously for the same k and ω. This happens whenever $k + pb = -(k+p'b) = \omega/c_0$. The first equality implies that $2k = (p'-p)b$, which is exactly the condition defining the crossings of the empty lattice model. Now if $|p' - p| > 1$ there is no off-diagonal coefficient in the matrix to couple the two amplitudes. Hence we can restrict the analysis to the case that $|p' - p| = 1$.

Let us specify the dispersion relation around the conditions $\omega = c_0 b/2$ and $k = b/2$, i.e. for $p' = 0$ and $p = -1$ (an equivalent possibility is to select any pair of integers such that $p' - p = 1$). In the infinite matrix, the only diagonal terms close to zero are those for p and p'. Then $\tilde{u}_p$ and $\tilde{u}_{p'}$ can be expected to have dominant values compared to all the other amplitudes. We can thus restrict the analysis to these two partial waves only. Correspondingly, we extract the submatrix of size 2×2

$$\begin{pmatrix} c_0^2(k-b)^2 - \omega^2 & \frac{c_1^2}{2\imath}k^2 \\ -\frac{c_1^2}{2\imath}(k-b)^2 & c_0^2k^2 - \omega^2 \end{pmatrix} \begin{pmatrix} \tilde{u}_{-1} \\ \tilde{u}_0 \end{pmatrix} = 0,$$

which amounts to neglecting all other Fourier amplitudes. The dispersion relation is given by the vanishing determinant

$$(c_0^2(k-b)^2 - \omega^2)(c_0^2k^2 - \omega^2) - \frac{c_1^4}{4}k^2(k-b)^2 = 0, \tag{2.46}$$

providing us with an implicit dispersion relation. In order to ease calculations we introduce the notations

$$\begin{aligned} \alpha &= \frac{c_1^2}{2c_0^2}, \\ k &= \frac{b}{2}(1+q), \\ \omega/c_0 &= \frac{b}{2}(1+\eta), \end{aligned}$$

with q and η assumed small compared to unity. The dispersion relation can be rewritten

$$(q - \eta - 2)(q + \eta)(q - \eta)(q + 2 + \eta) = \alpha^2(q+1)^2(q-1)^2.$$

A Taylor expansion limited to second-order terms gives

$$\eta^2 = q^2\left(1 - \frac{\alpha^2}{2}\right) + \frac{\alpha^2}{4}. \tag{2.47}$$

The dispersion relation $\omega(k)$ is thus parabolic around $k = b/2$ (or $q = 0$) with two branches that are symmetric with respect to the horizontal line $\omega/c_0 = b/2$ and given by

$$\omega(q)/c_0 = b/2\left(1 \pm \sqrt{q^2\left(1 - \frac{\alpha^2}{2}\right) + \frac{\alpha^2}{4}}\right). \tag{2.48}$$

The width of the band gap is $\Delta\omega/c_0 = \alpha b/2$, since the entrance and exit frequencies of the band gap are given by $\omega(q = 0)/c_0 = b/2(1 \pm \alpha/2)$ in (2.48).

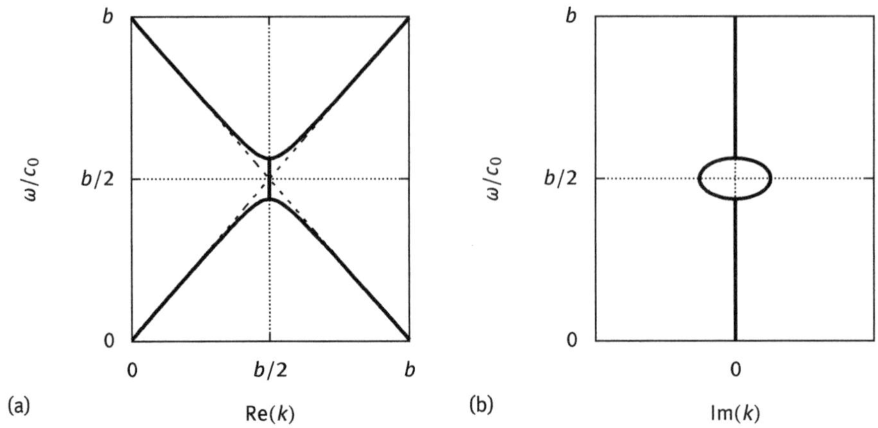

Fig. 2.16: Complex band structure $k(\omega)$ of the 1D sinusoidal grating. (a) The real part of the dispersion relation follows the empty lattice dispersion relation away from crossings where a band gap is formed. The band gap width is $\alpha b/2$, a measure of the modulation depth of the sinusoidal grating. (b) The imaginary part of the dispersion relation approximately forms a circle in the band gap range.

Let us analyze the complex band structure $k(\omega)$ in the vicinity of the band gap. We obtain from (2.47)

$$q(\eta) = \pm\frac{\sqrt{\eta^2 - \frac{\alpha^2}{4}}}{\sqrt{1 - \frac{\alpha^2}{2}}}. \tag{2.49}$$

This complex dispersion relation is plotted in Figure 2.16. Expression (2.49) shows that within the band gap the wavevector becomes complex imaginary since $\eta^2 - \frac{\alpha^2}{4} \leq 0$. The largest imaginary value is obtained at the center of the band gap (for $\eta = 0$) and is approximately $q = \pm\imath\frac{\alpha}{2}$. The two imaginary branches thus approximately form a circle connecting the lower and the upper real bands.

The solution technique we have just used can be termed a coupling of modes technique. The coupled equations actually relate the amplitudes $\tilde{u}_p \exp(-\imath(k+pb)x)$ of the Bloch waves of the empty lattice or unperturbed periodic medium. The significance of the result we have obtained is that the coupling of these amplitudes is only important around the crossings of the dispersion relations of the unperturbed Bloch waves. In this picture, we can say that the band gap opens as a result of the interference of a right-propagating Bloch wave with a left-propagating Bloch wave.

One-dimensional general grating. We can generalize the previous sinusoidal grating by considering an arbitrary periodic modulation of the celerity in the 1D wave equation. We will only outline the derivation, which is straightforwardly extended from the purely sinusoidal case, and give the main conclusions. Expression (2.44) is generalized to

$$\frac{\partial^2 u}{\partial t^2} - c^2(x)\frac{\partial^2 u}{\partial x^2} = 0, \quad c^2(x) = \sum_{q=-\infty}^{\infty} c_q^2 \exp(\imath qbx) \tag{2.50}$$

with $b = 2\pi/a$ as before. We can look for the Bloch waves of this equation by inserting the Fourier series expansion of $c^2(x)$ in the wave equation. We will end up with a linear system of coupled equations for the amplitudes of the unperturbed Bloch waves of the 1D lattice

$$\sum_{q=-\infty}^{\infty} c_q^2(k + (p+q)b)^2 \tilde{u}_{p+q} = \omega^2 \tilde{u}_p.$$

In matrix form, this system is

$$\begin{pmatrix} \ddots & \ddots & \ddots & \ddots & \ddots \\ & c_0^2(k-b)^2 - \omega^2 & c_1^2k^2 & c_2^2(k+b)^2 & \\ \cdots & c_{-1}^2(k-b)^2 & c_0^2k^2 - \omega^2 & c_1^2(k+b)^2 & \cdots \\ & c_{-2}^2(k-b)^2 & c_{-1}^2k^2 & c_0^2(k+b)^2 - \omega^2 & \\ \ddots & \ddots & \ddots & \ddots & \ddots \end{pmatrix} \begin{pmatrix} \vdots \\ \tilde{u}_{-1} \\ \tilde{u}_0 \\ \tilde{u}_1 \\ \vdots \end{pmatrix} = 0.$$

For any pair of integers p and $p' = p + q$ such that two diagonal terms with different indices vanish simultaneously, we can expect the corresponding unperturbed Bloch waves to become coupled. Coupling is then provided through the off-diagonal matrix elements containing $c_{\pm q}^2$. This happens whenever $k + pb = -(k + p'b) = \omega/c_0$, further implying $2k = -(p' + p)b = -(2p + q)b$ and $\omega/c_0 = -qb/2$. As a result, band gaps are expected to appear at all crossings of the empty lattice model, with band gap widths directly depending on the value of $c_{\pm q}^2/c_0^2$. The edges of the Brillouin zones are reached for odd values of q; their centers are reached for even values of q other than zero; no band gap opens at the center of the first Brillouin zone and $\omega = 0$ ($q = 0$ is not allowed).

2.3.2 Two- and three-dimensional cases

Wave propagation in 2D and 3D periodic media in general cannot be treated analytically as we have just done in a few 1D examples. In the subsequent chapters we will introduce theoretical and numerical tools to tackle them in the case of sonic and phononic crystals. Here we want to emphasize some physical concepts specific to wave propagation in two and three dimensions, and to introduce the concept of the Bragg band gap in a heuristic fashion based on them.

Figure 2.17 (a) shows an example calculation of the 2D scattering of a plane harmonic wave on a small cylindrical object. The practical implementation of this computation is explained in Chapter 3; it is based on solving the acoustic Helmholtz equation for pressure waves using the scattered field method (see Section 3.3). A plane harmonic

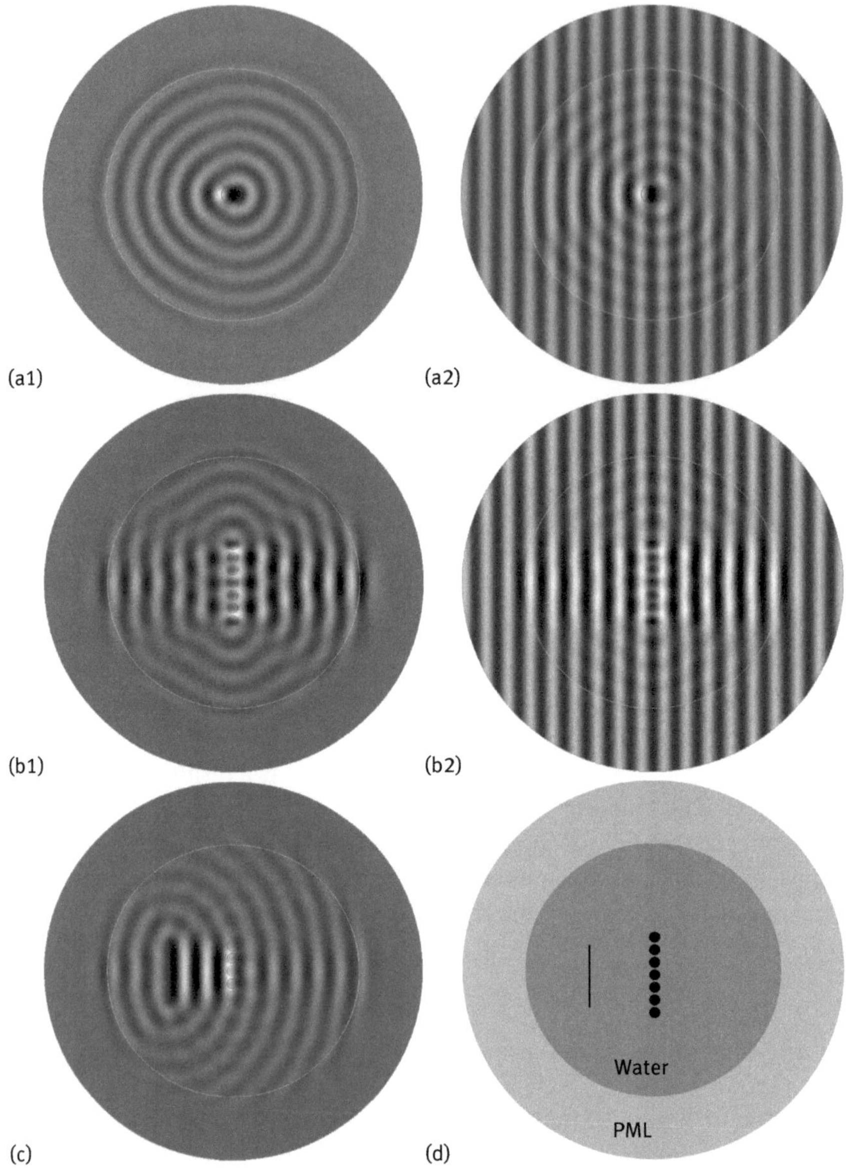

Fig. 2.17: Scattering of a harmonic plane wave by cylindrical inclusions. Inclusions are made of steel and placed in water. The frequency is 850 kHz in all examples. Incidence is horizontal and from the left to the right. A perfectly matched layer (PML) terminates the computation domain. (a) A single cylinder with radius 0.8 mm scatters the incident wave in all space, re-emitting cylindrical waves in the far field. Both the scattered field (a1) and the total field (a2) are shown. (b) 7 cylinders form a row with separation $a = 1$ mm. The radius of the cylinders is now 0.4 mm. Both the scattered field (b1) and the total field (b2) are shown. (c) A line source with height $5a$ is added to the left of the 7-cylinder row. A unit pressure at the line source is enforced as an internal boundary condition. (d) Sketch of the computational domain in case (c).

wave of the form $u_i = \exp(\imath(\omega t - k_1 x_1))$ is incident along direction x_1 in a homogeneous and isotropic medium with velocity c_1. A cylindrical object with a different velocity c_2 is placed somewhere in space. The diameter of the object is roughly of the order of one wavelength as measured in the incident medium. The small object introduces a heterogeneity that redistributes the incident wave in all space. The scattered field $u_s = u - u_i$ plotted in Figure 2.17 (a1) measures this change, with u the actual wavefield plotted in Figure 2.17 (a2). Of course, $u = u_i + u_s$ can also be said to be the sum of the incident and of the scattered fields. In the near field, the scattered field appears mostly as a dipole with an excess of pressure in front of the cylinder, caused by reflection on its boundary, and low pressure inside it. But in the far field, the wavefronts become more and more circular in shape as they escape from the scatterer. This is consistent with Property 2.5 in Section 2.1.2, considering the scatterer a secondary source of waves.

Next we form a row of seven cylinders parallel to the incident wavefronts, as shown in Figure 2.17 (b). The diameter of the cylinders is now about half a wavelength. The distance between cylinders, a, is chosen in anticipation of introducing a Bragg band gap in a sonic crystal composed of such steel cylinders in water. Being smaller, the individual scatterers are less efficient than the single scatterer case we considered before, but their combination as a row is able to create a scattered field forming vertical lines similar to those of the incident plane harmonic wave.

Instead of an incident plane harmonic wave, which is after all only a mathematical idealization, it is tempting to try to include a source of pressure waves that would resemble an actual ultrasonic transducer. In Figure 2.17 (c) this is implemented by introducing a line source along which a unit pressure is imposed. The height of the line source is only $5a$, or a bit less than 3 wavelengths in water. For such a small height, the waves emitted by the source line undergo a strong natural diffraction, diverging fast as they propagate away. Compared to the incident harmonic plane wave case in Figure 2.17 (b), a standing wave appears between the source and partial reflection on the row of cylinders; in the transmitted wavefield after the cylinders there is to some degree a re-collimation of the waves emitted by the small line source. Diffraction and collimation in sonic and phononic crystals will be discussed specifically in Chapter 12.

A square-lattice sonic crystal is next formed by considering four rows of seven steel cylinders, as shown in Figure 2.18. With the lattice constant and the material combination chosen, there is in principle a band gap around the frequency of 850 kHz, as we will show in Chapter 4. The existence of the band gap will be demonstrated for an infinite periodic crystal, but how does it manifest in a finite crystal with only a small number of rows? The scattered field shown in Figure 2.18 (a1) is concentrated before the entrance interface and after the exit interface between plain water and the crystal. On the reflection side, the scattered waves are approximately plane and in phase with the incident field: interference is constructive and reflection is enhanced. On the transmission side, the scattered waves are also approximately plane but in phase opposition with the incident field: interference is destructive and transmission tends to

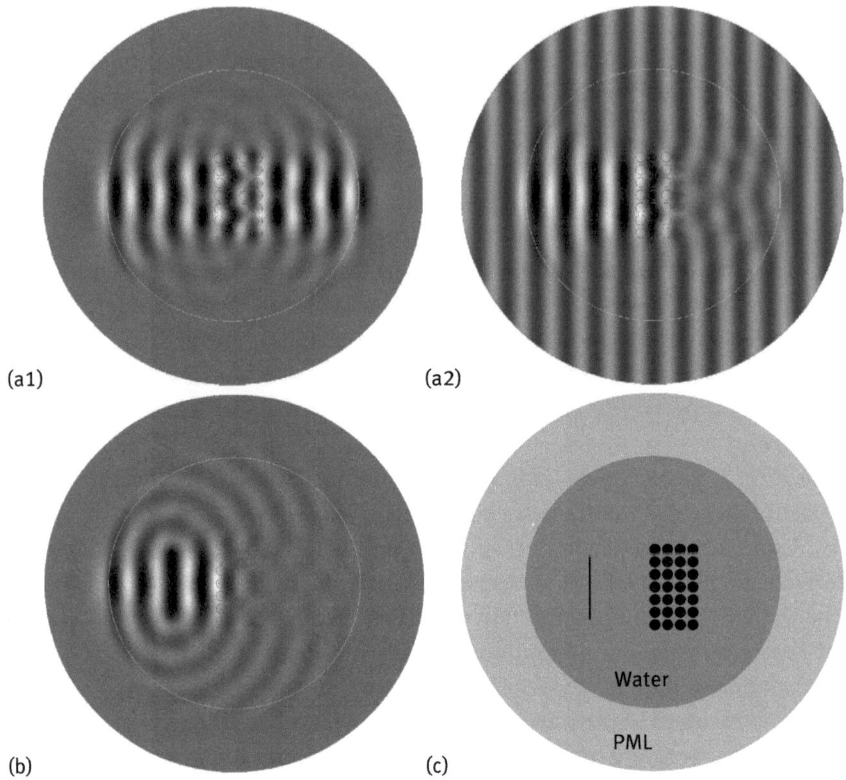

Fig. 2.18: Appearance of a Bragg band gap in a finite-size square-lattice sonic crystal. A 7×4 array of steel cylinders in placed in water. The radius of the cylinders is 0.4 mm and the lattice constant $a = 1$ mm. The frequency is 850 kHz. Incidence is horizontal and from the left to the right. A perfectly matched layer (PML) terminates the computation domain. (a) Numerical simulation is first performed in the scattered field formulation with an incident harmonic plane wave. Both the scattered field (a1) and the total field (a2) are shown. (b) Numerical simulation is then performed by adding a line source with height $5a$. (c) Sketch of the computational domain.

be canceled. This is a direct illustration of the mechanism of Bragg interference. The addition of more crystal rows reinforces reflection and accordingly decreases transmission. Various illustrations of the Bragg band gap mechanism will be shown in later chapters.

2.3.3 Local resonance

We have already discussed at length the Bragg mechanism for opening band gaps. Local resonances can also open band gaps, as we will now illustrate with a simple one-dimensional example that is adapted from reference [161]. A more in-depth discussion of local resonance is proposed in Chapter 10.

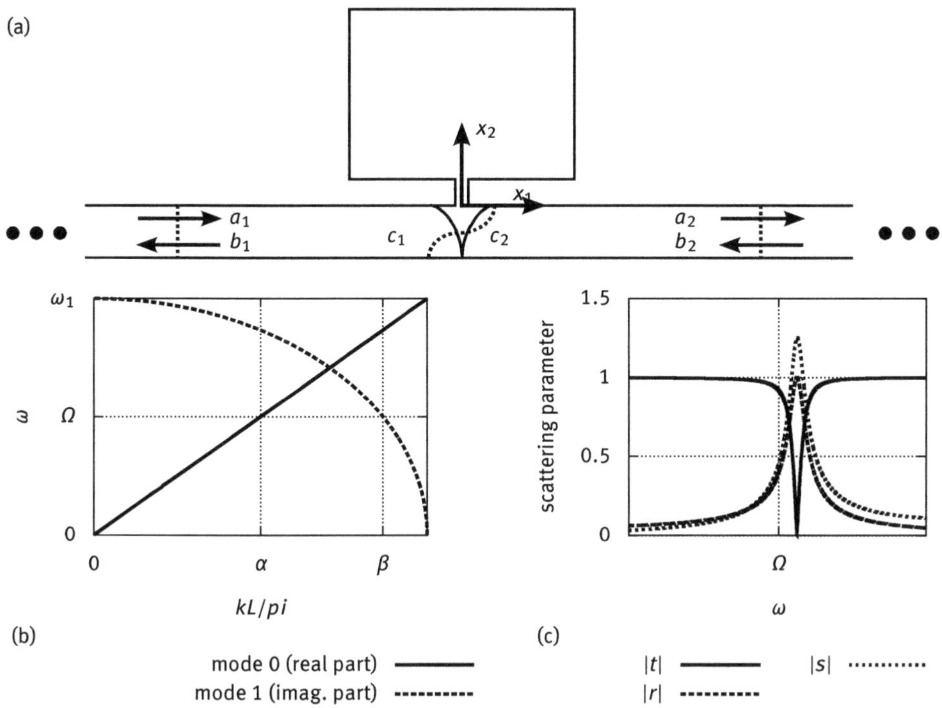

Fig. 2.19: A single resonator grafted on a waveguide. (a) Schematic with definitions of transmission channels (see text). The two-dimensional waveguide is single-mode within the frequency range of interest. Neumann boundary conditions apply on the waveguide walls. The width of the waveguide is L. (b) Dispersion relation of the fundamental guided mode (mode 0) and of the first evanescent guided mode (mode 1). ω_1 is the cut-off frequency of the first guided mode, Ω is the resonance frequency of the isolated resonator in (2.52). α and β are propagation constants. (c) Plot of the reflection ($|r|$), transmission ($|t|$), and evanescence ($|s|$) coefficients as a function of frequency.

A single resonator grafted on a waveguide. Let us consider a generic waveguide supporting only one propagating mode in the frequency range of interest. We also say that the waveguide is single-mode. As an example, the waveguide in Figure 2.11 supports only the propagation of the fundamental mode for angular frequencies $\omega < \pi c/L$, with L the width of the waveguide. We consider Neumann boundary conditions on the waveguide walls, so that the fundamental guided mode is nondispersive. A practical implementation is the air tube [161].

As depicted in Figure 2.19, there is a resonator connected to the side of the waveguide. For our purpose, we will not need to precisely explicate the details of the resonator. Its resonant behavior can be simply understood by stating that even a vanishingly small excitation can set it into motion, providing it is applied with the correct frequency. A model of this behavior is the forced harmonic oscillator with equation

$$Ku + M\frac{\partial^2 u}{\partial t^2} = F\exp(\imath\omega t), \tag{2.51}$$

with M a mass, K a stiffness, and F a force. u here represents displacement of the center of mass with respect to an equilibrium position. The same simple model is found with optical or electrical resonators, for instance, with a different physical meaning attached to the parameters of the model (i.e. dielectric constant, or capacitance and inductance). Under harmonic motion, $u = \bar{u}\exp(\imath\omega t)$ and the resonator equation becomes

$$(K - \omega^2 M)\bar{u} = F.$$

With $\Omega = \sqrt{\frac{K}{M}}$ the oscillation amplitude $\bar{u}$ has the Lorentzian frequency dependence

$$\bar{u} = \frac{F}{M}\frac{1}{\Omega^2 - \omega^2} = \frac{F}{M}L(\omega). \tag{2.52}$$

Obviously, $\bar{u}$ becomes infinite at the resonance condition $\omega = \Omega$.

Now, we wish to describe the interaction of the waveguide with the resonator, and ultimately to compute a reflection and a transmission for waves inside the waveguide. As the waveguide is single-mode for the frequency range of interest, the fundamental guided mode is a propagation channel for waves traveling to the right and to the left. The field to the left of the resonator and far from it can be written as the superposition of a right-propagating and of a left-propagating guided wave

$$p_0(\mathbf{x}) = \psi_0(x_2)\left(a_1 e^{-\imath\beta x_1} + b_1 e^{+\imath\beta x_1}\right), \tag{2.53}$$

with wavenumber $k_1 = \pm\beta = \pm\omega/c$. Similarly, the field to the right of the resonator and far from it can also be written

$$p_0(\mathbf{x}) = \psi_0(x_2)\left(a_2 e^{-\imath\beta x_1} + b_2 e^{+\imath\beta x_1}\right). \tag{2.54}$$

The function $\psi_0(x_2)$ is the transverse distribution of the fundamental guided mode; in the 2D case it can be taken as a constant. In close proximity to the resonator, however, the situation is different. Far from the connection between waveguide and resonator, the force density is identically zero at equilibrium. In the vicinity of the connection, however, the boundary conditions are modified and the field inside the waveguide induces a force that can drive the resonator. Accordingly, energy from the waveguide will be dragged by the resonator and will set it into motion. Reciprocally, the motion of the resonator acts back on the waveguide and changes its equilibrium very locally. As a useful idealization, we will consider that the connection can be reduced to a single point at $x_1 = 0$ where the force density is proportional to a Dirac delta function $\delta(x_1)$. Such a localized excitation of the waveguide results in the excitation of bound modes composed of evanescent waves of the waveguide. Coming back to the dispersion relation of the waveguide depicted in Figure 2.11, all modes with number $n \geq 1$ are evanescent with $k_{1n}(\omega) = \pm\imath\sqrt{(n\pi/L)^2 - \omega^2/c^2}$. For simplicity, we will only retain the least evanescent partial wave with $n = 1$ and introduce the notation $\alpha = \sqrt{(\pi/L)^2 - \omega^2/c^2}$. Figure 2.19 (b) shows the dispersion relations of the fundamental and of the evanescent guided modes. There is thus a second channel for transmission through the waveguide, centered on the connection point and such that

$$p_1(\mathbf{x}) = \begin{cases} c_1\psi_1(x_2)e^{\alpha x_1} & \text{if } x_1 < 0, \\ c_2\psi_1(x_2)e^{-\alpha x_1} & \text{if } x_1 > 0. \end{cases} \tag{2.55}$$

The function $\psi_1(x_2)$ is the transverse distribution of the first evanescent guided mode; in the 2D case it can be taken as proportional to $\cos(\pi x_2/L)$. The appearance of evanescent waves inside the waveguide can be understood as arising from the Green's function of the waveguide under the excitation $\delta(x_1)$. At any position along the waveguide except at $x_1 = 0$, the total field is the superposition $p(\mathbf{x}) = p_0(\mathbf{x}) + p_1(\mathbf{x})$.

At the connection point, we will assume that the field p is continuous. Because the two waveguide modes are orthogonal there result the two conditions

$$a_1 + b_1 = a_2 + b_2, \tag{2.56}$$

$$c_1 = c_2. \tag{2.57}$$

If the field p is continuous at the junction, its first derivative $\partial p/\partial x_1$ may not necessarily be. Actually, p_1 in (2.55) is clearly discontinuous, and so should the superposition be. We write the jump of $\partial p/\partial x_1$ as

$$\left[\frac{\partial p}{\partial x_1}\right]_{0^-}^{0^+} = L(\omega)\kappa(x_2)p(0, x_2) \tag{2.58}$$

which can be obtained by integration of the wave equation in the waveguide subject to a force density $L(\omega)\kappa(x_2)p(\mathbf{x})\delta(x_1)$, meaning that the resonator was initially set into motion by the field distribution $p(\mathbf{x})$ at the junction, with a resonant line shape given by the Lorentzian $L(\omega)$, and creates back a stress distribution with some cross-section dependence indicated by $\kappa(x_2)$, some function of x_2. Projecting this equation on the two modes, we obtain the two equations

$$\imath\beta(-a_2 + b_2) - \imath\beta(-a_1 + b_1) = L(\omega)(\kappa_{00}(a_1 + b_1) + \kappa_{01}c_2),$$
$$-\alpha(c_1 + c_2) = L(\omega)(\kappa_{10}(a_1 + b_1) + \kappa_{11}c_2),$$

with $\kappa_{ij} = \int dx_2\kappa(x_2)\psi_i(x_2)\psi_j(x_2)$. The presence of the discontinuity thus acts as a mixer of the two types of modes, which are otherwise completely uncoupled along the waveguide. In obtaining this set of equations, however, we have created two equations for the amplitudes of $\psi_0(x_2)$ and $\psi_1(x_2)$ from only one, so we must add the compatibility relation that the determinant of matrix κ_{ij} vanishes, or $\kappa_{00}\kappa_{11} - \kappa_{01}\kappa_{10} = 0$.

We define the coefficients $\gamma_{0j} = L(\omega)\kappa_{0j}/(2\imath\beta)$ and $\gamma_{1j} = L(\omega)\kappa_{1j}/(2\alpha)$. The equation system relating outgoing amplitudes (a_2 and b_1) to incoming amplitudes (a_1 and

b_2) can then be written

$$\begin{pmatrix} 1 & -1 & 0 \\ \frac{1}{2} & \frac{1}{2}+\gamma_{00} & \gamma_{01} \\ 0 & \gamma_{10} & 1+\gamma_{11} \end{pmatrix} \begin{pmatrix} a_2 \\ b_1 \\ c_2 \end{pmatrix} = \begin{pmatrix} 1 & -1 \\ \frac{1}{2}-\gamma_{00} & \frac{1}{2} \\ -\gamma_{10} & 0 \end{pmatrix} \begin{pmatrix} a_1 \\ b_2 \end{pmatrix}. \tag{2.59}$$

The solution to this linear system can be obtained exactly as

$$\begin{pmatrix} a_2 \\ b_1 \\ c_2 \end{pmatrix} = \begin{pmatrix} t & r \\ r & t \\ s & s \end{pmatrix} \begin{pmatrix} a_1 \\ b_2 \end{pmatrix} \tag{2.60}$$

with

$$t = \frac{1+\gamma_{11}}{D}, \qquad r = t-1,$$
$$s = \frac{-\gamma_{10}}{D}, \quad D = (1+\gamma_{00})(1+\gamma_{11}) - \gamma_{01}\gamma_{10}\,.$$

r and t have the meaning of a reflection coefficient and of a transmission coefficient, respectively. s measures the part of the incoming amplitudes that is stored in the evanescent wave. Because of the compatibility relation, $D = 1+\gamma_{00}+\gamma_{11}$. Far from resonance, the γ_{ij} coefficients are negligible and it follows that $t = 1$, i.e. waves are simply transmitted through the waveguide without reflection and significant excitation of the resonator. r and t are seen to be related by the simple formula $r = t-1$, implying that if the transmission vanishes, then the incident wave is fully reflected (with a phase shift of π). The condition for $t = 0$ is given by $1+\gamma_{11} = 0$, which can happen at the angular frequency

$$\omega_0^2 = \Omega^2 + \frac{\kappa_{11}}{2\alpha}. \tag{2.61}$$

Transmission canceling mostly occurs thanks to the Lorentzian line shape: even if the coupling strength is small, since an undamped resonator will be able to exactly cancel the transmission through the waveguide at a frequency close to but different from the natural frequency Ω.

Figure 2.19 (c) shows the variation of the coefficients $|t|$, $|r|$, and $|s|$ as a function of frequency. The coupling coefficients are given the arbitrary values $\kappa_{00} = 0.05$, $\kappa_{11} = 0.2$, and $\kappa_{01} = \kappa_{10} = 0.1$. Note the sharp drop to zero of the transmission coefficient at locally-resonant frequency ω_0, in agreement with formula (2.61) (upward frequency shift). At that frequency, transmission is completely canceled and a standing wave is formed as a result of the interference of the incident wave with the reflected wave, as Figure 2.20 illustrates. The numerical computation in this figure is conducted with the finite element and for pressure waves in air as explained in Chapter 3. In particular, radiation boundary conditions are implemented at both ends of the waveguide in order to cancel spurious reflections.

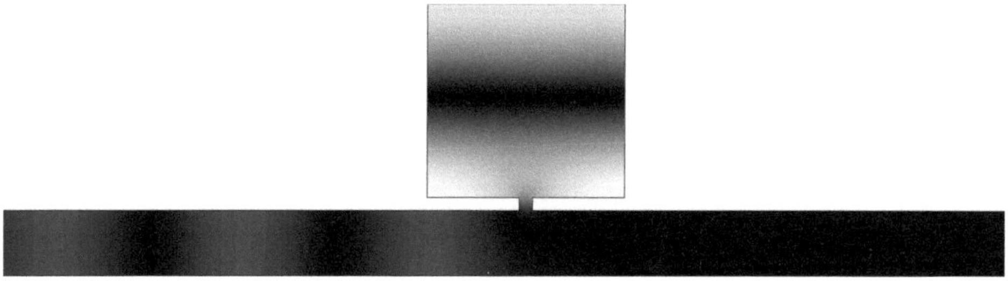

Fig. 2.20: A single resonator grafted on a waveguide. Numerical simulation of the pressure distribution in an air tube with a grafted resonator at exactly the locally-resonant condition (2.61). For this frequency, transmission is completely canceled. Note the much stronger pressure distribution inside the resonator compared to the waveguide.

A periodic array of resonators grafted on a waveguide. Let us now consider a one-dimensional crystal of resonators grafted on a waveguide, as depicted in Figure 2.21. We rewrite the scattering matrix of (2.60), limited to the amplitudes of the fundamental mode, as a transmission matrix.

$$\begin{pmatrix} a_2 \\ b_2 \end{pmatrix} = \begin{pmatrix} \frac{t^2-r^2}{t} & \frac{r}{t} \\ -\frac{r}{t} & \frac{1}{t} \end{pmatrix} \begin{pmatrix} a_1 \\ b_1 \end{pmatrix}. \tag{2.62}$$

It can be easily verified that the determinant of this transmission matrix is unity. We wish to relate the modal amplitudes at both ends of the unit cell of the crystal. We thus use the propagators

$$\begin{pmatrix} a_2' \\ b_2' \end{pmatrix} = \begin{pmatrix} e^{-\imath\beta a/2} & 0 \\ 0 & e^{+\imath\beta a/2} \end{pmatrix} \begin{pmatrix} a_2 \\ b_2 \end{pmatrix} \quad \text{and} \quad \begin{pmatrix} a_1 \\ b_1 \end{pmatrix} = \begin{pmatrix} e^{-\imath\beta a/2} & 0 \\ 0 & e^{+\imath\beta a/2} \end{pmatrix} \begin{pmatrix} a_1' \\ b_1' \end{pmatrix}$$

to obtain

$$\begin{pmatrix} a_2' \\ b_2' \end{pmatrix} = \begin{pmatrix} \frac{t^2-r^2}{t} e^{-\imath\beta a} & \frac{r}{t} \\ -\frac{r}{t} & \frac{1}{t} e^{+\imath\beta a} \end{pmatrix} \begin{pmatrix} a_1' \\ b_1' \end{pmatrix} = T \begin{pmatrix} a_1' \\ b_1' \end{pmatrix}.$$

We can now apply Bloch's theorem. As the problem is scalar, we can write Bloch waves as $p_0(\mathbf{x}) = \bar{p}_0(\mathbf{x}) \exp(-\imath k x_1)$ with $\bar{p}_0(\mathbf{x})$ periodic. In particular, we have the periodic Bloch boundary conditions

$$p_0(a/2, x_2) = \exp(-\imath k a) p_0(-a/2, x_2).$$

Such periodic boundary conditions are somehow weaker than the original Bloch theorem, but are sufficient to determine the solution of the problem. Indeed, they show that $\lambda = \exp(-\imath k a)$ is an eigenvalue of the transmission matrix T. In order to obtain the eigenvalues, we can look for the zeros of the characteristic polynomial $\det(T - \lambda I)$, with I the identity matrix. A direct calculation shows that

$$\det(T - \lambda I) = \lambda^2 + 1 - \frac{\lambda}{t}[(t^2 - r^2)e^{-\imath\beta a} + e^{+\imath\beta a}].$$

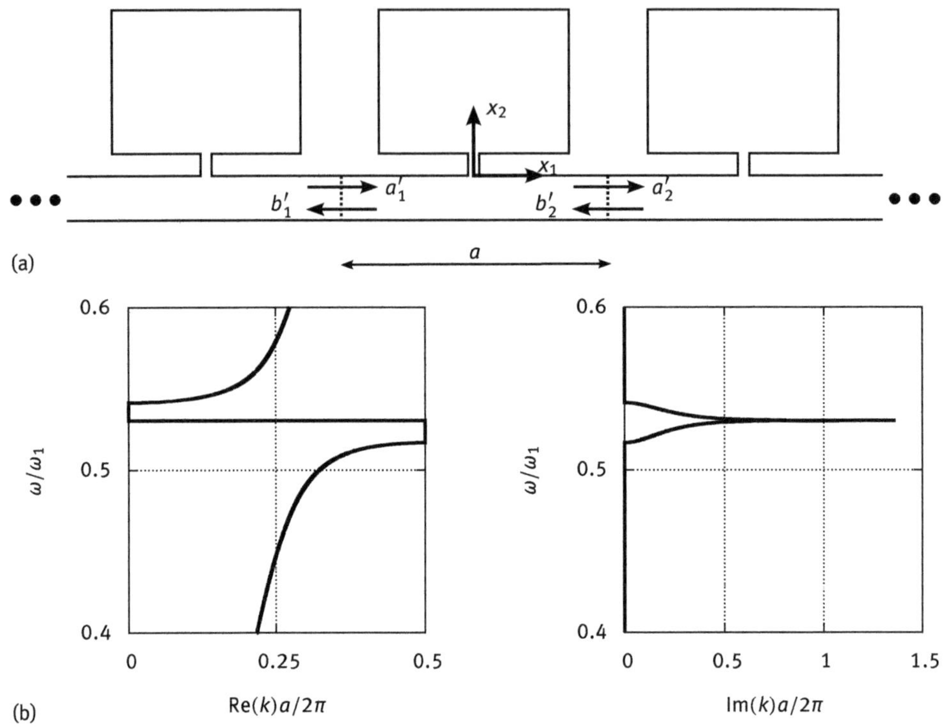

Fig. 2.21: A periodic array of resonators grafted on a waveguide. (a) The resonators are arranged periodically along the waveguide with a lattice constant a. (b) The complex band structure shows that around the locally-resonant frequency ω_0 the imaginary part of the wavenumber becomes nonzero, indicating evanescence within a band gap range. The real part of the wavenumber concurrently undergoes a π-phase shift at frequency ω_0, in connection with the zero of transmission at this frequency.

Next we note that $t - r = 1$ and $t^2 - r^2 = t + r$, but also that

$$\lambda + \frac{1}{\lambda} = 2\cos(ka).$$

$\det(T - \lambda I) = 0$ then leads to the dispersion relation

$$\cos(ka) = \cos(\beta a) + \imath \frac{r}{t} \sin(\beta a).$$

A smooth dependence with frequency enters via the definition of β, but a sharp dependence around the resonance frequency enters via r and t.

To understand the physical meaning of this dispersion relation, let us first remark that whenever reflection is negligible ($r = 0$), transmission is unity ($t = 1$) and the dispersion relation reduces to $k = \pm\beta$, the dispersion of the fundamental guided mode. Physically, the waves propagating along the waveguide do not see the local resonator and nothing particular happens.

Now, as we approach resonance, t becomes smaller and r increases. As a result the second term (in $\sin(\beta a)$) will dominate the first one ($\cos(\beta a)$) as resonance is approached. It is actually useful to express

$$\frac{r}{t} = \frac{t-1}{t} = \frac{-\gamma_{11}}{1+\gamma_{22}} = \imath \frac{L(\omega)\alpha\kappa_{11}}{2\alpha\beta + L(\omega)\beta\kappa_{22}}.$$

Finally, we can express the dispersion relation as

$$\cos(ka) = \cos(\beta a) - \frac{L(\omega)\alpha\kappa_{11}}{2\alpha\beta + L(\omega)\beta\kappa_{22}} \sin(\beta a).$$

In the absence of loss, $L(\omega)$ is purely real. Assuming further that the coupling constants κ_{ij} are real, the right-hand side is real also. The resonant term in factor of $\sin(\beta a)$ goes to $\pm\infty$ at frequency ω_0. Around the resonant frequency, there is a range of frequencies where k becomes complex, defined by the condition $|\cos(ka)| > 1$. A band gap is thus introduced by the periodic array of resonators grafted on the waveguide.

Figure 2.21 displays the shape of the locally-resonant band gap in the complex band structure. It can be noticed that this shape is quite different from those obtained previously for the bi-atomic chain (Figure 2.14) and for the 1D sinusoidal modulation (Figure 2.16). Here, the real part of the wavenumber reaches 0 at the band gap entrance (defined by $\cos(ka) = \pm 1$), stays constant until the resonant frequency ω_0 where it suddenly flips to π/a, reflecting a sign change as the resonance is crossed. It remains constant again until the band gap exit (defined by $\cos(ka) = \mp 1$) is reached. Concurrently, the imaginary part of the wavenumber forms an asymmetric peak centered on the resonant frequency ω_0.

2.4 Lattices, Brillouin zones and the band structure

In this section we give some definitions relative to the mathematical and physical description of crystals and more conceptually of periodic geometries. The presentation is not intended to be comprehensive and we only describe the lattices that will be most useful for subsequent chapters.

2.4.1 Bravais lattice

A Bravais lattice is an infinite array of discrete points generated by a set of discrete translation operations described by vectors

$$\mathbf{R} = n_1\mathbf{a}_1 + n_2\mathbf{a}_2 + n_3\mathbf{a}_3,$$

a notation we have already used in Section 2.2. The integers n_i are arbitrary and the vectors $\mathbf{a}_i$ are the primitive vectors. It is clear that seen from any position $\mathbf{R}$ the lattice looks exactly the same. In other words, it is translationally invariant.

A natural or artificial crystal is made up of a periodic arrangement of one or more atoms, or of a given unit cell with an internal structure. This basis is repeated at each lattice point, which is similar to a convolution of the punctual lattice with the unit cell. Consequently, the crystal is translationally invariant in the same discrete translations as the lattice.

Based on symmetry considerations,[2] there is only a finite number of distinct Bravais lattices in each dimension.

- In one-dimensional space there is just one type of Bravais lattice, obviously.
- In two-dimensional space, there are five Bravais lattices: oblique, rectangular, centered rectangular (rhombic), hexagonal, and square. They are depicted in Figure 2.22, together with a primitive cell (see next section).
- There are a total of 14 possible Bravais lattices in three-dimensional space. They are conventionally grouped in 7 lattice systems and possibly differentiated by lattice centerings according to the definitions:
  - Primitive (P): lattice points on the cell corners only.
  - Body (I): one additional lattice point at the center of the cell.
  - Face (F): one additional lattice point at the center of each of the faces of the cell.
  - Base (A, B or C): one additional lattice point at the center of each of one pair of the cell faces.

  Bravais lattices classified by lattice system and lattice centerings are listed in Table 2.1. The example of the cubic lattice system is depicted in Figure 2.23.

Tab. 2.1: The 14 three-dimensional Bravais lattices according to the seven lattice systems and four different lattice centerings. Each Bravais lattice is named according to the combination of lattice centering and system, in that order. For instance, the simple cubic (SC), the body-centered cubic (BCC), and the face-centered cubic (FCC) lattices belong to the cubic lattice system.

	Lattice centering			
Lattice system	**Simple (P)**	**Body-centered (I)**	**Face-centered (F)**	**Base-centered (A, B or C)**
Triclinic	P	–	–	–
Monoclinic	P	–	–	C
Orthorhombic	P	I	F	C
Tetragonal	P	I	–	–
Rhombohedral	P	–	–	–
Hexagonal	P	–	–	–
Cubic	P	I	F	–

2 Two Bravais lattices are said to be equivalent if they have isomorphic symmetry groups.

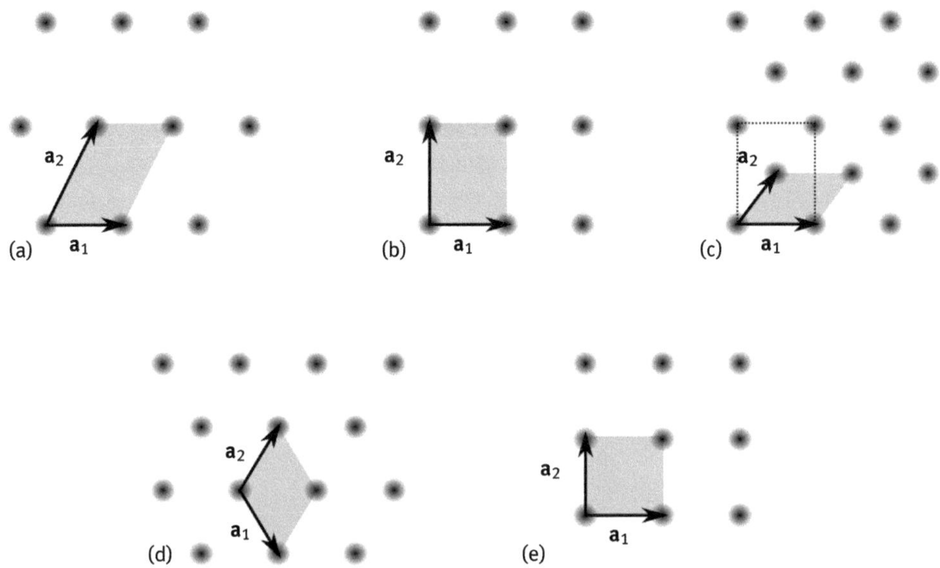

Fig. 2.22: The five two-dimensional Bravais lattices. The primitive cell is shown as the gray parallelogram. The angle between primitive vectors $\mathbf{a}_1$ and $\mathbf{a}_2$ is denoted ϕ. (a) The oblique lattice is such that $|\mathbf{a}_1| \neq |\mathbf{a}_2|$ and $\phi \neq \pi/2$. (b) The rectangular lattice is such that $|\mathbf{a}_1| \neq |\mathbf{a}_2|$ and $\phi = \pi/2$. (c) The centered rectangular lattice is such that $|\mathbf{a}_1| \neq |\mathbf{a}_2|$ and $\phi \neq \pi/2$, but $(2\mathbf{a}_2 - \mathbf{a}_1) \cdot \mathbf{a}_1 = 0$ (the initial rectangle is shown with the dotted line). (d) The hexagonal lattice is such that $|\mathbf{a}_1| = |\mathbf{a}_2|$ and $\phi = 2\pi/3$. (e) The square lattice is such that $|\mathbf{a}_1| = |\mathbf{a}_2|$ and $\phi = \pi/2$.

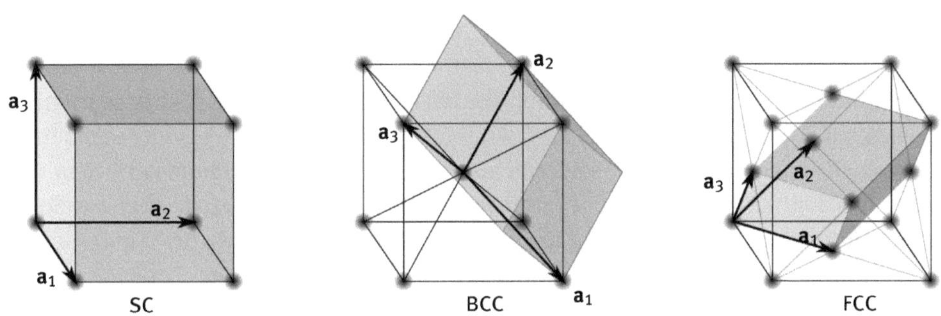

Fig. 2.23: Three-dimensional cubic Bravais lattices. Primitive vectors and primitive cell are depicted for the simple cubic (SC), the body-centered cubic (BCC), and the face-centered cubic (FCC) lattices.

The lattice system should not be confused with the crystal system which is a class of point groups. There are 7 different crystal systems that gather the 32 point groups in three-dimensional space. The concepts of crystal system and point group will be used to introduce the tensorial notation for elastic constants in Chapter 5. Sonic and phononic crystals will be explicitly classified according to their Bravais lattice and their symmetry properties will be discussed separately.

2.4.2 Primitive cell

The crystal structure of a material, be it natural or artificial, can be described in terms of its unit cell. The unit cell is any geometric "box" containing one or more "atoms" arranged in two- or three-dimensions. Unit cells stacked periodically form the crystal without leaving any empty space. There are actually an infinite number of possible unit cells in general.

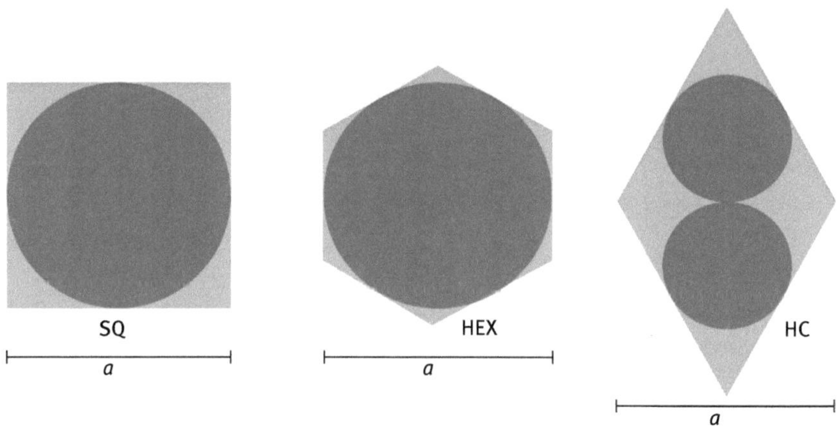

Fig. 2.24: Wigner–Seitz cells for the square lattice, the hexagonal lattice and the honeycomb lattice. The inclusions or "atoms" inside the cell are drawn for the close packing condition where they are touching but nonoverlapping between adjacent cells.

The primitive cell is a unit cell built on the basis of the primitive vectors, as we have depicted in Figures 2.12, 2.22, and 2.23. It has the minimum surface or volume required to pave all space. The Wigner–Seitz cell also has exactly this minimum surface or volume. It is defined around a lattice point as the locus of points in space that are closer to that lattice point than to any of the other lattice points. To construct it graphically, one may first pick a lattice point. Then, lines are drawn to all closest lattice points. At the midpoint of each line, another line (or plane) is drawn normal to each of the first set of lines. As an example, Figure 2.24 shows the Wigner–Seitz for the square lattice, the hexagonal lattice, and the honeycomb lattice, in two dimensions. The so-called honeycomb lattice is actually a hexagonal lattice with two "atoms" in the primitive cell, centered at positions $\pm(\mathbf{a}_1 - \mathbf{a}_2)/3$. The "atoms" are drawn at the maximum filling fraction, just at the condition of close packing. Note that the Wigner–Seitz cells of the square and the honeycomb lattices are identical to the primitive cells, but that it is a hexagon for the hexagonal lattice (compare with Figure 2.22). The Wigner–Seitz cells of 3D Bravais lattices should ideally be practiced with 3D drawing software in order to gain some intuition with them. Representations can be found in specialized books and on the internet (e.g. Wikipedia).

Tab. 2.2: Some useful geometrical quantities for some Bravais lattices. Coordinates are given with respect to a reference Cartesian coordinate system. $a = |\mathbf{a}_i|$ for 2D lattices; a is the side length of the original cube for 3D cubic lattices. Filling fractions (FF) are defined with d the diameter of a cylindrical or spherical inclusion. The maximum filling fraction is given for the close packing condition.

Lattice	$\mathbf{a}_1$	$\mathbf{a}_2$	$\mathbf{a}_3$	S or V	FF	max. FF
SQ	$(a, 0)$	$(0, a)$	–	a^2	$\frac{\pi}{4}\frac{d^2}{a^2}$	$\frac{\pi}{4}$
HEX	$(\frac{a}{2}, -\frac{\sqrt{3}a}{2})$	$(\frac{a}{2}, \frac{\sqrt{3}a}{2})$	–	$\frac{\sqrt{3}a^2}{2}$	$\frac{\pi}{2\sqrt{3}}\frac{d^2}{a^2}$	$\frac{\pi}{2\sqrt{3}}$
HC	$(\frac{a}{2}, -\frac{\sqrt{3}a}{2})$	$(\frac{a}{2}, \frac{\sqrt{3}a}{2})$	–	$\frac{\sqrt{3}a^2}{2}$	$\frac{\pi}{\sqrt{3}}\frac{d^2}{a^2}$	$\frac{\pi}{3\sqrt{3}}$
SC	$(a, 0, 0)$	$(0, a, 0)$	$(0, 0, a)$	a^3	$\frac{\pi}{6}\frac{d^3}{a^3}$	$\frac{\pi}{6}$
BCC	$(\frac{a}{2}, \frac{a}{2}, -\frac{a}{2})$	$(-\frac{a}{2}, \frac{a}{2}, \frac{a}{2})$	$(\frac{a}{2}, -\frac{a}{2}, \frac{a}{2})$	$\frac{a^3}{2}$	$\frac{\pi}{3}\frac{d^3}{a^3}$	$\frac{\pi\sqrt{3}}{8}$
FCC	$(\frac{a}{2}, \frac{a}{2}, 0)$	$(0, \frac{a}{2}, \frac{a}{2})$	$(\frac{a}{2}, 0, \frac{a}{2})$	$\frac{a^3}{4}$	$\frac{2\pi}{3}\frac{d^3}{a^3}$	$\frac{\pi}{3\sqrt{2}}$

Table 2.2 lists some useful geometrical values for some usual Bravais lattices we will consider with sonic and phononic crystals.

2.4.3 Reciprocal lattice

The reciprocal lattice of a Bravais lattice is by definition the lattice in which the Fourier transform of the wavefield of the original lattice is represented. The reciprocal lattice is itself a Bravais lattice, and the reciprocal of the reciprocal lattice is the original lattice. Mathematically, the reciprocal lattice can be described as the set of all vectors **K** that satisfy the following identity for all lattice point position vectors **R**:

$$\exp(-i\mathbf{K}\cdot\mathbf{R}) = 1.$$

In the derivation of Bloch's theorem we have already met the required reciprocal lattice primitive vectors, defined from the relations

$$\mathbf{b}_j \cdot \mathbf{a}_i = 2\pi\delta_{ij},$$

which form the natural basis for reciprocal lattice vectors written as

$$\mathbf{K} = m_1\mathbf{b}_1 + m_2\mathbf{b}_2 + m_3\mathbf{b}_3.$$

The reciprocal lattice of the square lattice is also a square lattice. The reciprocal lattice of the hexagonal lattice is also hexagonal, but with an additional $\pi/6$ rotation in reciprocal space. The simple cubic Bravais lattice, with cubic primitive cell of side a, has for its reciprocal a simple cubic lattice with a cubic primitive cell of side $b = \frac{2\pi}{a}$.

The reciprocal lattice to a face-centered cubic lattice is the body-centered cubic lattice, and reciprocally.

The first Brillouin zone can be simply defined as the Wigner–Seitz cell of the reciprocal lattice. Knowledge of the Bloch waves inside it is thus sufficient to construct all Bloch waves of a periodic medium. Any wavevector outside the first Brillouin zone can be folded back inside because of periodicity, using some vector K with integer coordinates (m_1, m_2, m_3). This is the reason why band structures are plotted with respect to the first Brillouin zone. Further considering all of the symmetries of the unit cell in the point group of the lattice, the first Brillouin zone can be reduced to the irreducible Brillouin zone.

Figure 2.25 depicts the first Brillouin zones for the square and the hexagonal lattices in two dimensions. Some conventional high symmetry points are indicated in each case. Point Γ represents the zero wavevector and all its translations by a reciprocal lattice vector; it is conventionally taken at the center of the first Brillouin zone.

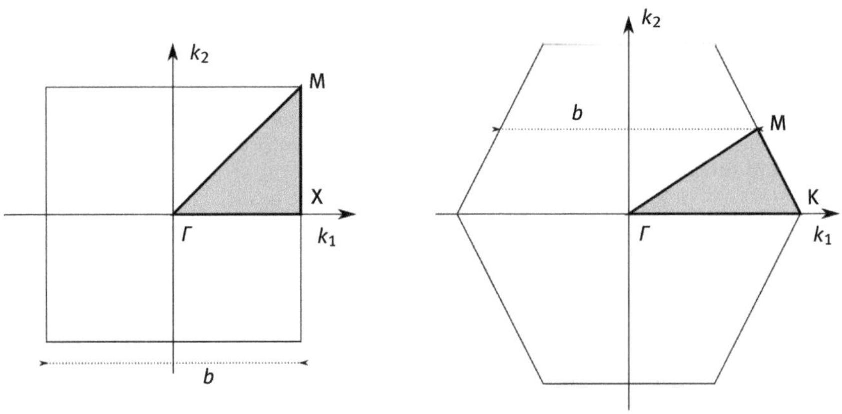

Fig. 2.25: First Brillouin zones for the square and the hexagonal lattices. $a = |\mathbf{a}_i|$; $b = 2\pi/a$. The irreducible Brillouin zone for a centro-symmetric unit cell is shaded in gray. The letters define high symmetry points.

Tab. 2.3: Coordinates of high symmetry points of the first Brillouin zone for some Bravais lattices. Coordinates are given with respect to a reference Cartesian coordinate system of reciprocal space. $a = |\mathbf{a}_i|$ for 2D lattices; a is the side length of the original cube for 3D cubic lattices; $b = 2\pi/a$.

SQ	X $(\frac{b}{2}, 0)$	M $(\frac{b}{2}, \frac{b}{2})$			
HEX	K $(\frac{2b}{3}, 0)$	M $(\frac{b}{2}, \frac{b}{2\sqrt{3}})$			
SC	X $(\frac{b}{2}, 0, 0)$	M $(\frac{b}{2}, \frac{b}{2}, 0)$	R $(\frac{b}{2}, \frac{b}{2}, \frac{b}{2})$		
BCC	H $(0, 0, b)$	P $(\frac{b}{2}, \frac{b}{2}, \frac{b}{2})$	N $(\frac{b}{2}, \frac{b}{2}, 0)$		
FCC	X $(0, b, 0)$	L $(\frac{b}{2}, \frac{b}{2}, \frac{b}{2})$	W $(\frac{b}{2}, b, 0)$	U $(\frac{b}{4}, b, \frac{b}{4})$	K $(\frac{3b}{4}, \frac{3b}{4}, 0)$

Points X and M for the square lattice, and K and M for the hexagonal lattice, are particular vertices of the first Brillouin zone boundary. Except for point K, they are midway between two adjacent nodes of the reciprocal lattice. According to our discussion of the empty lattice model, they are points in the band structure where the dispersion of oppositely propagating Bloch waves cross; they are thus possible loci for the appearance of Bragg band gaps. Many examples of this property will be met later; it was clearly apparent in the band structure in Figure 1.2 for a square-lattice sonic crystal. In addition, Figure 2.25 also shows the irreducible Brillouin zones for the case that the inclusions in the unit cell do not reduce the symmetries of the square and of the hexagonal lattices, as for instance drawn in Figure 2.24. Band structures are conventionally plotted for a path in reciprocal space that goes along the boundary of the irreducible Brillouin zone.

We do not specifically plot the Brillouin zone for 3D lattices, but representations can easily be found in specialized books or internet webpages. For reference, we give the coordinates of the high symmetry points of the Brillouin zones for the square, the hexagonal, the simple cubic, the body-centered cubic, and the face-centered cubic lattices in Table 2.3. These coordinates are typically useful to write computer programs to obtain band structures.

3 Acoustic waves

This chapter proposes a synthetic presentation of the subject of acoustic waves in fluid media, limited to linear elasticity. The contents are purposely limited to the necessary background for understanding sonic crystals.

3.1 Dynamical equations of acoustic waves

3.1.1 1D acoustic equations

In this section, we present a basic derivation of linear acoustic equations, limited to small deformations. For simplicity, we start with the one-dimensional case.

Lagrangian and Eulerian descriptions. Let us consider a continuous, isotropic, and perfectly compressible fluid. The Lagrange variables are attached to the material point, for instance a particular molecule of the fluid. We denote $\mathbf{a}(t)$ the equilibrium position at time t. Any physical quantity can be thought of as a function of this position and of time: $G(t, \mathbf{a})$. The Euler variables, in contrast, only pay attention to geometrical points of a given coordinate system: coordinate $\mathbf{x}$ and time t. The same physical quantity as before will then be denoted $g(t, \mathbf{x})$. Of course, the two descriptions aim at describing the same reality and they must be connected. The position of any material point is a certain trajectory in the Eulerian coordinate system that we write as $\mathbf{x} = \mathbf{X}(t, \mathbf{a})$. It follows that $G(t, \mathbf{a}) = g(t, \mathbf{X}(t, \mathbf{a}))$. This basic relation can be specialized to the displacement by writing

$$\mathbf{U}(t, \mathbf{a}) = \mathbf{X}(t, \mathbf{a}) - \mathbf{a} = \mathbf{u}(t, \mathbf{X}(t, \mathbf{a})). \tag{3.1}$$

This relation is illustrated by Figure 3.1.

The particle velocity is $\mathbf{V}_p = \frac{\partial \mathbf{U}}{\partial t} = \frac{\partial \mathbf{X}}{\partial t}$ in the Lagrangian description, while the local velocity is $\mathbf{v} = \frac{\partial \mathbf{u}}{\partial t}$ in the Eulerian description. Given the relation (3.1) between Lagrangian and Eulerian descriptions, we have by the rules of differentiation of functions of several variables

$$(V_p)_i = \frac{\partial u_i(t, \mathbf{X}(t, \mathbf{a}))}{\partial t} = v_i + (V_p)_j \frac{\partial u_i}{\partial x_j}. \tag{3.2}$$

The approximation of linear acoustics is that every component of the gradient of the displacement vector is much smaller than unity, or $|\frac{\partial u_i}{\partial x_j}| \ll 1$. Under this approximation, the strain $S = \nabla \cdot \mathbf{u} = \frac{\partial u_i}{\partial x_i}$ is also much smaller than unity, or $|S| \ll 1$, and the particle and the local velocities are equal to the first order of approximation, or

$$\mathbf{V}_p \approx \mathbf{v}.$$

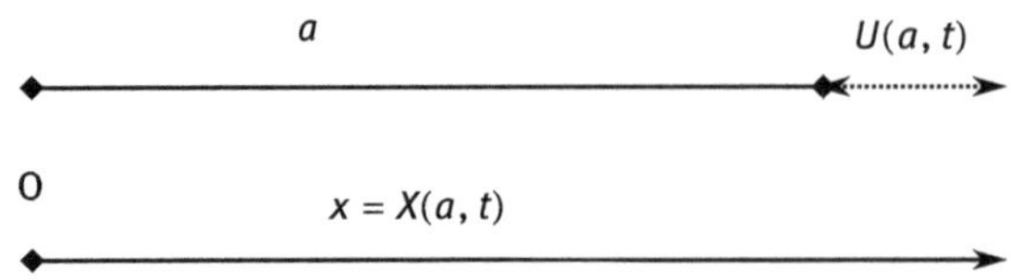

Fig. 3.1: Relation between Lagrangian and Eulerian coordinates.

There is then no longer any need to distinguish between the Lagrangian and the Eulerian descriptions.

Let us consider the motion of a slice of fluid of surface σ and extending from x to $x + \mathrm{d}x$, such as depicted in Figure 3.2. We assume that $\mathrm{d}x$ is very small and will eventually let it go to zero as a limit. The left-hand side of the slice experiences a displacement u, while the right-hand side experiences a displacement $u + \mathrm{d}u$. We note that

$$\mathrm{d}u = \frac{\partial u(t, x)}{\partial x}\,\mathrm{d}x = S\,\mathrm{d}x$$

is much smaller than $\mathrm{d}x$ because of the approximation of linear acoustics. The coordinate of the left-hand side is $x + u$, while the coordinate of the right-hand side is $x + \mathrm{d}x + u + \mathrm{d}u \approx x + \mathrm{d}x + u$.

We assume that an acoustic wave is characterized by the pressure field $P(t, x)$ it induces within the fluid. The elementary pressure force acting on the slice is

$$\mathrm{d}F = \sigma(P(t, x + u) - P(t, x + u + \mathrm{d}x)),$$

where care has been taken that the force on each side points inward. To first order, the elementary pressure force is then approximated by

$$\mathrm{d}F \approx -\sigma\frac{\partial P(t, x + u)}{\partial x}\,\mathrm{d}x = -\frac{\partial P(t, x + u)}{\partial x}\,\mathrm{d}V,$$

with $\mathrm{d}V = \sigma\,\mathrm{d}x$ the volume of the fluid slice. The units of pressure are Pa, or $\mathrm{N/m^2}$.

We can now apply Newton's first law to the slice of fluid. We neglect any force other than pressure, including gravity. We have

$$-\frac{\partial P(t, x)}{\partial x} = \rho\frac{\partial^2 u(t, x)}{\partial t^2}, \tag{3.3}$$

where ρ is the (static) density of the fluid, with units $\mathrm{kg/m^3}$. In writing this relation, we have approximated the mass of the slice as $\rho\,\mathrm{d}V$, thus neglecting any (first-order and above) variation of the fluid density. For the acceleration of the slice, the time derivative of the local velocity has been used.

Pressure is in fact the sum of a static pressure and of a dynamical pressure

$$P(t, x) = P_0 + p(t, x). \tag{3.4}$$

For instance, P_0 could be the atmospheric pressure for acoustic waves in air. If seeking a dynamic equation for the acoustic wave, we will only be interested in the variable

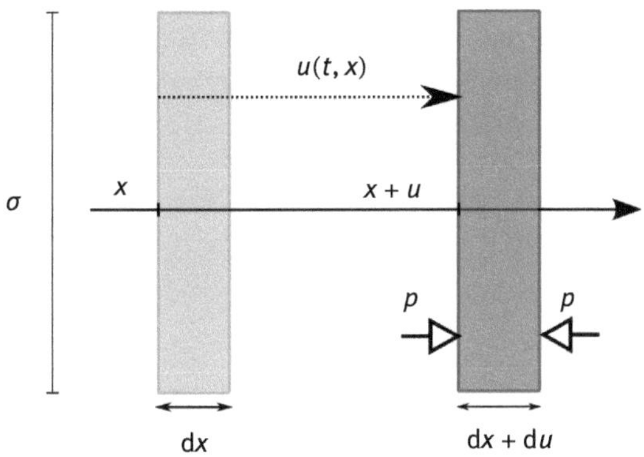

Fig. 3.2: Deformation of a slice of a fluid under pressure. In its equilibrium state, the slice has position x and thickness dx. Under the action of the pressure field $P(t, x) = p_0 + p(t, x)$, the slice is moving in time, reaching position $x + u(t, x)$, while its thickness is changed by $dx + du$ with $du \approx S\,dx$ and S the strain.

part $p(t, x)$. Indeed, P_0 does not contribute to the dynamics of the fluid slice, as prescribed by (3.3), if it is independent of space. For the compressible fluid model, we have the condition that the dynamical pressure is proportional to the local volume variation. Hence, a constitutive relation is written as

$$p = -B\frac{\delta(\mathrm{d}V)}{\mathrm{d}V} = -B\frac{\partial u}{\partial x} = -BS, \tag{3.5}$$

where B is an elastic modulus of compression (units: Pa); its inverse B^{-1} is the compressibility coefficient. It is implicit in this formula that the strain is considered a measure of the local dilation or compression of the fluid. In deriving (3.5), we have approximated $\delta(\mathrm{d}V) \approx \sigma((\mathrm{d}x + \mathrm{d}u) - \mathrm{d}x) = \sigma\,\mathrm{d}u$. The choice of sign is such that a positive elastic modulus implies that the volume of the fluid is diminished ($S < 0$) when it is compressed ($p > 0$).

1D homogeneous wave equation. Suppose first that both ρ and B are constants anywhere in the spatial domain. Then combining (3.3) and (3.5), we obtain the 1D homogeneous wave equation

$$\frac{\partial^2 u}{\partial t^2} - c^2\frac{\partial^2 u}{\partial x^2} = 0 \quad \text{or} \quad \frac{\partial^2 p}{\partial t^2} - c^2\frac{\partial^2 p}{\partial x^2} = 0, \tag{3.6}$$

with the acoustic velocity $c = (B/\rho)^{1/2}$ (units: m/s). The local velocity and the strain satisfy exactly the same scalar wave equation.

1D inhomogeneous wave equation. Suppose now that ρ and B are both functions of position x. They could even be discontinuous functions. This is the general case for sonic crystals, because the unit cell is inhomogeneous since the material constants change abruptly when passing from one material to another. The different physical quantities describing wave propagation then follow slightly different differential equations, though they are all still of the wave equation type.

Pressure satisfies

$$\frac{\partial}{\partial x}\left(\frac{1}{\rho}\frac{\partial p}{\partial x}\right) = \frac{\partial^2}{\partial t^2}\left(\frac{p}{B}\right), \tag{3.7}$$

while the displacement satisfies

$$\frac{\partial}{\partial x}\left(B\frac{\partial u}{\partial x}\right) = \rho\frac{\partial^2 u}{\partial t^2}. \tag{3.8}$$

The main difference between the two equations comes from the fact that the material constants cannot be factored out of the parentheses enclosing partial derivatives since ρ and B depend on x.

In addition, it is seen that the strain follows yet another slightly different equation

$$\frac{\partial}{\partial x}\left(\frac{1}{\rho}\frac{\partial}{\partial x}(BS)\right) = \frac{\partial^2 S}{\partial t^2}. \tag{3.9}$$

Acoustic impedance. Assuming again a homogeneous acoustic medium, we can come back to the general 1D solution to the wave equation outlined in Chapter 2:

$$u(t,x) = F(t - x/c) + G(t + x/c),$$

with F and G two arbitrary functions (arbitrary, though at least continuous and twice differentiable). We have

$$v(t,x) = \frac{\partial u}{\partial t} = \dot{F}(t - x/c) + \dot{G}(t + x/c)$$

and

$$p(t,x) = -B\frac{\partial u}{\partial x} = Z(\dot{F}(t - x/c) - \dot{G}(t + x/c)),$$

with the acoustic impedance $Z = \rho c = B/c = \sqrt{\rho B}$ (units: N.s/m^3). As a result, pressure and velocity are proportional for waves propagating to the right, $p^+ = Zv^+$, and for waves propagating to the left $p^- = -Zv^-$. It is important to notice that the sign changes before the acoustic impedance when propagating from left to right, and conversely. This remark will be important in the derivation of reflection and transmission coefficients at the interface between two homogeneous media.

3.1.2 3D acoustic equations

In this section, we reconsider the equations previously obtained for 1D acoustic wave propagation, in order to extend them to 3D propagation.

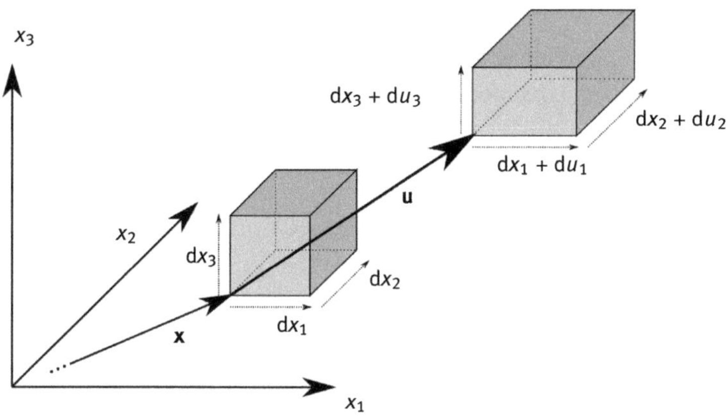

Fig. 3.3: Displacement and deformation of a small parallelepipedic volume $\mathrm{d}V$ under the action of an acoustic wave.

The total pressure is

$$P(t, \mathbf{x}) = P_0 + p(t, \mathbf{x}) \tag{3.10}$$

and is now a function of time t and of the position vector $\mathbf{x} = (x_1, x_2, x_3)^T$. Strain is still defined as the volume variation caused by the passing of the acoustic wave, or

$$S(t, \mathbf{x}) = \frac{\delta(\mathrm{d}V)}{\mathrm{d}V}.$$

With the linear acoustics frame, strain is approximated as

$$S(t, \mathbf{x}) \approx \nabla \cdot \mathbf{u} = u_{i,i}, \quad i = 1, 2, 3. \tag{3.11}$$

As an elementary proof, let us consider a parallelepiped defining volume $\mathrm{d}V$, as depicted in Figure 3.3. One corner is at position $\mathbf{x}$ and side lengths are $\mathrm{d}x_1$, $\mathrm{d}x_2$, and $\mathrm{d}x_3$. As before, the parallelepiped moves in time and space. The origin corner is displaced to position $\mathbf{x} + \mathbf{u}$ and the side lengths are changed to $\mathrm{d}x_i + \mathrm{d}u_i$, with $\mathrm{d}u_i \approx u_{i,i}\,\mathrm{d}x_i$ (no repeated index summation in the last formula). The volume of the parallelepiped is $\mathrm{d}V = \mathrm{d}x_1\,\mathrm{d}x_2\,\mathrm{d}x_3$. The variation of this volume is

$$\mathrm{d}V + \delta(\mathrm{d}V) = (\mathrm{d}x_1 + \mathrm{d}u_1)(\mathrm{d}x_2 + \mathrm{d}u_2)(\mathrm{d}x_3 + \mathrm{d}u_3),$$

so that to first order

$$\delta(\mathrm{d}V) \approx \mathrm{d}u_1\,\mathrm{d}x_2\,\mathrm{d}x_3 + \mathrm{d}x_1\,\mathrm{d}u_2\,\mathrm{d}x_3 + \mathrm{d}x_1\,\mathrm{d}x_2\,\mathrm{d}u_3,$$

and finally

$$S = \frac{\delta(\mathrm{d}V)}{\mathrm{d}V} \approx \frac{\mathrm{d}u_1}{\mathrm{d}x_1} + \frac{\mathrm{d}u_2}{\mathrm{d}x_2} + \frac{\mathrm{d}u_3}{\mathrm{d}x_3} \approx u_{i,i}.$$

Newton's first law writes

$$\rho\frac{\partial^2 \mathbf{u}}{\partial t^2} = -\nabla p \quad \text{or} \quad \rho\frac{\partial^2 u_i}{\partial t^2} = -p_{,i}. \tag{3.12}$$

This vector equation is a direct generalization of (3.3). In particular, this equation shows that the polarization of a plane wave is longitudinal: displacements are in the direction of propagation. Indeed, consider the plane wave described by pressure $p = p_0 F(t - s_i x_i)$, then

$$\rho \frac{\partial^2 u_i}{\partial t^2} = p_0 s_i F'(t - s_j x_j)$$

and the acceleration is aligned with the slowness vector **s**.

For a linear and compressible fluid, we have again the constitutive relation $p = -BS$. In the general case that the material constants are allowed to be discontinuous with the domain of definition, pressure and displacement vector satisfy the following wave equations

$$\nabla \cdot \left(\frac{1}{\rho}\nabla p\right) = \frac{\partial^2}{\partial t^2}\left(\frac{p}{B}\right) \tag{3.13}$$

and

$$\nabla \left(B \nabla \cdot \mathbf{u}\right) = \rho \frac{\partial^2 \mathbf{u}}{\partial t^2}. \tag{3.14}$$

The equation for pressure is a scalar wave equation while the equation for the displacement is a vector wave equation. The former is generally easier to solve than the latter, this is the reason why pressure is most often used as the independent variable for acoustic wave problems. In case the propagation medium is homogeneous, both equations simplify to a scalar or vector form of the wave equation

$$\frac{\partial^2 p}{\partial t^2} - c^2 \Delta p = 0 \quad \text{or} \quad \frac{\partial^2 p}{\partial t^2} - c^2 p_{,ii} = 0, \tag{3.15}$$

and

$$\frac{\partial^2}{\partial t^2}\mathbf{u} - c^2 \nabla(\nabla \cdot \mathbf{u}) = 0 \quad \text{or} \quad \frac{\partial^2 u_i}{\partial t^2} - c^2 u_{j,ji} = 0. \tag{3.16}$$

Both equations are straightforward generalizations of (3.6). The velocity of acoustic waves in the homogeneous fluid is still $c = (B/\rho)^{\frac{1}{2}}$.

3.1.3 Poynting's theorem for acoustic waves

Let us outline a result that is very useful for the analysis of the energy content of an acoustic wave, Poynting's theorem. We consider again (3.12), but now adding a body force term that drives the solution

$$\rho \frac{\partial \mathbf{v}}{\partial t} + \nabla p = \mathbf{f}.$$

Taking the scalar product with the velocity, we get

$$\rho \frac{\partial \mathbf{v}}{\partial t} \cdot \mathbf{v} + \nabla p \cdot \mathbf{v} = \mathbf{f} \cdot \mathbf{v}.$$

The right-hand side is the time derivative of the density of work of the body force. The first term of the left-hand side is the time derivative of the density of kinetic energy, or

$$\rho \frac{\partial \mathbf{v}}{\partial t} \cdot \mathbf{v} = \frac{\partial(\frac{1}{2}\rho \mathbf{v} \cdot \mathbf{v})}{\partial t}.$$

The second term can be transformed using the divergence relation

$$\nabla \cdot (p\mathbf{v}) = (\nabla p) \cdot \mathbf{v} + p(\nabla \cdot \mathbf{v})$$

to give

$$\nabla p \cdot \mathbf{v} = \nabla \cdot (p\mathbf{v}) - p(\nabla \cdot \mathbf{v}) = \nabla \cdot (p\mathbf{v}) + BS\frac{\partial S}{\partial t}.$$

Gathering all the terms, we obtain the relation

$$\mathbf{f} \cdot \mathbf{v} = \frac{\partial(\frac{1}{2}\rho \mathbf{v} \cdot \mathbf{v})}{\partial t} + \frac{\partial(\frac{1}{2}BS^2)}{\partial t} + \nabla \cdot (p\mathbf{v}).$$

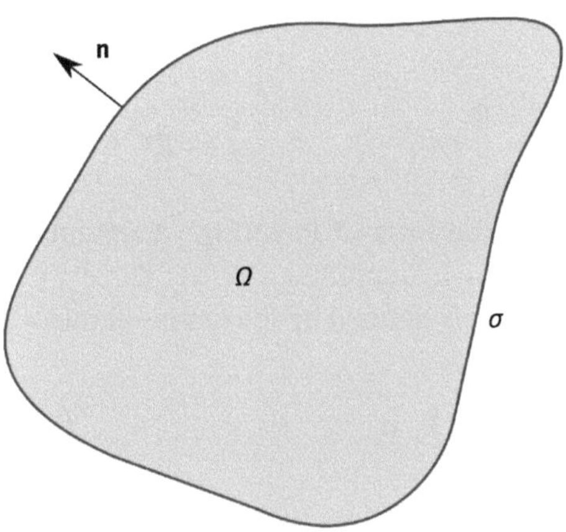

Fig. 3.4: Schematic representation of Poynting's theorem. The temporal variation of the internal energy of domain Ω is given by the work of internal forces, and is a balance of kinetic energy, potential (elastic) energy, and Poynting's vector flux through surface σ.

Figure 3.4 depicts some domain of space Ω that is enclosed by surface σ. Integrating the previous relation on Ω allows us to assign physical meanings to its different terms and leads to the following property.

Property 3.1 (Poynting's theorem). The temporal variation of the internal energy of the closed domain Ω is given by the work of internal forces W, and is a balance of kinetic energy E_k, potential (elastic) energy E_p, and the flux of Poynting's vector $\mathbf{P}$ through the enclosing surface σ. This relation reads mathematically

$$\frac{\partial w}{\partial t} = \frac{\partial e_k}{\partial t} + \frac{\partial e_p}{\partial t} + \nabla \cdot \mathbf{P} \quad \text{or} \quad \frac{\partial W}{\partial t} = \frac{\partial E_k}{\partial t} + \frac{\partial E_p}{\partial t} + \int_{\sigma} \mathbf{P} \cdot \mathbf{n}, \tag{3.17}$$

with

$$W = \int_{\Omega} \mathrm{d}\mathbf{x} w, \qquad \frac{\partial w}{\partial t} = \mathbf{f} \cdot \mathbf{v}, \tag{3.18}$$

$$E_k = \int_{\Omega} \mathrm{d}\mathbf{x} e_k, \qquad e_k = \frac{1}{2} \rho \mathbf{v} \cdot \mathbf{v}, \tag{3.19}$$

$$E_p = \int_{\Omega} \mathrm{d}\mathbf{x} e_p, \qquad e_p = \frac{1}{2} \frac{p^2}{B} = \frac{1}{2} B S^2, \tag{3.20}$$

$$\mathbf{P} = p\mathbf{v}. \tag{3.21}$$

The small letters w, e_k, and e_p are for the densities of the relevant quantities. The flux of the Poynting vector $\mathbf{P}$ actually measures the power transported by the wave. Note that $\mathbf{P}$ itself does not have a clear and unique physical meaning, contrary to its divergence at any position in space or its flux through a closed surface.

Furthermore, for complex harmonic plane waves with a $\exp(\imath(\omega(t - x/c)))$ dependence, one should instead use the formulas

$$e_k = \frac{1}{4} \rho \operatorname{Re}(\mathbf{v}^* \cdot \mathbf{v}), \quad e_p = \frac{1}{4B} \operatorname{Re}(p^* p) = \frac{B}{4} \operatorname{Re}(S^* S), \quad \mathbf{P} = \frac{1}{2} \operatorname{Re}(p\mathbf{v}^*).$$

Case of plane waves, intensity. Let us specify the form of Poynting's theorem for plane waves. While in this case $\mathbf{P} \cdot \mathbf{n}$ represents the instantaneous power density per unit surface transported by the wave, the intensity is defined by the temporal mean

$$I = \langle \mathbf{P} \cdot \mathbf{n} \rangle = \lim_{T \to \infty} \frac{1}{T} \int_0^T \mathrm{d}t \, \mathbf{P} \cdot \mathbf{n} \tag{3.22}$$

For a plane wave propagating in direction $\mathbf{n}$, we know that the displacements are longitudinal and we have

$$u_n = F(t - x/c), \quad v_n = \dot{F}(t - x/c), \quad p = Z\dot{F}(t - x/c) = Zv_n,$$

with $x = \mathbf{x} \cdot \mathbf{n}$. As a result,

$$\begin{aligned} e_k = e_p &= \frac{1}{2} \rho \dot{F}^2(t - x/c), \\ \mathbf{P} \cdot \mathbf{n} &= Z\dot{F}^2(t - x/c), \\ \mathbf{P} \cdot \mathbf{n} &= c(e_k + e_p), \end{aligned}$$

since $Z = \sqrt{\rho B} = \rho c$. The intensity is

$$I = Z \langle \dot{F}^2(t - x/c) \rangle .$$

For a given waveform, intensity is then mostly dependent on the acoustic impedance.

Case of harmonic plane waves. For harmonic plane waves of the form $F(t - x/c) = u_0 \sin(\omega(t-x/c))$, we have $v_n = \omega u_0 \cos(\omega(t-x/c)) = v_0 \cos(\omega(t-x/c))$. Accordingly,

$$e_k = e_p = \frac{1}{2}\rho\omega^2 u_0^2 \cos^2(\omega(t - x/c)).$$

As a result

$$\langle e_k \rangle = \langle e_p \rangle = \frac{1}{4}\rho\omega^2 u_0^2 = \frac{1}{4}\rho v_0^2.$$

Furthermore

$$\mathbf{P} \cdot \mathbf{n} = Z v_0^2 \cos^2(\omega(t - x/c)).$$

Hence

$$I = \frac{Z}{2} v_0^2 = \frac{1}{2Z} p_0^2.$$

3.1.4 Constants of fluids, loss

As the previous equations show, the relevant material properties of acoustic waves described using linear pressure waves are the mass density ρ and the bulk modulus B. If these quantities are known for every fluid material in the domain considered, then wave problems can be solved in principle. Table 3.1 lists a few representative numerical values for different fluids. These values will be used for the different examples in this book and they assume standard and controlled conditions. Actual values depend on temperature, static pressure, altitude, humidity (for air), salinity (for water), and so on. For experiments, consequently, it is generally safer to trust experimentally determined results, as acoustic velocities can be rather easily and precisely measured by delay-on-transmission experiments. The table also gives effective values for some isotropic solids. Solids of course do not support exclusively longitudinal waves, as we will see in Chapter 5. There are some situations, however, where it is practical to assume it is so to simplify calculations, and thus to treat a solid as an equivalent fluid. This is often the case in the literature about certain sonic crystals. In Chapter 8 we will discuss the limits of this approximation.

A simple model for acoustic loss. Let us introduce a simple phenomenological model for acoustic loss in fluids. It seems intuitive that a fluid cannot react instantly to a solicitation. Phenomenologically, and to first order in time, the constitutive relation can be modified to read

$$p = -B\left(S + \tau\frac{\partial S}{\partial t}\right), \tag{3.23}$$

where τ is some time constant.

For instance, if pressure were made abruptly discontinuous at $t = 0$, jumping from 0 to p_0, we could write $p(t) = p_0 H(t)$ where H is the Heaviside distribution. It is not difficult to check that strain then responds as

$$S(t) = -\frac{p_0}{B}(1 - \exp(-t/\tau))H(t).$$

Tab. 3.1: Material constants for acoustic waves in fluids, at room temperature ($T = 293$ K). The second part of the table gives values for isotropic solids considered as effective fluids, i.e. values for longitudinal waves are given ($B = c_{11}$).

	Density ρ kg/m^3	**Bulk modulus B** GPa	**Sound velocity c** m/s	**Impedance Z** N.s/m^3
Air	1.2041	1.42 10^{-4}	343.4	413.5
Water	1 000	2.2	1 483	1.483 10^6
Propanol	786	1.076	1 170	1.16 10^6
Mercury	13 500	28.38	1 450	19.6 10^6
Nylon	1 150	6.6	2 400	2.76 10^6
Plexiglas	1 190	9.0	2 750	3.27 10^6
Steel	7 780	264	5 825	45.3 10^6
PVC	1 560	7.8	2 236	3.49 10^6

Hence $S(t)$ continuously decreases from 0 to the minimum value $-p_0/B$ in roughly time τ and asymptotically becomes a constant.

Now what happens to the acoustic wave equation? The propagation equation (3.14) becomes

$$\rho \frac{\partial^2 \mathbf{u}}{\partial t^2} = \nabla \left(B \nabla \cdot \mathbf{u} + \tau \frac{\partial \nabla \cdot \mathbf{u}}{\partial t} \right). \tag{3.24}$$

This is no longer a wave equation, obviously, because of the mixed partial derivative with respect to time and space coordinates. Anyway, the Helmholtz equation for monochromatic waves can still be given a simple and meaningful form. Consider propagation in the x direction (all directions are equivalent, since the medium is assumed isotropic) of a harmonic plane wave $\exp(\imath(\omega t - kx)$. Substituting in (3.24) we obtain the complex dispersion relation

$$\omega^2 = c^2(1 + \imath\omega\tau)k^2. \tag{3.25}$$

The appearance of complex coefficients suggests to solve for the complex wavenumber and thus to write $k = k_r + \imath k_i$. This describes a harmonic plane wave whose amplitude depends on the propagation distance, as

$$\exp(\imath(\omega t - kx) = \exp(k_i x) \exp(\imath(\omega t - k_r x).$$

As a result, the amplitude of the plane wave is either exponentially increasing or decreasing, a situation that represents either gain or loss. Providing $\omega\tau \ll 1$, the wavenumber reads to second order

$$k_i \approx -\frac{\omega^2 \tau}{2c}, \tag{3.26}$$

$$k_r \approx \omega/c \left(1 - \frac{3}{8}(\omega\tau)^2 \right). \tag{3.27}$$

Hence, in the low frequency range, where low refers to frequencies much smaller than the inverse of the time constant τ, losses scale with the square of the frequency. The units of k_i are Np/m. The minus sign guarantees that the amplitude of the wave is indeed exponentially attenuated.

In practice, the dispersion relation (3.25) suggests that we complexify B as $B(1 + \imath\omega\tau)$. This defines a complex-valued and frequency-dependent bulk modulus that can readily be used in the Helmholtz equation derived from (3.13) or (3.14) for harmonic acoustic waves.

As a note, the simple model we have outlined gives the same exponential attenuation as the Stokes law of sound attenuation, providing we take

$$\tau = \frac{\frac{4}{3}\eta + \eta^V}{B} \tag{3.28}$$

with η the dynamic viscosity coefficient of the fluid and η^V its volume viscosity coefficient. For water at $T = 293$ K, $\eta \approx 1$ mPa.s and $\eta^V \approx 3$ mPa.s. As a result $\tau \approx 2$ ps. Viscosity values for gases and liquids can be found in the literature. They very significantly depend on temperature and other factors.

3.2 Reflection and refraction

We propose a brief explanation of the phenomena of reflection and refraction at an interface separating two nonviscous fluids, as depicted in Figure 3.5. This is a situation that frequently arises in inhomogeneous media, including sonic crystals.

Suppose a pressure wave is incident on the interface. We first want to write boundary conditions that connect the wavefields inside the two fluids. It is natural to consider that pressure is continuous across the interface if the latter is supposed to remain

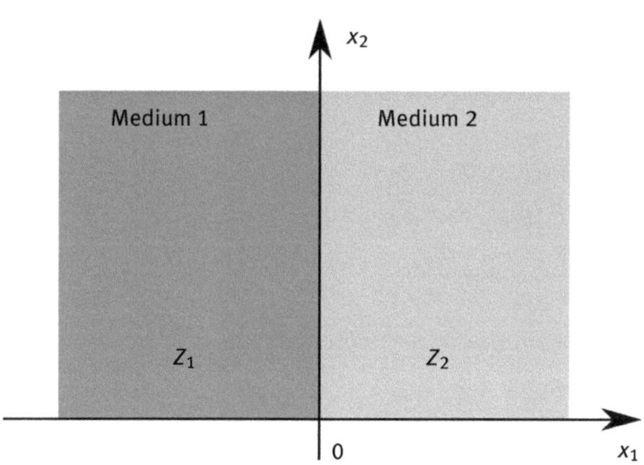

Fig. 3.5: Interface between two fluid media with acoustic impedances Z_1 and Z_2.

at equilibrium. It can furthermore be shown (see also Section 3.3.2) that the normal acceleration is further continuous. Mathematically, this boundary condition is obtained by application of the divergence theorem to (3.13) upon introducing a suitable closed contour around the interface. Assuming the interface is at coordinate $x_1 = 0$ and extends normally to the x_1 axis, we can write

- the continuity of acoustic pressure at the interface,

$$p^{(1)}(t; x_1 = 0, x_2, x_3) = p^{(2)}(t; x_1 = 0, x_2, x_3), \tag{3.29}$$

- and the continuity of the normal component of displacement at the interface,

$$u_1^{(1)}(t; x_1 = 0, x_2, x_3) = u_1^{(2)}(t; x_1 = 0, x_2, x_3). \tag{3.30}$$

  Similar relations hold for the normal components of velocity and acceleration.

For the numerical solution of scattering problems, these two boundary conditions are enough. Further analytical results – the Fresnel coefficients for reflection and transmission – can nonetheless be obtained in the case of plane waves.

Normal incidence for plane waves. A plane wave normally incident on the interface gives rise to both reflected and transmitted plane waves. The normal displacements at the interface are then

$$u_1^{(1)}(t; r) = F_i(t - x_1/c_1) + F_r(t + x_1/c_1)$$

and

$$u_1^{(2)}(t; r) = F_t(t - x_1/c_2).$$

At the interface ($x_1 = 0$), (3.29) implies

$$\dot{F}_i(t) + \dot{F}_r(t) = \dot{F}_t(t)$$

and (3.30) implies

$$Z_1(\dot{F}_i(t) - \dot{F}_r(t)) = Z_2\dot{F}_t(t).$$

From these equations, we obtain the reflection and transmission velocity coefficients as

$$r_v = \frac{\dot{F}_r(t)}{\dot{F}_i(t)} = \frac{Z_1 - Z_2}{Z_1 + Z_2}, \tag{3.31}$$

$$t_v = \frac{\dot{F}_t(t)}{\dot{F}_i(t)} = \frac{2Z_1}{Z_1 + Z_2}. \tag{3.32}$$

Similarly, the reflection and transmission pressure coefficients are

$$r_p = -\frac{\dot{F}_r(t)}{\dot{F}_i(t)} = \frac{Z_2 - Z_1}{Z_1 + Z_2}, \tag{3.33}$$

$$t_p = \frac{Z_2\dot{F}_t(t)}{Z_1\dot{F}_i(t)} = \frac{2Z_2}{Z_1 + Z_2}. \tag{3.34}$$

The velocity and pressure coefficients are connected by the very simple relations $r_p = -r_v$ and $Z_1 t_p = Z_2 t_v$. Though this may be puzzling at first sight, these relations simply remind us that acoustic waves can be represented either using their velocity or their pressure distributions, but that these cannot be independent. A practical pitfall would be to assume that a concept such as the reflection and transmission coefficients can be uniquely defined; as the formulas above show, whether the coefficients are defined for velocity or for pressure has to be specified.

As a note, it can easily be checked that $t_v = 1 + r_v$ and $t_p = 1 + r_p$. Overall, only one of the four amplitude coefficients we have defined contains all the necessary information on reflection and transmission at the interface.

Considering that acoustic power was previously related to the flux of Poynting's vector through a given surface, we can further define reflection and transmission acoustic power coefficients as

$$R = \frac{|P_r|}{|P_i|} = -r_v r_p = \left(\frac{Z_1 - Z_2}{Z_1 + Z_2}\right)^2 , \tag{3.35}$$

$$T = \frac{|P_t|}{|P_i|} = t_v t_p = \frac{4Z_1 Z_2}{(Z_1 + Z_2)^2} . \tag{3.36}$$

It is easily verified that $T + R = 1$, meaning that acoustic power is conserved upon reflection and transmission. This is fortunate, since there is no source of loss in the present problem, so that any departure from this relation would indicate a faulty derivation. $R = (1 - z)^2/(1 + z)^2$ is a function of the impedance contrast $z = Z_2/Z_1$ only and is plotted in Figure 3.6. It can be seen that total reflection ($R = 1$) is obtained as the impedance contrast goes either to zero or to infinity, which is just a matter of exchanging the roles of the incident and of the transmission medium. Reflection vanishes only if the impedance contrast $z = 1$. This can happen for media with different densities and bulk moduli, providing their acoustic impedances are equal. As a note, R is positive since r_v and r_p have opposite signs.

Almost total reflection is obtained for instance at the interface between air and water, for which the impedance contrast $z \approx 3600$ according to Table 3.1. For such a value, the reflection acoustic power coefficient $R \approx 0.998\,89$. Accordingly, the conversion of acoustic waves from air to water or the converse is very inefficient, and almost total reflection is the rule.

For acoustic waves incident from air onto water, the velocity and pressure induced into water are almost zero ($t_v = 0$ and $t_p = 0$). Concurrently, $r_v = -1$, meaning that the reflected velocity is in antiphase with the incident velocity (there is a π-phase shift between them). Normal velocities inside the fluid then vanish at the interface, as a result of the interference of incident and reflected waves. In addition, $r_p = 1$, meaning that the reflected pressure is in phase with the incident pressure; as a result, the total pressure is doubled at the interface as seen from air. The corresponding equivalent boundary condition is that of a "wall" causing the normal velocity to vanish at the interface ($v_1 = 0$). For the pressure acoustic equation (3.13) this is a Neumann boundary condition.

For acoustic waves incident from water onto air, the velocity and pressure induced into air are again almost zero ($t_v = 0$ and $t_p = 0$). In contrast, however, $r_v = 1$ and $r_p = -1$, meaning that the reflected pressure is in antiphase with the incident pressure. The total pressure in the incident medium then vanishes at the interface. The corresponding equivalent boundary condition is that of a "soft boundary" causing pressure to vanish at the interface ($p = 0$). For the pressure acoustic equation (3.13) this is a Dirichlet boundary condition.

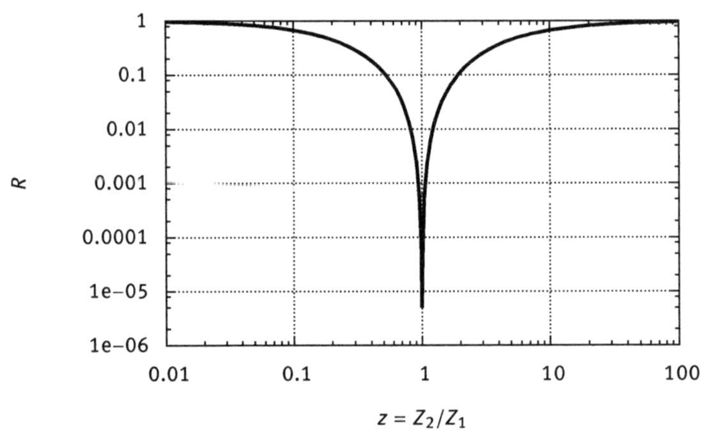

Fig. 3.6: Reflection acoustic power coefficient as a function of the impedance contrast Z_2/Z_1, for normal incidence.

Arbitrary incidence for an harmonic plane wave. Let us now consider an arbitrary incidence angle ϕ_i, defined with respect to the normal to the interface. For harmonic plane waves, the equality of pressures along the interface implies

$$A_i \exp(\imath(\omega_i t - k_{i2}x_2)) + A_r \exp(\imath(\omega_r t - k_{r2}x_2)) = A_t \exp(\imath(\omega_t t - k_{t2}x_2)).$$

This relation must hold at all times and for all points of the interface, which implies the following properties.

- Reflection and transmission on a static (nonmoving) interface occur without a change in frequency ($\omega_i = \omega_r = \omega_t$).
- The Snell–Descartes law applies: components of the wavevector tangential to the interface are conserved ($k_{i\parallel} = k_{r\parallel} = k_{t\parallel}$).

A graphical construction of the reflected and the transmitted wavevectors that is obtained from the second condition is depicted in Figure 3.7. This second condition indeed implies

$$\phi_r = -\phi_i, \tag{3.37}$$

$$\frac{\sin\phi_t}{c_2} = \frac{\sin\phi_i}{c_1}. \tag{3.38}$$

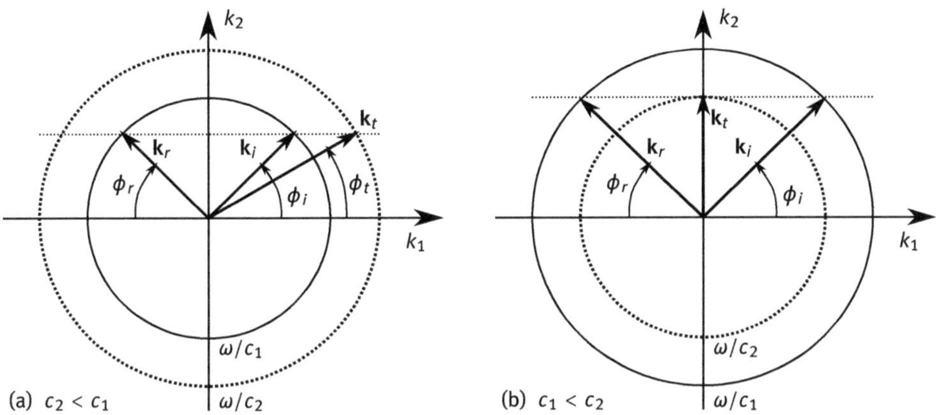

Fig. 3.7: The Snell–Descartes law. The wavevectors of the reflected and of the transmitted harmonic plane waves can be constructed from the dispersion relations in both fluid media and knowledge of the incident wavevector. The projection of the incident wavevector onto the interface axis is a conserved quantity. Diagrams are drawn in the incidence plane containing all wavevectors. (a) If $c_2 < c_1$, then a transmitted plane wave exists for all angles of incidence ϕ_i ($\phi_t < \phi_i$). (b) If $c_1 < c_2$, then there is a critical incidence angle beyond which the transmitted plane wave becomes evanescent. The figure is drawn for that condition ($\phi_t = \pi/2$).

The continuity of pressures implies

$$A_i + A_r = A_t.$$

Normal velocities are further continuous at the interface, or

$$\frac{\cos\phi_i}{Z_1}A_i - \frac{\cos\phi_r}{Z_1}A_r = \frac{\cos\phi_t}{Z_2}A_t.$$

From these two relations, the reflection and transmission pressure coefficients can be readily obtained as

$$r_p = \frac{A_r}{A_i} = \frac{Z_2\cos\phi_i - Z_1\cos\phi_t}{Z_2\cos\phi_i + Z_1\cos\phi_t}, \tag{3.39}$$

$$t_p = \frac{A_t}{A_i} = \frac{2Z_2\cos\phi_i}{Z_2\cos\phi_i + Z_1\cos\phi_t}. \tag{3.40}$$

The reflection and transmission velocity coefficients are

$$r_v = -r_p, \tag{3.41}$$

$$t_v = \frac{Z_1 \cos\phi_t}{Z_2 \cos\phi_i} t_p = \frac{2Z_1 \cos\phi_t}{Z_2 \cos\phi_i + Z_1 \cos\phi_t}. \tag{3.42}$$

The reflection and transmission acoustic power coefficients are simply given by $R = |r_v|^2$ and $T = 1 - R$.

The formulas (3.37) to (3.42) are valid for all values of the angle of incidence, of the impedance contrast z, and of the ratio of velocities c_2/c_1. Figure 3.8 displays the variations of reflection and transmission coefficients as a function of the angle of incidence for different cases. In the particular case that $c_1 < c_2$, there exists a critical

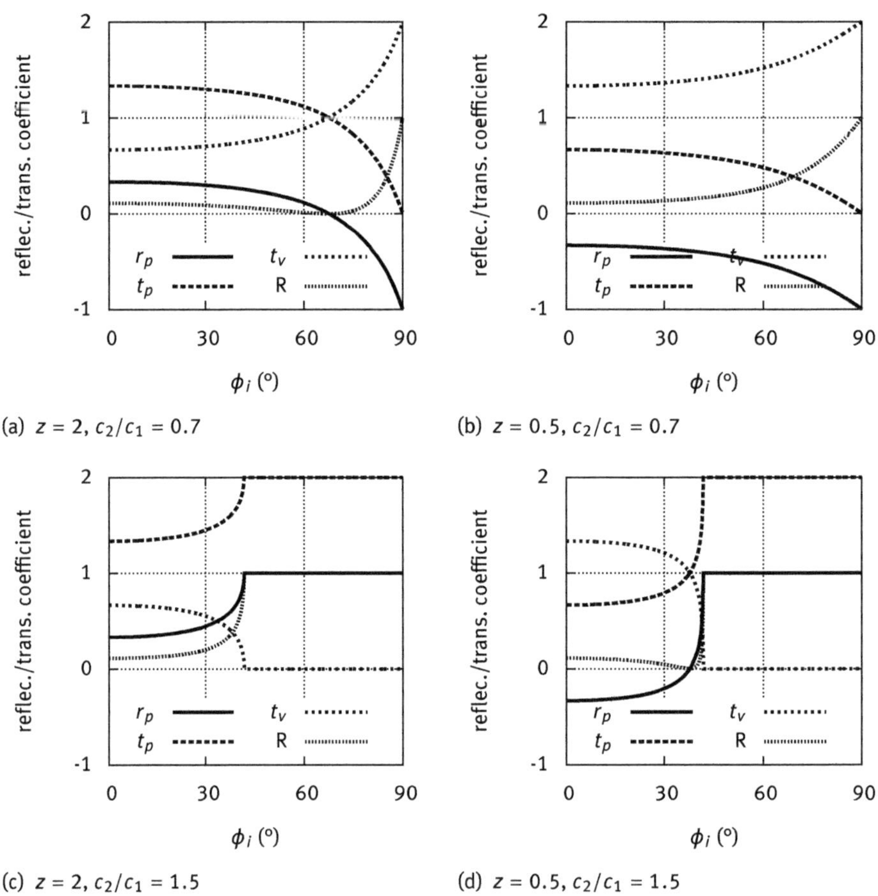

Fig. 3.8: Reflection and transmission coefficients as a function of the angle of incidence ϕ_i. Four different cases are represented depending on whether the impedance contrast $z = Z_2/Z_1$ and the ratio of velocities c_2/c_1 are larger or smaller than one.

incidence angle such that $\phi_t = \pi/2$ in (3.38),

$$\phi_c = \sin^{-1}\left(\frac{c_1}{c_2}\right).$$

For ϕ_i larger than this critical angle, ϕ_t becomes complex valued. Letting $\phi_t = \pi/2 + \imath\psi$, (3.38) is solved for $\phi_i > \phi_c$ by

$$\psi = \cosh^{-1}\left(\frac{c_2}{c_1}\sin\phi_i\right)$$

and $\cos\phi_t = -\imath\sinh\psi$. Hence, r_p and r_p are complex numbers of unit modulus, and then $R = 1$. The transmitted harmonic plane wave is evanescent and the real part of the Poynting's vector flux is zero, consistently with $T = 1 - R = 0$. This phenomenon is called total internal reflection (TIR).

3.3 Finite element modeling of scattering acoustic problems

In this section we describe how the partial differential equations of linear acoustic (pressure) waves can be solved in practice for a rather arbitrary domain of definition using the finite element method. The section also serves the purpose of introducing the finite element method that has been used to obtain most of the computational results illustrating this book, at first for the relatively simple case of the scalar Helmholtz equation. The presentation in particular includes the methods that were used to obtain the computation results shown in Figures 2.17 and 2.18.

3.3.1 Mesh, finite element space, weak form, problem solving

Figure 3.9 illustrates different useful concepts. The domain is a finite piece of space, here a square, inside which we seek to solve a partial differential equation given appropriate boundary conditions. In the example, it is the union of two nonoverlapping subdomains, $\Omega = \Omega_1 \bigcup \Omega_2$. Subdomain Ω_2 is a disk here.

A boundary is a line in two dimensions, a surface in three dimensions. Different boundaries are shown in Figure 3.9. $\sigma = \sigma_1 \bigcup \sigma_2$ is a closed line made of two nonoverlapping open lines with two common points, which will be used later to impose either Neumann or Dirichlet boundary conditions. σ_i is the internal boundary separating subdomains Ω_1 and Ω_2, and is a closed circle. Actually, $\Omega_1 \bigcap \Omega_2 = \sigma_i$.

A mesh Th generated for domain Ω is shown in Figure 3.9 (b). The mesh closely follows the boundaries, replacing them with sequences of straight segments. The inner parts of the subdomains are meshed with triangles that are placed freely by the meshing program. Such a mesh is termed unstructured because the internal points are not fixed beforehand.

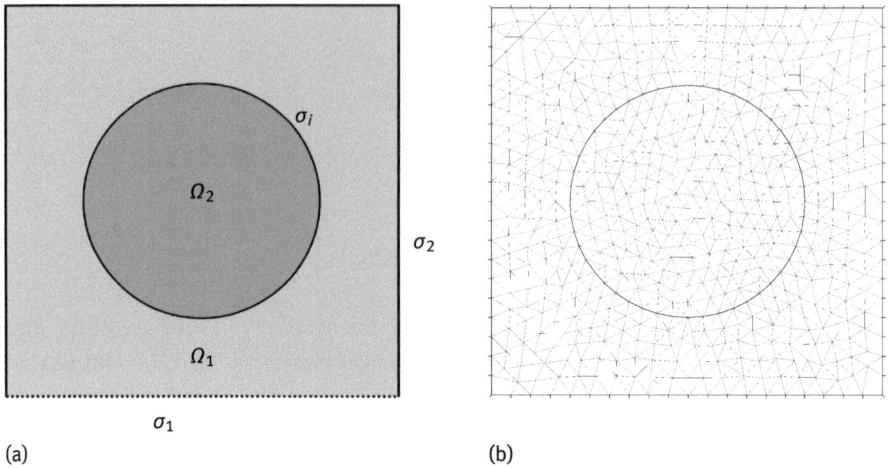

Fig. 3.9: A computation domain and its mesh. (a) The domain Ω is the union of sub-domains Ω_1 and Ω_2. The outer boundary is $\sigma = \sigma_1 \bigcup \sigma_2$. σ_i is the internal boundary separating sub-domains Ω_1 and Ω_2,. (b) The mesh follows closely the boundaries, replacing them with sequences of segments. The inner parts of the sub-domains are meshed with triangles.

Each of the triangles in the mesh is the support of a finite element. Given the domain Ω and its mesh Th, any function of space coordinates $u(\mathbf{x})$ is approximated using only a finite number of degrees of freedom (dof), for instance the nodal values u_j^e for the j nodes of element e. As an example, Lagrange elements in 2D are defined on a triangle by a number of nodes defined with respect to the vertices. The P1 element uses 3 dof placed at each vertex of the triangle. The P2 element uses 6 dof placed at each vertex plus in the middle of each side, defining a total of 6 nodes. Inside each element, $u(\mathbf{x})$ is approximated by

$$u^e(\mathbf{x}) = \sum_j N_j^e(\mathbf{x}) u_j^e$$

where the $N_j^e(\mathbf{x})$ are basis functions. For Pn elements ($n = 1, 2, \ldots$), polynomials of degree n are used. In 2D, polynomials of degree 1 can be generated from 3 independent basis functions (obtained from 1, x_1, and x_2), hence the necessary 3 dof and 3 nodes of element P1. Still in 2D, polynomials of degree 2 can be generated from 6 independent basis functions (obtained from 1, x_1, x_2, x_1^2, x_2^2, and $x_1 x_2$), hence the necessary 6 dof and 6 nodes of element P2. In 3D, triangles can be replaced by tetrahedrons; the P1 element has 4 nodes; the P2 element has 10 nodes. Note that almost all nodes are common to at least two different elements. The common nodes can be used to enforce continuity of $u^e(\mathbf{x})$ between elements or not, depending on the type of elements that are chosen. Pn elements enforce this continuity. P2 elements in particular are used almost exclusively in this book. Finally, the finite element space Wh is the set of all functions defined piecewise by $u(\mathbf{x}) = u^e(\mathbf{x})$ if $\mathbf{x} \in e$. A finite element space is a functional space of finite size, contrary to the initial partial differential problem. The finite size is the total number of dofs, with each dof belonging to several elements counted only once.

In two-dimensional space a partial differential equation is any equation of the type

$$\mathcal{L}u = f$$

with $u(x_1, x_2)$ a function and $\mathcal{L}$ is a differential operator containing x_1, x_2, u, $\frac{\partial u}{\partial x_1}$, $\frac{\partial u}{\partial x_2}$, ... and functions of these quantities. The definition is easily extended to 3D. Classical examples include:

- Laplace's equation, $-\nabla \cdot (\nabla u) = 0$,
- Poisson's equation, $-\nabla \cdot (\epsilon \nabla \phi) = \rho$,
- Helmholtz equation, $-\nabla \cdot (c\nabla u) - k^2 u = 0$.

Finite element models can be built using Galerkin's method, as outlined next. Given a partial differential equation $\mathcal{L}u = f$ defined on domain Ω, solution u is expanded on the functional basis w_j of the finite element space Wh,

$$u = \sum_{j=1}^{n} a_j w_j,$$

where a_j are unknown coefficients (real or complex). The differential equation is projected on all possible functions w_i in Wh:

$$\int_{\Omega} w_i \mathcal{L}u = \int_{\Omega} w_i f, \quad \forall i = 1 \dots n,$$

with n the dimension of the finite element space. Inserting the expression for u, we obtain a linear equation of the form

$$A\mathbf{a} = \mathbf{f}$$

with $A_{ij} = \int_{\Omega} w_i \mathcal{L} w_j$ and $f_i = \int_{\Omega} w_i f$. A is a square matrix of size $n \times n$; $\mathbf{f}$ is a vector of length n. The formal solution is then $\mathbf{a} = A^{-1}\mathbf{f}$. In practice, the solution to the linear problem is obtained numerically using a solver, which is a computer program specially optimized for this purpose. A significant advantage of finite element models is that the matrices that are obtained are sparse, reducing both storage requirement and computation time. There are many powerful FEM software applications that can be used, either commercial or open source. All examples in this book were obtained with FreeFem++ [48] completed with gmsh [41] for 3D meshing.

3.3.2 Weak form of the acoustic wave equation

Let us consider the acoustic wave equation written for pressure in (3.13)

$$-\nabla \cdot \left(\frac{1}{\rho}\nabla p\right) + \frac{1}{B}\frac{\partial^2 p}{\partial t^2} = f, \tag{3.43}$$

where we have explicitly considered that B is independent of time. It can, however, still be a function of position $\mathbf{x}$. It can moreover be a discontinuous function of position as the wave travels across the interface between media with different properties.

We consider that function $p(t, \mathbf{x})$ at fixed t belongs to a finite element space defined on a domain Ω, in the sense that we just explained in the previous subsection. We further consider all possible test functions $q(t, \mathbf{x})$ belonging to the same finite element space and we form their scalar products with (3.43)

$$-\int_{\Omega} \mathrm{d}\mathbf{x}\, q\nabla\cdot\left(\frac{1}{\rho}\nabla p\right) + \int_{\Omega} \mathrm{d}\mathbf{x}\, q\frac{1}{B}\frac{\partial^2 p}{\partial t^2} = \int_{\Omega} \mathrm{d}\mathbf{x}\, qf. \tag{3.44}$$

The weak form of (3.43) is then: instead of solving the original equation at every point in space (which means at infinitely many points), we solve (3.44) for all elements q of the finite element space. This leads to as many equations as degrees of freedom, and in any case to a finite number of equations.

Equation (3.44) contains second-degree derivatives in space, meaning that p should be at least continuous and once differentiable. This is a problem with Lagrange finite elements since the basis functions are only continuous. The problem is circumvented by transferring one of the spatial derivatives to the test function by using the divergence or Gauss theorem. This theorem states that

$$\int_{\Omega} \mathrm{d}\mathbf{x}\, \nabla\cdot\mathbf{g} = \int_{\sigma} \mathrm{d}s\, \mathbf{g}\cdot\mathbf{n}, \tag{3.45}$$

that is, the integral over the closed domain Ω of the divergence of a vector function equals the contour integral along the boundary σ of the normal component of the vector function. Considering $\mathbf{g} = q\left(\frac{1}{\rho}\nabla p\right)$, we have the chain rule derivation formula

$$\nabla\cdot\mathbf{g} = q\nabla\cdot\left(\frac{1}{\rho}\nabla p\right) + \nabla q\cdot\left(\frac{1}{\rho}\nabla p\right)$$

and as a result the weak form can be written as

$$\int_{\Omega} \mathrm{d}\mathbf{x}\, \nabla q\cdot\left(\frac{1}{\rho}\nabla p\right) - \int_{\sigma} \mathrm{d}s\, q\left(\frac{1}{\rho}\nabla p\right)\cdot\mathbf{n} + \int_{\Omega} \mathrm{d}\mathbf{x}\, q\frac{1}{B}\frac{\partial^2 p}{\partial t^2} = \int_{\Omega} \mathrm{d}\mathbf{x}\, qf. \tag{3.46}$$

There are two benefits to this formulation. First, we need only evaluate first spatial derivatives of the functions, which is compatible with Lagrange finite elements. Second, boundary conditions along σ can be explicitly considered.

Let us first consider boundaries σ_i internal to the domain Ω, along which the material constants can be discontinuous. To simplify the notation, we restrict our attention to the case that σ_i divides Ω in exactly two subdomains Ω_1 and Ω_2. The case of an arbitrary but finite number of internal boundaries follows obviously. All domain integrals in (3.44) are easily broken down as

$$\int_{\Omega} \mathrm{d}\mathbf{x} \cdots = \int_{\Omega_1} \mathrm{d}\mathbf{x} \cdots + \int_{\Omega_2} \mathrm{d}\mathbf{x} \cdots .$$

In the application of the Gauss theorem, however, every internal boundary will appear twice, whether we transform an integral on Ω_1 or on Ω_2. Since the normal to a boundary is oriented, we have $\mathbf{n}_2 = -\mathbf{n}_1$ on σ_i (the normal is reversed when seen "from" Ω_1 or Ω_2). Thus we can write the boundary integral in (3.46) as

$$\int_{\sigma_i} \mathrm{d}s\, q \left[\left(\frac{1}{\rho} \nabla p \right)_1 - \left(\frac{1}{\rho} \nabla p \right)_2 \right] \cdot \mathbf{n}_1 .$$

As a consequence, omitting the surface internal boundary integrals in the variational formulation implies that $\left(\frac{1}{\rho}\nabla p\right) \cdot \mathbf{n}$ is implicitly assumed to be continuous across σ_i. In physical terms, it means that the normal acceleration is continuous across the interface. This is the natural boundary condition for acoustic waves.

Let us now consider the boundaries enclosing domain Ω. There are two basic types of external boundary conditions that simplify the problem. Neumann or free boundary conditions are such that

$$\frac{1}{\rho} \nabla p = 0 \quad \text{on } \sigma_1, \tag{3.47}$$

where σ_1 is some subset of σ. Obviously, the boundary integral vanishes identically for a Neumann boundary condition, whatever the test function q,

$$\int_{\sigma_1} \mathrm{d}s\, q \left(\frac{1}{\rho} \nabla p \right) \cdot \mathbf{n} = 0.$$

Dirichlet or essential boundary conditions are of the form $p = p_0$ on a subset σ_2 of σ, with p_0 some constant function. In this case, the finite element space for q has to be modified to impose $q = 0$ on σ_2. The reason is that the degrees of freedom lying on σ_2 are all imposed beforehand, and hence an equal number of equations must be removed from the set of equations in (3.46). Accordingly, for the Dirichlet boundary condition,

$$\int_{\sigma_2} \mathrm{d}s\, q \left(\frac{1}{\rho} \nabla p \right) \cdot \mathbf{n} = 0.$$

Other boundary conditions can be considered as well. In general the boundary integral will not necessarily be zero, however.

3.3.3 Radiation boundary condition

Very often we would like to consider conditions where waves can propagate through all space without being reverberated by obstacles. In a numerical simulation, the typical case is that we have some internal source generating waves and that we want to observe their interaction with some object or structure. Waves should then be free to escape the computational domain Ω without significant reflection. A first approach to the simulation of an infinite domain with only a finite computational domain is the radiation boundary condition.

Suppose a monochromatic plane wave is incident on a boundary with local normal $\mathbf{n}$. The pressure distribution of the plane wave is written

$$p = p_0 \exp(\imath(\omega t - \mathbf{k}\cdot\mathbf{x}))$$

with p_0 a constant. Its gradient is

$$\nabla p = -\imath \mathbf{k} p_0 \exp(\imath(\omega t - \mathbf{k}\cdot\mathbf{x})) = -\imath \mathbf{k} p.$$

The boundary integral in (3.46) involves the quantity $(\frac{1}{\rho}\nabla p)\cdot\mathbf{n}$ that has the physical meaning of the normal acceleration at the boundary. In the present case, we have

$$\int_\sigma \mathrm{d}s\, q \left(\frac{1}{\rho}\nabla p\right)\cdot\mathbf{n} = \int_\sigma \mathrm{d}s\, q \left(-\imath\mathbf{k}\cdot\mathbf{n}\frac{p}{\rho}\right).$$

Suppose in addition that the wavevector is everywhere directed along the normal to the boundary, then $\mathbf{k}\cdot\mathbf{n} = k = \omega/c$ according to the dispersion relation in the homogeneous fluid. The boundary integral further simplifies to

$$\int_\sigma \mathrm{d}s\, q \left(\frac{1}{\rho}\nabla p\right)\cdot\mathbf{n} = \int_\sigma \mathrm{d}s\, q \left(-\imath\frac{\omega p}{c\rho}\right).$$

We have seen in Chapter 2 that sufficiently far away from all sources, the plane wave spectrum loses any evanescent component, but also that the wavevector becomes nearly orthogonal to the local wavefront. Under such conditions, the relation $\mathbf{k}\cdot\mathbf{n} = \omega/c$ can be considered as almost verified for plane waves (for 1D geometries or waveguides), for cylindrical waves (for 2D geometries), or for spherical waves (for 3D open space). The radiation boundary condition is thus simply expressed as

$$\left(\frac{1}{\rho}\nabla p\right)\cdot\mathbf{n} = -\imath\frac{\omega p}{\rho c}. \tag{3.48}$$

It is implied that because it is satisfied, plane (or cylindrical, or spherical) waves will not be reflected at the boundary. Then, since any solution to the wave equation can be considered a superposition of plane waves, reflections can be expected to vanish or at least to remain small.

Suppose further that in addition we want to let some plane wave be incident in the computation domain from the outside. The pressure of this incident plane wave can be written

$$p_i = p_0 \exp(\imath(\omega t - \mathbf{k}_0\cdot\mathbf{x})), \tag{3.49}$$

with $\mathbf{k}_0$ an incident wavevector. We now remove the condition that this wavevector should be directed along the boundary normal. Elaborating upon the previous radiation boundary condition, we write

$$\left(\frac{1}{\rho}\nabla p\right)\cdot\mathbf{n} = -\imath\frac{\omega p}{\rho c} + \imath\left(\frac{\omega}{c} - \mathbf{k}_0\cdot\mathbf{n}\right)\frac{p_i}{\rho}. \tag{3.50}$$

There are three terms in the right-hand side of this expression. The first one is as before added to avoid reflection of plane waves incident from the inside of the domain. The third one is added in view of taking into account that

$$\nabla p_i = -\imath \mathbf{k}_0 p_i$$

for the wave incident from the outside. The second term is then a correction to the first one so that the latter is canceled when $p = p_i$, i.e. when only the wave incident from the outside is considered. The present reasoning holds because of the linearity of the wave equation.

3.3.4 Representation of an internal source of waves

We have just seen that a source of waves incident from the outside of the boundary enclosing the computational domain can be considered using the radiation boundary condition. If we instead want to consider an internal source of waves, we have a number of options.

A first option would be to use the function f in (3.46) to represent a distribution of the excitation. Note that f is homogeneous to the divergence of the acceleration of the fluid, hence it is expressed in units of s^{-2}.

Second, we can use a Dirichlet boundary condition to impose the value of pressure along some internal boundary σ_i of Ω. For instance, Figure 3.10 shows the case of a line of excitation with an imposed pressure of $p = 1$ Pa.

As a third option, we could require that the normal acceleration of the fluid has a given value along the internal boundary σ_i. This condition would be expressed as

$$\left(\frac{1}{\rho}\nabla p\right) \cdot \mathbf{n} = a_n$$

with a_n some normal acceleration distribution in units of m/s^2. The corresponding boundary condition is then imposed via the boundary integral by setting

$$\int_{\sigma_i} ds\, q \left(\frac{1}{\rho}\nabla p\right) \cdot \mathbf{n} = \int_{\sigma_i} ds\, q a_n .$$

As a note, this boundary condition is enforced only weakly, hence it should not be expected that the normal acceleration of the solution equals exactly the requested value a_n on σ_i.

Figure 3.10 shows an example combining an internal source of prescribed pressure with a radiation boundary condition applied along a circular outer boundary. Note that the radiation boundary condition without an incoming wave in (3.48) is pretty flexible as there is no explicit dependence with position $\mathbf{x}$ or local normal $\mathbf{n}$.

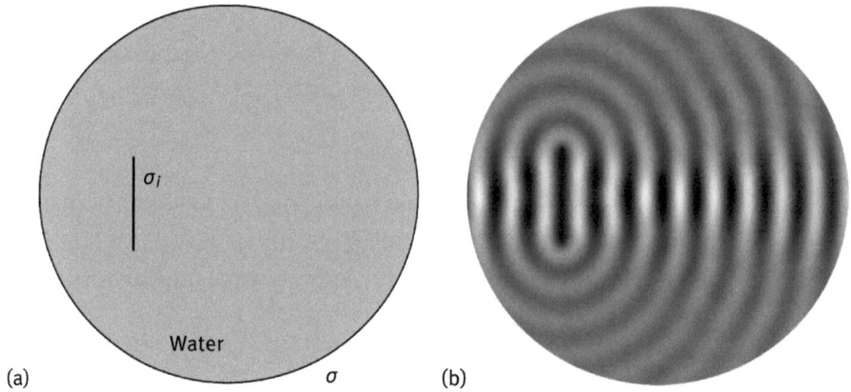

Fig. 3.10: Internal source and radiation boundary condition. (a) The computational domain is a disk of water inside which a linear source is added by prescribing $p = 1$ Pa along internal boundary σ_i. A radiation boundary condition (without an incoming wave) is applied at boundary σ. (b) The solution shows the natural diffraction of the acoustic beam radiated from the source. The source dimension is slightly less than 3 wavelengths in water.

3.3.5 Perfectly matched layer for monochromatic waves

As an alternative to radiation boundary conditions, one can consider adding an absorbing region around the computational domain to absorb outgoing waves. An absorbing region can be defined simply by adding an imaginary part to a material constant, as we discussed in Section 3.1.4. The magnitude of this imaginary part should be quite important if we do not want to extend the mesh more than necessary. On second thoughts, however, we remark that we have to define a boundary between the original computational domain and the absorbing region. Defining the transition carelessly must surely result in a reflection at the boundary by virtue of the Fresnel coefficients, since we will introduce an impedance mismatch between the two regions of space.

The principle of an absorbing region that does not introduce spurious reflections was introduced by Bérenger [13] and has become a widely popular topic in applied mathematics, electromagnetics, optics, and acoustics. It is known as a perfectly matched layer (PML). The concept was originally introduced for the (Maxwell) wave equation based on complex coordinate transformations, but can be more easily explained in the context of monochromatic waves by simple considerations. From the expression of the Fresnel reflection coefficients, it is clear that there is no reflection at the (plane) boundary between two media with identical impedances. Hence, the physical idea is to introduce an adiabatic increase of losses as the absorbing region is penetrated by the waves. Consider for instance a circular domain of radius R centered at the origin and enclosed by an additional cylindrical PML region with thickness D.

We define a complex loss function as

$$\gamma(\mathbf{x}) = 1 + \imath \frac{x^2 - R^2}{D^2} \alpha. \tag{3.51}$$

As a result, $\gamma(\mathbf{x})$ varies very smoothly from 1 at the entrance of the PML to $1 + \imath\alpha$ at its end. α is an dimensionless parameter that has to be adjusted to the problem considered.

The loss function is inserted in the original weak formulation as follows

$$\int_{\Omega} \mathrm{d}\mathbf{x}\, \nabla q \cdot \left(\frac{1}{\gamma\rho}\nabla p\right) - \int_{\sigma} \mathrm{d}s\, q \left(\frac{1}{\rho}\nabla p\right) \cdot \mathbf{n} + \int_{\Omega} \mathrm{d}\mathbf{x}\, q \frac{1}{B}\frac{\partial^2 p}{\partial t^2} = \int_{\Omega} \mathrm{d}\mathbf{x}\, q f. \tag{3.52}$$

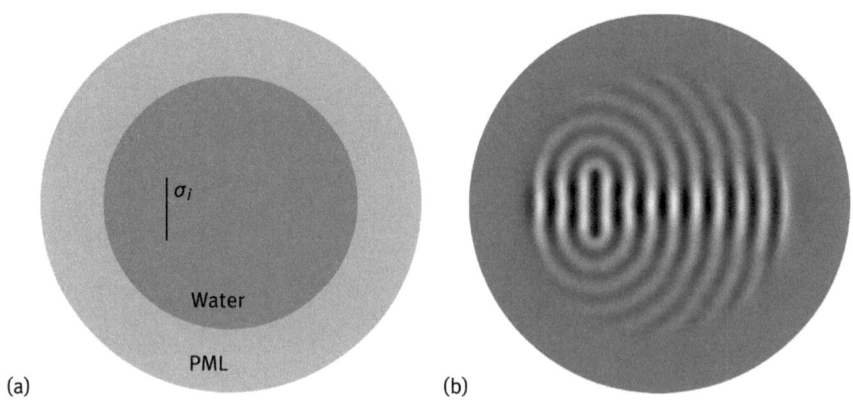

Fig. 3.11: Internal source and perfectly matched layer for harmonic waves. (a) The computational domain is initially a disk of water inside which a linear source is added by prescribing $p = 1$ Pa along internal boundary σ_i, similarly to Figure 3.10. An additional cylindrical region is added in order to implement a perfectly matched layer (PML) absorbing waves outgoing from the central region before they reach the outer boundary. (b) The solution again shows the natural diffraction of the acoustic beam radiated from the source that is slightly less than 3 wavelengths in water. Notice the close correspondence with the result in Figure 3.10.

As an example, Figure 3.11 considers the same radiation problem as in Figure 3.10, but replacing the radiation boundary condition with a perfectly matched layer. The computational result inside the central region is essentially the same. The computational result inside the PML is only displayed to show the effective absorption of outgoing waves.

3.3.6 Scattering of an incident plane wave

We have seen earlier that the radiation boundary condition allows for adding an external source of waves. With the PML, because of the addition of a lossy artificial region

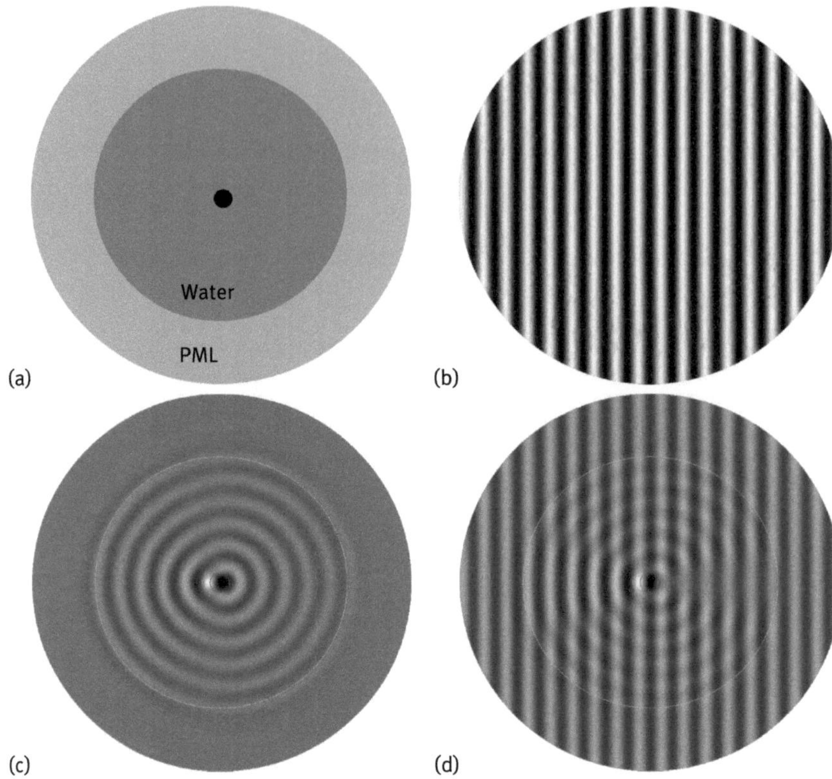

Fig. 3.12: Illustration of the scattered wave formulation. The problem considered is the scattering of an incident plane harmonic wave on a circular inclusion, with PML and external source. (a) The computational domain defines three different subdomains, one for the central scatterer, one for water, and one for the PML. (b) The incident harmonic plane wave p_i is a solution of the wave equation in water. Its real part is displayed. (c) The scattered wave p_s is obtained by solving (3.51). The PML only applies to this field. (d) The total pressure field is reconstructed artificially by summing p_i and p_s. Values inside the PML are not physically significant. (c) and (d) were already presented in Figure 2.17.

to the original domain, the same technique is not directly possible. It is convenient instead to implement a scattering wave formulation along the lines of the following ideas.

Consider the particular but often encountered case that we are interested in the scattering of an incident wave by some small region with different material properties. As an example, we could be interested in the scattering of an acoustic plane wave in water incident on a spherical air bubble. As depicted in Figure 3.12, the incident medium has material properties ρ_1 and B_1, while the scatterer has material properties ρ_2 and B_2. The scatterer defines a subdomain Ω_2 imbedded in the incident subdomain Ω_1. Obviously, $\Omega = \Omega_1 \cup \Omega_2$. ρ and B are piecewise continuous functions of position and they are discontinuous at the boundary between the two subdomains.

Suppose that the incident wave is a solution to the wave equation in the homogenous incident medium, that is

$$-\nabla \cdot \left(\frac{1}{\rho_1}\nabla p_i\right) + \frac{1}{B_1}\frac{\partial^2 p_i}{\partial t^2} = 0, \quad \text{in } \Omega_1. \tag{3.53}$$

Of course, the incident wave p_i is not a solution of the wave equation inside the scatterer. We write the total pressure field as a sum of the incident wave and of a scattered wave, $p = p_i + p_s$. Upon replacing this superposition in the original wave equation, the scattered wave is found to be the solution of the modified problem

$$-\nabla \cdot \left(\frac{1}{\rho}\nabla p_s\right) + \frac{1}{B}\frac{\partial^2 p_s}{\partial t^2} = \nabla \cdot \left(\left(\frac{1}{\rho} - \frac{1}{\rho_1}\right)\nabla p_i\right) - \left(\frac{1}{B} - \frac{1}{B_1}\right)\frac{\partial^2 p_i}{\partial t^2}. \tag{3.54}$$

Since p_i is assumed to be specified, the right-hand side can readily be evaluated and acts as a driving force for the scattered wave. The material functions $\frac{1}{\rho} - \frac{1}{\rho_1}$ and $\frac{1}{B} - \frac{1}{B_1}$ vanish in Ω_1 but not in Ω_2. As a result, the right-hand side is similar to a source of waves f placed inside the scatterer. The scattered field p_s can be solved for using (3.46).

4 Sonic crystals

In this chapter, we get a first glimpse at artificial crystals for acoustic waves. Sonic crystals are considered here as artificial crystals for sound or pressure waves. Strictly speaking, this definition requires that all materials composing the crystal are fluids. For obvious reasons of difficulty of fabricating a perfectly periodic fluidic structure, however, sonic crystal is a term adopted whenever the host material is a fluid, such as air or water, even though the inclusions can be solid. Waves propagating through a sonic crystal can then be considered as purely longitudinal as long as coupling of shear elastic waves existing inside the solid parts to pressure waves in the fluid can be completely ignored. Figure 4.1 presents photographs of some sonic crystals that have been studied in the literature. They most often consist of a stable solid crystal that is immersed in a host fluid.

4.1 Modeling of sonic crystals

In this section, we give a general treatment of sonic crystal calculations that is mostly centered around the problem of obtaining accurate band structures. Furthermore, the mathematics described are useful for other problems that will be tackled in subsequent chapters. The developments are made for pressure waves only, i.e. they apply only to scalar longitudinal waves propagating in fluids. Generalization to elastic waves in solids will proceed in Chapter 6, where the complications introduced by the polarization of waves will be added.

4.1.1 Dynamical equations

Before turning to the different methods that are used to describe wave propagation in sonic crystals, let us remark that one can encounter different kinds of computational problems. There is a first, obvious distinction of problems depending on whether they are considered in the spectral or in the time domain. In principle it is always possible to go from one type to the other with the help of Fourier transforms. A deeper separation of problems, arguably, stems from the boundary conditions that are considered, as these lead to different types of waves (i.e. bulk, surface, and plate waves); only bulk waves are considered in this chapter, for the sake of simplicity. Less fundamental in character, but very important for the understanding of the physics behind sonic and phononic crystals are the different problem types that can be considered in addition to band structures: waveguides, cavities, and scattering problems. The latter problems will be described in depth in Chapters 11 and 12.

Fig. 4.1: Some sonic crystals that have appeared in the literature. (a) A square-lattice crystal of acrylic rods in air [100]. (b) A sculpture by Eusebio Sempere (Fundación Juan March, Madrid) [96]. (c) A phononic crystal of steel rods in water [110].

Summary of equations for pressure waves. As we have set out in Chapter 3, plane wave propagation in fluids (gases, liquids) only involves longitudinal displacements, at least as long as viscoelasticity can be neglected. As a consequence, it might be thought that propagation in fluids could be treated as a limiting case of elastic wave propagation in isotropic solids with only the longitudinal displacements considered, but it turns out that this simplification is not correct in general. Indeed, as the velocity vector in the pressure wave formulation includes both longitudinal and shear components, it is not permitted to simply set the shear components to zero. In the absence of applied external forces, the basic equations for propagation of acoustic waves in a still fluid can be taken as the first-order differential equations (see Section 3.1.2)

$$-\frac{1}{\rho(\mathbf{x})}\nabla p(t,\mathbf{x}) = \frac{\partial \mathbf{v}(t,\mathbf{x})}{\partial t}, \tag{4.1}$$

$$\frac{1}{B(\mathbf{x})}\frac{\partial p(t,\mathbf{x})}{\partial t} = -\nabla\cdot\mathbf{v}(t,\mathbf{x}), \tag{4.2}$$

where variable dependence has been stressed and will not be repeated in the following. ρ is the mass density and B is a bulk modulus expressed with the same units as elastic constants. We have purposely employed vector notations instead of tensorial notations because it is the customary way to present these equations, but of course both are equivalent. The spatially-dependent material constants ρ and B are purposely placed close to the pressure terms rather than close to the velocity terms. This choice makes the subsequent derivations much simpler.

Velocity can be eliminated from (4.1–4.2) to get a scalar wave equation for the pressure only

$$\frac{1}{B}\frac{\partial^2 p}{\partial t^2} = \nabla\cdot\left(\frac{1}{\rho}\nabla p\right). \tag{4.3}$$

This has the form of a second-order differential equation. If we had eliminated the pressure instead, we would have obtained a vector wave equation for the velocity (or the displacement) similar (but not equivalent) to (4.3). Solving the pressure wave equation is the best choice in practice with the finite element method (FEM), in particular for band structure computations. Nevertheless, when surface and evanescent wave problems are considered, the set of first-order differential equations (4.1–4.2) relating pressure and velocity is the logical choice. It will also prove useful to derive the plane wave expansion (PWE) method.

When looking for Bloch waves, the pressure distribution has to be taken in the form

$$p(t,\mathbf{x}) = \tilde{p}(\mathbf{x})\exp(\imath(\omega t - \mathbf{k}\cdot\mathbf{x})) \tag{4.4}$$

with $\tilde{p}(\mathbf{x})$ a periodic function, i.e. $\tilde{p}(\mathbf{x}+\mathbf{R}) = \tilde{p}(\mathbf{x})$ with $\mathbf{R}$ any lattice vector. In this case, the gradient of pressure in the previous equations has to be replaced with

$$\nabla p = (\nabla\tilde{p} - \imath\mathbf{k}\tilde{p})\exp(\imath(\omega t - \mathbf{k}\cdot\mathbf{x})) \tag{4.5}$$

and a similar equation for velocity. Equations (4.1) and (4.2) become for Bloch waves

$$-\frac{1}{\rho}(\nabla\tilde{p} - \imath\mathbf{k}\tilde{p}) = \imath\omega\tilde{\mathbf{v}}, \tag{4.6}$$

$$\imath\omega\frac{1}{B}\tilde{p} = -(\nabla\cdot\tilde{\mathbf{v}} - \imath\mathbf{k}\cdot\tilde{\mathbf{v}}), \tag{4.7}$$

Boundary conditions. Boundary conditions are very important to define the solution of a partial differential equation. For pressure waves, Dirichlet or essential boundary conditions are of the form

$$p = p_0 \tag{4.8}$$

on some part of the boundary of the domain, with p_0 a constant or possibly a function of position. For sonic crystals, Dirichlet boundary conditions are mostly useful to represent a source of waves, but hardly for anything else.

Neumann or free boundary conditions are of the form

$$\left(\frac{1}{\rho}\nabla p\right)\cdot\mathbf{n} = 0 \tag{4.9}$$

with $\mathbf{n}$ a unit vector normal to the boundary. This condition means that the normal derivative of pressure vanishes on the boundary. Physically, this situation arises at the boundary between the fluid and a very massive wall. Specifying such a condition amounts to assuming that no elastic waves are generated inside the solid wall as pressure waves reflect on it. It is a good approximation for a low density fluid enclosed in a solid container, such as air in a closed room with concrete walls. It is not a valid approximation for a dense fluid enclosed in a container, such as water in a tank. As an example, let us consider the interface between water and Plexiglas. The impedances of water ($Z_1 = 1.483\,10^6$ N.s/m^3) and of Plexiglas ($Z_2 = 3.27\,10^6$ N.s/m^3) are in a ratio of $z = Z_2/Z_1 = 2.2$, or a power reflection coefficient of only $R = 0.14$ in normal incidence. As a comparison, for air in the same solid container we have $z = 7908$ and $R = 0.9995$.

At the boundary between two fluids, pressure is continuous. Note that we will neglect surface tension effects at the interface of two fluids such as water and air.

Periodic boundary conditions are further needed for the computation of band structures when the solution is defined on a unit cell. In such a case, the unit cell can be chosen to be bounded by d pairs of boundaries. With a primitive cell, d is also the number of periodicities. In this case, each pair of boundaries is such that one boundary (Σ_2) is obtained by a translation a_i of the other (Σ_1), with $i = 1, \ldots d$ (see Figures 2.22 and 2.23). If a different unit cell is used, for instance a Wigner–Seitz cell, d can be larger than the number of periodicities (for instance, the Wigner–Seitz cell of the hexagonal lattice has $d = 3$, see Figure 2.24). Still one boundary is obtained by a translation $\mathbf{R}$ of the other, with $\mathbf{R}$ some lattice vector. As a whole, the periodic boundary conditions write for each pair

$$\tilde{p}(\mathbf{x}+\mathbf{R})|_{\Sigma_2} = \tilde{p}(\mathbf{x})|_{\Sigma_1} \tag{4.10}$$

or

$$p(\mathbf{x}+\mathbf{R})|_{\Sigma_2} = \exp(-\imath\mathbf{k}\cdot\mathbf{R})\; p(\mathbf{x})|_{\Sigma_1}. \tag{4.11}$$

In this description, the constant $\exp(-\imath\mathbf{k}\cdot\mathbf{R})$ is a phase factor relating both sides of the unit cell. It must be noted that periodic boundary conditions (4.10) and (4.11) are weaker statements of periodicity than the original strong statement $\tilde{p}(\mathbf{x}+\mathbf{R}) = \tilde{p}(\mathbf{x})$ with $\mathbf{R}$ any lattice vector and $\mathbf{x}$ any position in space. In practice, however, solutions of the wave equation obtained with periodic boundary conditions can be used to pave all space and thus satisfy the strong periodicity statement individually.

We finish this discussion of boundary conditions for pressure waves by a note on the interface between a fluid and a solid material. The case of phononic crystals containing both solid and fluid materials has been considered quite often in the literature, especially in relation to experiments, for instance in cases of steel rods in water or air. The reason is that macroscopic realizations of phononic crystals are rather easy with these material systems. As will be apparent when we introduce elastic waves in solids in Chapter 5, it is not possible to use just one set of equations with coefficients assuming different values in different regions of space. The rigorous way to consider the problem is to solve the problem known as fluid-structure interaction, and more precisely for our needs the coupling of acoustic and elastic waves. Simpler but only approximate methods can be followed if the solid is considered very rigid: either the solid boundary can be considered to have no motion (hence it is assumed there are no elastic waves in the solid), or the solid material can be regarded as an equivalent fluid supporting only longitudinal waves (and thus having an equivalent bulk modulus). The latter approximate solution gives results less distant from the exact solution than the former. These are the two approaches we will follow in this chapter to describe fluid-solid sonic crystals, the more exact acoustic-elastic coupling formulation being the subject of Chapter 8.

4.1.2 Plane wave expansion (PWE) method

The PWE method is directly inspired by Bloch's theorem: as both the material properties and $\tilde{p}$ are periodic functions of position, their Fourier series expansions can be conveniently used to solve the problem.

Fourier series of material constants. In a perfect sonic crystal, the material constants distributions $1/\rho(\mathbf{x})$ and $1/B(\mathbf{x})$ are periodic functions of position. As such, they admit Fourier series expansions. In most actual situations, however, they can be considered only as piecewise continuous functions, because the unit cell contains more than one material. If the constituent are homogeneous, the material constants distributions are piecewise constant functions. Fourier series in contrast are intrinsically continuous, i.e.

$$\rho^{-1}(\mathbf{x}) = \sum_{m=-\infty}^{+\infty} \rho_m^{-1} \exp(-\imath \mathbf{G}_m \cdot \mathbf{x}), \tag{4.12}$$

$$B^{-1}(\mathbf{x}) = \sum_{m=-\infty}^{+\infty} B_m^{-1} \exp(-\imath \mathbf{G}_m \cdot \mathbf{x}) \tag{4.13}$$

define continuous material distributions. In the PWE method, it is implicitly assumed that if the series are truncated to an arbitrary number of coefficients, convergence to the exact solution can be achieved to a desired precision simply by increasing the number of harmonics. In the Fourier series, the index m runs over all possible reciprocal lattice vectors, according to $\mathbf{G}_m = m_1\mathbf{b}_1 + m_2\mathbf{b}_2 + m_3\mathbf{b}_3$. Thus m can be understood as a shorthand notation for the triplet of (m_1, m_2, m_3). The exact order in which the different indices (m_1, m_2, m_3) are classified to define the vector $(\mathbf{G}_m)$ is not important providing the relation is bijective.[1]

The Fourier series expansion of a continuous and regular function is in principle infinite. In practice, however, when writing a computer program to handle Fourier expansions, one has to consider some truncation of the Fourier coefficients. To be specific, let us consider Fourier series with $2N_i + 1$, $i = 1, \dots 3$ coefficients for each periodicity direction (a 2D sonic crystal is simply described by setting $N_3 = 0$). The total number of harmonics is then $2N + 1 = (2N_1 + 1)(2N_2 + 1)(2N_3 + 1)$ and we can write

$$\rho^{-1}(\mathbf{x}) = \sum_{m=-N}^{+N} \rho_m^{-1} \exp(-\imath \mathbf{G}_m \cdot \mathbf{x}),$$

$$B^{-1}(\mathbf{x}) = \sum_{m=-N}^{+N} B_m^{-1} \exp(-\imath \mathbf{G}_m \cdot \mathbf{x}).$$

For a function $\alpha(\mathbf{x})$, the coefficients α_m can be obtained by a simple Fourier integral over the unit cell as

$$\alpha_m = \frac{1}{V(\Omega)} \int_\Omega d\mathbf{x} \alpha(\mathbf{x}) \exp(\imath \mathbf{G}_m \cdot \mathbf{x}), \tag{4.14}$$

with $V(\Omega) = \int_\Omega d\mathbf{x}$ the volume (or surface $S(\Omega)$) of the unit cell. The orthonormality of Fourier exponentials writes

$$\frac{1}{V(\Omega)} \int_\Omega d\mathbf{x} \exp(\imath \mathbf{G}_m \cdot \mathbf{x}) \exp(-\imath \mathbf{G}_n \cdot \mathbf{x}) = \delta_{m-n}. \tag{4.15}$$

For some simple inclusion shapes, the Fourier coefficients are known analytically [74]. In 2D, a centered disk of radius r has the values

$$\alpha(\mathbf{x}) = \begin{cases} 1 & \text{if } |\mathbf{x}| \le r, \\ 0 & \text{if } |\mathbf{x}| > r, \end{cases}$$

$$\alpha_m = 2 \frac{\pi r^2}{S(\Omega)} \frac{J_1(|\mathbf{G}_m| r)}{|\mathbf{G}_m| r},$$

where J_1 is the Bessel function of first kind and first order. A centered rectangular inclusion in 2D with side lengths l_1 and l_2 has the values

$$\alpha(\mathbf{x}) = \begin{cases} 1 & \text{if } |x_1| \le l_1/2 \text{ and } |x_2| \le l_2/2, \\ 0 & \text{otherwise,} \end{cases}$$

$$\alpha_m = \frac{l_1 l_2}{S(\Omega)} \operatorname{sinc}(G_{m1} l_1/2) \operatorname{sinc}(G_{m2} l_2/2),$$

with an obvious generalization to the 3D case.

1 Bijective here means that (m_1, m_2, m_3) defines m uniquely, and reciprocally.

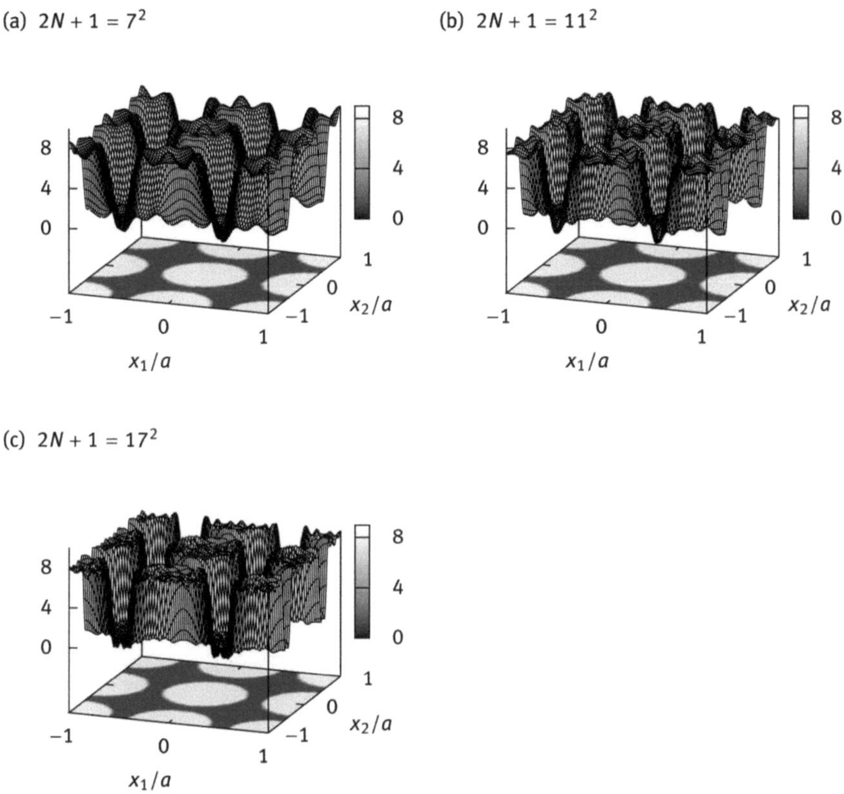

Fig. 4.2: Fourier series expansion of material constants. The figure illustrates the effect of increasing the number of Fourier harmonics.

As an illustration, let us examine the 2D representation of a disk inclusion with constant value $\alpha = 8$ inside the disk and $\alpha = 1$ outside of it. As shown in Figure 4.2, we consider a square lattice with $a_1 = a_2 = a$ and the radius of the inclusion is $r = 0.4a$. The figure shows the spatial distribution obtained by keeping exactly $2N+1 = 7^2, 11^2$, and 17^2 Fourier harmonics. Though the approximation of the sought step function clearly improves as N increases, it is also observed that high spatial frequency oscillations remain. The amplitude of these oscillations around the discontinuities actually converges to a finite value, which is known as Mach's phenomenon. As a consequence, it is illusory to think of approaching a discontinuous function with its Fourier series (uniform convergence), and this limitation should be accepted with the PWE method.

PWE equations. Let us now set out the PWE method explicitly for sonic crystals. The following derivation is simple but not completely rigorous. An alternative derivation using Laurent's theorem for Fourier series and explicitly considering discontinuities of the material constants is given in Appendix 4.A. Assuming monochromaticity (time harmonic waves) and making use of the Bloch–Floquet theorem, we have for the periodic part of the solution

$$-\frac{1}{\rho}(\tilde{p}_{,i} - \imath k_i \tilde{p}) = \imath\omega\tilde{v}_i, \tag{4.16}$$

$$\imath\omega\frac{1}{B}\tilde{p} = -(\tilde{v}_{i,i} - \imath k_i \tilde{v}_i). \tag{4.17}$$

The periodicity of $\tilde{p}$ and $\tilde{v}_i$ allows us to use the Fourier series expansions

$$\tilde{p}(\mathbf{x}) = \sum_{m=-N}^{+N} p_m \exp(-\imath\mathbf{G}_m \cdot \mathbf{x}),$$

$$\tilde{v}_i(\mathbf{x}) = \sum_{m=-N}^{+N} v_{im} \exp(-\imath\mathbf{G}_m \cdot \mathbf{x}).$$

The first derivatives are simply

$$\tilde{p}_{,i} - \imath k_i \tilde{p} = -\imath \sum_{m=-N}^{+N} (k_i + G_{im}) p_m \exp(-\imath\mathbf{G}_m \cdot \mathbf{x}),$$

$$\tilde{v}_{i,i} - \imath k_i \tilde{v}_i = -\imath \sum_{m=-N}^{+N} (k_i + G_{im}) v_{im} \exp(-\imath\mathbf{G}_m \cdot \mathbf{x}).$$

In order to make use of matrix algebra later on, we define the following vectors of Fourier coefficients with $2N + 1$ elements

$$P = (p_{-N}, \ldots, p_0, \ldots, p_N)^T, \qquad V_i = (v_{i-N}, \ldots, v_{i0}, \ldots, v_{iN})^T, \tag{4.18}$$

and the diagonal square matrix with $(2N + 1) \times (2N + 1)$ elements

$$(\Gamma_i)_{mn} = \delta_{m-n}(k_i + G_{im}); \quad m, n = -N, \ldots, N. \tag{4.19}$$

With these notations, we can rewrite

$$\tilde{p}_{,i} - \imath k_i \tilde{p} = -\imath \sum_{m=-N}^{+N} (\Gamma_i P)_m \exp(-\imath\mathbf{G}_m \cdot \mathbf{x}),$$

$$\tilde{v}_{i,i} - \imath k_i \tilde{v}_i = -\imath \sum_{m=-N}^{+N} (\Gamma_i V_i)_m \exp(-\imath\mathbf{G}_m \cdot \mathbf{x}),$$

where $\Gamma_i P$ and $\Gamma_i V_i$ should be understood as matrix-vector products. The meaning of these equations is that the Fourier series of the gradient of a function are simply obtained by left-multiplying its Fourier coefficients with matrix Γ_i.

The products of the material constants by the physical fields are a bit more involved but can again be written as matrix-vector products. For a reason that will be understood soon, we extend the Fourier series for the material constants to the range

$[-2N : 2N]$. Expressing first the left-hand side of equation (4.16)

$$B^{-1}\tilde{p}(\mathbf{x}) = \sum_{m=-2N}^{+2N} \sum_{n=-N}^{+N} B_m^{-1} p_n \exp(-\imath(\mathbf{G}_m + \mathbf{G}_n) \cdot \mathbf{x}),$$

$$B^{-1}\tilde{p}(\mathbf{x}) = \sum_{n=-N}^{+N} \sum_{m'=n-2N}^{n+2N} B_{m'-n}^{-1} p_n \exp(-\imath\mathbf{G}_{m'} \cdot \mathbf{x}),$$

$$B^{-1}\tilde{p}(\mathbf{x}) = \sum_{m'=-N}^{+N} \sum_{n=-N}^{N} B_{m'-n}^{-1} p_n \exp(-\imath\mathbf{G}_{m'} \cdot \mathbf{x}).$$

In the second line we have swapped both summations and considered the change of variables $m' = m+n$. In the last line we have truncated the sum over m' to the original range $[-N : N]$ and then swapped back the sums. The truncation introduces an error, obviously, but this error is limited to Fourier harmonics outside the range we have chosen for the field quantities. Finally we define the square band matrix with $(2N+1)\times(2N+1)$ elements

$$\mathcal{B}_{mn} = B_{m-n}^{-1}; \quad m, n = -N, \ldots, N, \tag{4.20}$$

with which we can write

$$B^{-1}\tilde{p}(\mathbf{x}) = \sum_{m=-N}^{+N} (\mathcal{B}P)_m \exp(-\imath\mathbf{G}_m \cdot \mathbf{x}).$$

Hence the Fourier coefficients of function $B^{-1}\tilde{p}(\mathbf{x})$ are just $(\mathcal{B}P)_m$. In order to transform equation (4.17), we further define the square band matrix with $(2N + 1) \times (2N + 1)$ elements

$$\mathcal{A}_{mn} = \rho_{m-n}^{-1}; \quad m, n = -N, \ldots, N, \tag{4.21}$$

and we can now write equations (4.16)–(4.17) in the matrix form

$$\mathcal{A}\Gamma_i P = \omega V_i, \tag{4.22}$$

$$\omega\mathcal{B}P = \Gamma_i V_i. \tag{4.23}$$

These equations relate the vectors of Fourier coefficients of pressure and velocity. They can be written because the Fourier series expansions are unique, so that equality of the functions can be replaced by equality of the Fourier coefficients.

The band structure can be adequately computed from the following eigenvalue problem, obtained by eliminating the velocity vectors

$$(\Gamma_i\mathcal{A}\Gamma_i)P = \omega^2\mathcal{B}P. \tag{4.24}$$

This is a generalized eigenvalue problem. $\Gamma_i\mathcal{A}\Gamma_i$ and $\mathcal{B}$ are square matrices with $(2N+1)\times(2N+1)$ elements. There are then $2N+1$ eigenvalues ω^2 and up to the same number of linearly independent eigenvectors P. Since the matrices Γ_i, $i = 1; \ldots, p$ are functions of the Bloch wavevector, solving the eigenvalue problem allows one to plot the band structure in the form $\omega(\mathbf{k})$. Appendix 4.B summarizes some useful properties of eigenvalue problems.

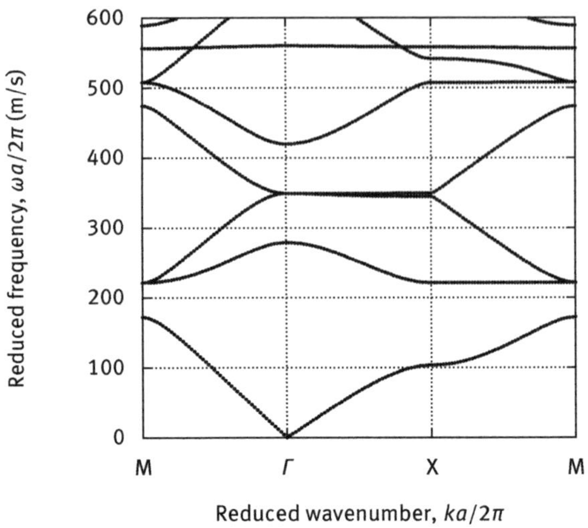

(a)

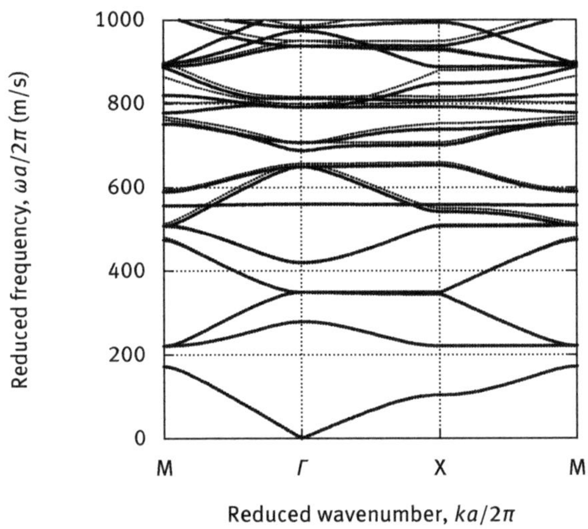

(b)

Fig. 4.3: Band structure of a square-lattice sonic crystal composed of steel rods in air. The radius of the rods is $r/a = 0.4$. (a) The PWE calculation is converged for $2N + 1 = 9^2$ Fourier harmonics, showing the existence of a complete band gap around $\frac{\omega a}{2\pi} = 200$ m/s. (b) The same calculation is shown for a wider frequency range (big dots) together with the calculation result for only $2N+1 = 5^2$ Fourier harmonics (small dots). Visual inspection shows that for a given number of harmonics, convergence is more easily achieved for lower bands than for higher bands.

Band structure convergence. Let us apply the previous results in the case of a square-lattice sonic crystal of steel cylinders in air or water. This choice of materials is mostly practical: it is quite difficult to actually build purely fluid sonic crystals, with the possible exception of crystals of air bubbles in water. Still, periodic arrays of steel cylinders in air or water are usually termed sonic crystals since acoustic waves are incident from the fluid side. The reverse situation, fluid inclusions inside a solid matrix, would generally be called a phononic crystal, for a similar reason. Of course, steel is not a fluid and a trick is needed here to account for acoustic wave propagation inside it. Elastic waves in solids will be treated in Chapter 5 and fluid-solid phononic crystals will only be accounted for satisfactorily in Chapter 8. Anyway, we will consider here that only longitudinal elastic waves are excited in the steel inclusions, so that we can take $B = c_{11} = 264\,\text{GPa}$. The mass density of steel is taken to be $\rho = 7780\,\text{kg/m}^3$. As a result, the longitudinal velocity in steel is taken as $c = \sqrt{B/\rho} = 5825\,\text{m/s}$. This velocity, as well as the impedance $Z = \sqrt{B\rho} = 45.3\ 10^6\,\text{N}\cdot\text{s}\cdot\text{m}^{-3}$, are markedly larger than the corresponding values in air and water given in Table 3.1. As a result, we can expect that the reflection coefficients at the interface between fluid and solid are quite high. For normal incidence, the power reflection coefficient is $R = 0.88$ at a water-steel interface.

Figure 4.3 (a) shows the band structure for a square-lattice sonic crystal of steel rods in air, with $r/a = 0.4$. The diagram is a plot of frequency as a function of wavenumber, considered along the enclosure of the irreducible Brillouin zone (IBZ). Only the lowest bands are shown, but the band structure extends to higher frequencies. In this conventional plot, both frequency and wavenumber are normalized against the lattice constant a. The reduced frequency, $\frac{\omega a}{2\pi}$, has the units of velocity, m/s. The reduced wavenumber, $\frac{ka}{2\pi}$, is a dimensionless coordinate with respect to the first Brillouin zone extension. Figure 4.3 (a) was plotted for $2N + 1 = 9^2$ Fourier harmonics. In the low frequency range, this number of harmonics is enough to obtain convergence of the PWE method. As an illustration, and as a warning to the reader, we plot in Figure 4.3 (b) an extended view of the band structure with more bands included. In addition to $2N + 1 = 9^2$, the plot for $2N + 1 = 5^2$ Fourier harmonics was also added. It can be seen in the latter case that convergence is achieved for the low frequency range but not for the high frequency range. This is a general trend: convergence of the band structure calculation generally degrades for the higher bands. For the frequency range shown in Figure 4.3 (b), convergence is actually obtained for $2N + 1 = 7^2$ and larger values.

In generating band structures, we are imposing the wavevector to lie in the first Brillouin zone. For instance, along direction ΓX, $k_1 a/2\pi$ varies between 0 and 1/2. We could have used any real value for the reduced wavenumber and we would expect the result to be the same because of periodicity. Because of truncation of the Fourier series, however, numerical errors can accumulate, degrading convergence of the band structure for increasing k. To see this, let us write the Bloch wave for $\mathbf{k}' = \mathbf{k} + \mathbf{G}_n$, i.e.

for an integer shift in reciprocal space

$$\begin{aligned} p(t,\mathbf{x}) &= \tilde{p}(\mathbf{x})\exp(\imath(\omega t - \mathbf{k}'\cdot\mathbf{x})) \\ &= \sum_{m=-N}^{+N} p_m \exp(-\imath(\mathbf{G}_m + \mathbf{G}_n)\cdot\mathbf{x})\exp(\imath(\omega t - \mathbf{k}\cdot\mathbf{x})) \\ &= \sum_{m=-N+n}^{+N+n} p_m \exp(-\imath\mathbf{G}_m\cdot\mathbf{x})\exp(\imath(\omega t - \mathbf{k}\cdot\mathbf{x})). \end{aligned}$$

Hence, the Bloch wave for $\mathbf{k}'$ is the same as the Bloch wave for $\mathbf{k}$, but the index of Fourier series must be shifted to account for the shift in reciprocal space. The choice we have made of considering Fourier harmonics ranging between $-N$ and $+N$ is adapted to the first Brillouin zone, but is decreasingly well adapted as we move away from small k values.

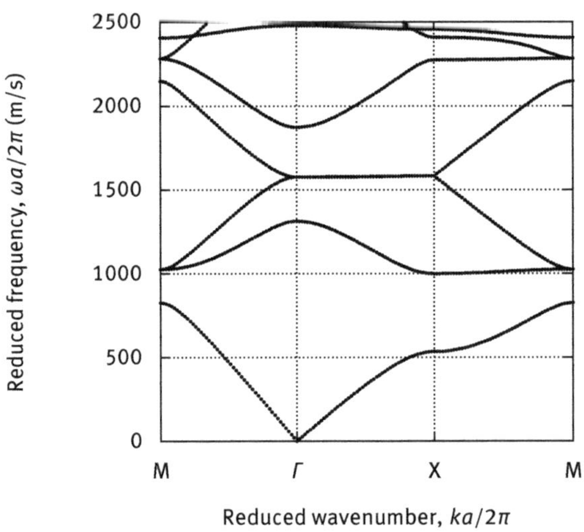

Fig. 4.4: Band structure of a square-lattice sonic crystal composed of steel rods in water. The radius of the rods is $r/a = 0.4$. The PWE calculation is shown for $2N + 1 = 9^2$ Fourier harmonics. A complete band gap extends around $\frac{\omega a}{2\pi} = 850$ m/s.

Figure 4.4 shows the band structure for the same square-lattice sonic crystal of steel rods, with $r/a = 0.4$, but now immersed in water instead of air. The resemblance between Figures 4.3 (a) and 4.4 is striking, and apart from small details in the bands, it almost seems as if they can be superposed by simply scaling the frequency axis. The complete band gap appearing around $\frac{\omega a}{2\pi} = 200$ m/s in air now appears around $\frac{\omega a}{2\pi} = 850$ m/s in water. The central value in both cases is given approximately by $c/2$, with c the longitudinal velocity in the fluid.

4.1.3 Multiple scattering theory (MST and LMS)

Multiple scattering theory (MST) type computation methods have a long history in condensed matter and particle physics. The methods used for phononic crystals have a close relationship with those used for the treatment of electron scattering in solids and for the computation of the band structure of photonic crystals and transmission through them. The basic idea is to make use of the superposition principle for identical nonoverlapping scatterers embedded in a homogeneous matrix.[2] Considering a single scatterer with a given shape in the matrix, one first obtains the field scattered from an incident plane wave with given angular frequency and wavevector. Then the scattered field from an assembly of scatterers is constructed by summing properly all individual contributions. The method thus works very efficiently if the single scatterer problem can be solved analytically, as is the case of isotropic spherical scatterers in a homogeneous isotropic medium (for which spheroidal expansions are known) or of infinite isotropic cylinders in a homogeneous isotropic medium (for which Bessel expansions are known). Multiple scattering methods have been employed with great success for the computation of band structures, density of states, and transmission through finite phononic crystals, though almost always for isotropic spheres or cylinders embedded in an isotropic background, as is the case for sonic crystals. Generalization of multiple scattering methods to arbitrary anisotropic periodic materials still remains a highly difficult problem.

Arguably the most developed and versatile MST method is the layer multiple scattering (LMS) method [126, 135, 136]. We refer the reader to these papers for a full mathematical description of the method. We will only outline it here. The basic principle of the LMS method is that the wave incident on a given scatterer can be viewed as the sum of waves outgoing from all the other scatterers and from the externally incident wave. A scattering matrix relating the amplitudes (which are expansion coefficients in a given basis) of the scattered wave with those of the incident wave is obtained for the composite system from the corresponding matrices of the individual scatterers and proper propagator functions in the host medium. The LMS method considers a phononic crystal to be constituted by a stack of phononic crystal layers. It proceeds layer by layer, evaluating the scattering properties of the phononic crystal from those of the constituent phononic crystal layers, for given frequency and wavevector. The layers can be either planes of 3D scatterers with the same 2D periodicity, or homogeneous layers. For each plane of scatterers, the method calculates the full multipole expansion of the total multiply scattered wavefield and deduces the corresponding transmission and reflection matrices in the plane wave basis for the surrounding media. For homogeneous layers, the transmission and reflection matrices are thus directly obtained in the plane wave basis of the homogenous material.

2 The fact that the matrix is homogeneous is very important as the method relies implicitly on the shift-invariance of the individual scattering problem (scattering on one isolated inclusion).

In any multiple scattering calculation, the angular frequency is a fixed quantity. The wavevector only appears as an eigenvalue in band structure computations or as a particular component of the wavevector of the wave incident on a finite phononic crystal. As such, MST and LMS methods give access to complex band structures (see Chapter 9) and to the consideration of propagation loss in the form of frequency-dependent viscous elastic constants [124]. An open-source LMS program is available.[3]

4.1.4 Finite-difference time-domain (FDTD)

There exist a variety of implementations of FDTD methods for pressure waves. We here refer specifically to the implementation presented by Miyashita [102] for sonic crystals. The derivation is only presented in the 2D case, but the 3D version is straightforward. The starting point is again equations (4.1) and (4.2). In the FDTD method, all fields are defined on a staggered and interleaved grid, as depicted in Figure 4.5. A time increment, Δt, and space increments, Δx_1 and Δx_2, are first defined. The pressure is evaluated at time and space points ($t_n = n\Delta t; x_{1i} = i\Delta x_1, x_{2j} = j\Delta x_2$) of the initial grid.

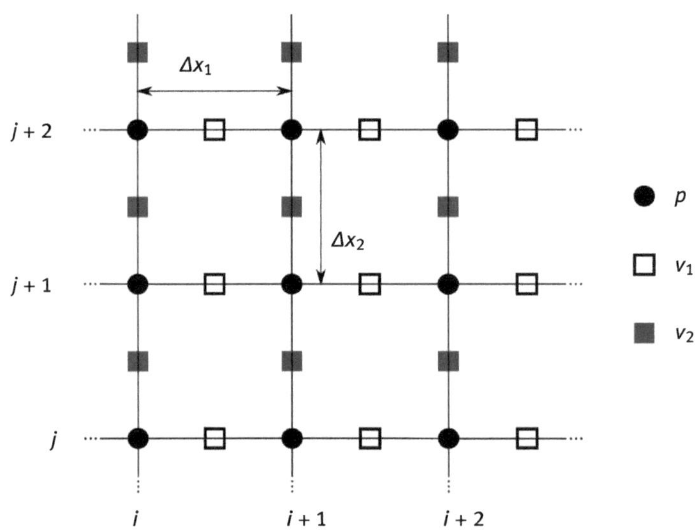

Fig. 4.5: Finite-difference time-domain (FDTD) space grid for pressure waves.

3 Program MULTEL, http://cpc.cs.qub.ac.uk/summaries/ADUT_v1_0.html.

First derivatives are estimated using centered differences, e.g.

$$\frac{\partial p}{\partial t}(n + 1/2; i, j) \simeq \frac{p(n+1; i, j) - p(n; i, j)}{\Delta t}$$
$$p_{,1}(n; i + 1/2, j) \simeq \frac{p(n; i+1, j) - p(n; i, j)}{\Delta x_1}$$
$$p_{,2}(n; i, j + 1/2) \simeq \frac{p(n; i, j+1) - p(n; i, j)}{\Delta x_2}.$$

As a result, first derivatives are displaced by half a grid increment in either time or space. Because of the particular form of waves equations (4.1) and (4.2) – pressure and velocity are related through their first derivatives – it is natural to define the velocity on shifted grids. v_1 is evaluated at time and space points $(t_n = (n+1/2)\Delta t; x_{1i} = (i+1/2)\Delta x_1, x_{2j} = j\Delta x_2)$ and v_2 is evaluated at $(t_n = (n + 1/2)\Delta t; x_{1i} = i\Delta x_1, x_{2j} = (j + 1/2)\Delta x_2)$. It thus takes three interleaved grids to write FDTD equations in 2D (and four interleaved grids in 3D).

It is then straightforward to work out the FDTD equations. With $Z = \rho c = B/c$ the acoustic impedance,

$$v_1(n + 1/2; i + 1/2, j) = v_1(n - 1/2; i + 1/2, j) - \frac{1}{Z}\frac{c\Delta t}{\Delta x_1}(p(n; i+1, j) - p(n; i, j)),$$
$$v_2(n + 1/2; i, j + 1/2) = v_2(n - 1/2; i, j + 1/2) - \frac{1}{Z}\frac{c\Delta t}{\Delta x_2}(p(n; i, j+1) - p(n; i, j)),$$
$$\begin{aligned} p(n+1; i, j) = p(n; i, j) & \\ & - Z\frac{c\Delta t}{\Delta x_1}(v_1(n + 1/2; i + 1/2, j) - v_1(n + 1/2; i - 1/2, j)) \\ & - Z\frac{c\Delta t}{\Delta x_2}(v_2(n + 1/2; i, j + 1/2) - v_2(n + 1/2; i, j - 1/2)). \end{aligned}$$

These formulas are forward steps in time: velocities on the grid are first updated given their previous values and those of the pressures, then pressures are updated in a similar fashion and the process can be repeated. There exist more elaborate time schemes which use an improved approximation for time derivatives and result in a sequence of forward and backward time steps, but the scheme above was used successfully [100–102]. Since the wave equations are solved both in time and space, the FDTD method closely matches usual experimental conditions under which wave propagation can be observed. It is, however, not too well suited to resonant problems with long lived excitations, since the simulation time has to be extended to capture the existence of all waves.

The grid size cannot be chosen arbitrarily for convergence to be achieved. The Courant–Friedrichs–Lewy (CFL) condition is generally used as a necessary condition for stability of the numerical solution of the above equations. The CFL condition reads in this case

$$\frac{c\Delta t}{\Delta x_1} + \frac{c\Delta t}{\Delta x_2} < 1.$$

Physically, it means that to correctly capture the amplitude of a wave traveling across a discrete spatial grid, the time increment must be less than the time it takes for the wave to travel to adjacent grid points. The velocity c in the CFL condition should be taken as the smallest in the whole crystal.

There is another limitation of FDTD that arises in relation with discontinuities in material constants. Actually, in the three FDTD equations above the impedance Z has to be estimated at space points $(i+1/2, j)$, $(i, j+1/2)$, and then (i, j). Depending on where grid points are positioned with respect to the internal boundaries, it is not always clear which impedance value should be used: host matrix, inclusion, or some combination of both? To improve upon this situation, alternative models can be constructed where the raw impedance value is replaced by an adequate weighting taking into account the position of grids points with respect to boundaries [90].

4.1.5 Finite element modeling (FEM)

The finite element method (FEM) was introduced in Chapter 3. Here we discuss how the band structure can be obtained by taking into account periodicity in application of Bloch's theorem.

The basic equations for propagation of acoustic waves in a fluid are (4.1) and (4.2). In the context of Bloch's theorem, they further simplify to (4.6) and (4.7), where $\tilde{p}$ and $\tilde{\mathbf{v}}$ become the unknown field distributions instead of the pressure p and of the velocity $\mathbf{v}$. Each solution is associated with a pair $(\omega, \mathbf{k})$.

The domain of definition of the periodic parts $\tilde{p}$ and $\tilde{\mathbf{v}}$ is naturally a unit cell of the phononic crystal. The domain can thus be a primitive cell or the Wigner–Seitz cell, for instance. Finite element meshes for some primitive unit cells for 2D phononic crystals are shown in Figure 4.6. All these meshes are bounded externally by exactly two pairs of boundaries. We apply periodic boundary conditions on each pair of external boundaries: the unknown fields are set equal at corresponding nodes of these boundaries. Such periodic boundary conditions are hard-coded in the very definition of the finite element space.

In order to obtain the band structure, it is enough here to consider a variational formulation of (4.3). Multiplying with the complex conjugate of a test function $\tilde{q}\exp(-\imath\mathbf{k}\cdot\mathbf{r})$, integrating over domain Ω (the unit cell), and performing an integration by parts, we obtain

$$\omega^2 \int_\Omega d\mathbf{r}\left(\tilde{q}^* \frac{1}{B}\tilde{p}\right) = \int_\Omega d\mathbf{r}\left((\nabla\tilde{q} - \imath\mathbf{k}\tilde{q})^\dagger \frac{1}{\rho}(\nabla\tilde{p} - \imath\mathbf{k}\tilde{p})\right), \quad \forall\tilde{q}. \tag{4.25}$$

As compared to the variational formulation (3.47), we note that there is no source term – since we are looking for modes and not at the response of a system under some force – and that the boundary integral vanishes identically because of the periodic boundary condition (the normal pressure gradient changes sign at either pair of boundaries and hence the total boundary integral vanishes). Another difference to (3.47) is that the wavevector **k** enters directly inside the variational formulation, and more precisely inside the stiffness matrix associated with the FEM problem.

The variational equation (4.25) is used to obtain band structures simply by varying the wavevector in the first Brillouin zone and computing the discrete set of eigenfrequencies. As a result of Hermitian symmetry in case the material parameters are purely real (no material loss), the resulting eigenvalue problem for ω^2 has only real and positive solutions. The FEM formulation just described only requires truly periodic boundary conditions; an alternative derivation with periodic boundary conditions involving complex phase factors can be equivalently used [63]. In the latter case, the wavevector dependence only enters the periodic boundary conditions, but not the FEM matrices.

With FEM, the mesh quality is important to obtain a converged solution. The basis functions on each element have few degrees of freedom, usually some coefficients of a polynomial of small degree, and thus can only represent functions of space that do

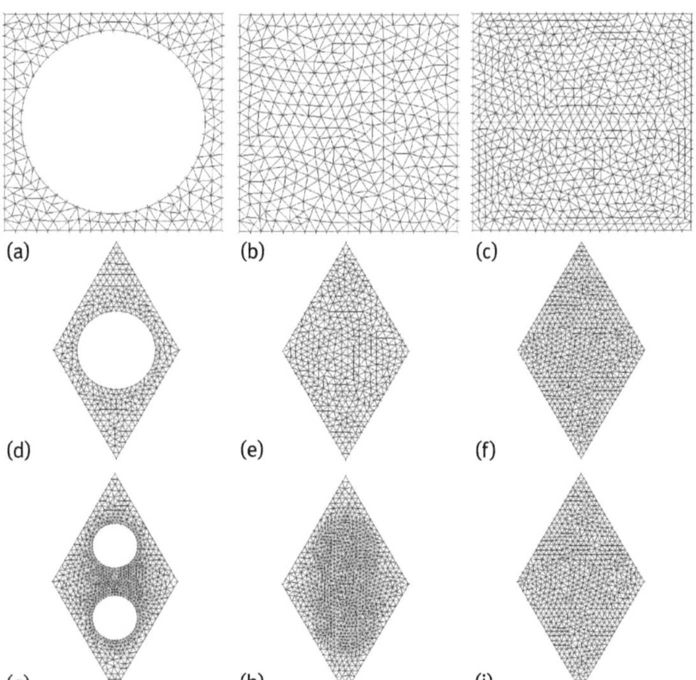

Fig. 4.6: Some examples of FEM meshes for various unit cells. 2D examples are given for the square (a)–(c), the hexagonal (d)–(f) and the honeycomb (g)–(i) lattices. In each case, examples are given for hollow circular inclusions (left), filled circular inclusions (middle), and an unstructured mesh of the unit cell (right). See text for a description of the use of each type of mesh. For the hexagonal lattice, the Wigner–Seitz cell can be used alternatively to the primitive cell.

not vary too fast within one element. In the context of waves, spatial variations are scaled by the wavelength, and it is generally considered that elements with longest dimension 5 to 10 times smaller than the wavelength are sufficient for a proper convergence. Numerical convergence should anyway always be checked for the particular case considered.

Independent of mesh refinement, discontinuity of the material constants at the boundaries between any inclusion and the matrix must be taken into account properly. Let us consider the very common case that functions of space $\frac{1}{B}$ and $\frac{1}{\rho}$ are piecewise constants: they have one value in subdomain Ω_1 (the matrix) and another value in subdomain Ω_2 (the inclusion). The union of the two subdomains is domain Ω. The integrals in (4.25) can be decomposed over the subdomains, for instance

$$\int_{\Omega} d\mathbf{r}\left(\tilde{q}^*\frac{1}{B}\tilde{p}\right) = \int_{\Omega_1} d\mathbf{r}\left(\tilde{q}^*\frac{1}{B}\tilde{p}\right) + \int_{\Omega_2} d\mathbf{r}\left(\tilde{q}^*\frac{1}{B}\tilde{p}\right).$$

In order to implement this decomposition in practice, the mesh must follow the boundary between subdomains as precisely as possible. If it can be performed, the domain decomposition method is probably the most accurate for a given mesh refinement. Figure 4.6 presents some examples of meshes for the square, the hexagonal, and the honeycomb lattices. For a sonic crystal of solid inclusions in air, it is sufficient to use the meshes shown in Figure 4.6 (a), (d) or (g). Indeed, since pressure can be assumed not to penetrate the solid, meshing of the solid part becomes unnecessary; the inclusions can simply be replaced with a free (Neumann) boundary condition on their boundaries, implying that pressure will be extremal at any point of the boundaries. For a sonic crystal mixing two different fluids, the meshes shown in Figure 4.6 (b), (e) or (h) should be used if possible. They contain well-defined internal boundaries in the form of polygons closely following the shape of the circular inclusions. Even though the mesh is composed of triangles and fills the whole domain, there are nodes that fall exactly on the inclusion boundary, so that the elementary integrals on each triangle can be evaluated only on homogeneous subdomains (material constants can be taken as numerical constants inside every triangle).

If domain decomposition of the mesh is difficult to achieve, for any reason, it is still possible to use the mesh of the unit cell together with projections of $\frac{1}{B}$ and $\frac{1}{\rho}$ as finite element functions. In this case, an arbitrary mesh of the unit cell can be considered, such as those depicted in Figure 4.6 (c), (f) or (i); the shape of the inclusions is not taken into account during the meshing process. For Lagrange finite elements, finite element functions are continuous between elements, hence the original discontinuous material functions have to be replaced by continuous approximations. For instance, for a 2D sonic crystal with circular inclusions, we can write

$$\frac{1}{B} \approx \frac{1}{B_2}w(\mathbf{x}) + \frac{1}{B_1}(1 - w(\mathbf{x})),$$
$$w(\mathbf{x}) = \frac{1}{2}\left(1 - \tanh\left(\frac{|\mathbf{x} - \mathbf{x}_0| - r}{h}\right)\right).$$

Function $w(\mathbf{x})$ equals 1 inside the inclusion of radius r centered at point $\mathbf{x}_0$ of the unit cell and equals 0 outside. The jump of B^{-1} at the boundary is replaced by a continuous transition occurring within distance h. Note that the definition of $w(\mathbf{x})$ above is equally valid to represent a sphere of radius r centered at point $\mathbf{x}_0$ in the unit cell of a 3D sonic crystal. Here we have used the hyperbolic tangent function to define a proper window function $w(\mathbf{x})$ for piecewise constant functions, but other choices are obviously possible.

4.1.6 Other methods

The methods we have described in this section are not exhaustive, and there are many different possible approaches that have been taken in the literature. We will only mention some, though they have generally not been used extensively by researchers other than the original authors.

We demonstrate in Appendix 4.A that the PWE method can be considered as a subcase of the Galerkin method, performed on a functional space spanned by Fourier exponentials functions. In this case, the Galerkin method is meshless and the degrees of freedom are not functional values at nodal positions, but expansion coefficients. Sánchez-Pérez et al. have devised a Galerkin method specific to sonic crystals [137]. For the functional basis, they considered a superposition of localized functions that are products of 1D cubic B-splines. A related approach is the use of a wavelet basis in a Galerkin method [171]. Though it was presented in the context of phononic crystals of solid/solid and solid/fluid composition, it should apply as well to sonic crystals.

Another alternative is the use of Green's functions. Such an approach is especially useful if there is only one homogeneous material, possibly within a complex geometry, for which the Green's function for free propagation is known. For instance, a boundary element method (BEM) for acoustic waves can be implemented efficiently for sonic crystals of rigid scatterers in air, since only pressure waves in air need to be described. An open-source BEM software is freely available.[4]

4.2 2D sonic crystal

In this section, we discuss some examples of 2D sonic crystals that have been studied experimentally in the literature. It should be stressed that there have not been reports about truly fluid/fluid compositions, because of the obvious experimental difficulty in obtaining a stable periodic structure composed only of gases and liquids. A close situation to the ideal sonic crystal is however the fluid/solid composition (matrix is

4 Program AcouSTO is available at http://acousto.sourceforge.net/.

fluid; inclusions are solid), since only pressure waves can propagate inside the matrix. A major distinction can be made regarding whether the inclusions can be considered as perfectly rigid or not. For solid cylinders in air, the rigid inclusion approximation holds firmly. As plotted in Figure 3.6, the reflection coefficient for a pressure wave at an interface separating two fluids with an impedance contrast larger than 100 can practically be considered as reaching 100 %. Hence, pressure waves in air will hardly set any solid material into motion. The situation is different if the impedance contrast is limited, as in the case of pressure waves in water impinging on solid materials such as polymers or plastics. In the context of this chapter, we will neglect elastic wave propagation in the solid parts but replace the solid by an equivalent effective fluid with the same density and longitudinal wave velocity. The validity of this approximation will be discussed later in Chapter 6.

4.2.1 Rigid cylinders in air

The case of a periodic array of solid cylinders in air attracted early experimental attention because of the ease of manufacturing sonic crystal samples in this case. The solid cylinders can be made of plastic, glass, metal or wood, and can be held together by frames ensuring periodicity. If the cylinders are long enough, the sonic crystal is approximately two-dimensional; long enough here means longer than the dimension of the source and the receiver. The source should emit quasi-plane waves in order to define a precise direction of propagation, i.e. a wavevector direction. Similarly, the receiver should give a measured value proportional to a plane wave amplitude if the result of an experiment is to be tested against a computed band structure. In air, the most practical arrangement is to use a combination of loudspeakers and microphones, which are readily available in the spectral range 40 Hz to 20 kHz. Signal processing of experimental data can be performed with the inexpensive digital audio equipment of personal computers. Both emitted and received signals will generally be expressed as a voltage as a function of time, loudspeakers and microphones acting as transducers converting electrical signals to pressure waves and vice versa.

There are different signals that can be used to measure a transmission as a function of frequency. One might simply think of playing one pure frequency after another, but this is a very long and inefficient procedure. Instead, one can rely on Fourier analysis and send a known temporal signal smoothly covering a given spectral range, for instance a Gaussian pulse. The different frequency components will travel with different speeds and suffer different attenuation while propagating through the sonic crystal. The spectral transmission is obtained by comparing the results with a reference measurement obtained in the absence of the sonic crystal.

Miyashita [100] studied a 2D square-lattice sonic crystal of acrylic cylinders in air and compared the transmission measurement to FDTD computations. The reported measurements are shown in Figure 4.7. A complete band gap is found experimentally

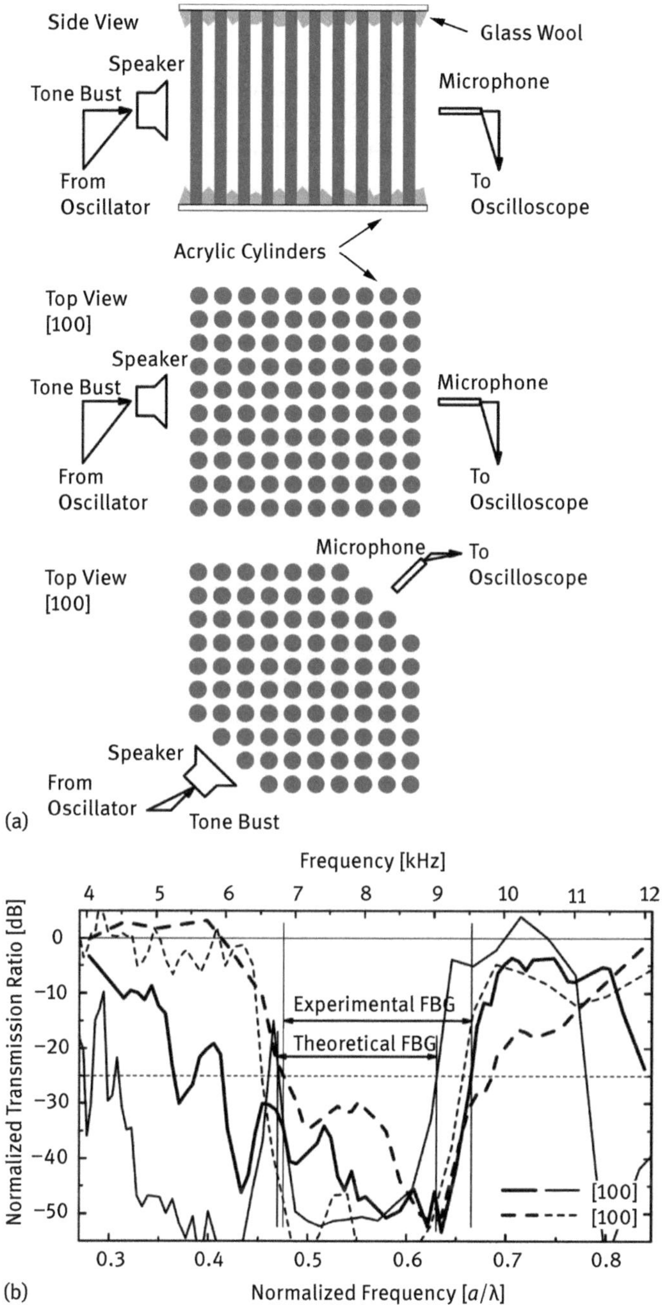

Fig. 4.7: A 2D sonic crystal of acrylic cylinders in air. The crystal is arranged according to a square lattice with lattice constant $a = 24$ mm and radius $r = 10.2$ mm. The filling fraction is 0.567. (a) Experimental set-up and (b) transmission measurement compared to an FDTD computation (adapted from Miyashita et al. [100], copyright 2002 The Japan Society of Applied Physics).

between 6.8 to 9.5 kHz, or between 0.475 to 0.67 in normalized frequency units $a/\lambda = \omega a/(2\pi c)$ with $c = 343$ m/s.

Since the recorded signal is a real-valued function of time, $V(t)$, its Fourier transform is complex valued and has Hermitian symmetry, $\tilde{V}(-\omega) = \tilde{V}^*(\omega)$. Hence, only positive values of frequency need to be represented in the transmission spectra. The intensity spectrum is customarily plotted as a function of frequency, as a measure of spectral transmission. One also has direct access to the spectral phase and to the group delay. Writing $\tilde{V}(\omega) = a(\omega)\exp(-\imath\varphi(\omega))$ with $a(\omega)$ the real-valued amplitude and $\varphi(\omega)$ the phase, it is easy to check that

$$t_g(\omega) = \frac{\mathrm{d}\varphi}{\mathrm{d}\omega} = -\operatorname{Im}\left(\frac{\mathrm{d}\tilde{V}}{\mathrm{d}\omega}\frac{1}{\tilde{V}}\right). \tag{4.26}$$

The latter expression of the group delay is well adapted to numerical approximation by finite differences, because the function $\tilde{V}(\omega)$ varies smoothly. In contrast, the former expression is ill-defined since the spectral phase $\varphi(\omega)$ is numerically reset every 2π.

Figure 4.8 shows the band structure computed with FEM using the rigid inclusion model for the parameters of Figure 4.7. The complete band gap outlined in gray extends from $\omega a/(2\pi) = 160$ to 228 m/s, or from 0.466 to 0.665 in normalized frequency units, in close agreement with the experimental result quoted above. Apart from the obvious complete band gap, reading a phononic band structure is not always easy but offers a lot of information in a synthetic plot. Below the band structure, we show the pressure distribution of Bloch waves belonging to the first five bands in the ΓX direction. The pressure distributions are obtained from the eigenvectors of (4.25) at the X point of the Brillouin zone. Bloch waves 1, 2, and 5 actually belong to a subset of bands forming Bragg band gaps at the Γ and at the X points, where they undergo band foldings. Bloch waves 3 and 4 belong to a different subset which we will later identify in Chapter 9 as arising from Bragg diffraction inside the sonic crystal. It suffices to remark at this point that these Bloch waves have a sinusoidal spatial variation in the x_2 direction, contrary to Bloch waves 1, 2, and 5 (propagation is along the x_1 direction in this case). Furthermore, Bloch wave 4 is deaf, meaning that for symmetry reason it cannot be excited by a plane wave source emitting waves along axis x_1. The concept of deafness will be clarified in Section 4.2.3.

The case of the hexagonal lattice and its derivatives, including the honeycomb lattice, was discussed by Caballero et al. with an emphasis of the effect of symmetry breaking on band gap width [20]. These authors used a frame to hold 1 meter long cylinders with variable diameter according to a hexagonal lattice with lattice constant $a = 63.5$ mm. When the unit cell contains two cylinders with different diameters, as depicted in Figure 4.9, the lattice constant is larger by a factor $\sqrt{3}$, or $a = 110$ mm. Two complete band gaps were found experimentally, with the additional surprise of transmission dips wider than expected from the band structure alone, which the authors attributed to the existence of deaf bands, whereby introducing this concept.

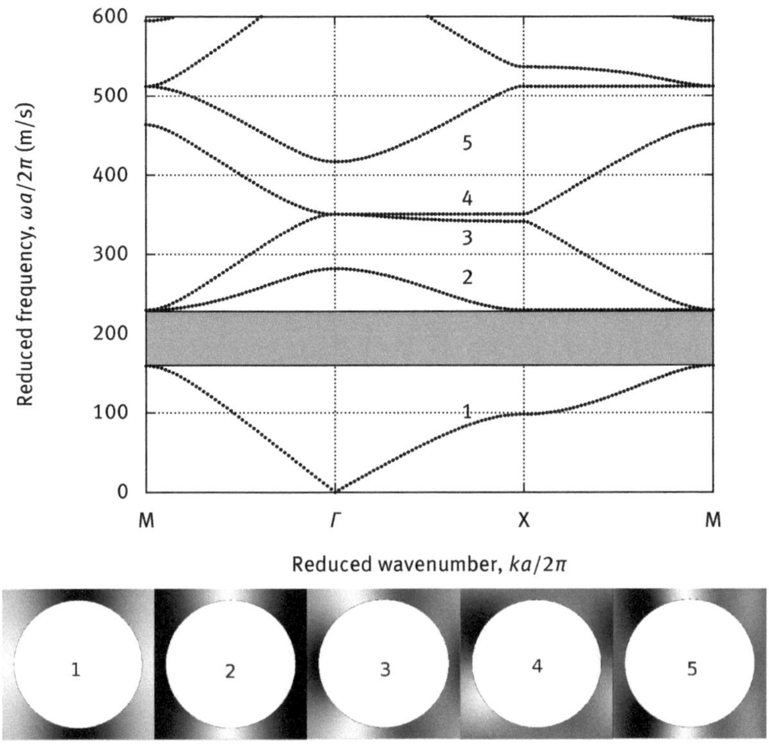

Fig. 4.8: Band structure of a 2D sonic crystal of steel cylinders in air. The crystal is arranged according to a square lattice with $d/a = 0.85$, similarly to Figure 4.7. The FEM computation assumes a rigid boundary condition on the cylinders. The real part of the pressure field of the first five Bloch waves at point X are depicted, with number corresponding to the band number. The gray scale is such that negative values of pressure appear darker, $p = 0$ is 50 % gray, and positive values appear lighter.

Let us consider the band structure for the hexagonal lattice of rigid cylinders in air, shown in Figure 4.10 with the dimensions considered in Caballero's experiment. It can be seen that no complete band gap is opened. In particular, a partial band gap opens between the first and the second band at the M point of the Brillouin zone, but remains closed at the K point. The plot in the ΓK direction was also extended beyond the boundary of the first Brillouin, by simply letting the wavenumber span twice a larger range. Indeed, as k increases from its value at the K point, $4\pi/3a$, it will reach a point M′ equivalent to the M point at $k = 6\pi/3a$ and a point K′ equivalent to K at $k = 8\pi/3a$. Hence, an experiment performed in the ΓK direction will reveal the existence (or not) of a complete band gap without having to scan to ΓM and the MK directions. It is however customary to plot the band structure along a path enclosing the irreducible Brillouin zone, and we will stick with this custom in the following.

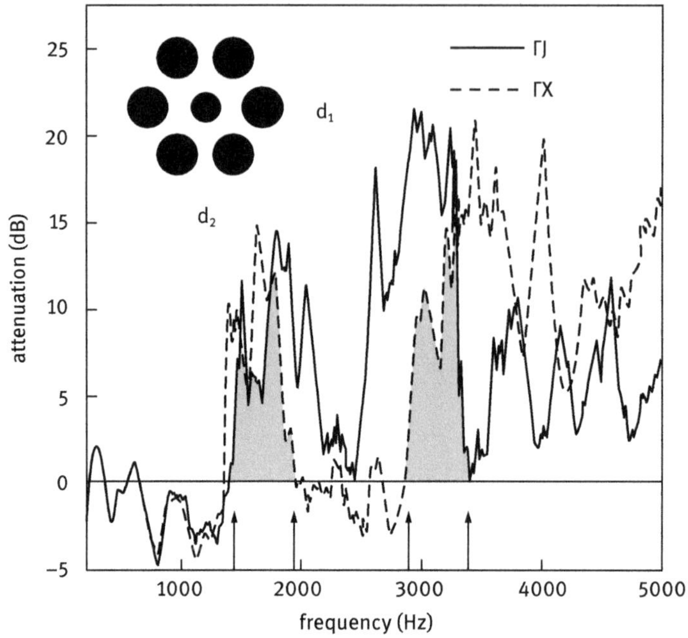

Fig. 4.9: A 2D hexagonal-lattice sonic crystal of acrylic cylinders in air. The lattice constant is a = 110 mm, and there are two different cylinders with diameters d_1 = 40 mm and d_2 = 10 mm. The transmission measurement indicate the existence of two complete band gaps, indicated by arrows (from Caballero et al. [20], copyright 1999 American Physical Society).

The simple hexagonal lattice is not very well adapted to the production of complete band gaps. Indeed, such a complete band opens as d/a approaches the close-packing value, but it is more efficient for a given d/a value to invoke the symmetry-breaking principle. For instance, if a second inclusion with the same diameter is inserted in the hexagonal unit cell, the honeycomb lattice results. As shown in Figure 4.11, two complete band gaps now open, between the first and the second bands, and between the third and the fourth bands.

4.2.2 Steel cylinders in water

The case of a 2D array of solid inclusions in water is similar to sonic crystals in air with regards to fabrication. Again, the sonic crystal can be assembled mechanically before it is immersed in a water tank for measurements. Because the velocity of pressure waves in water is larger than in air, operation can be achieved at ultrasonic frequencies, typically around a few MHz for millimeter-size lattice constants. In this range, commercial transducers are readily available, together with dedicated signal forming electronics sold for nondestructive evaluation (NDE) applications as well as for medi-

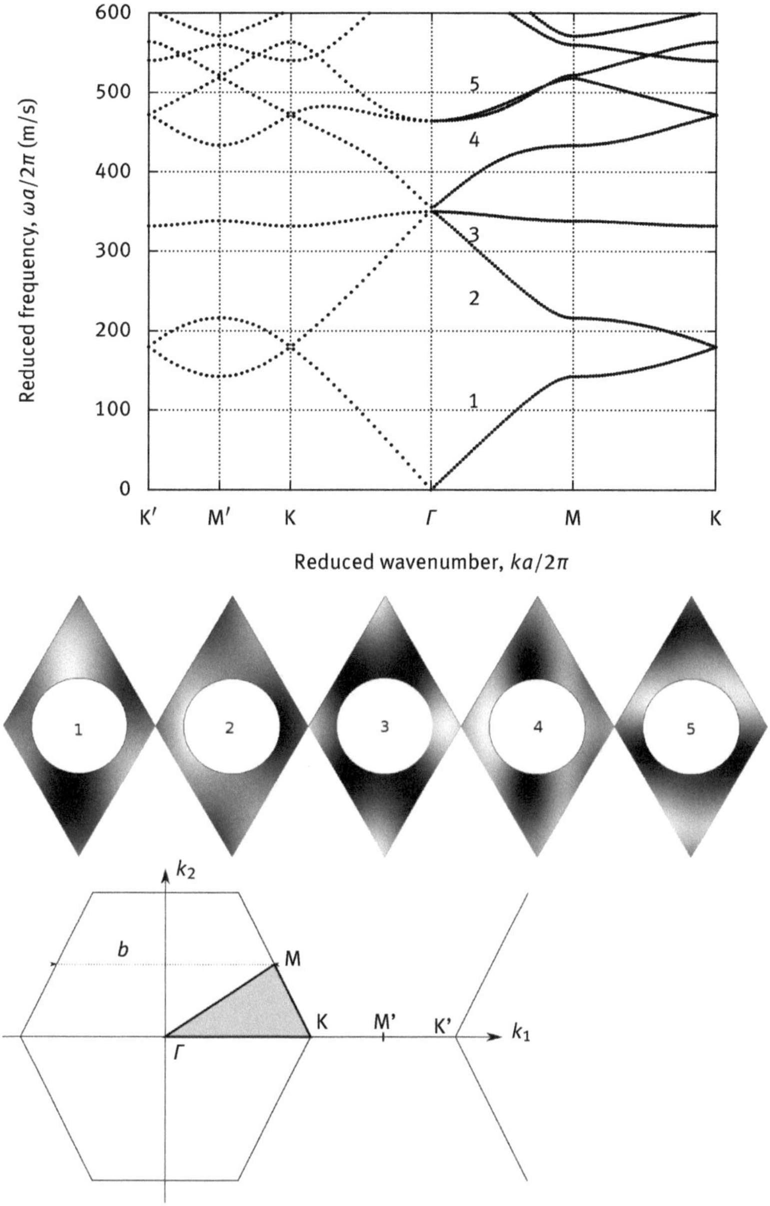

Fig. 4.10: A 2D hexagonal-lattice sonic crystal of acrylic cylinders in air. The diameter of the cylinders normalized to the lattice constant is $d/a = 0.63$. The FEM computation assumes a rigid boundary condition on the cylinders. The real part of the pressure field of the first five Bloch waves at point K are depicted, with number corresponding to the band number. The extended Brillouin zone view shows the definition of points M′ and K′, which are equivalent to the original M and K points in a reciprocal lattice translation ($b = 2\pi/a$).

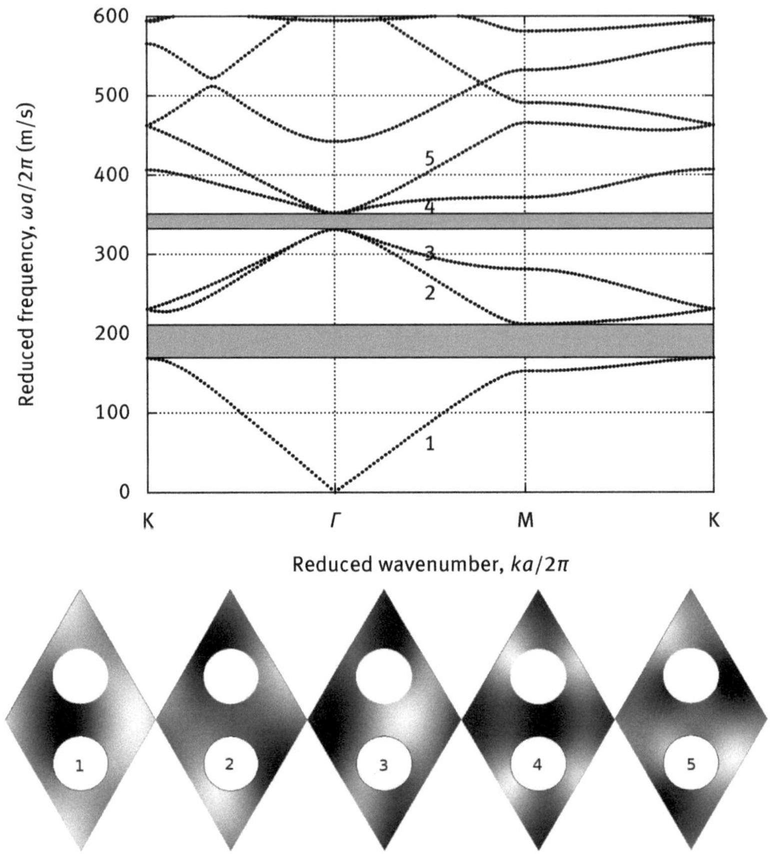

Fig. 4.11: A 2D honeycomb-lattice sonic crystal of acrylic cylinders in air. The diameter of the cylinders normalized to the lattice constant is $d/a = 0.3636$. The FEM computation assumes a rigid boundary condition on the cylinders. The real part of the pressure field of the first five Bloch waves at point K are depicted, with number corresponding to the band number.

cal imaging applications. In particular, transducers emitting intense short pulses with plane wavefronts are available.

Khelif et al. explored the properties of square-lattice 2D sonic crystals of cylindrical steel rods in water [65]. Steel is chosen as the material for the inclusion because of the strong contrast it offers with respect to water. The density of steel is about 7.8 times greater than the density of water, while its elastic modulus for longitudinal waves is about 73 times greater. The rigid approximation used for air is obviously no longer valid and we use the "liquid" steel model for the inclusions.

Figure 4.12 shows the experimental arrangement used to measure the transmission through the square-lattice water/steel sonic crystal. The input short pulse as well as the dispersed signal detected by the receiver are shown. Combining measurements

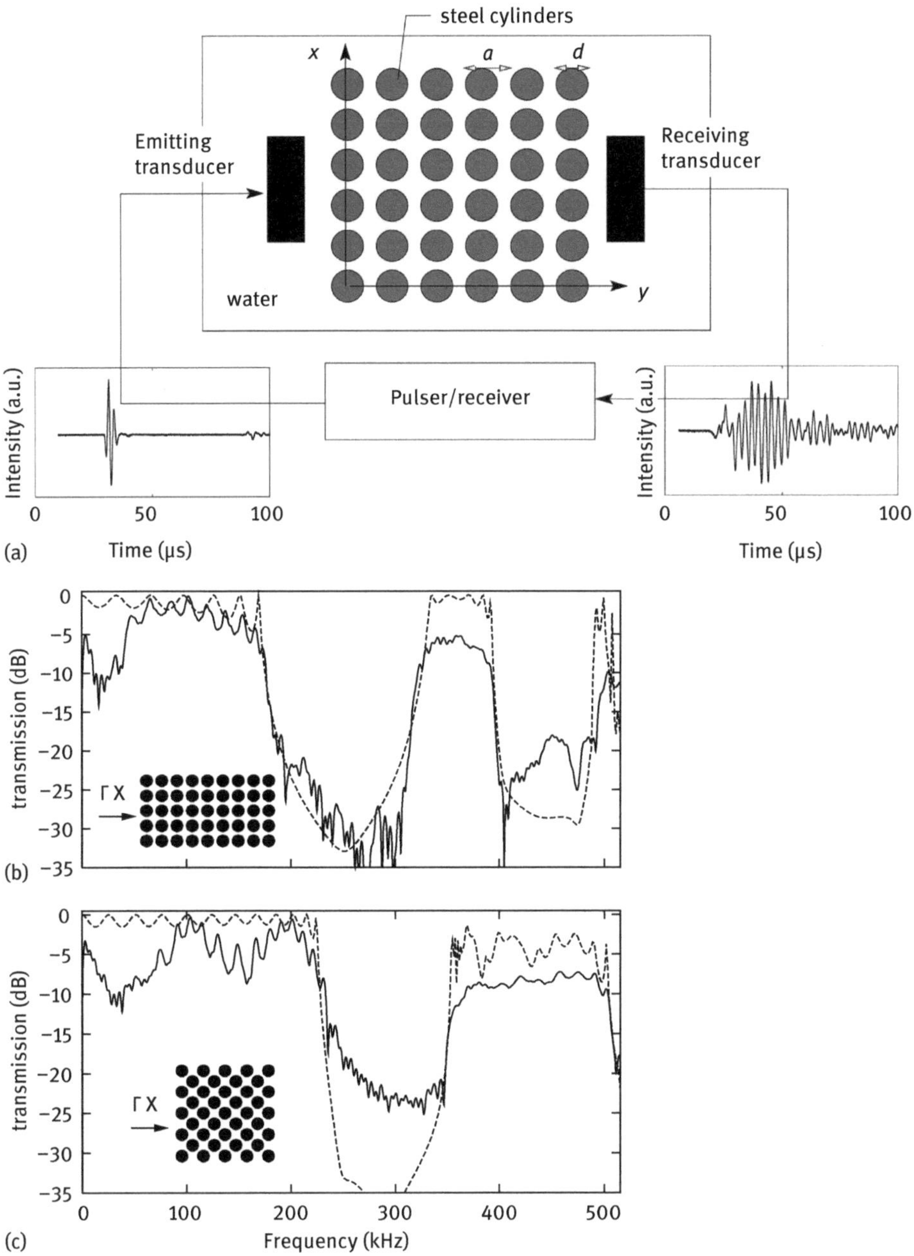

Fig. 4.12: A square-lattice sonic crystal of steel rods in water. The lattice constant is $a = 3$ mm and the diameter of the rods is $d = 2.5$ mm (filling fraction 0.54). (a) The experimental set-up involves two transducers that are controlled by a pulser/receiver. Both the generated and the received waveforms are shown for one particular measurement. (b) The transmission is shown along the ΓX direction and (c) along the ΓM direction. Both experimental values (solid line) and FDTD simulated values (dashed line) are presented (after Khelif et al. [65], copyright 2003 American Physical Society).

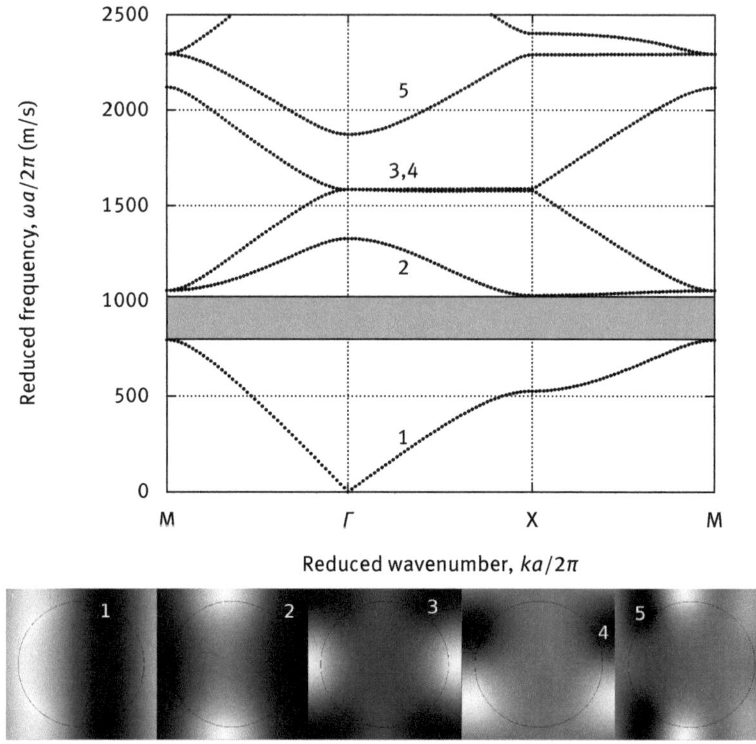

Fig. 4.13: 2D square-lattice sonic crystal of steel cylinders in water. The diameter of the cylinders normalized to the lattice constant is $d/a = 0.833$, as in Figure 4.12. The FEM computation assumes pressure waves only inside the steel rods. The real part of the pressure field of the first five Bloch waves at point X are depicted, with number corresponding to the band number. Bloch mode 4 is deaf.

along directions ΓX and ΓM, a complete band gap is found to extend from 260 to 312 kHz. Along the ΓX direction, there is an additional partial band gap between 400 and 487 kHz. In the figure, measurements are compared with an FDTD computation, with overall good agreement, particularly regarding band gap positions.

The band structure for the 2D square-lattice water/steel sonic crystal is shown in Figure 4.13, for dimensions corresponding to the dimensions of the experiment in Figure 4.12. The complete band gap extends from $\omega a/(2\pi) = 800$ to 1023 m/s. Given that the lattice constant $a = 3$ mm, the complete band gap is predicted between 267 and 341 kHz, and is thus a bit overestimated compared to the measurements. We will see in Chapter 8 that taking elastic wave propagation inside the steel inclusions into account indeed improves the numerical result. Pressure distributions for the first five Bloch waves at the X point of the Brillouin zone are also shown in the figure. The main difference to the air/rigid sonic crystal of Figure 4.8 is that pressure now extends continuously inside the steel inclusion. The absolute values of pressure in water and in steel are of the same order of magnitude, but it should be kept in mind that the kinetic energy density and the potential energy density are proportional to ρ^{-1} and B^{-1} and are thus clearly larger inside water. Bloch waves 1, 2, and 5 again belong to a subset of bands forming Bragg band gaps at the Γ and at the X points, where they undergo band foldings. Bloch waves 3 and 4 again belong to a different subset of Bragg diffraction inside the sonic crystal. Bloch wave 4 is again deaf.

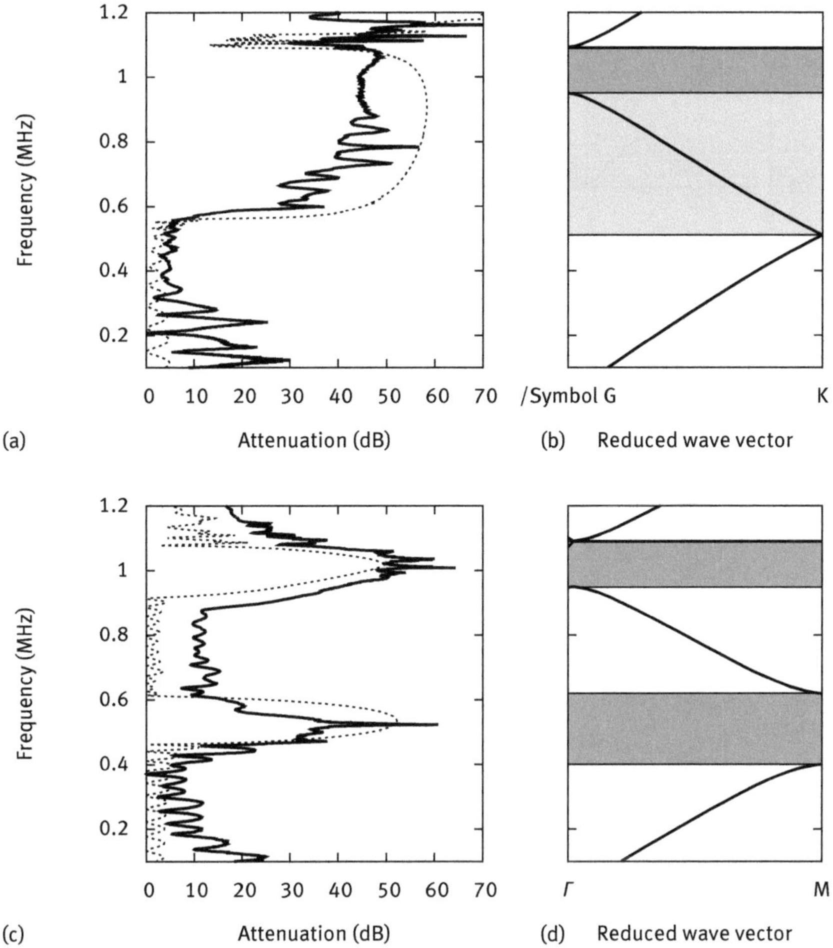

Fig. 4.14: A hexagonal-lattice sonic crystal of steel rods in water. The lattice constant is $a = 1.5$ mm and the diameter of the rods is $d = 1.2$ mm (filling fraction 0.58). Transmission measurements (solid line) along directions (a) ΓK and (c) ΓM are compared to FEM band structures in (b) and (d). FDTD simulations of transmission are shown in (a) and (c) in dashed line. Band gaps are outlined in dark gray and deaf bands are outlined in light gray (after Hsiao et al. [53]).

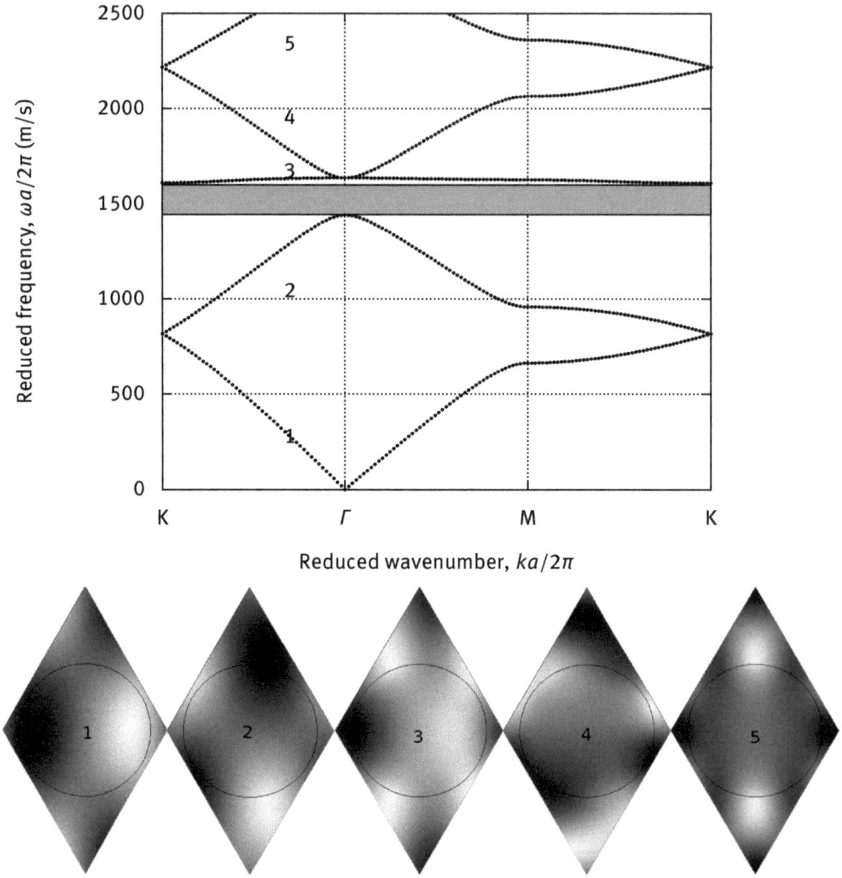

Fig. 4.15: 2D hexagonal-lattice sonic crystal of steel cylinders in water. The diameter of the cylinders normalized to the lattice constant is $d/a = 0.8$, as in Figure 4.14. The FEM computation assumes pressure waves only inside the steel rods. The real part of the pressure field of the first five Bloch waves at point K are depicted, with number corresponding to the band number. Bloch modes 2 and 4 are deaf.

The case of the hexagonal and honeycomb-lattice 2D water/steel sonic crystal was studied by Hsiao et al. [53]. Their experimental results are shown in Figure 4.14 for the hexagonal lattice and in Figure 4.16 for the honeycomb lattice. In each case, measurements are shown along the ΓK and the ΓM directions, and are compared with the theoretical band structure computed with the coupled acoustic and elastic FEM that we will describe in Chapter 8. The presence of Bragg band gaps and of deaf bands is specified in each case. It can be noticed that the presence of the deaf bands has a strong influence on transmission dips. Without the information on deafness obtained from the theoretical model it would be almost impossible to experimentally separate

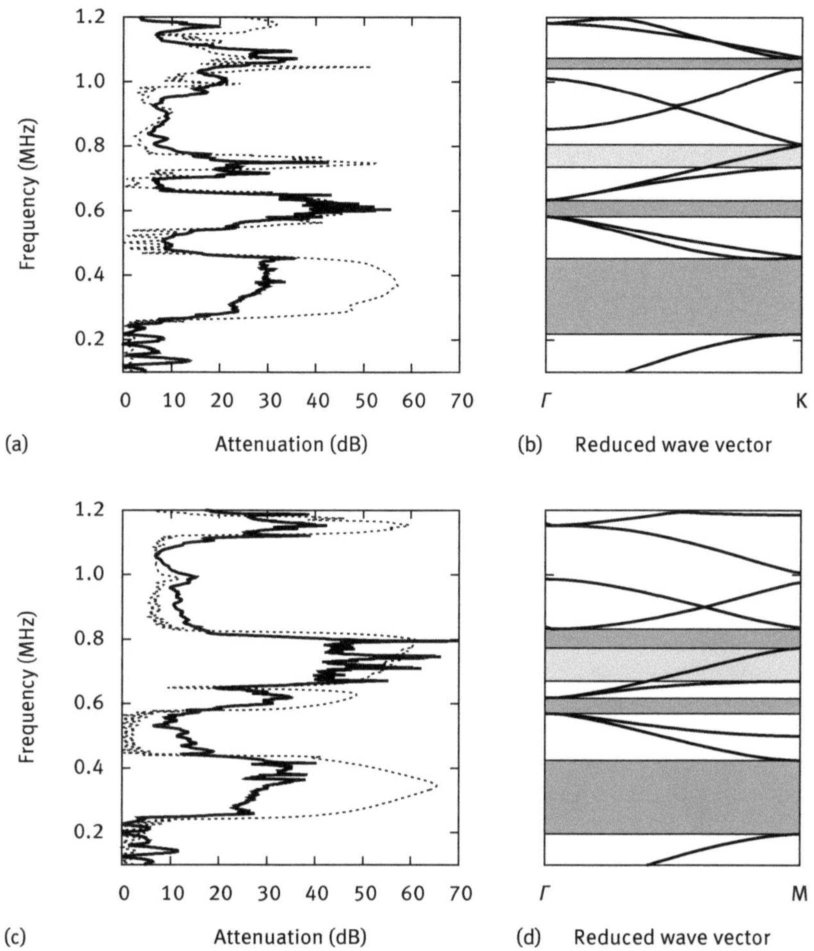

Fig. 4.16: A honeycomb-lattice sonic crystal of steel rods in water. The lattice constant is $a = 2.6$ mm and the diameter of the rods is $d = 1.2$ mm (filling fraction 0.387). Transmission measurements (solid line) along directions (a) ΓK and (c) ΓM are compared to FEM band structures in (b) and (d). FDTD simulations of transmission are shown in (a) and (c) in dashed line. Band gaps are outlined in dark gray and deaf bands are outlined in light gray (after Hsiao et al. [53]).

Bragg band gaps from deaf frequency ranges. Compared to the air/rigid sonic crystal, a band gap is opened between the second and the third band for the hexagonal lattice.

Band structures for the hexagonal and the honeycomb lattices are presented in Figures 4.15 and 4.17 for the geometrical parameters of Figures 4.14 and 4.16. The computation is performed with steel represented as an equivalent fluid medium. Most of the features of the coupled acoustic and elastic computation are reproduced, at least for the lowest bands, although frequencies are generally slightly overestimated. In the

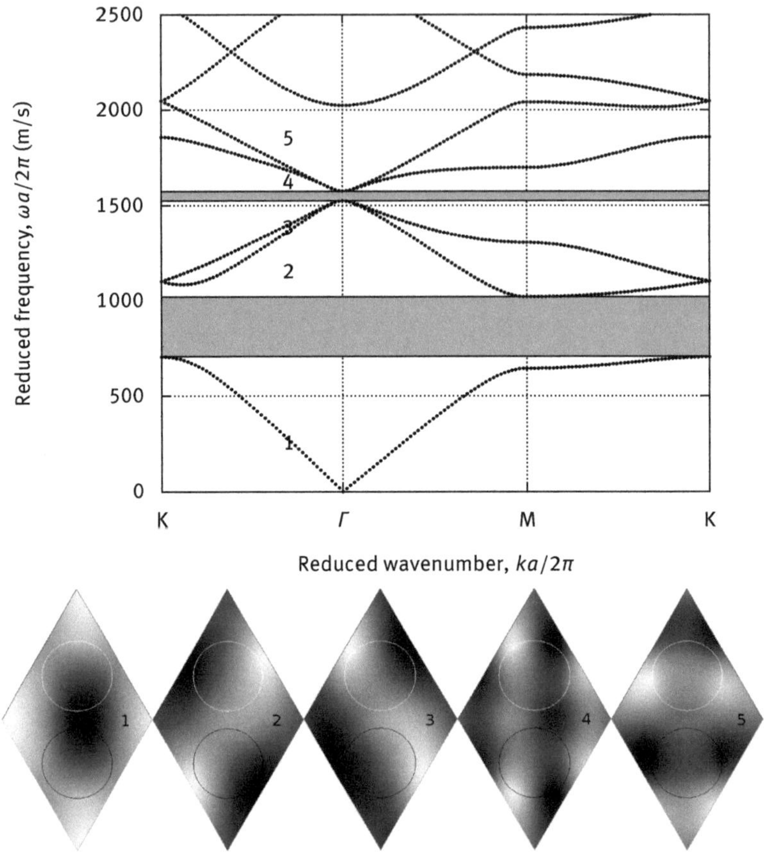

Fig. 4.17: 2D honeycomb-lattice sonic crystal of steel cylinders in water. The diameter of the cylinders normalized to the lattice constant is $d/a = 0.46$, as in Figure 4.16. The FEM computation assumes pressure waves only inside the steel rods. The real part of the pressure field of the first five Bloch waves at point K are depicted, with number corresponding to the band number. Bloch waves 2 and 5 are deaf.

ΓK direction, both bands 2 and 4 are deaf for the hexagonal lattice. For the honeycomb lattice, bands 2 and 5 are deaf.

4.2.3 Deaf bands and unit cell symmetry

We now discuss the relation between symmetries of the unit cell of a sonic crystal and the existence of deaf bands. Our discussion follows the presentation in Laude et al. [82].

Experimental investigations of the dispersion of a sonic crystal or of a phononic crystal usually rely on the excitation of Bloch waves by an external source, for instance by monitoring the transmission of a particular mode of the medium surrounding a finite size crystal sample, as we have illustrated previously. It is thus clear that the intrinsic properties of the crystal are only explored through modal conversions that occur at the boundary between the surrounding medium and the crystal. Obviously, modal conversion to different Bloch waves results in different conversion efficiencies, which we will specify in Chapter 12. Here we only discuss the nonexcitation of certain Bloch waves because of symmetry reasons.

Deaf Bloch waves were originally observed in two-dimensional (2D) sonic crystal of rigid rods in air arranged according to a square lattice [20, 132, 137, 138]. They were afterward also found in three-dimensional (3D) crystals of solid spheres in a solid [126, 127] or a fluid matrix [125], and in 2D sonic crystal of steel rods in water with hexagonal-type lattices [53]. They were observed in phononic crystal slabs as well [44, 148, 170]. Deaf Bloch waves manifest themselves as dips in the transmission, much as band gaps do, and thus lead to apparent discrepancies between the theoretical band structure and the experimental result. These discrepancies are soon explained by observing that a plane wave incident normally on the crystal is symmetric with respect to the propagation direction; if a particular Bloch wave is antisymmetric with respect to the same axis, then it cannot be excited and it cannot contribute to transmission through the crystal. In the band structure, bands formed by deaf Bloch waves are by extension termed deaf bands.

Deafness and unit cell symmetry. Let us consider 2D crystals such as depicted in Figure 4.18. Two different lattices are shown, square (SQ) and hexagonal (HEX). The SQ lattice has symmetries with respect to axes x_1 and x_2 and with respect to both diagonals, i.e. every $n\pi/4$ with n an integer. The HEX lattice has symmetries every $n\pi/6$. Two different inclusions are further considered, either a circle or a semicircle. While the circular inclusion preserves all lattice symmetries, the semicircular inclusion only preserves symmetry along the x_2 axis. It is thus the combination of the symmetries of lattice and inclusions that defines the symmetry of the unit cell of the crystal and an inclusion of arbitrary shape generally removes any symmetry.

Let us now consider wave propagation in a sonic crystal. The Bloch–Floquet theorem states that for a fixed angular frequency ω, the modes of the phononic crystal have the form (4.4). The symmetry directions of the crystal are shared by Bloch waves. It can further be inferred that if the unit cell has a symmetry direction, then Bloch waves with their wavevector pointing in that direction part into two groups, either symmetric or antisymmetric. Antisymmetric Bloch waves are then deaf for this wavevector. Even in the case of a symmetric lattice and of a symmetric inclusion, there is consequently a simple way to break symmetry: it is sufficient to consider a wavevector that is not along a direction of symmetry. As a result, changing the angle of incidence even slightly generally removes the deafness character from all Bloch waves.

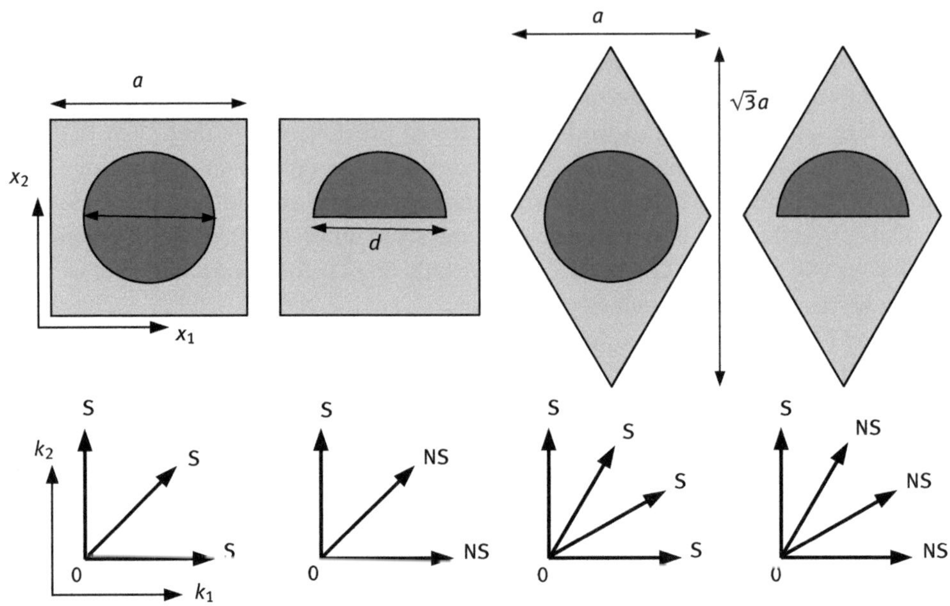

Fig. 4.18: Primitive cells for square- and hexagonal-lattice 2D phononic crystals, containing either a circular or a semicircular inclusion. Regarding wave propagation, the circular inclusion preserves all symmetry directions of the lattice, while the semicircular inclusion only preserves symmetry along the x_2 axis. Symmetric (S) and nonsymmetric (NS) directions are shown for each case.

A criterion for deafness. Identifying deaf Bloch waves by plotting their modal distribution band by band is cumbersome. Can we obtain a simple criterion to judge whether they are deaf or not? Bloch waves are intrinsic modes of the periodic crystal. In practice, the crystal is of finite extent and has boundaries separating it from the surrounding medium. Figure 4.19 depicts such a boundary for a square-lattice crystal. Two cases have been illustrated, with the crystal cut along symmetry direction X or M. It can be noted that the periodicity constant along the crystal boundary, $a_\parallel$, is different in both cases. A plane wave with angular frequency ω and wavevector $\mathbf{k}_0$ is incident on the crystal boundary. The incident pressure field is

$$p_i(\mathbf{r}, t) = \exp(\imath(\omega t - k_{01}x_1 - k_{01}x_2)), \tag{4.27}$$

with the dispersion relation $k_{01}^2 + k_{02}^2 = \omega^2/c^2$ and c the velocity in the incident medium. There is a certain degree of arbitrariness in the exact choice of boundary σ; in Figure 4.19 we have chosen it to lie exactly on the boundary of the unit cell, but other choices can be made without changing the result of the analysis.

We next examine the conversion of the incident wave to Bloch waves. Conservation of energy and momentum requests conservation of the frequency and of the k_1 component of the wavevector. Thus, the relevant Bloch waves in the PC are given as solutions of a problem of the form $k_2(\omega; k_{01})$ that yields a discrete set of complex-valued

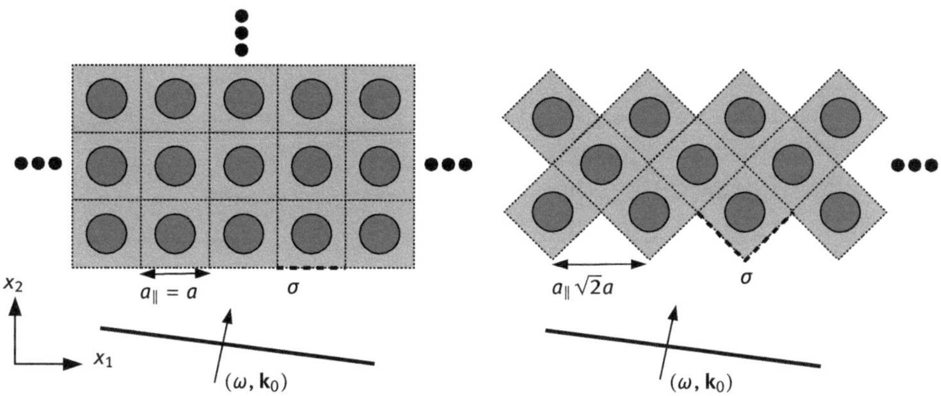

Fig. 4.19: Modal conversion problem for a plane wave incident on the boundary of a sonic or phononic crystal. The two cases shown are for a square-lattice 2D crystal cut either along (a) direction X or (b) direction M.

solutions k_{n2}, with n the mode index [78]. Anticipating the results of Chapter 12, we consider the scalar product between the incident wave and the n-th Bloch wave, measured along the boundary σ, as a measure of modal conversion. This quantity reduces to a line integral of the periodic part of the Bloch wave solution

$$< p_i(\omega; k_{01}) | p_n(\omega; k_{01}) > = \int_\sigma ds \exp(\imath(k_{02} - k_{n2})x_2) \tilde{p}_n(\mathbf{r}). \tag{4.28}$$

In the particular case that σ is a simple straight line, as in Figure 4.19 (a), we can take the integral along $x_2 = 0$, in which case the above equation reduces to

$$< p_i(\omega; k_{01}) | p_n(\omega; k_{01}) > = \int_\sigma ds\, \tilde{p}_n(\mathbf{r}). \tag{4.29}$$

Even though the incident wavevector does not appear explicitly in this expression, both k_{n2} and $\tilde{p}_n(\mathbf{r})$ depend on the pair $(\omega; k_{01})$. Obviously, if the periodic part of the Bloch wave, $\tilde{p}_n(\mathbf{r})$, is antisymmetric, then the scalar product in (4.29) is zero. For normal incidence, i.e. $k_{01} = 0$, this property remains true whatever the value of k_{ny} and deaf Bloch waves form deaf bands. When incidence is not normal, i.e. $k_{01} \neq 0$, deafness is lost immediately since the wavevector breaks the symmetry. Deafness is indeed highly sensitive to the incidence angle.

The line integral defined above provides a simple means of evaluating the symmetry or antisymmetry of Bloch waves for normal incidence. It can thus be used to investigate the intrinsic symmetry properties of band structures. Figures 4.20 and 4.21 display band structures for propagating Bloch waves for a square-lattice 2D sonic crystal of steels rods in water, in the case of the circular and of the semicircular inclusion depicted in Figure 4.18 (a) and (b), respectively. For the circular inclusion, Figure 4.20, the three symmetry directions X, M, and Y show deaf bands, and directions X and Y

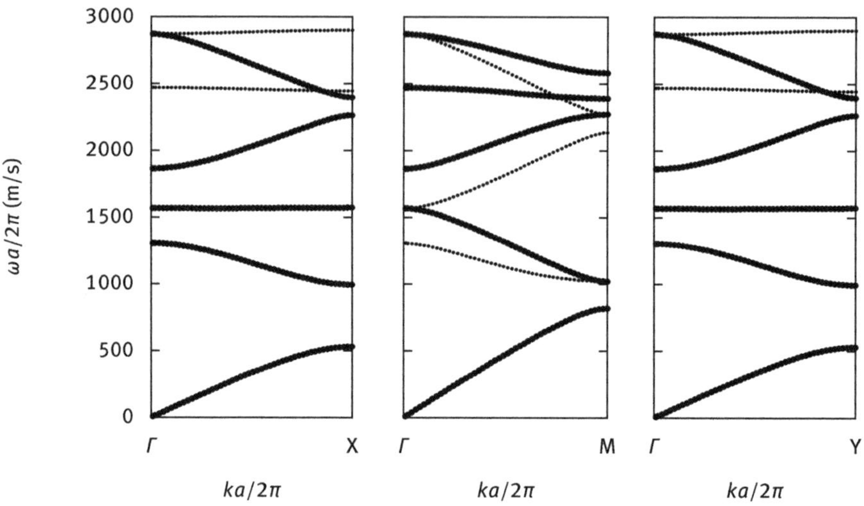

Fig. 4.20: Band structure for a square-lattice 2D sonic of circular steels rods in water with $d/a - 0.8$, shown along direction X, M and Y in the first Brillouin zone. Antisymmetric (deaf) bands are plotted with small dots, while nondeaf bands are plotted with large dots. Note that bands 3 and 4 in directions X and Y are degenerate in frequency, but that band 3 is deaf while band 4 is not.

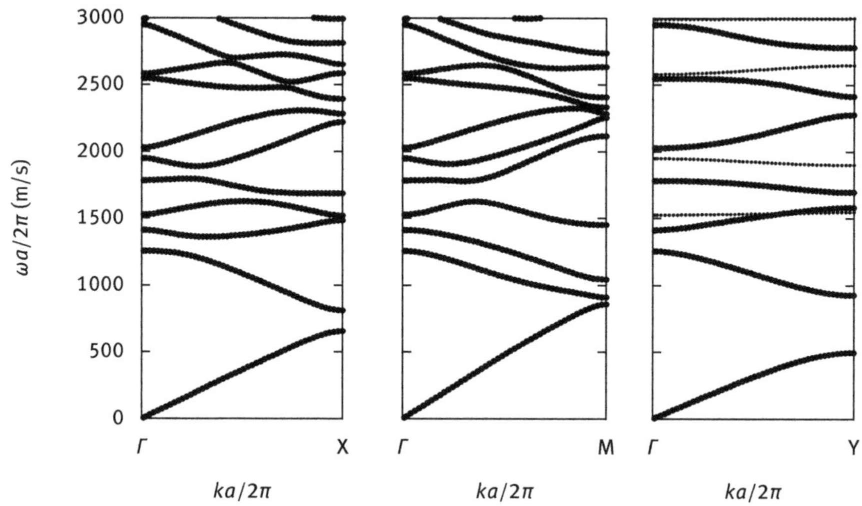

Fig. 4.21: Band structure for a square-lattice 2D PC of semi-circular steels rods in water with $d/a = 0.8$, shown along direction X, M and Y in the first Brillouin zone. Antisymmetric (deaf) bands are plotted with small dots, while nondeaf bands are plotted with large dots.

are equivalent. For the semicircular inclusions that only preserve symmetry along direction Y, deaf bands only appear along this same direction, as Figure 4.21 illustrates. It can further be observed that the complete band gap observed in the case of a circular inclusion disappears for the semicircular inclusion.

4.2.4 Sonic crystal design

Let us finish our presentation of 2D sonic crystals with some general remarks on the design of sonic crystals with large band gaps. The widest band gaps that were predicted are for water/air sonic crystals, understood in the form of air "bubbles" in water [72]. Kushwaha et al. have provided gap maps for 2D bubble sonic crystals, which particularly show that the widest band gaps are obtained at the lowest filling fractions [72]. Of course, this particular type of crystal is probably difficult to obtain as a two-dimensional array of bubbles with a very long third length. It remains that except for localized frequency bands arising from internal resonances of the bubbles, such a sonic crystal is in theory able to block the propagation of sound in a very wide frequency range. Experimental demonstrations are still missing, though there have been some attempts in 3D (see next section).

The bubble provides very soft boundary conditions around the inclusions that scatter acoustic waves in water. Considering perfectly rigid boundary conditions instead, such as the air/rigid sonic crystal, leads to relatively moderately wide band gaps. As a note, gap maps for the air/rigid composition were provided by Caballero et al. [20]. Band gaps generally open as the filling fraction tends to its maximum value. This trend remains the same if the inclusion is chosen to be more rigid and more massive than the matrix, such as with water/steel sonic crystals.

4.3 3D sonic crystals

We now turn our attention to 3D sonic crystals. Figure 4.22 displays 3D FEM meshes for the simple cubic (SC) and the face-centered cubic (FCC) lattice sonic crystals. Again following the strategy introduced for 2D FEM meshes in Figure 4.6, we differentiate the different cases of a totally unstructured mesh, to be used with material constants described as space-dependent functions of space coordinates; of a mesh with an internal hollow inclusion, to be used with the rigid boundary condition; and of a mesh with an internal plain inclusion, to be used with a domain decomposition method. The meshes shown are for the primitive cell, but the Wigner–Seitz cell can be considered as well in the FCC case, resulting in a polyhedron with 12 faces.

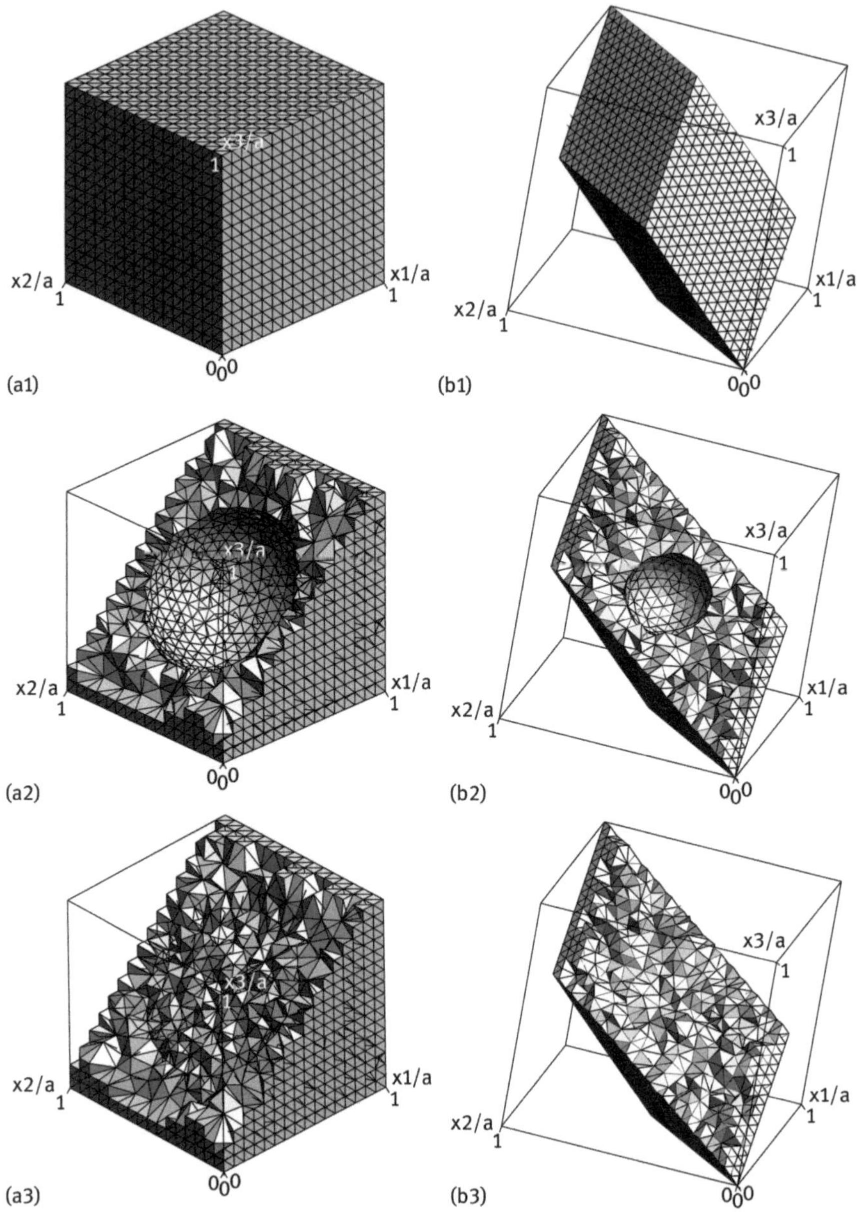

Fig. 4.22: 3D meshes for simple cubic (SC, left) and face-centered cubic (FCC, right) lattice sonic crystals. From top to bottom, the unstructured 3D meshes of the primitive cell, of the primitive cell with a hollow spherical inclusion, and of the primitive cell with a plain spherical inclusion are shown. The meshes with an inclusion are clipped to show their internal structure.

4.3.1 Air bubbles in water

Water/air sonic crystals, or bubble crystals, possess the widest complete band gaps of all sonic and phononic crystals, both in 2D [72] and in 3D [73]. We are aware, however, of only one experimental report of a periodic structure approaching this theoretically ideal case [84]. In contrast, the theory is rather well developed.

Figure 4.23 shows the band structure for the simple cubic (SC) lattice for a filling fraction of 0.1, corresponding to a case discussed by Kushwaha et al. [73]. The first band is really extremely flat and a series of three very wide complete band gaps until a reduced frequency $\omega a/(2\pi) = 749$ m/s. In-between there are two groups of flat bands, corresponding to resonant modes of the air bubbles: three resonant modes around $\omega a/(2\pi) = 402$ m/s and four resonant modes around $\omega a/(2\pi) = 649$ m/s. Further groups of resonant bubble modes can be identified on the figure for higher frequencies. Note that the band starting at $\omega a/(2\pi) = 749$ m/s at the Γ point is a Bragg order of diffraction.

The case of a face-centered cubic (FCC) lattice water/air sonic crystal is presented in Figure 4.24. Again we consider a filling fraction of 0.1, following Kushwaha et al. [73]. In the latter paper, the authors used a PWE method to obtain the band structure. The same problem was considered by Kafesaki et al. using an MST method [60]. Their results are essentially identical to our Figure 4.24 that is obtained with FEM. Similar

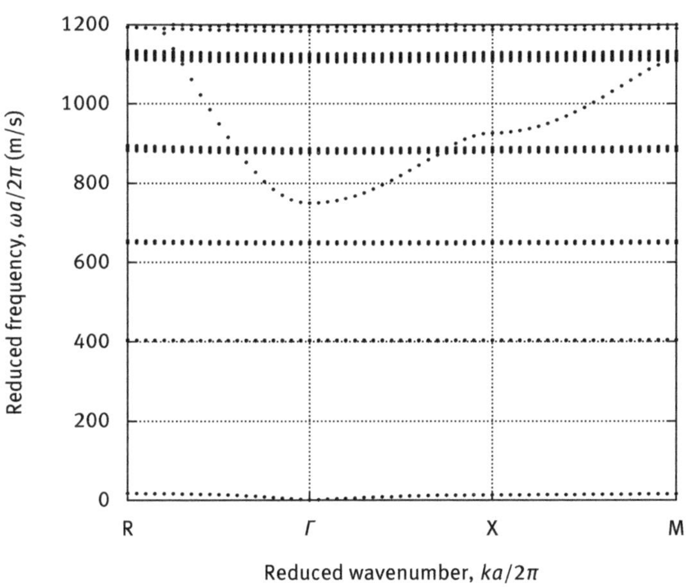

Fig. 4.23: Band structure of simple cubic-lattice sonic crystal of air bubbles in water. The bubbles have a normalized diameter $d/a = 0.576$ (the filling fraction is 0.1).

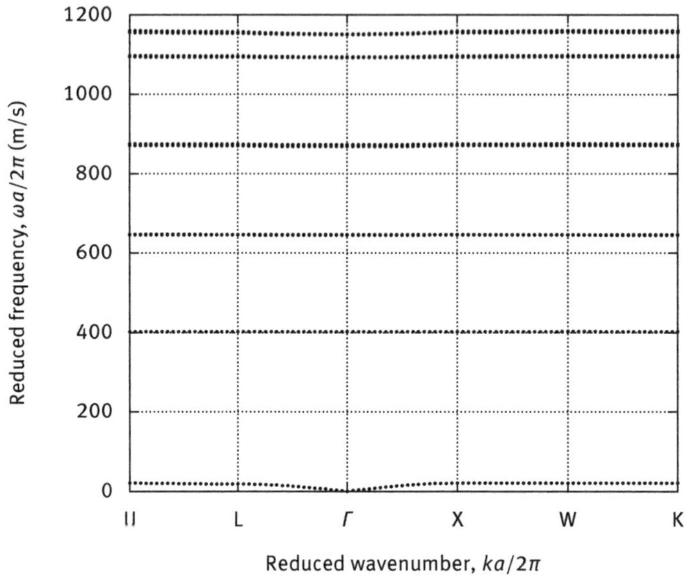

Fig. 4.24: Band structure of face-centered cubic lattice sonic crystal of air bubbles in water. The bubbles have a normalized diameter $d/a = 0.3628$ (the filling fraction is 0.1).

to the SC case, the first band is very flat and a sequence of complete band gaps starts almost immediately, only interrupted by flat bands corresponding to groups of resonant modes of the air bubble. There is also a band corresponding to the first Bragg order of diffraction, but its onset frequency is above the range of Figure 4.24, since the effective grating constant of any Bragg plane is significantly smaller for the FCC lattice compared to the SC lattice.

As we noted before, an experimental demonstration of bubble sonic crystals is still missing, and their huge complete band gaps remain a theoretical conjecture at the time of writing this book.

4.3.2 Tungsten carbide beads in water

A 3D sonic crystal of tungsten carbide spheres arranged according to the FCC lattice was used by Yang et al. [172] to explore tunneling of acoustic waves within a Bragg band gap. We will discuss the associated results in Chapter 12 and only present here the band structure computed with the FEM method and the unstructured mesh shown in Figure 4.22. Since the crystal is constructed by hand by piling 0.8 mm diameter beads, it can only respect the close-packing condition that every sphere is in contact with its six closest neighbors. For the FCC lattice, the close-packing condition is ob-

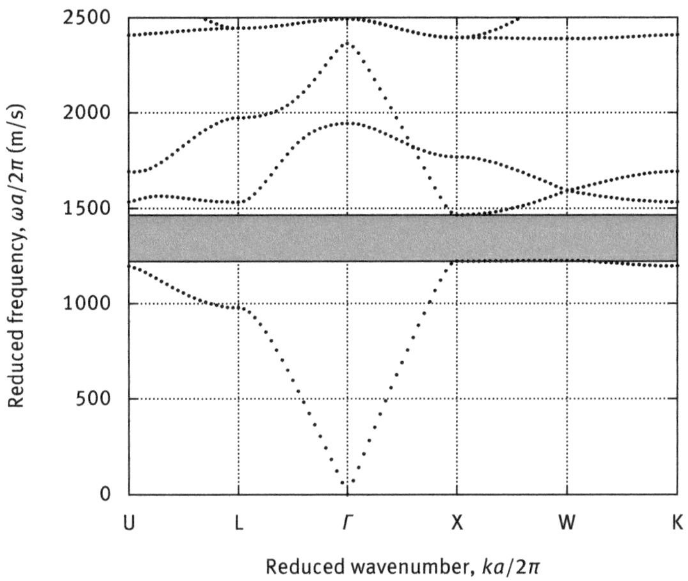

Fig. 4.25: Band structure of face-centered cubic lattice sonic crystal of tungsten carbide beads in water. The close-packed beads have a normalized diameter $d/a = 0.707$ (the filling fraction is 0.74).

tained for $d/a = 1/\sqrt{2}$, and the filling fraction is then 0.74. If we were to attempt a domain decomposition method, we would need to clip the central sphere at the six faces of the FCC primitive parallelepiped and to incorporate the six clipped sphere caps emerging from the six neighboring spheres. To avoid these complications, we simply describe the material constants by space-dependent functions of space coordinates. An alternative is to use the Wigner–Seitz cell, since the latter fully contains the inclusion even at the close-packed condition.

The band structure is shown in Figure 4.25. A rather large complete band gap can be observed, but it should be noted that it only forms as the sphere diameter closely approaches the close-packing value. For small filling fractions as considered previously with bubble crystals, no complete band gap would be observed. It can also be remarked that there are only dispersive bands, i.e. no resonant modes of the tungsten carbide beads appear.

4.A Derivation of PWE equations

In this appendix, we give a more rigorous derivation of the PWE equations than the one presented in Section 4.1.2. Laurent's rule for Fourier series states that if function $h(x) = f(x)g(x)$ is the product of two functions f and g, then the Fourier series coefficients of

h are obtained as the convolution of the coefficients of f and g, or

$$h_n = \sum_{m=-\infty}^{+\infty} f_{n-m}\, g_m.$$

Since we made a direct derivation of the PWE equations, we did not have to use Laurent's rule explicitly, but it is clearly behind definitions (4.20) and (4.21). Of course, there are conditions that the various functions should satisfy for Laurent's rule to lead to a useful result; in the case of sonic crystals, these conditions are related to discontinuities of the material constants and of the physical fields.

To simplify the analysis, let us consider univariate functions of x with period 2π. Let us denote the Fourier series expansion of function f as

$$F[f](x) = \sum_{m=-\infty}^{+\infty} f_m \exp(-\imath m x).$$

$F[f](x)$ is a continuous function of x, which is not necessarily the case of the original function $f(x)$. If $f(x)$ is continuous and left- and right-differentiable at point x, then the Fourier series converges to the function value, i.e. $F[f](x) = f(x)$. This is the case for the sonic crystals we considered, for any point x not exactly on the discontinuity between two material regions. At a staircase discontinuity, however, we already mentioned the Gibbs phenomenon: $F[f](x)$ does not converge to $f(x)$ but instead oscillates very fast around the mean value $\frac{1}{2}(f(x^+) + f(x^-))$, as N goes to infinity. So, does Laurent's rule always lead to a meaningful result? The question has recently been debated extensively in the optical grating and photonic crystal literature. Li [85] gave mathematical arguments as to when one should select the Fourier series (f_m), or the inverse of the Fourier series of the inverse function $((f_m^{-1})^{-1})$ when evaluating the Fourier coefficients of $h(x)$. As an example, Li considers

$$f(x) = \begin{cases} 1 & \text{if } |\mathbf{x}| < \frac{\pi}{2} \\ \frac{1}{2} & \text{if } \frac{\pi}{2} < |\mathbf{x} \leq \pi \end{cases}$$

and $g(x) = f^{-1}(x)$. It is then obvious that $h(x) = f(x)g(x) = 1$. Anyway, when Laurent's rule is used, $h_n = \sum_{m=-\infty}^{+\infty} f_{n-m}\, g_m$, $F[h](x)$ oscillates very fast around 1 at the original discontinuity points. Of course, in this case, using $h_n = \sum_{m=-\infty}^{+\infty} (f_{n-m}^{-1})^{-1}\, g_m$ removes the problem, since $g_m = f_m^{-1}$. What the present example says is that we should avoid considering the product of two functions that are simultaneously discontinuous at the same points.

Coming back to the sonic crystal problem, we know that pressure p is continuous everywhere, but that only the normal velocity is continuous across a discontinuity between material regions. The material constants ρ and B, as well as their inverses, are clearly discontinuous. This is the reason why we wrote (4.16) and (4.17) as they stand: discontinuities are present on either side of the equations, but not twice on the same side. If instead we had written

$$-(\tilde{p}_{,i} - \imath k_i \tilde{p}) = \imath\omega\rho\tilde{v}_i,$$
$$\imath\omega\tilde{p} = -B(\tilde{v}_{i,i} - \imath k_i \tilde{v}_i),$$

then we would have introduced a convergence problem in the resulting PWE equations since both right-hand sides would now be twice discontinuous at every material boundary.

Let us now show that the PWE method is actually a Galerkin method on the basis of Fourier exponential functions. In this respect, this property implies that the PWE method has a well-defined meaning: it has a weak form, or variational formulation on some functional space. This functional space is, however, not a finite element space in that no mesh is required. The PWE method is thus a meshless method.

We consider the space spanned by the Fourier exponentials on the periodic unit cell, which we term the PWE space. The degrees of freedom of functions f in this space are just the Fourier coefficient f_m in the Fourier series expansion. The Fourier exponentials $\exp(-\imath\mathbf{G}_m \cdot \mathbf{x})$ are orthogonal on the domain Ω since

$$\frac{1}{V(\Omega)} \int\limits_{\Omega} \exp(\imath\mathbf{G}_n \cdot \mathbf{x}) \exp(-\imath\mathbf{G}_m \cdot \mathbf{x}) = \delta_{n-m}.$$

Let us build the weak form of (4.16) and (4.17). We consider the test functions $\tilde{q}$ (dual to $\tilde{p}$) and $\tilde{w}_i$, $i = 1, 2, 3$ (dual to $\tilde{v}_i$). The test functions belong to the PWE space

$$\tilde{q}(\mathbf{x}) = \sum_m q_m \exp(-\imath\mathbf{G}_m \cdot \mathbf{x}),$$
$$\tilde{w}_i(\mathbf{x}) = \sum_m w_{im} \exp(-\imath\mathbf{G}_m \cdot \mathbf{x}).$$

The scalar equation (4.17) is left-multiplied by $\tilde{q}^*$ and integrated over the unit cell. We have for the left-hand side

$$\begin{aligned} \imath\omega \frac{1}{V(\Omega)} \int\limits_{\Omega} \tilde{q}^* \frac{1}{B}\tilde{p} &= \imath\omega \sum_n \sum_m q_n^* p_m \frac{1}{V(\Omega)} \int\limits_{\Omega} \frac{1}{B} \exp(\imath(\mathbf{G}_n - \mathbf{G}_m) \cdot \mathbf{x}) \\ &= \imath\omega \sum_n \sum_m q_n^* p_m \left(\frac{1}{B}\right)_{n-m} \\ &= \imath\omega Q^\dagger \mathcal{B} P. \end{aligned}$$

The matrix $\mathcal{B}$ obtained from the Fourier coefficients of $\frac{1}{B}$ thus introduces naturally. Q is obviously the vector of Fourier coefficients q_n and $Q^\dagger$ denotes its transpose conjugate.

Furthermore, the right-hand side is

$$
\begin{aligned}
-\frac{1}{V(\Omega)}\int_{\Omega}\tilde{q}^*(\tilde{v}_{i,i}-\imath k_i\tilde{v}_i) &= \imath\sum_n\sum_m q_n^*(k_i+G_{im})v_{im} \\
&\quad\times\frac{1}{V(\Omega)}\int_{\Omega}\exp(\imath(\mathbf{G}_n-\mathbf{G}_m)\cdot\mathbf{x}) \\
&= \imath\sum_m q_m^*(k_i+G_{im})v_{im} \\
&= \imath Q^\dagger\Gamma_i V_i.
\end{aligned}
$$

The equality of the two bilinear forms holds whatever the vector Q, from which the PWE equation (4.17) follows: $\omega\mathcal{B}P=\Gamma_i V_i$.

Each of the three equations (4.16) is left-multiplied by $\tilde{w}_i^*$ $(i=1,2,3)$ and integrated over the unit cell. We have for the left-hand side (the repeated index summation rule is not applied here)

$$
\begin{aligned}
-\frac{1}{V(\Omega)}\int_{\Omega}\tilde{w}_i^*\frac{1}{\rho}(\tilde{p}_{,i}-\imath k_i\tilde{p}) &= \imath\sum_n\sum_m w_{in}^*(k_i+G_{im})p_m \\
&\quad\times\frac{1}{V(\Omega)}\int_{\Omega}\frac{1}{\rho}\exp(\imath(\mathbf{G}_n-\mathbf{G}_m)\cdot\mathbf{x}) \\
&= \imath\sum_n\sum_m w_{in}^*(k_i+G_{im})p_m\left(\frac{1}{\rho}\right)_{n-m} \\
&= \imath W_i^\dagger\mathcal{B}\Gamma_i P.
\end{aligned}
$$

Furthermore, the right-hand side is

$$
\begin{aligned}
\imath\omega\frac{1}{V(\Omega)}\int_{\Omega}\tilde{w}_i^*\tilde{v}_i &= \imath\omega\sum_n\sum_m w_{in}^*v_{im}\frac{1}{V(\Omega)}\int_{\Omega}\exp(\imath(\mathbf{G}_n-\mathbf{G}_m)\cdot\mathbf{x}) \\
&= \imath\omega W_i^\dagger V_i.
\end{aligned}
$$

The equality of the three pairs of bilinear forms holds whatever the vector W_i, from which the PWE equation (4.16) follows: $\mathcal{B}\Gamma_i P=\omega V_i$.

4.B Some properties of eigenvalue problems

Let us consider a square matrix M_{ij} with dimension $n\times n$, whose elements can be either real or complex valued. An eigenvalue problem for the eigenvalue λ and the eigenvector u_i is of the form

$$M_{ij}u_j=\lambda u_i \tag{4.30}$$

Eigenvalues are roots of the characteristic polynomial obtained with the determinant $|M_{ij}-\lambda\delta_{ij}|=0$.

There are exactly n eigenvalues $\lambda^{(k)}$ and at most n eigenvectors $u_i^{(k)}$, which are a priori complex valued. Eigenvectors are nonvanishing and can be renormalized ($u_i^{(k)} u_i^{(k)} = 1$). They can be arranged as columns in a square matrix $X_{ik} = u_i^{(k)}$ so that (4.30) becomes

$$M_{ij} X_{jk} = X_{ij} \Lambda_{jk} \quad \text{with} \quad \Lambda_{jk} = \lambda^{(k)} \delta_{jk}. \tag{4.31}$$

If X is nonsingular, then $M = X \Lambda X^{-1}$.

If M is real and symmetrical, eigenvalues are real and eigenvectors are mutually orthogonal: $X^{-1} = X^T$.

In practice, there exist very efficient algorithms to obtain numerically eigenvalues and eigenvectors, and one should not try to obtain the roots of the characteristic polynomial for this purpose.

Part II: **Elastic waves in phononic crystals**

I come from fields of fractured ice,
Whose wounds are cured by squeezing,
Melting they cool, but in a trice,
Get warm again by freezing.
Here, in the frosty air, the sprays
With fern-like hoar-frost bristle,
There, liquid stars their watery rays
Shoot through the solid crystal.

James Clerk Maxwell
To the Chief Musician upon Nabla: A Tyndallic Ode

5 Elastic waves

In this chapter, we give an overview of the theory of elastic waves in solids, including piezoelectricity, plus an original part on finite element modeling of elementary elastic wave problems. This presentation aims at providing some basics required for the presentation of phononic crystals that follows in the next chapters.

5.1 Elastodynamic equations

Let us first introduce the basic concepts and relations for linear elastic waves.

Strain tensor. Let us consider some solid material. The solid is certainly composed of a very large number of atoms that are firmly bonded together but are still relatively free to vibrate about their mean equilibrium positions. In crystalline solids, the amplitude of vibration of an individual atom can safely be assumed to remain much smaller than the mean interatomic distance, because of the very stiff bonds that tie the atom to its neighbors. We will assume this linear elastic approximation to apply in the following. We want to describe these tiny vibrations which can ultimately form elastic waves when they act collectively. To obtain a tractable problem, we have to abandon the idea of describing the individual motion of each and every atom. Instead, in the frame of continuum mechanics, we consider there is a rigid reference frame that is attached to the solid body and that can be used to describe mechanical motion. Let us thus consider some point $\mathbf{x}$ with coordinates (x_1, x_2, x_3) inside a solid material. As a measure of the local deformation around $\mathbf{x}$, we introduce the displacement vector $\mathbf{u}$. The displacement vector and its components, the displacements u_i, are functions of the position $\mathbf{x}$. The first-order Taylor expansion of the displacements is

$$u_i(\mathbf{x} + \mathrm{d}\mathbf{x}) = u_i(\mathbf{x}) + \frac{\partial u_i}{\partial x_j} \mathrm{d}x_j.$$

By definition, $\frac{\partial u_i}{\partial x_j}$ is the displacement gradient, a tensor of order 2. We can separate this gradient into its symmetrical part, S_{ij}, and its antisymmetrical part, Ω_{ij}:

$$\begin{aligned} \frac{\partial u_i}{\partial x_j} &= S_{ij} + \Omega_{ij}, \\ S_{ij} &= \frac{1}{2}\left(\frac{\partial u_i}{\partial x_j} + \frac{\partial u_j}{\partial x_i}\right), \\ \Omega_{ij} &= \frac{1}{2}\left(\frac{\partial u_i}{\partial x_j} - \frac{\partial u_j}{\partial x_i}\right). \end{aligned} \tag{5.1}$$

S_{ij} is the (symmetrical) strain tensor that measures the local deformation of the solid structure. The antisymmetrical tensor Ω_{ij} measures local rotations. The expansion, or local variation of the volume, is

$$S_{ii} = S_{11} + S_{22} + S_{33} = \nabla.\mathbf{u},$$

or the trace of the strain tensor. Diagonal tensor elements S_{11}, S_{22}, S_{33} are associated with longitudinal motion, while antidiagonal tensor elements S_{ij} with $i \neq j$ are associated with shear motion.

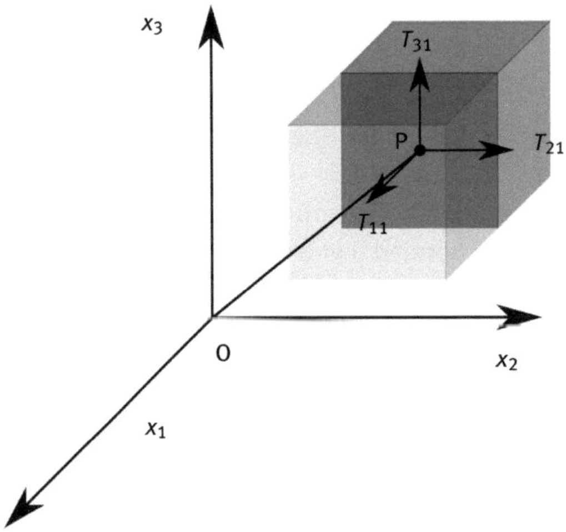

Fig. 5.1: Representation of the stress vector on a surface containing the center P of an elementary solid cube. The case depicted is for a surface orthogonal to axis x_1. The stress orthogonal to the surface is $T_{11} + T_{21} + T_{31}$ in the direction of positive x_1, with T_{11} the longitudinal stress and $T_{21} + T_{31}$ the shear stress.

Stress tensor. In contrast to fluids, shear forces are transmitted through a solid and at an interface between two bonded solids. Three independent forces can be applied on a surface: a traction-compression stress and two shear stresses. Let us consider that the small elementary cube depicted in Figure 5.1 is subject to a deformation, for instance caused by an elastic wave. In addition to possible body forces f_i, for instance gravitation, stresses appear inside the bulk of the solid because the strong atomic cohesion resists local deformations. Considering a surface orthogonal to axis x_1 and passing by the cube center P with coordinates $\mathbf{x}$, the traction per unit surface, or stress, in the positive x_1 direction can be written as $T_{11} + T_{21} + T_{31}$. By the principle of action and reaction, or Newton's third law, the stress in the negative x_1 direction is exactly opposite. Similar considerations of course apply to the surfaces orthogonal to axes x_2 and x_3, and more generally to any surface passing by point P. This means that there is a linear relationship between the surface traction and the normal to the surface: for an elementary surface with normal $\mathbf{n}$, the stress is $T_{ik}n_k$. Overall, this linear relation defines a symmetrical rank-two tensor T_{ij}, which is called the (Cauchy) stress tensor.

Tab. 5.1: Contracted (or Voigt) notation for symmetric pairs of indices.

(ij)	11	22	33	23 or 32	13 or 31	12 or 21
I	1	2	3	4	5	6

The units are those of pressure, Pa or N/m^2. The linear relation between stress and local normal follows from the fundamental laws of conservation of linear momentum and static equilibrium of forces. Moreover, the principle of conservation of angular momentum implies that the stress tensor is symmetric.

The fundamental equation of elastodynamics can be written

$$\frac{\partial T_{ik}}{\partial x_k} + f_i = \rho \frac{\partial^2 u_i}{\partial t^2}, \tag{5.2}$$

with f_i the body forces and ρ the mass density. The first term is the divergence of the stress tensor, $\nabla \cdot T$. This equation is obtained by writing the equilibrium condition for a small region of 3D space, like the cube in Figure 5.1, and then letting its volume go to zero.

Hooke's law and elastic constants. Experience shows that in the limit of small deformations, most solids obey Hooke's law, which stipulates that the Cauchy stress tensor is proportional to the strain, or

$$T_{ij} = c_{ijkl} S_{kl}. \tag{5.3}$$

In the case of tensors, proportionality means that a tensor with constant elements relates the two proportional tensors. Since both the strain tensor and the stress tensor are second-order tensors, the tensor relating them linearly must be of fourth order, hence the four indices indicated for the elastic tensor c_{ijkl} in Hooke's law (see Appendix 5.A).

The elastic tensor has in principle $3^4 = 81$ elements. The symmetry of both T_{ij} and S_{kl} however implies that

$$c_{jikl} = c_{ijlk} = c_{jilk} = c_{ijkl}. \tag{5.4}$$

As a consequence, there are at most 36 independent tensor elements. This property is used to reduce the tensor notation of Hooke's law to a 6×6 matrix notation. First, the pair of symmetric indices (ij) (i and j running from 1 to 3) is replaced by a single number I running from 1 to 6, according to the rule summarized in Table 5.1. Similarly, the symmetric pair (kl) is replaced by J. The contracted notation is then used to write the second-order stress tensor as a 6-component vector and the fourth-order elastic tensor as a 6×6 matrix

$$T_I = T_{ij}, \tag{5.5}$$

$$c_{IJ} = c_{ijkl}. \tag{5.6}$$

Hooke's law then has the matrix-vector product form

$$T_I = c_{IJ} S_J, \tag{5.7}$$

with the vector S_J having components

$$\begin{aligned} S_1 &= S_{11}; \quad & S_2 &= S_{22}; \quad & S_3 &= S_{33}; \\ S_4 &= 2S_{23}; \quad & S_5 &= 2S_{31}; \quad & S_6 &= 2S_{12}. \end{aligned} \tag{5.8}$$

The factors 2 in the shear strains S_4, S_5, and S_6 must be added so that the matrix-vector product gives the same result as the original tensor product.

Tab. 5.2: Elastic constants for some isotropic solids and anisotropic crystalline solids. The crystal system is indicated in bold in the first column. This table should be used together with Figure 5.2.

Material	Point group	Elastic constants (10^{10} N/m^2)						ρ (10^3 kg/m^3)
Iso.		c_{11}	c_{44}					
SiO_2	–	7.85	3.12					2.203
Cubic		c_{11}	c_{12}	c_{44}				
AsGa	$\bar{4}3m$	11.88	5.38	2.83				5.307
Si	$m3m$	16.56	6.39	7.95				2.329
Hexa.		c_{11}	c_{12}	c_{13}	c_{33}	c_{44}		
PZT-4	–	13.9	7.8	7.4	11.5	2.6		7.5
ZnO	$6mm$	21.0	12.1	10.5	21.1	4.2		5.676
Tetra.		c_{11}	c_{12}	c_{13}	c_{33}	c_{44}	c_{66}	
TiO_2	4/mmm	26.6	17.3	13.6	47.0	12.4	18.9	4.28
Trigonal		c_{11}	c_{12}	c_{13}	c_{33}	c_{44}	c_{14}	
Al_2O_3	$\bar{3}m$	49.7	16.3	11.1	49.8	14.7	−2.3	3.986
$LiNbO_3$	$3m$	20.3	5.3	7.5	24.5	6.0	0.9	4.7
quartz-α	32	8.7	0.7	1.2	10.7	5.8	−1.8	2.648

The final number of truly independent elastic constants, however, depends on the point group of the crystal that is considered. Isotropic materials like glasses have only two independent elastic constants, cubic crystals have three, and so on. Figure 5.2 gives the shape of matrix c_{IJ} depending on the crystal system and on the point group. Further, in Table 5.2 we give a list of the elastic constants and of the mass densities that are used in the examples in this book. Values for many materials and crystals as well as a detailed explanation of the shape of c_{IJ} can be found in specialized books.

Energy relations and Poynting's theorem for elastic waves. Let us neglect body forces in the elastodynamic relation (5.2). Combined with Hooke's law (5.3), we are led at once

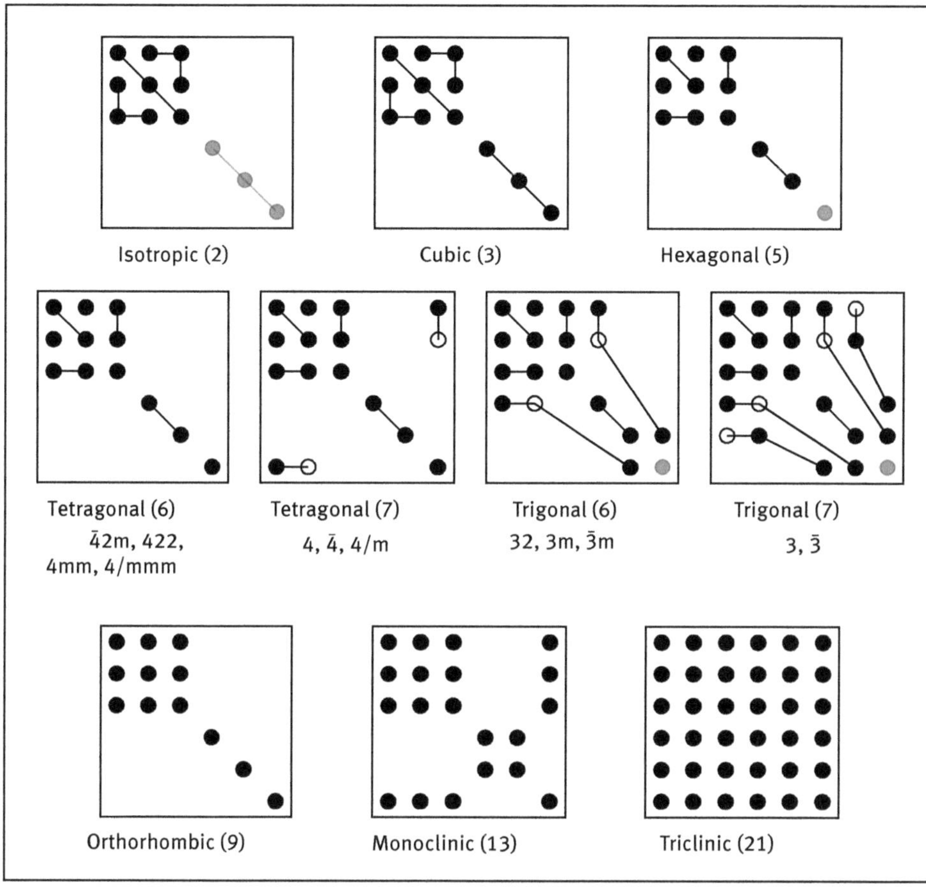

Fig. 5.2: Independent elements of the elastic tensor represented in contracted notation c_{IJ} and in the frame of the crystallographic axes for the different crystal systems and point groups. Missing symbols indicate zero values. Equal values are connected with lines. Empty circles indicate minus the value of connected plain circles. Gray circles indicate $c_{66} = (c_{11} - c_{12})/2$. The number of independent elastic constants is indicated inside parentheses in each case. Note that c_{IJ} is symmetric.

to an anisotropic wave equation

$$\rho \frac{\partial^2 u_i}{\partial t^2} = c_{ijkl} \frac{\partial^2 u_l}{\partial x_j \partial x_k}. \tag{5.9}$$

This equation is the fundamental equation of elastodynamics.

Let us give an energetic interpretation for the propagation of elastic waves in general. In this section, we assume that all physical quantities are real valued. We contract the elastodynamic equation with the velocity vector **v**:

$$\rho \frac{\partial v_i}{\partial t} v_i = \frac{\partial T_{ij}}{\partial x_j} v_i.$$

The left-hand side is the time derivative of the density of kinetic energy,

$$e_k = \frac{1}{2}\rho v_i v_i. \tag{5.10}$$

To transform the right-hand side, we use the chain-rule relation for derivatives:

$$\begin{aligned}\frac{\partial}{\partial x_j}(T_{ij}v_i) &= \frac{\partial T_{ij}}{\partial x_j}v_i + T_{ij}\frac{\partial v_i}{\partial x_j}\\ &= \frac{\partial T_{ij}}{\partial x_j}v_i + T_{ij}\frac{\partial S_{ij}}{\partial t}\\ &= \frac{\partial e_k}{\partial t} + \frac{\partial e_p}{\partial t},\end{aligned}$$

with the density of elastic energy defined as

$$e_p = \frac{1}{2}T_{ij}S_{ij} = \frac{1}{2}c_{ijkl}S_{ij}S_{kl}. \tag{5.11}$$

Note that the interchange of the time derivative from the strain to the density of elastic energy in the relations above is made possible by the symmetries of the elastic tensor. Finally, defining the Poynting vector as

$$P_j = -T_{ij}v_i, \tag{5.12}$$

we obtain the local form of Poynting's theorem as

$$\frac{\partial e_k}{\partial t} + \frac{\partial e_p}{\partial t} = -\nabla \cdot \mathbf{P} = -\frac{\partial P_j}{\partial x_j}. \tag{5.13}$$

From the local form of the Poynting theorem, a global form can be obtained by invoking the Gauss (or divergence) theorem. Let Ω denote a closed domain with enclosing boundary σ; $\mathbf{n}$ is the outward normal to boundary σ. Integrating (5.13) over domain Ω and swapping time and space derivatives we have the Poynting theorem

$$\frac{\partial E_k}{\partial t} + \frac{\partial E_p}{\partial t} = -\int_\sigma \mathbf{P}\cdot\mathbf{n}, \tag{5.14}$$

with

$$E_k = \int_\Omega e_k, \tag{5.15}$$

$$E_p = \int_\Omega e_p. \tag{5.16}$$

The physical meaning of this result is that the time variation of the internal energy (the sum of the kinetic energy and of the elastic energy) equals minus the flux of the Poynting vector through the boundary enclosing the domain of integration. The flux is thus understood as measuring the power flow leaving the system if its internal energy decreases. This statement justifies the sign convention used when defining Poynting's vector.

5.2 Bulk waves in elastic solids

In this section we discuss the anisotropy of elastic wave propagation in crystalline solids. This will lead us to define the slowness curves and wave surfaces for elastic waves. The relation between group velocity, Poynting's theorem, and energy conservation will be made apparent in this nondispersive case.

Christoffel equation – For harmonic plane waves of the form

$$u_i(t, \mathbf{r}) = \hat{u}_i \exp(\imath\omega(t - s\,\mathbf{n}.\mathbf{r})), \tag{5.17}$$

we are led to the Christoffel equation

$$\rho\hat{u}_i = s^2 c_{ijkl} n_j n_k \hat{u}_l. \tag{5.18}$$

Here, the slowness $s(\mathbf{n}) = k(\mathbf{n})/\omega$ (with units s/m) is a function of the propagation direction as given by the unit vector $\mathbf{n}$. As a note, any quantity of the form $\sqrt{c/\rho}$ with c some elastic constant has units of velocity. The Christoffel equation is independent of frequency, meaning that propagation of bulk acoustic waves is anisotropic but not dispersive.

In elastic solids, the velocities of bulk elastic waves can be very strongly anisotropic. We introduce the second-order symmetric Christoffel tensor

$$\Gamma_{il} = c_{ijkl} n_j n_k, \tag{5.19}$$

which is the contraction of the elastic tensor with the unit vector $\mathbf{n}$ (twice). We emphasize that the Christoffel tensor depends on the selected direction of propagation. The Christoffel equation can now be cast as an eigenvalue problem for the square of the velocity

$$\Gamma_{il}\hat{u}_l = \rho v^2 \hat{u}_i. \tag{5.20}$$

Solving the eigenvalue problem we can obtain the velocities of all bulk waves propagating in direction $\mathbf{n}$ together with their polarization $\hat{u}_i$. Since Γ_{il} is a second-order tensor and the indices i and j vary from 1 to 3, there are exactly 3 eigenvalues. There are then exactly three bulk waves with orthogonal polarizations. Let us now discuss some simple cases and then give a general procedure to easily obtain all bulk velocities with a computer program.

Bulk elastic waves in isotropic solids. In the isotropic case, bulk wave velocities must be invariant with the propagation direction. Let us consider, for instance, x_1 as the direction of propagation. Then $n_1 = 1$, $n_2 = 0$, and $n_3 = 0$. In general, the Christoffel tensor for this direction of propagation is

$$\Gamma = \begin{pmatrix} c_{1111} & c_{1112} & c_{1113} \\ c_{2111} & c_{2112} & c_{2113} \\ c_{3111} & c_{3112} & c_{3113} \end{pmatrix} = \begin{pmatrix} c_{11} & c_{16} & c_{15} \\ c_{61} & c_{66} & c_{65} \\ c_{51} & c_{56} & c_{55} \end{pmatrix} = \begin{pmatrix} c_{11} & 0 & 0 \\ 0 & c_{44} & 0 \\ 0 & 0 & c_{44} \end{pmatrix}$$

with $c_{44} = \frac{c_{11}-c_{12}}{2}$. The first equality is just the definition (5.19), the second equality uses the contracted notation, and the third equality uses Figure 5.2 to write the result with only independent elastic constants for isotropic solids. The resulting matrix is diagonal; it has a simple eigenvalue and a pair of degenerate eigenvalues.

- The bulk wave with velocity $c_L = \sqrt{c_{11}/\rho}$ is associated with the eigenvector $\hat{u} = (1, 0, 0)^T$. This is a longitudinal wave since the displacement is in the direction of propagation.
- The two bulk waves with velocity $c_S = \sqrt{c_{44}/\rho}$ are associated with the eigenvectors $\hat{u} = (0, 1, 0)^T$ et $\hat{u} = (0, 0, 1)^T$. They are shear waves since their displacement is transverse.
- Since $c_{12} > 0$, $V_S < V_L/\sqrt{2}$ and it can be concluded that the longitudinal bulk velocity is always larger than the shear wave velocity.

The direct calculation above can be repeated for an arbitrary direction of propagation **n** and we would find that the eigenvalues remain exactly the same. The polarization vector $\hat{\mathbf{u}}$, however, would change since for the longitudinal wave the displacement would be along **n** while it would be orthogonal to it for the two shear waves. It can further be concluded that the properties above remain true for any solution of the wave equation in isotropic media, by considering its plane wave spectrum as in Chapter 2.

Bulk elastic waves in cubic crystals. The Christoffel tensor for a cubic crystal in general is

$$\Gamma = \begin{pmatrix} c_{11}n_1^2 + c_{44}(n_2^2 + n_3^2) & (c_{12} + c_{44})n_1 n_2 & (c_{12} + c_{44})n_1 n_3 \\ \cdot & c_{11}n_2^2 + c_{44}(n_1^2 + n_3^2) & (c_{12} + c_{44})n_2 n_3 \\ \cdot & \cdot & c_{11}n_3^2 + c_{44}(n_1^2 + n_2^2) \end{pmatrix},$$

which can be obtained using Figure 5.2 to express it with only the three independent elastic constants c_{11}, c_{12}, and c_{44}.

Let us first consider propagation along the [100] direction ($n_1 = 1$, $n_2 = 0$, and $n_3 = 0$). The Christoffel tensor Γ is diagonal, with a simple eigenvalue, c_{11}, and two degenerate eigenvalues, c_{44}. Exactly as in the isotropic case, there are one longitudinal bulk wave with velocity $\sqrt{c_{11}/\rho}$ and two shear bulk waves with velocity $\sqrt{c_{44}/\rho}$.

Let us then consider propagation along direction [110] ($n_1 = 1/\sqrt{2}$, $n_2 = 1/\sqrt{2}$, and $n_3 = 0$). There are now three distinct eigenvalues: c_{44}, $\frac{1}{2}(c_{11} - c_{12})$ and $\frac{1}{2}(c_{11} + c_{12}) + c_{44}$. There are then

- one purely shear bulk wave polarized along x_3, with velocity $\sqrt{c_{44}/\rho}$,
- one quasi-shear bulk wave with velocity $\sqrt{(c_{11} - c_{12})/2\rho}$,
- and one quasi-longitudinal bulk wave with velocity $\sqrt{(2c_{44} + c_{11} + c_{12})/2\rho}$.

It can be observed that the two shear waves are not degenerate because $c_{44} \neq \frac{c_{11}-c_{12}}{2}$, marking the distinction between isotropic and cubic solids. The courageous reader might attempt to analytically obtain the eigenvalues of the Christoffel tensor for all values of the direction vector **n**. This is not any easy task, however, and it is more efficient to write a computer program since indeed all that is required is to solve for the eigenvalues of a 3×3 square matrix. We give examples later on when we describe the characteristic surfaces for bulk elastic waves in anisotropic solids. See also Appendix 5.B for a simple finite element implementation of the problem.

Energy relations for harmonic plane waves. For harmonic plane waves we have the following results from the Christoffel equation and the Poynting theorem introduced in the previous section.

The kinetic energy density is

$$e_c = \frac{1}{4}\rho\omega^2 \hat{u}_i^* \hat{u}_i. \tag{5.21}$$

A note of caution is necessary here since this relation cannot be derived directly from (5.10). The Poynting theorem was derived for real-valued physical quantities and we now want to invoke it with complex-valued quantities. In the context of complex signals, we have to consider time-averaged energy quantities [8]. The elastic energy density is

$$e_p = \frac{1}{4}c_{ijkl}\hat{S}_{ij}^* \hat{S}_{kl} = \frac{1}{4}\omega^2 s^2 \Gamma_{il}\hat{u}_i^* \hat{u}_l. \tag{5.22}$$

Contracting the Christoffel equation with $\hat{u}_i$, it can be seen that $e_p = e_c$: the kinetic and elastic energy densities are equal for harmonic plane waves. The total energy density is then

$$e = e_c + e_p = \frac{1}{2}\rho\omega^2 \hat{u}_i^* \hat{u}_i = \frac{1}{2}\omega^2 s^2 \Gamma_{il}\hat{u}_i^* \hat{u}_l.$$

Poynting's vector is

$$P_i = \frac{1}{2}\operatorname{Re}(\imath\omega \hat{T}_{ij}\hat{u}_j^*) = \frac{1}{2}s\omega^2 \operatorname{Re}(c_{ijkl}\hat{u}_j^* \hat{u}_k n_l).$$

The energy velocity is by definition the power flow divided by the energy of the wave. In the case of harmonic plane waves, it is sufficient to consider the Poynting vector divided by the energy density, which defines the energy velocity vector

$$V_i^e = P_i/e.$$

By direct inspection of the relations, it is easy to check that

$$V_i^e n_i = v.$$

As a result, there is a strong relation between the phase velocity and the energy velocity, which is clearly reminiscent of Property 2.5. In particular, the energy velocity (as the modulus of the energy velocity vector) is always larger than the phase velocity for a given propagation direction. Furthermore, the equality of the energy velocity and of the spatial group velocity can be demonstrated. This demonstration will be given explicitly in Section 5.3 for the more general case of harmonic plane waves in piezoelectric media.

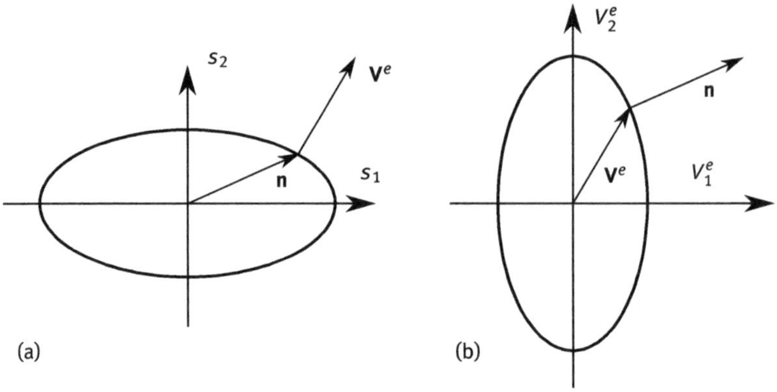

Fig. 5.3: Characteristic surfaces for waves in anisotropic media. The slowness surface is a 3D plot of the slowness as a function of the direction in space. The slowness curve in (a) is a cross section of the slowness surface by some 2D plane. The wave surface, or group velocity surface, is a 3D plot of the extremity of the energy velocity vector: it represents the surface attained by the energy front of the wave after a unit propagation time, as if it had been emitted from the origin. A 2D cross section of the wave surface is shown in (b).

Characteristic surfaces. By definition, the slowness surface depicted in Figure 5.3 is defined by the locus of the slowness vector $\mathbf{s} = s\mathbf{n}$ as a function of the direction $\mathbf{n}$. The slowness surface is then a graphical representation of the dispersion relation $\mathbf{k}(\omega, \mathbf{n})/\omega$ exploiting the property that harmonic plane wave propagation is not dispersive in homogeneous solids. The energy velocity vector V^e (or the spatial group velocity) is orthogonal to the slowness surface. This is a direct consequence of the far-field property of solutions of the wave equation that we presented in Chapter 2.

By definition, the wave surface is the locus of the energy velocity vector $\mathbf{V}^e$ as the unit direction vector $\mathbf{n}$ varies. Physically, it represents the far-field surface that is reached in 3D space after a unitary time by a wave emitted from a single point source placed at the origin. It is then also an equiphase surface. $\mathbf{n}$ is orthogonal to the wave surface: wavefronts are tangent to it.

In the case of bulk elastic waves, there exist three different slowness surfaces, one being quasi-longitudinal and the other two being quasi-shear. These surfaces are symmetric with respect to the origin.

In Figures 5.4, 5.5 and 5.6 we plot slowness curves and wave surfaces for the cases of silicon (Si), rutile (TiO_2) and sapphire (Al_2O_3). These crystalline materials are readily available in the form of thin wafers with various diameters that can be used for technological processing. They will serve to illustrate the influence of the anisotropy of the elastic tensor on the characteristic surfaces.

Silicon as a crystal is ordered as a diamond cubic lattice, i.e. a face-centered cubic lattice with two Si atoms per unit cell, positioned at the origin and at $1/4\,(\mathbf{a}_1 + \mathbf{a}_2 + \mathbf{a}_3)$, with the m3m point group symmetry. We plot in Figure 5.4 (a) a cut of the slowness

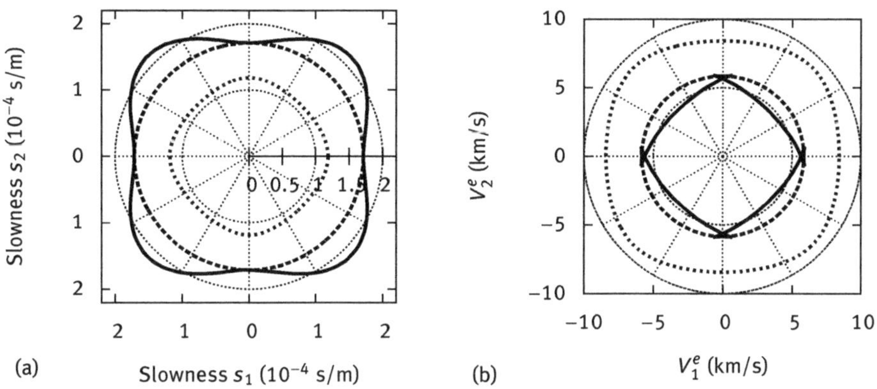

Fig. 5.4: Slowness curves and wave surfaces for silicon. (a) Slowness curves for bulk elastic wave propagation are plotted in the (x_1, x_2) plane for silicon (Si, cubic crystal system, m3m point group). The three slowness curves are for a quasi-longitudinal (short dashed line) wave, a pure shear wave (long dashed line) and a quasi-shear wave (solid line). (b) Corresponding wave surface cross sections.

surfaces orthogonal to the Z axis. One of the two shear waves is pure, i.e. the corresponding slowness curve is a circle and this wave is isotropic; its displacement is along the Z axis only. The other quasi-shear wave combines a dominant shear displacement with some longitudinal displacement. Note that the concept of shear or longitudinal displacement rotates with the propagation direction in the (s_1, s_2) plane. The slowness curves are obtained by solving the eigenvalue problem (5.13) as a function of the propagation angle in the plane. The three eigenvalues are classified as pure shear, quasi-shear and quasi-longitudinal by considering the components of the eigenvectors. Figure 5.4 (b) displays cuts of the wave surface orthogonal to the Z axis, in correspondence with Figure 5.4 (a). These curves are obtained by plotting the first two components of the energy velocity vector (V_1^e, V_2^e) computed from the eigenvalue problem solution. Double points for the quasi-shear wave can be observed at the intersections with the X and the Y axes: at these points the wave surface cut forms a loop with two cusps. These particular accidents in the wave surface are known to be related to the phenomenon of phonon focusing in solid state physics.

Atoms in rutile or titanium dioxide are ordered according to a body-centered tetragonal lattice with the 4/mmm point group symmetry. Its slowness curves are very strongly anisotropic, as Figure 5.5 (a) shows in the (s_1, s_2) plane. There is again a pure shear wave. The quasi-shear wave slowness depends very acutely on the propagation angle: it can come close to the longitudinal slowness along the [100] and [010] directions (X and Y axes) as well as increase strongly in the [110] and [1$\bar{1}$0] directions. Correspondingly, the wave surface cuts in Figure 5.5 (b) clearly show marked double points, loops and cusps.

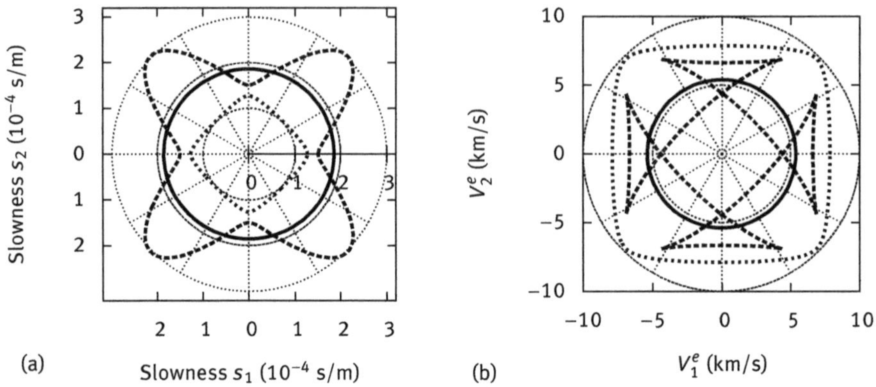

Fig. 5.5: Slowness curves and wave surfaces for rutile. (a) Slowness curves for bulk elastic wave propagation are plotted in the (x_1, x_2) plane for rutile (TiO_2, tetragonal crystal system, 4/mmm point group). The three slowness curves are for a quasi-longitudinal (short dashed line) wave, a pure shear wave (solid line) and a quasi-shear wave (long dashed line). (b) Corresponding wave surface cross sections. Note the pronounced double points and cusps in the quasi-shear wave case.

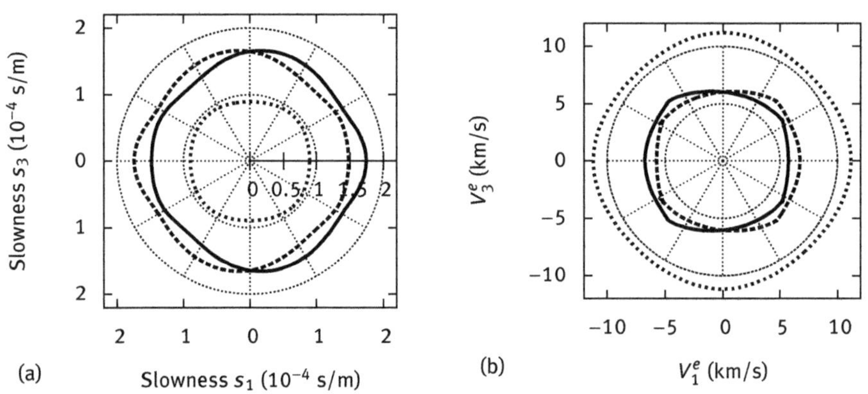

Fig. 5.6: Slowness curves and wave surfaces for sapphire. (a) Slowness curves for bulk elastic wave propagation are plotted in the (x_1, x_3) plane for sapphire (Al_2O_3, trigonal crystal system, $\bar{3}$m point group). The three slowness curves are for a quasi-longitudinal (short dashed line) wave and two quasi-shear waves (solid and long dashed line). (b) Corresponding wave surface cross sections.

Sapphire belongs to the trigonal crystal system but has a rhombohedral lattice system. In Figure 5.6 (a) we display the slowness curves in the (x_1, x_3) plane, that is, orthogonal to the [010] direction. The slowness curves are mildly anisotropic, at least in comparison with rutile, but it should be noticed that there are no purely polarized waves in this case. Also, the two shear wave slowness curves are images in mirror symmetry with respect to axis [001]. The same symmetry also applies to the wave surface cuts in Figure 5.6 (b).

Attenuation. Losses of elastic waves propagating in solids can have various origins: thermal conduction, interaction with thermal phonons, scattering by defects, etc. Losses are generally larger in metals than in insulators, and in polycrystals than in monocrystals. For most natural and synthetic crystals, losses can be taken as roughly proportional to ω^2, as experiment shows. Let us outline how this behavior can be reproduced with a simple model, valid in the limit of small loss, where the elastic constants become complex with a small imaginary part proportional to frequency. Specifically, we suppose the following dependency of the elastic tensor on frequency

$$c'_{ijkl} = c_{ijkl} + \imath\omega\eta_{ijkl}, \tag{5.23}$$

where η_{ijkl} is the phonon viscosity tensor, with the same symmetries as the elastic tensor. c_{ijkl} is the value of the elastic tensor without loss. From the Christoffel equation, we have

$$s'^2 = \frac{\rho}{\Gamma'},$$
$$\Gamma' = \frac{\hat{u}_i^* \Gamma'_{il} \hat{u}_l}{\hat{u}_i^* \hat{u}_l}.$$

The second relation defines an effective complex elastic constant $\Gamma' = \Gamma + \imath\omega\eta$. The first-order Taylor expansion of s' is

$$s' \approx s\left(1 - \frac{1}{2}\imath\omega\eta/\Gamma\right).$$

After propagation along a length L, the amplitude of the wave will be reduced by a factor

$$\exp(\mathrm{Im}(k')L) = \exp(\omega\,\mathrm{Im}(s')L) = \exp\left(-\frac{1}{2}\omega^2 sL\eta/\Gamma\right).$$

The ω^2-law for loss on propagation is thus explicitly apparent. Table 5.3 lists some values for phonon viscosity coefficients.

5.3 Piezoelectric media

In this section, we first briefly describe the physical principles of piezoelectricity. We then generalize the concepts of elastic wave propagation in the previous section to the case of piezoelectric media.

Physical origin of piezoelectricity. Piezoelectricity globally means that the mechanical properties of a dielectric solid are coupled with its electrical properties. It is conventional to distinguish between the direct and the inverse piezoelectric effects. This

Tab. 5.3: Phonon viscosity tensor for some materials and crystals.

Material	Point group	Phonon viscosity constants (10^{-3} N/m^2.s)					
Iso.		η_{11}	η_{44}				
SiO_2 [75]	—	5.9	≈ 0.6				
Cubic		η_{11}	η_{12}	η_{44}			
AsGa [49]	$\bar{4}3m$	7.49	6.57	0.72			
Si [49]	$m3m$	5.9	5.16	0.62			
Trigonal		η_{11}	η_{12}	η_{13}	η_{33}	η_{44}	η_{14}
$LiNbO_3$ [9]	$3m$	0.6547	0.2275	0.2499	0.3377	0.1765	−0.0687
quartz-α [75]	32	1.37	0.73	0.71	0.96	0.36	0.01

distinction, however, is based on experiment but will not appear in the theory. The direct piezoelectric effect means that under an applied stress or strain, an electrical polarization appears inside the solid. The inverse piezoelectric effect, or Lippman effect, means that an applied electrical field induces a deformation of the solid structure or the appearance of a stress distribution. At the microscopic level, this can happen because the deformation of the solid structure induces an asymmetry of the electronic distribution with respect to the atomic distribution. In an ordered crystal, this can only happen for noncentrosymmetric crystal structures, and hence not all crystals are piezoelectric.

Piezoelectricity is mostly useful to elaborate transducers that can convert electrical signals to mechanical waves and conversely. Common piezoelectrics include α-quartz, in widespread use for making watches, television sets, and very precise microbalances; lithium tantalate, which is ubiquitous in mobile phones; and lithium niobate, which is used in many scientific applications including acousto-optics. All these crystals belong to the trigonal crystal system, but piezoelectricity is not limited to this particular crystal system as we will see later. Isotropic and disordered dielectric crystals are generally not piezoelectric, for symmetry reasons.

Constitutive relations. Considering only the linear regime, the constitutive relations of piezoelectricity can be taken as

$$T_{ij} = c_{ijkl}S_{kl} - e_{kij}E_k, \tag{5.24}$$

$$D_i = e_{ikl}S_{kl} + \varepsilon_{ij}E_j. \tag{5.25}$$

In these equations, **E** is the electric field vector and **D** is the electric displacement that includes the polarizability of the piezoelectric solid. These quantities appear from Maxwell's equations as in the classical theory of dielectrics; if there were no coupling to elastic deformations of the solid they would only be related through the order-2 dielectric tensor ε_{ij}. Hooke's law is seen to be contained in the constitutive relations, for their purely elastic part. The order-3 tensor e_{kij} in (5.24) is the piezoelectric tensor and is responsible for the coupling of elastic and electric quantities. This tensor is symmetric with respect to its second and third indices, because these indices relate to the symmetric strain tensor. The same tensor enters (5.24), which can be shown based on thermodynamical considerations.

Tab. 5.4: Representation of the tensors of the piezoelectric constitutive relations as a matrix. There are at most 21 independent elastic constants, 18 independent piezoelectric constants, and 6 independent dielectric constants. The table is symmetric with respect to the diagonal of the matrix.

c_{11}	c_{12}	.	.	.	c_{16}	e_{11}	e_{21}	e_{31}
.	c_{22}	.	.	.	.	e_{12}	.	.
.	.	.	.	.	.	.	.	.
.	.	.	.	.	.	.	.	.
.	.	.	.	.	.	.	.	.
.	.	.	.	.	c_{66}	e_{16}	.	e_{36}
e_{11}	e_{12}	.	.	.	e_{16}	ε_{11}	ε_{12}	ε_{13}
e_{21}	.	.	.	.	.	.	ε_{22}	ε_{23}
e_{31}	.	.	.	.	e_{36}	.	.	ε_{33}

Using the contracted notation for pairs of symmetric indices, we have

$$T_I = c_{IJ}S_J - e_{kI}E_k, \tag{5.26}$$

$$D_i = e_{iJ}S_J + \varepsilon_{ij}E_j. \tag{5.27}$$

Note that the dielectric tensor is also symmetric and has thus at most 6 independent coefficients. Table 5.4 presents a very practical table storage for tensor coefficients to be used with the piezoelectric constitutive relations. As a whole, all coefficients can be presented as in a 9 × 9 symmetrical matrix, which is very practical for numerical simulation programs. The shape of this matrix is depicted in Figure 5.7 depending on the crystal system and point group. Table 5.5 lists the independent piezoelectric and dielectric constants for some piezoelectric solids.

Quasi-static approximation. Because of piezoelectric coupling, if an elastic wave travels in a solid it is accompanied by an electromagnetic wave. Let us specify the properties of the electromagnetic part of the wave. Maxwell's equations for a dielectric medium are

$$\nabla \times \mathbf{E} = -\frac{\partial \mathbf{B}}{\partial t}, \tag{5.28}$$

$$\nabla \times \mathbf{H} = \frac{\partial \mathbf{D}}{\partial t}, \tag{5.29}$$

$$\nabla \cdot \mathbf{D} = 0, \tag{5.30}$$

$$\nabla . \mathbf{H} = 0. \tag{5.31}$$

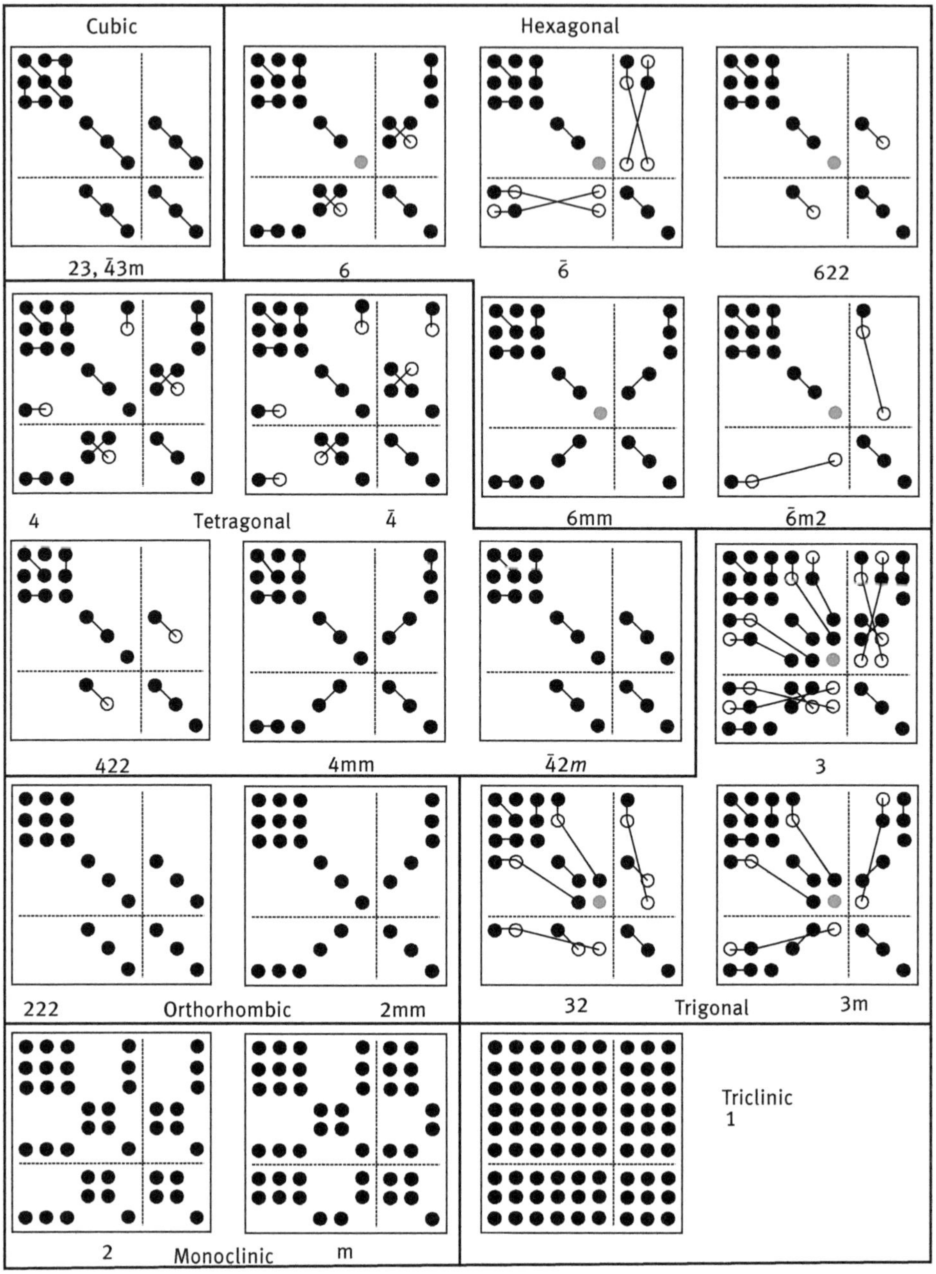

Fig. 5.7: Independent elements of the elastic, dielectric, and piezoelectric tensors. They are represented in contracted notation in the frame of the crystallographic axes for the different crystal systems and point groups, following the format in Table 5.4. Missing symbols indicate zero values. Equal values are connected with lines. Empty circles indicate minus the value of connected plain circles. Gray circles indicate $c_{66} = (c_{11} - c_{12})/2$.

Tab. 5.5: Independent piezoelectric and dielectric constants. Values are taken from Royer et al. [131]. This table should be used in conjunction with Figure 5.7.

Material	Piezoelectric constants (C/m^2)						Diel. constants (10^{-11} F/m)	
Cubic	e_{14}						ε_{11}	
AsGa	−0.16						9.73	
Hexa.	e_{13}	e_{31}	e_{33}				ε_{11}	ε_{33}
PZT-4	12.7	−5.2	15.1				650	560
ZnO	−0.59	−0.61	1.14				7.38	7.83
Trigonal	e_{11}	e_{14}	e_{13}	e_{22}	e_{31}	e_{33}	ε_{11}	ε_{33}
$LiNbO_3$	0	0	3.7	2.5	0.2	1.3	38.9	25.7
quartz-α	0.717	−0.0406	0	0	0	0	3.92	4.10

They are supplemented with the constitutive equation (nonmagnetic medium)

$$\mathbf{B} = \mu_0 \mathbf{H} \tag{5.32}$$

and the piezoelectric constitutive equations (5.24) and (5.25).

These equations can be simplified by observing that the usual frequencies of elastic waves (say, at most a few GHz) are sufficiently small that temporal derivatives can be safely ignored in a first approximation. To be more precise, the simplification arises from the huge difference in the magnitude of elastic wave velocities (a few km/s) as compared to electromagnetic wave velocities ($3\,10^8$ m/s in a vacuum), so that spatial derivatives (scaled by the elastic wavelength) dominate temporal derivatives in Maxwell's equations. As a result, both the electric field vector and the magnetic field vector are irrotational, or

$$\nabla \times \mathbf{E} = -\imath\omega \mathbf{B} \simeq 0,$$
$$\nabla \times \mathbf{H} = \imath\omega \mathbf{D} \simeq 0.$$

Consequently, the electric and the magnetic field vectors are completely decoupled in Maxwell's equations. Since magnetic quantities do not enter the piezoelectric equations, they can be ignored for our purpose. Moreover, **E** is irrotational and thus there exists a scalar potential ϕ such that

$$\mathbf{E} = -\nabla \Phi. \tag{5.33}$$

In this so-called quasi-static approximation the piezoelectric constitutive relations simplify to

$$T_I = c_{IJ} S_J + e_{kI} \Phi_{,k}, \tag{5.34}$$

$$D_i = e_{iJ} S_J - \varepsilon_{ij} \Phi_{,j}. \tag{5.35}$$

Together with the dynamical equations (5.2) and (5.30), they form the basic set of equations for elastic wave propagation in piezoelectric media.

Boundary conditions. Boundary conditions at the interface separating two piezoelectric media can be specified independently for mechanical and electromagnetic quantities.

Mechanical boundary conditions are easily specified if the two media are perfectly bonded, or at a free or clamped surface.

- If the solids are rigidly bonded, the displacements are continuous at every point of the boundary Σ separating media M_1 and M_2:

$$[u_i] = 0. \tag{5.36}$$

Furthermore, the mechanical tension is continuous across the interface:

$$[T_{ij}n_j] = 0 \tag{5.37}$$

with $\mathbf{n}$ the normal to Σ.

- At a free surface, displacements are left completely unspecified but the mechanical tension vanishes:

$$T_{ij}n_j = 0. \tag{5.38}$$

- At a clamped surface, stresses are left unspecified but displacements vanish:

$$u_i = 0. \tag{5.39}$$

In the frame of the quasi-static approximation, electrical boundary conditions are of the electrostatics type.

- At an interface Σ separating media M_1 and M_2, both the electric potential and the tangential electric field are continuous,

$$[\Phi] = 0, \tag{5.40}$$

$$[\mathbf{E} \times \mathbf{n}] = 0. \tag{5.41}$$

- If the surface can support a surface charge density σ_e, which can happen at a dielectric-metal interface, for instance, then

$$[D_i n_i] = \sigma_e. \tag{5.42}$$

At the interface between two dielectrics, the normal component of $\mathbf{D}$ is continuous.

- If the surface is shorted, for instance because of the presence of a thin metal film that is connected to the ground, then $\Phi = 0$.

Poynting's theorem for piezoelectric media. Expressing the work done by mechanical forces and electrical forces, following the procedure outlined in Section 5.1, Poynting's theorem for piezoelectric media has the form

$$\frac{\mathrm{d}W}{\mathrm{d}t} = \frac{\mathrm{d}}{\mathrm{d}t}(E_k + E_p) + \int_{\Sigma} P_j n_j \tag{5.43}$$

with

- the kinetic energy $E_k = \int e_k \,\mathrm{d}V$ and the kinetic energy density $e_k = \frac{1}{2}\rho v_i^2$,
- the potential energy $E_p = \int e_p \,\mathrm{d}V$ and the potential energy density

$$e_p = \frac{1}{2}(S_{ij}T_{ij} + E_k D_k), \tag{5.44}$$

- and Poynting's vector

$$P_j = -T_{ij}v_i + (\mathbf{E} \times \mathbf{H})_j. \tag{5.45}$$

Again, Poynting's theorem expresses that the work done by internal sources is partly stored in the form of kinetic and potential energy, and partly radiated through the boundary enclosing the domain. The flux of Poynting's vector represents this radiated power.

5.4 Bulk waves in piezoelectric media

We examine again the propagation of harmonic plane waves, but with piezoelectricity now taken into account.

Stiffened elastic constants for plane harmonic waves. For a harmonic plane wave, the elastodynamic equation and Poisson's equation

$$\begin{aligned} -\imath\omega s \hat{T}_{ij} n_j &= -\rho\omega^2 \hat{u}_i, \\ \hat{D}_j n_j &= 0, \end{aligned}$$

together with the constitutive relations

$$\begin{aligned} \hat{T}_{ij} &= -\imath\omega s(c_{ijkl} n_k \hat{u}_l + e_{kij} n_k \hat{\Phi}), \\ \hat{D}_j &= -\imath\omega s(e_{jkl} n_k \hat{u}_l - \varepsilon_{jk} n_k \hat{\Phi}), \end{aligned}$$

lead to

$$\begin{aligned} \rho \hat{u}_i &= s^2(\Gamma_{il}\hat{u}_l + \gamma_i \hat{\Phi}), \\ \gamma_l \hat{u}_l &= \varepsilon \hat{\Phi}, \\ \gamma_i &= e_{kij} n_j n_k, \\ \varepsilon &= \varepsilon_{jk} n_j n_k. \end{aligned}$$

Eliminating the electric potential from these equations, we are led to the Christoffel equation with stiffened constants

$$\rho \hat{u}_i = s^2 \bar{\Gamma}_{il} \hat{u}_l, \tag{5.46}$$

$$\bar{\Gamma}_{il} = \Gamma_{il} + \frac{\gamma_i \gamma_l}{\varepsilon}. \tag{5.47}$$

This equation is a practical means of obtaining the elastic components of harmonic plane waves in piezoelectric media, simply by solving an eigenvalue problem.

Slowness curves and electromechanical coupling. As an example, let us consider propagation along axis [010] ($n_1 = 0$, $n_2 = 1$, and $n_3 = 0$) inside lithium niobate ($LiNbO_3$, trigonal 3m). Using Figure 5.7, we can write the Christoffel tensor as

$$\Gamma = \begin{pmatrix} c_{66} & 0 & 0 \\ \cdot & c_{11} & -c_{14} \\ \cdot & \cdot & c_{44} \end{pmatrix},$$

with $c_{66} = \frac{c_{11}-c_{12}}{2}$. Neglecting piezoelectricity, we then find a pure shear wave with velocity $\sqrt{c_{66}/\rho}$, and a quasi-shear and a quasi-longitudinal waves with velocities $2\rho v^2 = \Gamma_{22} + \Gamma_{33} \pm \sqrt{(\Gamma_{22} - \Gamma_{33})^2 + 4\Gamma_{23}^2}$. Furthermore, from their definitions, we have $\gamma_1 = 0$, $\gamma_2 = e_{22}$, $\gamma_3 = e_{15}$, and $\varepsilon = \varepsilon_{11}$. As a result,

$$\bar{\Gamma} = \begin{pmatrix} c_{66} & 0 & 0 \\ \cdot & c_{11} + e_{22}^2/\varepsilon_{11} & -c_{14} + e_{22}e_{15}/\varepsilon_{11} \\ \cdot & \cdot & c_{44} + e_{15}^2/\varepsilon_{11} \end{pmatrix}.$$

Piezoelectricity then only affects the velocities of the quasi-shear and the quasi-longitudinal waves, which become

$$v = \sqrt{\frac{1}{2\rho}\left(\bar{\Gamma}_{22} + \bar{\Gamma}_{33} \pm \sqrt{(\bar{\Gamma}_{22} - \bar{\Gamma}_{33})^2 + 4\bar{\Gamma}_{23}^2}\right)},$$

with the minus sign for the quasi-shear wave and the plus sign for the quasi-longitudinal wave.

We could repeat the previous calculation for any direction of propagation if necessary, though it would obviously be tedious. It is easier to write a small computer program to plot the slowness curves. For the case of lithium niobate, we plot these slowness curves in the (s_2, s_3) plane in Figure 5.8, both with and without piezoelectricity taken into account. The uncoupled quasi-shear wave is unaffected by piezoelectricity. For the other quasi-shear wave and the quasi-longitudinal wave, however, it can be seen that the slowness change between piezoelectric and nonpiezoelectric depends acutely on the propagation direction. The stiffened elastic constants always lead to smaller slowness for a given propagation angle, or faster waves. Hence, it can be said that piezoelectricity increases the (effective) elastic constants, causing the solid to appear stiffer. In engineering terms, it is understood that this change is caused by the coupling of the mechanical wave with its accompanying electromagnetic wave: the larger the change in the slowness, the larger the coupling. It is then customary to define the electromagnetic coupling coefficient K^2 as the dimensionless quotient

$$K^2 = 2\frac{\Delta v}{v} \approx 2\frac{\Delta s}{s}. \tag{5.48}$$

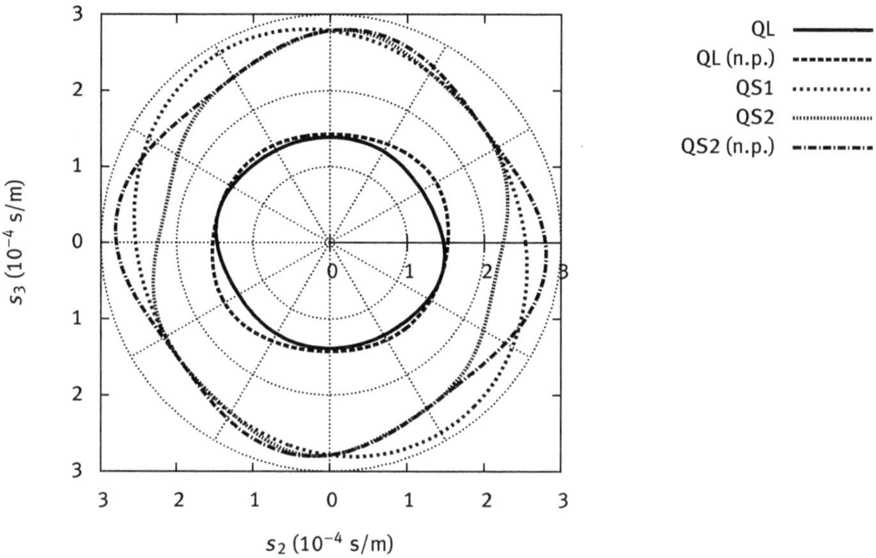

Fig. 5.8: Slowness curves for lithium niobate ($LiNbO_3$, trigonal 3m). Slowness curves are presented in the (s_2, s_3) plane. One quasi-shear wave (QS1) is unaffected by piezoelectricity. In contrast, the other quasi-shear wave (QS2) and the quasi-longitudinal wave (QL) are quite strongly affected.

For X-cut lithium niobate, as apparent in Figure 5.8, the largest K^2 is obtained for a propagation angle around 30° for the quasi-L wave, and for a propagation angle of about −15° for the quasi-S wave.

Figure 5.9 shows the slowness curves for Y-cut zinc oxide, ZnO. Again, there is a pure shear wave unaffected by piezoelectricity, polarized along the Y axis, and a quasi-S and a quasi-L waves that are sensitive to it. The electromagnetic coupling coefficient K^2 is optimized for propagation along the Z axis for the quasi-L wave, and for a propagation angle of about 50° for the quasi-S wave. Finally, we show the case of Y-cut quartz in Figure 5.10. Quartz has rather small piezoelectric constants and thus the electromagnetic coupling coefficient K^2 always remains small, though it does not vanish for any of the three elastic waves.

Energy velocity. For complex-valued physical quantities, we have that the time-averaged generalized energy density is given by [8]

$$W = \frac{1}{2}\operatorname{Re}(T_{ij}S_{ij}^* + D_k E_k^*), \tag{5.49}$$

and that the time-averaged generalized Poynting vector is

$$P_j = \frac{1}{2}\operatorname{Re}(-T_{ij}v_i^* + (\mathbf{E} \times \mathbf{H}^*)_j). \tag{5.50}$$

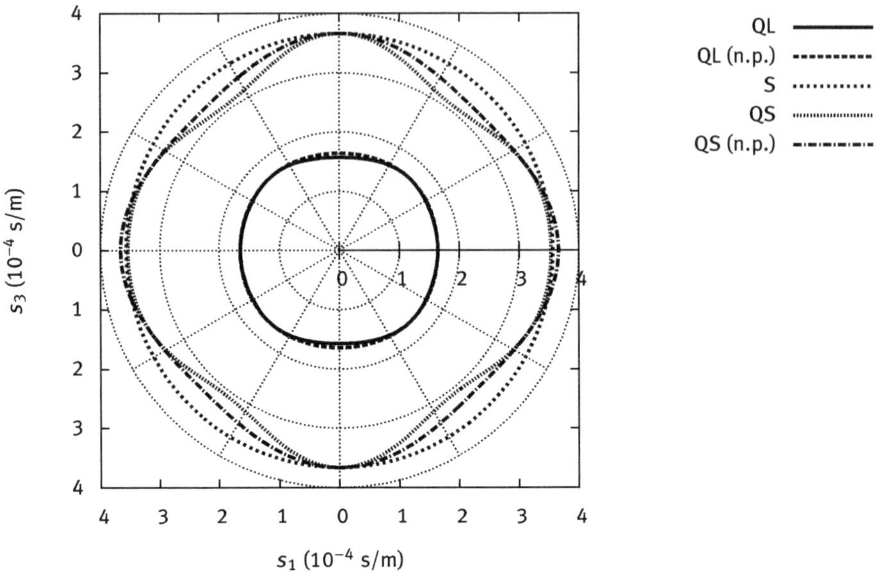

Fig. 5.9: Slowness curves for zinc oxide (ZnO, hexagonal 6mm). Slowness curves are presented in the (s_1, s_3) plane. The pure shear wave (S) is unaffected by piezoelectricity. In contrast, the other quasi-shear wave (QS) and the quasi-longitudinal wave (QL) are affected.

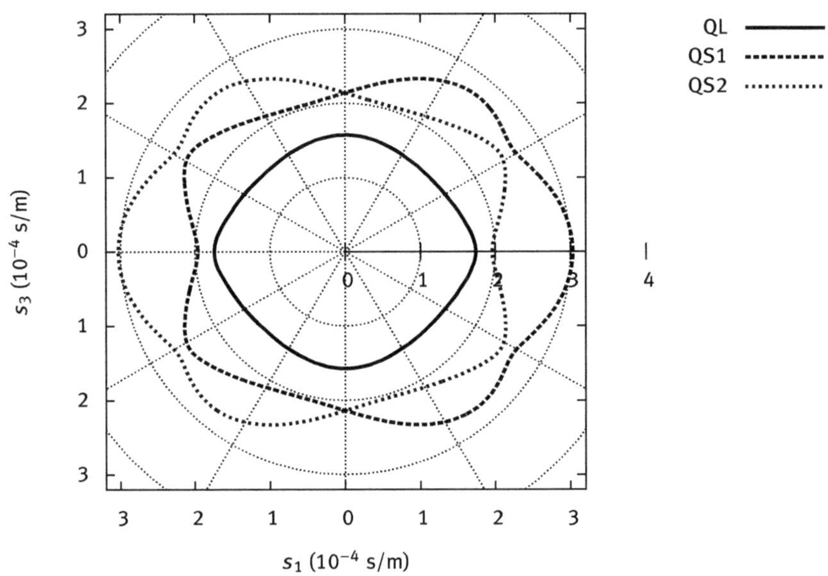

Fig. 5.10: Slowness curves for quartz (SiO_2, trigonal 32). Slowness curves are presented in the (s_1, s_3) plane. All three bulk waves are sensitive to piezoelectricity, though coupling is small for all propagation angles.

Furthermore, for harmonic plane waves and within the quasi-static approximation, the generalized energy density simplifies to

$$W = \frac{1}{2}\,\mathrm{Re}(\jmath\omega T_{ij} s_j u_i^*). \tag{5.51}$$

It can be seen that the electromagnetic part of the energy density vanishes because of Poisson's equation. The generalized Poynting vector becomes

$$P_j = \frac{1}{2}\,\mathrm{Re}\left(\jmath\omega(T_{ij}u_i^* + D_j\Phi^*)\right). \tag{5.52}$$

We now introduce the generalized notation

$$\bar{c}_{ijkl} = \begin{cases} c_{ijkl} & i,\ l = 1,2,3 \\ e_{kij} & l = 4,\ i = 1,2,3 \\ e_{jkl} & i = 4,\ l = 1,2,3 \\ -\epsilon_{jk} & i,\ l = 4 \end{cases}. \tag{5.53}$$

With the generalized displacements, $\bar{u}_i$, and stresses, $\bar{T}_{ij}$, defined as

$$\bar{u}_i = u_i,\ i = 1,2,3, \tag{5.54}$$

$$\bar{u}_4 = \Phi, \tag{5.55}$$

$$\bar{T}_{ij} = T_{ij},\ i = 1,2,3, \tag{5.56}$$

$$\bar{T}_{4j} = D_j, \tag{5.57}$$

the constitutive relations can be written in a compact way as

$$\bar{T}_{ij} = -\jmath\omega\bar{c}_{ijkl}s_k\bar{u}_l, \tag{5.58}$$

and the Christoffel equation [8] governing the wave dynamics assumes the form

$$\bar{c}_{ijkl}s_js_k\bar{u}_l = \rho_{il}\bar{u}_l, \tag{5.59}$$

with

$$\rho_{il} = \rho\begin{pmatrix} 1 & 0 & 0 & 0 \\ 0 & 1 & 0 & 0 \\ 0 & 0 & 1 & 0 \\ 0 & 0 & 0 & 0 \end{pmatrix}. \tag{5.60}$$

Contracting the Christoffel equation (5.59) with $\bar{u}_i^*$ yields the relation

$$\bar{c}_{ijkl}s_js_k\bar{u}_l\bar{u}_i^* = \rho_{il}\bar{u}_l\bar{u}_i^* = \rho u_i^* u_i. \tag{5.61}$$

It is also seen that

$$W = \frac{\omega^2}{2}\,\mathrm{Re}(\bar{c}_{ijkl}s_js_k\bar{u}_l\bar{u}_i^*). \tag{5.62}$$

Using (5.61), we observe that this quantity is real and thus

$$W = \frac{\omega^2}{2}\bar{c}_{ijkl}s_j s_k \bar{u}_l \bar{u}_i^* = \frac{\omega^2}{2}\rho u_i^* u_i. \tag{5.63}$$

This relation proves the equality of potential and kinetic energies for bulk acoustic waves as in the case of elastic media [131].

The expression for the Poynting vector can similarly be transformed by inserting the constitutive relations into (5.52), yielding

$$P_j = \frac{\omega^2}{2}\operatorname{Re}(c_{ijkl}s_k u_l u_i^* + e_{kij}s_k \Phi u_i^* + e_{jkl}s_k u_l \Phi^* - \epsilon_{jk}s_k \Phi\Phi^*),$$

or simply

$$P_j = \frac{\omega^2}{2}\operatorname{Re}(\bar{c}_{ijkl}s_k \bar{u}_l \bar{u}_i^*). \tag{5.64}$$

Defining the energy velocity as for bulk acoustic waves in elastic media by the ratio of the Poynting vector to the energy density, we obtain at once

$$(V_e)_j = \frac{P_j}{W} = \frac{\operatorname{Re}(\bar{c}_{ijkl}s_k \bar{u}_l \bar{u}_i^*)}{\rho u_i^* u_i}. \tag{5.65}$$

Furthermore, contracting this expression with s_j, the following useful relation is obtained

$$s_j (V_e)_j = 1, \tag{5.66}$$

as in the case of elastic media.

Group velocity. The Christoffel equation (5.59) is a generalized eigenvalue equation for the square of the phase velocity,

$$\bar{c}_{ijkl}n_j n_k \bar{u}_l = v^2 \rho_{il}\bar{u}_l. \tag{5.67}$$

The group velocity is defined as

$$(V_g)_j = \frac{\partial v}{\partial n_j}, \tag{5.68}$$

which implies that the group velocity vector is normal to the slowness surface by construction. The group velocity vector can be obtained by differentiating with respect to n_j the identity formed by contracting (5.59) with $\bar{u}_i^*$, or

$$\bar{c}_{ijkl}n_j n_k \bar{u}_l \bar{u}_i^* = v^2 \rho_{il}\bar{u}_l \bar{u}_i^*. \tag{5.69}$$

In this equation, $\bar{u}_l$ is a function of the unit vector $\mathbf{n}$ since it is the eigenvector associated with the eigenvalue v^2 of the Christoffel equation for a fixed propagation direction. The differentiation of (5.69) with respect to n_j results in

$$\begin{aligned}2\bar{c}_{ijkl}n_k \bar{u}_l \bar{u}_i^* &+ \bar{c}_{i\beta kl}n_\beta n_k \frac{\partial \bar{u}_l}{\partial n_j}\bar{u}_i^* + \bar{c}_{i\beta kl}n_\beta n_k \bar{u}_l \frac{\partial \bar{u}_i^*}{\partial n_j} \\ &= 2v(V_g)_j \rho_{il}\bar{u}_l \bar{u}_i^* + v^2 \rho_{il}\frac{\partial \bar{u}_l}{\partial n_j}\bar{u}_i^* + v^2 \rho_{il}\bar{u}_l \frac{\partial \bar{u}_i^*}{\partial n_j}. \end{aligned} \tag{5.70}$$

This equation can be simplified by considering two different contractions of the Christoffel equation (5.59). First, the contraction by $\frac{\partial \bar{u}_i^*}{\partial n_j}$ yields

$$\bar{c}_{i\beta kl} n_\beta n_k \bar{u}_l \frac{\partial \bar{u}_i^*}{\partial n_j} = v^2 \rho_{il} \bar{u}_l \frac{\partial \bar{u}_i^*}{\partial n_j}.$$

Second, complex conjugation of (5.59) and subsequent contraction by $\frac{\partial \bar{u}_i}{\partial n_j}$ results in

$$\bar{c}_{i\beta kl} n_\beta n_k \bar{u}_l^* \frac{\partial \bar{u}_i}{\partial n_j} = v^2 \rho_{il} \bar{u}_l^* \frac{\partial \bar{u}_i}{\partial n_j}.$$

Permutation of indices i and l, and β and k, respectively, and consideration of the symmetries $\bar{c}_{i\beta kl} = \bar{c}_{lk\beta i}$ and $\rho_{il} = \rho_{li}$ lead to

$$\bar{c}_{i\beta kl} n_\beta n_k \bar{u}_i^* \frac{\partial \bar{u}_l}{\partial n_j} = v^2 \rho_{il} \bar{u}_i^* \frac{\partial \bar{u}_l}{\partial n_j}.$$

Eventually, (5.70) simplifies to

$$(V_g)_j = \frac{\bar{c}_{ijkl} n_k \bar{u}_l \bar{u}_i^*}{v \rho u_i u_i^*} = \frac{\bar{c}_{ijkl} s_k \bar{u}_l \bar{u}_i^*}{\rho u_i u_i^*}. \tag{5.71}$$

This is identical with the expression (5.65) for the energy velocity and incidentally shows that the Re(.) operators in the expression of the Poynting's vector can be dropped. We have thus shown the equality of energy and group velocities for bulk elastic waves in a lossless piezoelectric medium. This is a particular case of a more general result in classical wave physics [14].

5.5 Reflection and refraction

Let us consider the interface between two media, which can be elastic solids or piezoelectric elastic solids. The following physical properties hold for harmonic plane wave propagation, generalizing the discussion of reflection and refraction presented in Chapter 2 for acoustic waves.

- The polarization of waves in an elastic solid is composed of the three displacements, but the polarization of waves in a piezoelectric elastic solid has an additional degree of freedom, the electric potential (Φ).
- An elastic solid supports three bulk waves whose polarizations are orthogonal in 3D space. Two of them have mostly shear polarization, one has mostly longitudinal polarization. A piezoelectric elastic solid additionally supports a mostly electromagnetic wave which is always evanescent. This can be seen from the shape of the Christoffel equation written in (5.59), because of the singular mass matrix (5.60); this is hidden in the stiffened form of Christoffel equation (5.46).

- A harmonic plane wave incident on the interface will give rise to three reflected and three transmitted harmonic plane waves if propagation media are elastic solids. If, in addition, the propagation media are piezoelectric, four reflected and four transmitted harmonic plane waves can be excited.
- Modal conversion is such that both frequency and the projection of the wavevector along the interface are conserved. If the media are homogeneous, the projection of the slowness vector along the interface is conserved.

For definiteness, let us consider a plane interface between two homogeneous media normal to axis x_1. Three quantities are conserved: the angular frequency ω, and slownesses s_2 and s_3. The question is then: what are the possible values of s_1? The answer to this problem actually gives the full solution to the reflection/refraction problem, as we shall see.

Remembering that the physical quantities that are conserved at the interface are the displacements and the normal stresses, it is useful to form an 8-vector of these conserved quantities as $(\bar{u}_i \tau_{i1})^T$, with i running from 1 to 4. We write all subsequent equations for the more general case of piezoelectric media, hence with eight independent variables (four generalized displacements and four generalized stresses), but it is easy to rewrite the same equations for nonpiezoelectric media (keeping only three mechanical displacements and three mechanical stresses in the vectors). After some algebra, it is possible to rewrite the dynamical equations, further taking account of the constitutive relations, as

$$\begin{pmatrix} -\bar{c}_{i12l}s_2 - \bar{c}_{i13l}s_3 & \delta_{il} \\ \sum_{j,k=2}^{3} \bar{c}_{ijkl}s_j s_k + \bar{\rho}_{il} & 0 \end{pmatrix} \begin{pmatrix} \bar{u}_l \\ \tau_{l1} \end{pmatrix} = s_1 \begin{pmatrix} \bar{c}_{i11l} & 0 \\ \bar{c}_{i21l}s_2 + \bar{c}_{i31l}s_3 & \delta_{il} \end{pmatrix} \begin{pmatrix} \bar{u}_l \\ \tau_{l1} \end{pmatrix}, \tag{5.72}$$

with $\tau_{ij} = T_{ij}/(-\imath\omega)$. Equation (5.72) has the form of a generalized eigenvalue problem

$$A\mathbf{h} = s_1 B\mathbf{h},$$

where the 8×8 square matrices A and B are functions of s_2 and s_3, and of the material constants. Vector $\mathbf{h}$ has 8 components, the 4 $\bar{u}_l$ and the 4 τ_{l1}.

Let us detail the algebraic properties of this eigenvalue system in the absence of loss, i.e. for real elastic constants. As matrices A and B are real valued but nonsymmetrical, the eigenvalues and eigenvectors are then generally complex valued. There are eight eigenvalues, giving eight possible values for the slowness s_1. These eigenvalues are either real or they belong to a pair of complex conjugate eigenvalues. They belong in pairs to each of four slowness curves: when they are real they belong to the slowness curves for bulk elastic waves that we discussed in the previous section; when they are complex they belong to complex branches of the same slowness curves, indicating that they are evanescent waves. The eight eigenvectors are called partial waves. There are exactly four reflected partial waves and four transmitted partial waves.

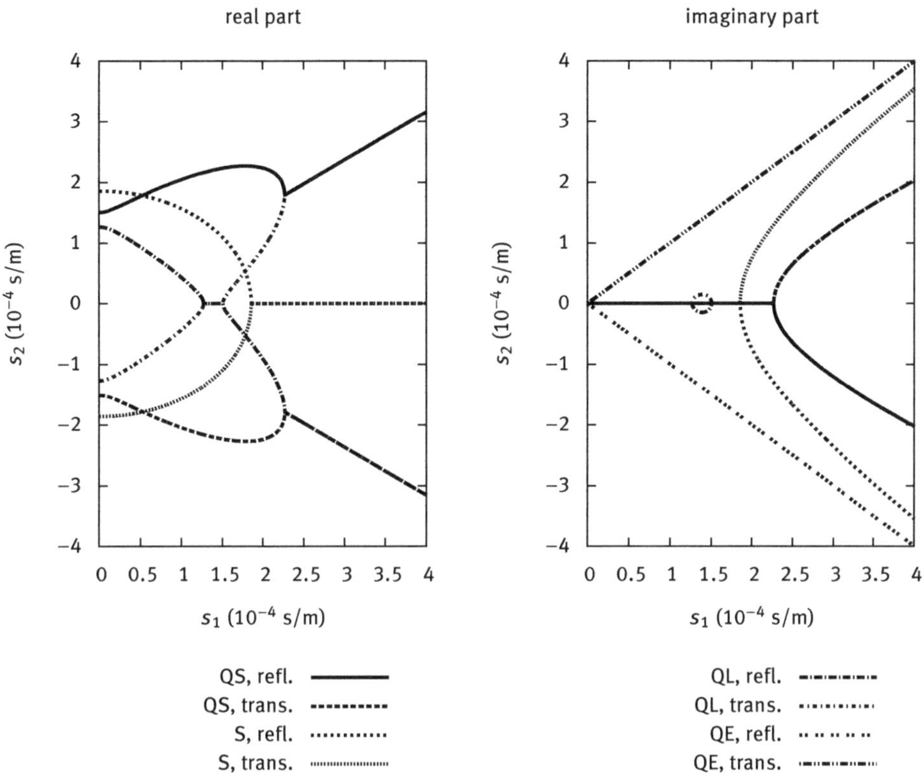

Fig. 5.11: Complex slowness curves for harmonic plane waves in rutile (TiO_2). $s_2(s_1)$ is plotted with $s_3 = 0$. The reference frame is aligned with crystallographic axes. Note the natural continuity between quasi-shear and quasi-longitudinal real and complex branches. Real branches are exactly identical to the slowness curves in Figure 5.5.

As an example, in Figures 5.11 and 5.12 we plot complex slowness curves for rutile and lithium niobate. For rutile, we have considered that the interface is defined by $x_1 = 0$ and that incidence is in the (x_1, x_2) plane ($s_3 = 0$ is imposed). The plots show $s_2(s_1)$ as s_1 varies along the real axis. We can see in Figure 5.11 that as s_1 increases from 0, the three slowness curves for bulk propagating waves in rutile, shown in Figure 5.5, are recovered in the real part of s_2. However, this is no longer a polar plot, and additional complex branches appear. As s_1 becomes larger than the maximum allowed purely real slowness for a bulk wave, the slowness curves appear to be continued by pairs of complex conjugate branches arising from extremal points. For the purely shear wave, there is only one pair of complex conjugate branches, but the quasi-shear wave has two different pairs, since there are two extremal points. We can also see that the quasi-longitudinal slowness curve is connected to the quasi-shear slowness curve by a pair of complex conjugate branches. The pure electromagnetic slowness curves (they are

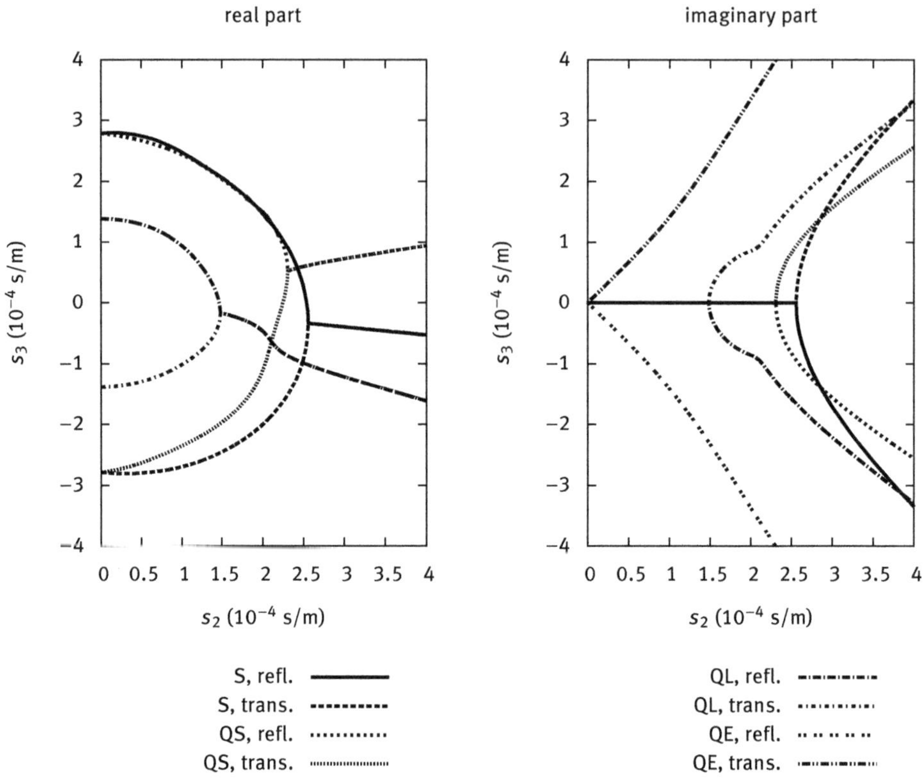

Fig. 5.12: Complex slowness curves for harmonic plane waves in lithium niobate ($LiNbO_3$). $s_3(s_2)$ is plotted with $s_1 = 0$. The reference frame is aligned with crystallographic axes. Real branches are exactly identical to the slowness curves in Figure 5.8.

pure EM waves since rutile is nonpiezoelectric) form a complex conjugate pair and are purely imaginary. As we noted before, they are evanescent EM waves traveling along the interface.

In the case of lithium niobate, shown in Figure 5.12, similar features are observed. The plots show $s_3(s_2)$ as s_2 varies along the real axis ($s_1 = 0$ is imposed). The three bulk slowness curves all have a single extremal point and are each continued by a pair of complex conjugate branches. The quasi-electromagnetic slowness curves again form a complex conjugate pair and are purely imaginary. They are evanescent quasi-EM waves traveling along the interface.

Finally, let us outline a general numerical procedure to solve a reflection/refraction problem at an interface between two solids, according to Figure 3.5 with fluid media replaced by solid media. Axis x_1 is orthogonal to the plane interface between the two solids. The first step is to solve the generalized eigenvalue problem (5.72) in both media 1 and 2. For each medium this gives us eight eigenvalues ($s_{1r}^{(1)}$ and $s_{1r}^{(2)}$) and eight eigenvectors ($\mathbf{h}_r^{(1)}$ et $\mathbf{h}_r^{(2)}$). According to the superposition principle, the general solution in each medium is a superposition of the eight partial waves

$$\mathbf{h}(t, \mathbf{x}) = \sum_{r=1}^{8} a_r \mathbf{h}_r^{(q)} \exp(\imath\omega(t - s_{1r}^{(q)} x_1 - s_2 x_2 - s_3 x_3)), \tag{5.73}$$

with $q = 1$ or 2. In this equation, the complex amplitudes a_r are still arbitrary. They can be identified from the incident conditions and the continuity relations at the interface. We write the 8×8 eigenvector matrix as a stack of two 4×8 matrices U and T. A transmission matrix is obtained by writing the continuity equations as

$$\begin{pmatrix} U^{(1)} \\ T^{(1)} \end{pmatrix} \mathbf{a}^{(1)} = \begin{pmatrix} U^{(2)} \\ T^{(2)} \end{pmatrix} \mathbf{a}^{(2)}, \tag{5.74}$$

which can be inverted to

$$\mathbf{a}^{(2)} = \mathcal{T}\mathbf{a}^{(1)} \quad \text{with} \quad \mathcal{T} = \begin{pmatrix} U^{(2)} \\ T^{(2)} \end{pmatrix}^{-1} \begin{pmatrix} U^{(1)} \\ T^{(1)} \end{pmatrix}. \tag{5.75}$$

The transmission matrix allows one to obtain the complex amplitudes $\mathbf{a}^{(2)}$ if the $\mathbf{a}^{(1)}$ are known. This is, however, not the usual form of a reflection/refraction problem and a scattering matrix can be obtained instead as follows. Partial waves are sorted in each medium into four right-propagating (R) partial waves and four left-propagating (L) partial waves. Assuming incidence from medium 1 and from left to right, incident and transmitted partial waves are right-propagating, while reflected partial waves are left-propagating. Conversely, assuming incidence from medium 2 and from right to left, incident and transmitted partial waves are right-propagating, while reflected partial waves are left-propagating. Relation (5.74) is rewritten with 4×4 block matrices as

$$\begin{pmatrix} U_L^{(1)} & U_R^{(1)} \\ T_L^{(1)} & T_R^{(1)} \end{pmatrix} \begin{pmatrix} \mathbf{a}_L^{(1)} \\ \mathbf{a}_R^{(1)} \end{pmatrix} = \begin{pmatrix} U_L^{(2)} & U_R^{(2)} \\ T_L^{(2)} & T_R^{(2)} \end{pmatrix} \begin{pmatrix} \mathbf{a}_L^{(2)} \\ \mathbf{a}_R^{(2)} \end{pmatrix}. \tag{5.76}$$

A scattering matrix relating the complex amplitudes of source and scattered partial waves is then

$$\begin{pmatrix} \mathbf{a}_L^{(1)} \\ \mathbf{a}_R^{(2)} \end{pmatrix} = \mathcal{S} \begin{pmatrix} \mathbf{a}_L^{(2)} \\ \mathbf{a}_R^{(1)} \end{pmatrix} \quad \text{with} \quad \mathcal{S} = \begin{pmatrix} U_L^{(1)} & -U_R^{(2)} \\ T_L^{(1)} & -T_R^{(2)} \end{pmatrix}^{-1} \begin{pmatrix} U_L^{(2)} & -U_R^{(1)} \\ T_L^{(2)} & -T_R^{(1)} \end{pmatrix}. \tag{5.77}$$

The different matrix elements of $\mathcal{S}$ can be given obvious physical interpretations as reflection and transmission coefficients between the four available propagation modes (the quasi-L, the two quasi-S and the quasi-EM bulk waves).

5.6 Plate waves

Elastic wave propagation in a homogeneous plate limited by two plane and parallel surfaces is depicted in Figure 5.13. This geometry is called either a plate, a membrane, or a slab in the literature. For the particular case of isotropic solids and for elastic waves polarized in the plane of incidence (a plane containing both x_3 and the direction of propagation), it is customary to speak of Lamb waves. For isotropic solids, there always additionally exist pure shear waves polarized orthogonally to the plane of incidence, or sagittal plane. In the general case of anisotropic solids or of piezoelectric solids, plate waves are generally polarized, with a x_3-dependent combination of all three displacements in space, so the distinction between Lamb waves and pure shear waves is no longer effective except for propagation along certain symmetry axes.

The most classical approach to obtaining the dispersion of plate waves is to use their expansion in partial waves

$$\mathbf{h}(t, \mathbf{x}) = \sum_{r=1}^{8} a_r \mathbf{h}_r \exp(\imath\omega(t - s_1 x_1 - s_2 x_2 - s_{3r} x_3)). \tag{5.78}$$

Actually, both frequency ω and the slownesses s_1 and s_2 are conserved quantities under propagation in the plate, and it is natural to consider s_3 as a function of all of them. Even though the eight (for piezoelectric solids) or six (for elastic solids) partial waves are not solutions of the elastodynamic equations, their superposition can be. For instance, in the case of elastic solids, the boundary conditions on the top and the bottom surfaces are such that the surface traction vanishes, resulting in the six equations

$$(T_{I3})_r a_r = 0,$$
$$(T_{I3})_r \exp(\imath\omega s_{3r} h) a_r = 0.$$

Since there are six unknown complex amplitudes a_r, the determinant of the matrix in these equations must vanish, which yields an implicit dispersion relation for plate waves. In the case of piezoelectric solids, two additional electrical boundary conditions must be added, for a similar result.

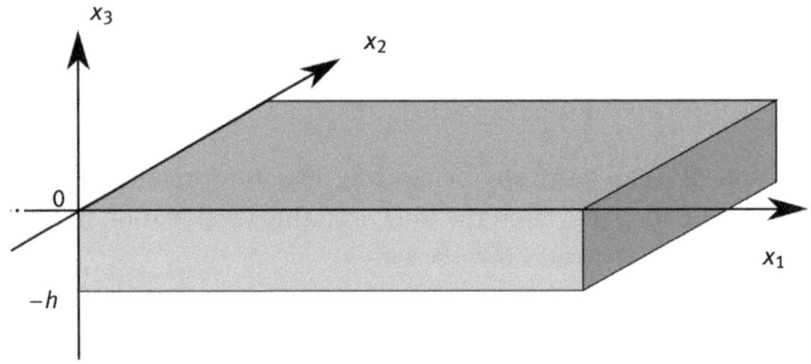

Fig. 5.13: Schematic representation of elastic wave propagation inside a plate of thickness h. Propagation is in the plane (x_1, x_2) and the plate depth h extends along axis x_3. The wave distribution is inhomogeneous along axis x_3.

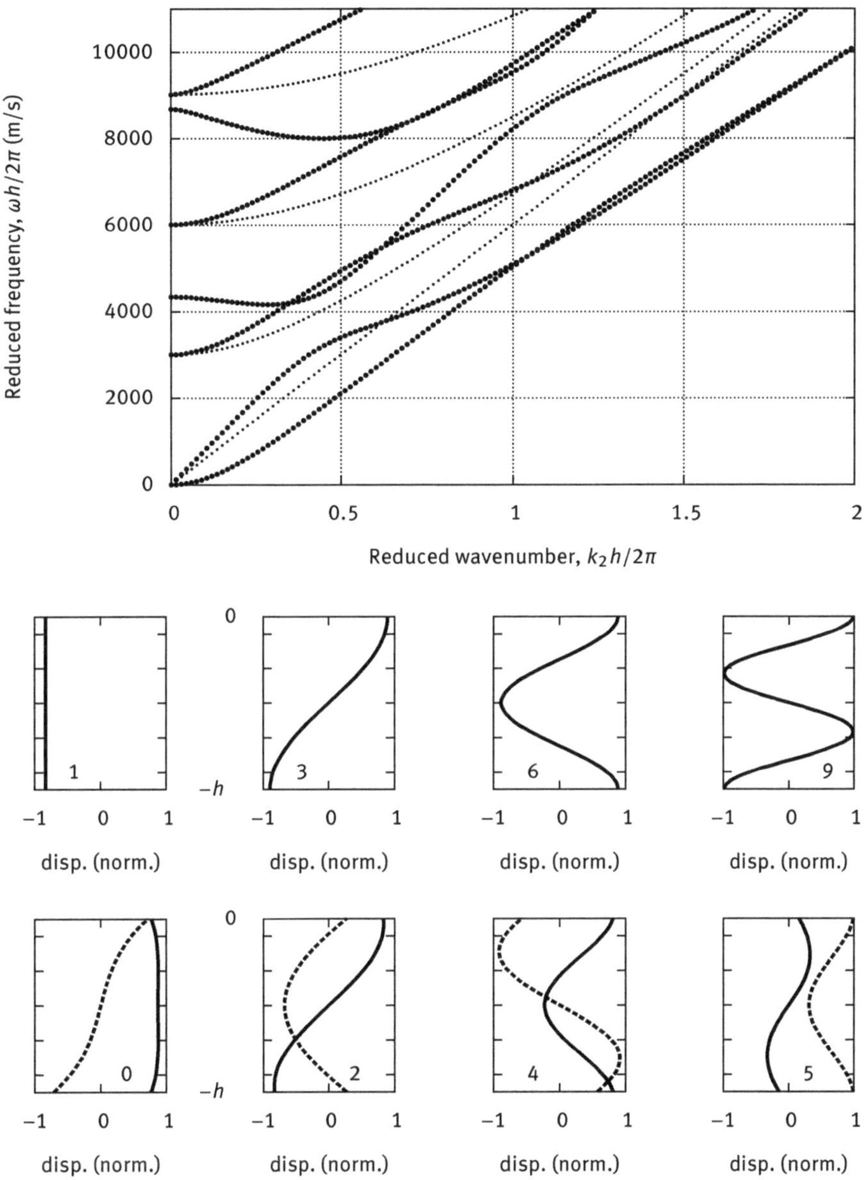

Fig. 5.14: Band structure for plate waves in silicon in its crystallographic axes. The reduced frequency is plotted as a function of the reduced wavenumber in the direction of propagation x_2 ($s_1 = 0$). Pure shear waves polarized along x_1 are shown with small dots. Lamb waves polarized in the sagittal plane, (x_2, x_3), are shown with large dots. The second line displays the first four pure shear modal distributions u_1 at $k_3h/2\pi = 1/2$. The third line displays the first four Lamb wave modal distributions, with $\text{Im}(u_2)$ in solid line and $\text{Re}(u_3)$ in dashed line.

This boundary condition determinant approach will be further explored in the next section to obtain the dispersion of surface elastic waves. Let us remark as a complement that the dispersion relation of plate waves can be most efficiently obtained using FEM as an eigenvalue problem. In this respect, the polarization of elastic waves in the plate can be written as

$$\bar{u}_i(t, \mathbf{x}) = \bar{u}_i(x_3) \exp(\imath\omega(t - s_1 x_1 - s_2 x_2)). \tag{5.79}$$

Since the unknown displacements $\bar{u}_i(x_3)$ are only a function of x_3, it is sufficient to employ a 1D-mesh of the segment from $x_3 = 0$ to $x_3 = -h$. Slownesses s_1 and s_2 can be incorporated inside the finite element method as described in Appendix 5.B in order to obtain the dispersion relation in the form $\omega(s_1, s_2)$. As an example, Figure 5.14 shows the dispersion relation for plate waves of silicon in its crystallographic axes. For this case, the separation between pure shear waves and Lamb waves is effective.

5.7 Surface waves

Put very boldly, a surface wave is a solution to the wave equation that satisfies boundary conditions at the surface of a semi-infinite medium, generally placed in a vacuum. The surface wave is guided if the modal amplitude decreases exponentially in the depth. As a consequence, the partial waves into which a lossless surface wave can be decomposed are all evanescent.

Let us assume that the surface slowness is imposed, i.e. both s_1 and s_2 are given. A surface wave can be written as a linear superposition of partial waves of the form (5.73), as in the case of plate waves. Not all partial waves, however, are allowed in the superposition. We know that partial waves can be divided between propagating

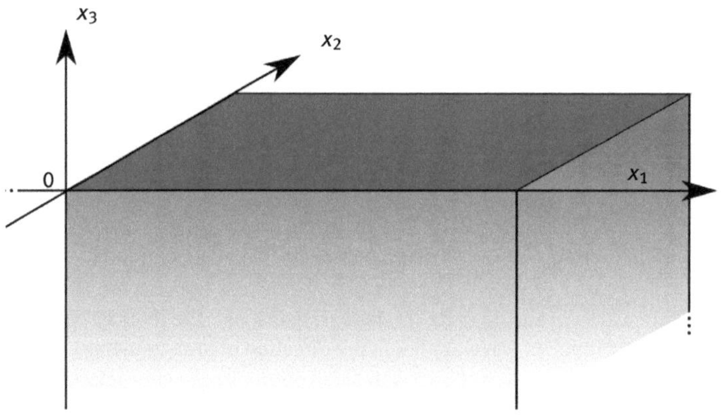

Fig. 5.15: Schematic representation of elastic wave propagation at the surface of a semi-infinite substrate. Propagation is in the plane (x_1, x_2). The plate depth extends to $-\infty$. The wave distribution is inhomogeneous along axis x_3.

and evanescent. Propagating partial waves inside the semi-infinite substrate can only propagate to its interior. Evanescent partial waves must have decreasing amplitude as x_3 decreases to $-\infty$, because of radiation conditions at infinity. The rule to decide whether a partial wave is eligible in the decomposition of a surface wave is then the following:

- Propagating partial waves (with s_3 real valued) are selected if their Poynting's vector enters the semi-infinite substrate, or $P_3 < 0$ with the reference frame of Figure 5.15.
- Evanescent partial waves (with s_3 complex valued) are selected if $\mathrm{Im}(s_3) > 0$, so that the amplitude of the partial wave decreases in the depth of the substrate (for $x_3 < 0$).

On the whole, permitted partial waves amount to three for an elastic solid and to four for a piezoelectric solid. Once the selection is performed, the surface wave can be written as the superposition over the eligible partial waves

$$\mathbf{h}(t, \mathbf{x}) = \left(\sum_{r=1}^{3 \text{ or } 4} a_r \mathbf{h}_r \exp(-\imath \omega s_{3r} x_3) \right) \exp(\imath \omega (t - s_1 x_1 - s_2 x_2), \tag{5.80}$$

with $h_l = (u_l, \tau_{l3})$ the vector gathering generalized displacements and stresses, s_{3r} and $\mathbf{h}_r$ are the eigenvalues and the eigenvectors of the r-th partial wave, and a_r are the modal amplitudes in the superposition.

Elastic boundary conditions for a free surface are that $T_{i3} = 0$. Electrical boundary conditions matter only in the piezoelectric case. For a free surface, i.e. at the interface between a dielectric and air or a vacuum, there is no surface charge density and the normal component of the dielectric displacement is continuous. At a metalized and short-circuited surface, i.e. for a short thin metal layer, $\phi = 0$.

Surface waves on nonpiezoelectric solids. The boundary conditions are $T_{i3} = 0$, $i = 1, 2, 3$, for $x_3 = 0$. There are three boundary conditions implying the three unknown modal amplitudes

$$\tau_{i3} = \sum_{r=1}^{3} a_r \tau_{i3}^{(r)} = 0, \quad i = 1, 2, 3. \tag{5.81}$$

There exists a solution if and only if the determinant of the boundary conditions vanishes, or

$$\Delta = |\tau_{i3}^{(r)}| = \begin{vmatrix} \tau_{13}^{(1)} & \tau_{13}^{(2)} & \tau_{13}^{(3)} \\ \tau_{23}^{(1)} & \tau_{23}^{(2)} & \tau_{23}^{(3)} \\ \tau_{33}^{(1)} & \tau_{33}^{(2)} & \tau_{33}^{(3)} \end{vmatrix} = 0. \tag{5.82}$$

This determinant is a function of the surface slownesses s_1 and s_2. The above secular equation in turn defines a slowness curve for surface waves in the plane of the surface. The surface wave is perfectly guided if all partial waves are inhomogeneous. As

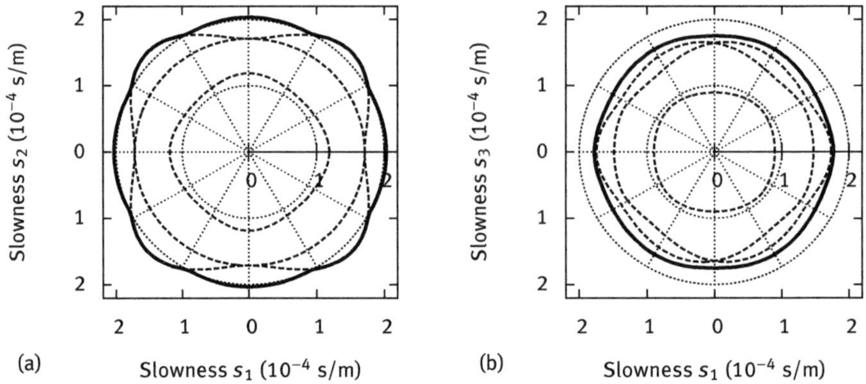

Fig. 5.16: Slowness curves for surface elastic waves of silicon and sapphire. (a) Propagation on the planar surface (001) is considered for silicon. (b) Propagation on the planar surface (010) is considered for sapphire. In each case, the slowness curve for surface elastic waves is shown with the thick solid line and the slowness curves for the three bulk elastic waves are shown with the thin dotted line.

a consequence, the velocity of a perfectly guided wave is necessarily smaller than the velocities of all partial waves that enter the superposition.

Figure 5.16 shows the slowness curves for surface elastic waves on the (001) plane of silicon and on the (010) plane of sapphire. These slowness curves are obtained by looking for zeros of the determinant (5.82) as a function of the propagation direction on the surface. The slowness curves of the three bulk elastic waves are also shown for comparison. It can be seen that the surface elastic waves are always slower than any bulk wave.

Surface waves on piezoelectric solids. Let us first consider the free surface. There are now four eligible partial waves. In addition to the vanishing mechanical traction at the surface, there are no charges accumulated on the surface and the surface jump of the normal electric displacement vanishes, $[D_3] = 0$. On the vacuum side, the electric potential satisfies Poisson's equation $s_i s_i \phi = 0$, so that

$$s_3 = -\imath\sqrt{s_1^2 + s_2^2},$$

where the sign is chosen so that the solution is evanescent and decreasing away from the surface ($x_3 > 0$). As a result, exactly at the surface ($x_3 = 0$), $\frac{D_3}{-\imath\omega} = -s_3\varepsilon_0\Phi = \imath\sqrt{s_1^2 + s_2^2}\varepsilon_0\Phi$. As a result, the electrical boundary condition is

$$\tau_{43} - \imath s\varepsilon_0\Phi = 0,$$

where we have written $s = \sqrt{s_1^2 + s_2^2}$. Gathering all boundary conditions, the surface boundary conditions determinant is

$$\Delta_F = \begin{vmatrix} \tau_{13}^{(1)} & \tau_{13}^{(2)} & \tau_{13}^{(3)} & \tau_{13}^{(4)} \\ \tau_{23}^{(1)} & \tau_{23}^{(2)} & \tau_{23}^{(3)} & \tau_{23}^{(4)} \\ \tau_{33}^{(1)} & \tau_{33}^{(2)} & \tau_{33}^{(3)} & \tau_{33}^{(4)} \\ \bar{\tau}_{43}^{(1)} & \bar{\tau}_{43}^{(2)} & \bar{\tau}_{43}^{(3)} & \bar{\tau}_{43}^{(4)} \end{vmatrix}, \tag{5.83}$$

with $\bar{\tau}_{43}^{(r)} = \tau_{43}^{(r)} - \imath s \varepsilon_0 \Phi^{(r)}$. The secular equation $\Delta_F(s_1, s_2) = 0$ gives the slowness curve for guided surface waves propagating on the free surface.

Let us now consider that the surface is shorted. In addition to vanishing surface traction, the electric potential vanishes at $x_3 = 0$

$$\Phi = \sum_{r=1}^{4} a_r \Phi^{(r)} = 0. \tag{5.84}$$

The surface boundary conditions determinant becomes

$$\Delta_S = \begin{vmatrix} \tau_{13}^{(1)} & \tau_{13}^{(2)} & \tau_{13}^{(3)} & \tau_{13}^{(4)} \\ \tau_{23}^{(1)} & \tau_{23}^{(2)} & \tau_{23}^{(3)} & \tau_{23}^{(4)} \\ \tau_{33}^{(1)} & \tau_{33}^{(2)} & \tau_{33}^{(3)} & \tau_{33}^{(4)} \\ \Phi^{(1)} & \Phi^{(2)} & \Phi^{(3)} & \Phi^{(4)} \end{vmatrix} = 0. \tag{5.85}$$

The secular equation $\Delta_S(s_1, s_2) = 0$ gives the slowness curve for guided surface waves propagating on the shorted surface.

The surface boundary conditions determinants Δ_F and Δ_S each give a different slowness curve. In practice we find that the slowness on the shorted surface is always larger than the slowness on the free surface. As in the case of bulk waves, we could think that if the velocity of the wave changes as it passes from a metalized region to an unmetalized region, and vice versa, then its kinetic energy content can also change as it either provides electrical energy to an external circuit or draws energy from it. Though this argument is very loose, if no other type of energy is exchanged, then the rate of energy exchange in the transduction by metal electrodes patterned on a piezoelectric surface is related to the velocity change. It is thus customary to define the electromechanical coupling by the formula

$$K^2 = 2\frac{\Delta v}{v} = 2\frac{\Delta s}{s}, \tag{5.86}$$

with v (respectively, s) the mean of the velocity (resp., slowness) on the free and the shorted surface.

When the slowness giving a zero in either determinant (5.83) or (5.85) has been located, the relative amplitudes a_r can be obtained by fixing the value of one of them,

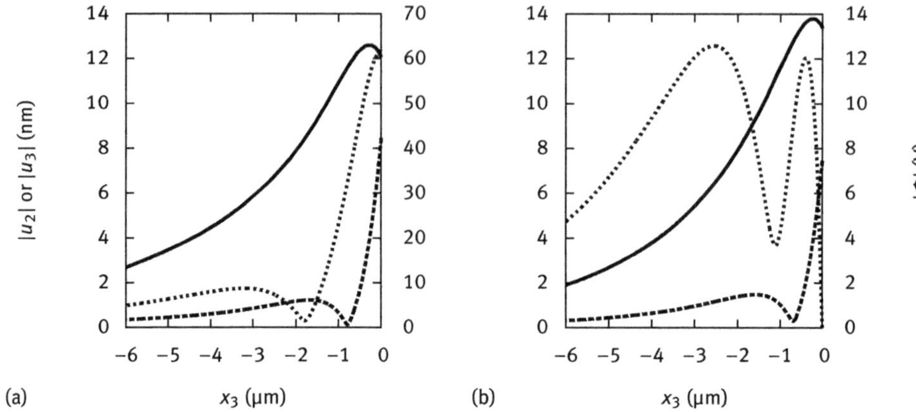

Fig. 5.17: Polarization of the surface elastic wave of Y-cut lithium niobate, for Z propagation. Both the case of the free surface (a) and of the shorted surface (b) are depicted. Displacements u_3 (along Y axis) and u_1 (along Z axis) are shown with solid and long dashed lines, respectively. The electric potential Φ is shown with the short dashed line. The frequency is 1 GHz.

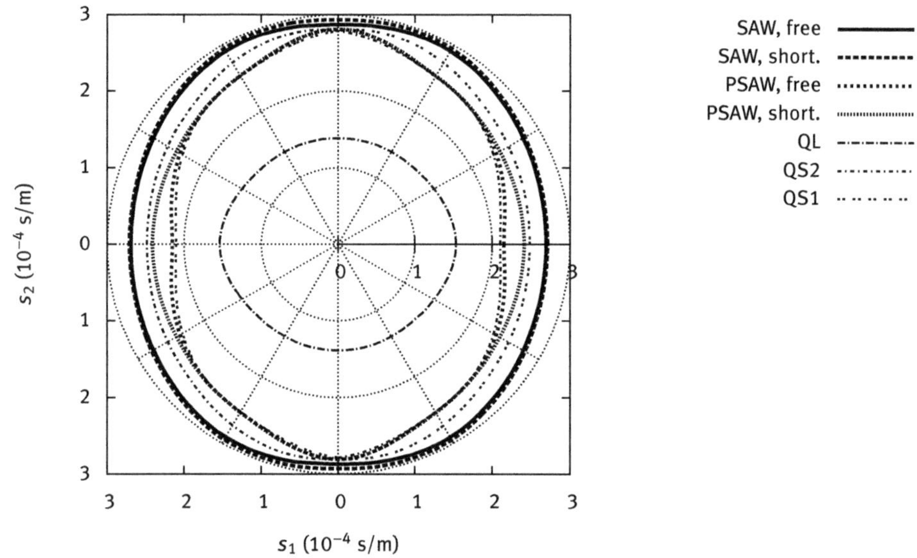

Fig. 5.18: Slowness curves for the surface elastic waves of Y-cut lithium niobate. Orientation is such that x_1 is along crystallographic axis X and x_2 is along crystallographic axis −Z. Both the lossless surface elastic wave (SAW) and the pseudo-surface acoustic wave (PSAW) are shown for the free and the shorted surface boundary conditions. The three bulk elastic waves are also depicted.

eliminating one equation, and solving the resulting linear equation. The polarization of the surface wave is then obtained from the superposition (5.80). As an example, Figure 5.17 displays the polarization of the surface elastic waves of Y-cut lithium niobate propagating in the Z direction. Y-cut means that the surface is orthogonal to the Y crystallographic axis and the outward normal points in the positive Y direction. With reference to Figure 5.15, x_1 = Z or [001], x_2 = X or [100], and x_3 = Y or [010]. It can be seen that changing the surface boundary condition from free to shorted slightly changes the distribution of the displacements and strongly changes the distribution of the electric potential. Propagation is here considered along symmetry axis Z and the horizontal shear displacements $u_2 = 0$, but for a general propagation direction all three displacements are nonzero.

Figure 5.18 shows the slowness curves for the surface elastic waves of Y-cut lithium niobate. With reference to Figure 5.15, x_1 = X or [100], x_2 = −Z or [00$\bar{1}$], and x_3 = Y or [010]. It can be seen that the shorted lossless surface wave is always slower then the free lossless surface wave, and that both of them are slower than the three bulk waves. The pseudo-surface wave (PSAW, see definition on page 172) for free and shorted boundary conditions is also shown. It appears in-between the two quasi-shear bulk waves.

Green's function and effective surface permittivity. Whatever the exact surface boundary conditions, the number of linear relations in the partial wave superposition is exactly twice the number of partial waves (once only eligible partial waves are retained). We can thus eliminate the modal amplitudes to obtain a linear (matrix) relation

$$\begin{pmatrix} u_1 \\ u_2 \\ u_3 \\ \phi \end{pmatrix} = G \begin{pmatrix} T_{13} \\ T_{23} \\ T_{33} \\ \frac{D_3}{-\imath\omega} \end{pmatrix}, \tag{5.87}$$

with the Green's matrix

$$G = \left(\bar{u}_i^{(r)}\right)\left(\bar{T}_{i3}^{(r)}\right)^{-1}. \tag{5.88}$$

The Green's matrix depends only on the eligible partial waves, or eigenvectors. With this linear relation, it is perfectly possible to consider general types of surface boundary conditions, involving generalized displacements and stresses and any linear combination of these.

When the surface is mechanically free, all normal mechanical stresses in (5.87) are zero, and thus $\Phi = G_{44}\frac{D_3}{-\imath\omega}$. It is customary to define an effective surface permittivity as the ratio of the surface charge density (which is the jump of the normal electric displacement) to the tangential electric field, or

$$\varepsilon_S = \frac{[D_3]}{|E_\parallel|} = \frac{D_3 - \omega s \epsilon_0 \Phi}{\omega s \Phi} = \frac{-\imath}{s G_{44}} - \epsilon_0. \tag{5.89}$$

This definition is chosen so that $\varepsilon_S = 0$ for a free surface, while $\varepsilon_S = \infty$ for a shorted surface.

Pseudo-surface acoustic waves (PSAW). As an example, Figure 5.19 shows the effective surface permittivity as a function of slowness for three rotated Y-cuts of lithium niobate and for waves propagating in the X or [100] direction. The notation Y+128°, for instance, means that the surface normal is at an angle of 128° from axis Y in the plane (Y, Z).

For large values of the surface slowness, $\imath\varepsilon_S$ remains almost constant and is complex imaginary; this indicates that there are no losses. The asymptotic value ε_∞ is a constant for any crystal cut. Whenever the real part of $\imath\varepsilon_S$ is nonzero, losses are present. This is generally the case for slownesses up to the slowest bulk elastic wave.

For the Y+128° cut in Figure 5.19 (a), the main feature is a pole: around $s = 2.6\ 10^{-4}$ s/m, ε_S drops to $-\infty$ before switching to $+\infty$ and dropping down to the asymptotic value ε_∞. The imaginary part remains zero in this slowness range and the pole is the signature of a perfectly guided surface wave, termed a Rayleigh wave or simply a surface acoustic wave (SAW). Similar poles are also present for the Y and the Y+36° cuts. For the latter cuts, however, in the intermediate slowness range around $s = 2.4\ 10^{-4}$ s/m there is an additional damped pole, or Lorentzian function. The sharp variation of the real part of ε_S is continuous and is accompanied by a peak in the imaginary part. The damped pole is a signature of a pseudo-surface wave, having partly the character of a surface wave but presenting some amount of loss.

Looking at the slowness curves of Figure 5.18, it can be seen that the Rayleigh wave has a slowness that is always larger than the slownesses of bulk elastic waves in the X direction. As a result, the partial wave superposition only includes evanescent partial waves and the surface wave energy cannot leak inside the substrate. In contrast, the PSAW slowness is located in-between the two quasi-shear bulk slownesses. The slowest of these waves is thus included in the partial wave superposition and is the origin of loss: part of the surface wave energy is radiated toward the interior of the substrate. The actual level of loss is directly related to the complex amplitude of the propagating partial wave. There are particular cases where this complex amplitude can vanish, leading to the existence of a perfectly surface-guided wave that is faster than the slowest shear bulk wave, but this is not the general case. Particularly useful PSAW can be found in lithium niobate and more importantly for applications in lithium tantalate. Table 5.6 lists a few cuts and orientations of piezoelectric substrates that are used in practice. Table 5.7 summarizes various names that are given to special cases of surface elastic waves on a semi-infinite substrate. For completeness, let us mention that in the case of a thin layer deposited on a semi-infinite substrate, it is conventional to refer to Love waves if the polarization is purely shear, to Sezawa waves if the polarization is in the sagittal plane, and to generalized Sezawa waves if the polarization is general. Depending on the thickness of the layer, there can be several Love and Sezawa waves.

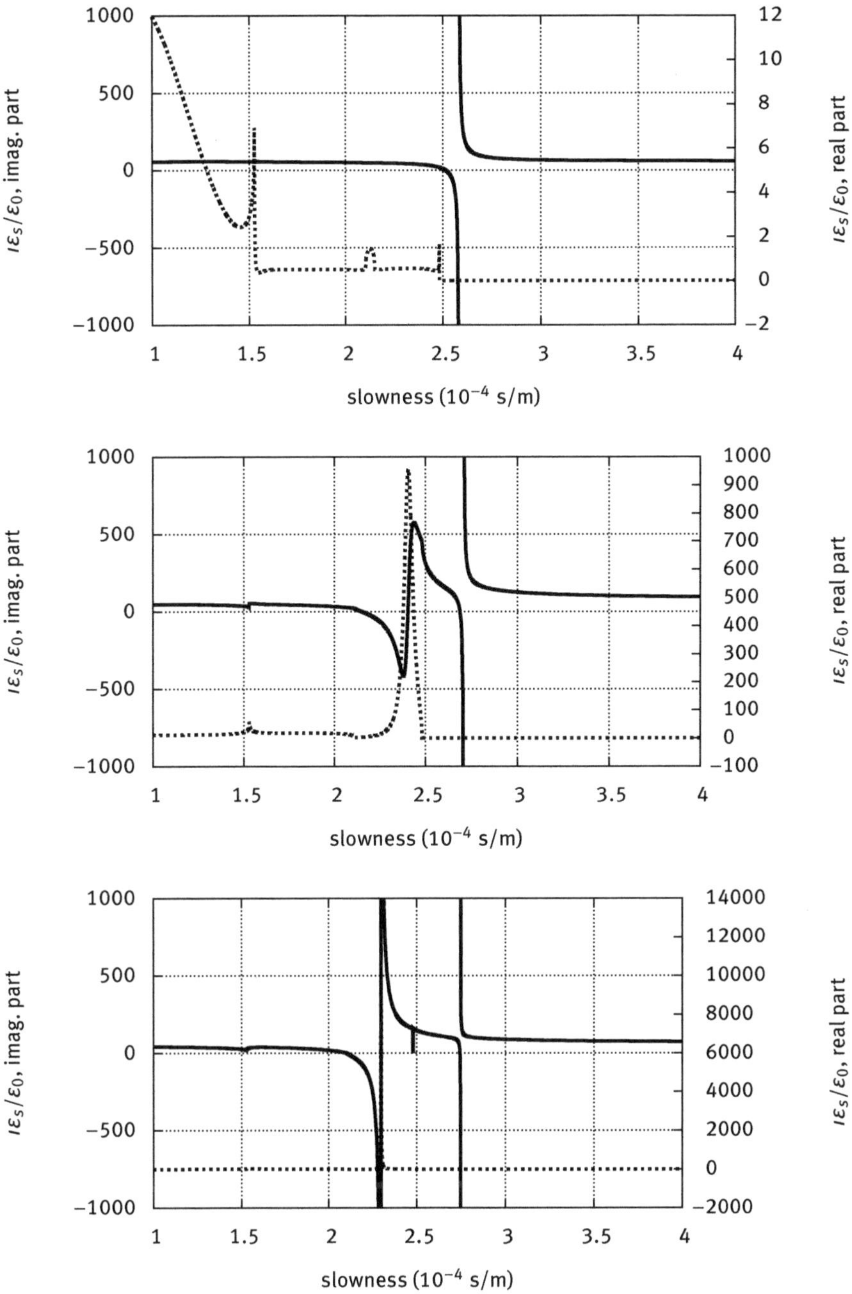

Fig. 5.19: Effective surface permittivity for lithium niobate. Three different orientations are considered: (a) Y+128°, (b) Y and (c) Y+36°. The propagation direction is always the X crystallographic axis.

Tab. 5.6: Characteristics of some cuts and orientations of piezoelectric crystal substrates. Velocities v_s are given for the shorted electrical boundary condition. R: Rayleigh wave; SAW: generally polarized SAW; BG: Bleustein–Gulyaev wave; PSAW: pseudo-surface wave.

Crystal	Cut	Direction	Type	v_s (m/s)	α (dB/λ)	K^2 (%)	$\frac{\varepsilon_\infty}{\varepsilon_0}$
$LiNbO_3$	Y	Z	R	3390	0	4.5	46
(−)	Y+128	X	SAW	3870	0	5	56
(−)	Y+128	X	BG	4030	0	$2.5\ 10^{-4}$	56
(−)	Y+41	X	PSAW	4380	$2\ 10^{-2}$	16	63
$LiTaO_3$	Y	Z	R	3210	0	0.9	48
(−)	Y+36	X	PSAW	4110	$3\ 10^{-4}$	5.5	50
Quartz	Y+42	X	SAW	3158	0	0.11	6
ZnO	X	Y	BG	2823	0	0.5	10

Tab. 5.7: Classification of surface acoustic waves.

Wave type	Polarization
Rayleigh (R)	Sagittal plane displacements only
Generalized Rayleigh (SAW)	All three displacements
Bleustein–Gulyaev (BG)	One pure shear evanescent partial wave
Pseudo-SAW (PSAW)	Leakage induced by radiation inside the substrate

5.A Tensors

Linear relations in anisotropic media. In anisotropic media, and especially in crystals, a cause applied in a given direction does not in general induce an effect oriented in the same direction. For instance, the linear relation between electric displacement and electric field vector is

$$\mathbf{D} = \varepsilon \mathbf{E}, \quad \text{or} \quad \begin{pmatrix} D_1 \\ D_2 \\ D_3 \end{pmatrix} = \begin{pmatrix} \varepsilon_{11} & \varepsilon_{12} & \varepsilon_{13} \\ \varepsilon_{21} & \varepsilon_{22} & \varepsilon_{23} \\ \varepsilon_{31} & \varepsilon_{32} & \varepsilon_{33} \end{pmatrix} \begin{pmatrix} E_1 \\ E_2 \\ E_3 \end{pmatrix}, \quad \text{or} \quad D_i = \varepsilon_{ij} E_j .$$

The first two equations use notations relative to vectors and matrices. The last equation is typically tensorial and uses the convention of implied summation on repeated indices (Einstein convention). D_i and E_j are first-order tensors; ε_{ij} is a second-order tensor. The order is simply the number of indices of the tensor.

The matrix expression is generally quite cumbersome, while the tensorial expression is both compact and much clearer. There is a much more fundamental reason to use tensors instead of matrices with anisotropic media: the physical meaning of a tensor is not modified in any arbitrary rotation of reference axes. Indeed, a rotation of the reference axes does not change the physical meaning while it necessarily deeply modifies the matrix representation. Let us specify this statement quantitatively.

Reference axes rotation. Let us denote the vectors defining two reference frames $\mathbf{e}_1, \mathbf{e}_2, \mathbf{e}_3$ and $\mathbf{e}'_1, \mathbf{e}'_2, \mathbf{e}'_3$. The components of the new reference frame vectors as a function of the old reference frame vectors are given by some matrix α such that

$$\mathbf{e}'_i = \alpha_i^k \mathbf{e}_k \quad \text{with} \quad \alpha = \begin{pmatrix} \alpha_1^1 & \alpha_1^2 & \alpha_1^3 \\ \alpha_2^1 & \alpha_2^2 & \alpha_2^3 \\ \alpha_3^1 & \alpha_3^2 & \alpha_3^3 \end{pmatrix}.$$

We stress that α is a matrix, not a tensor. Reciprocally, $\mathbf{e}_k = \beta_k^j \mathbf{e}'_j$ with $\alpha_i^k \beta_k^j = \delta_{ij}$, δ_{ij} being the Kronecker symbol. Matrix β is the inverse of matrix α. Furthermore, if both reference frames are orthonormal, then $\beta = \alpha^T$.

The coordinates of a vector $\mathbf{x}$ obey the transformation rule

$$x'_i = \beta_k^i x_k \quad \text{with} \quad \mathbf{x} = x_k \mathbf{e}_k = x'_i \mathbf{e}'_i.$$

In the case where the reference frames are orthonormal, the coordinates of a vector follow exactly the same transformation rule as the basis vectors

$$x'_i = \alpha_i^k x_k.$$

Properties of tensors. The following elementary properties are given without demonstration.

- A scalar physical quantity f (e.g. temperature, energy, and so on), is invariant of the chosen reference frame, or $f(x_1, x_2, x_3) = f(x'_1, x'_2, x'_3)$; it is an order-0 tensor.
- An order-1 tensor (or vector) transforms as the basis vectors

$$A'_i = \alpha_i^k A_k.$$

- An order-2 tensor is a set of nine scalars A_{ij} that transforms according to

$$A'_{ij} = \alpha_i^k \alpha_j^l A_{kl}.$$

- The definition extends flawlessly to an arbitrary order, with for instance the order-3 transformation rule

$$A'_{ijk} = \alpha_i^l \alpha_j^p \alpha_k^q A_{lpq}.$$

- The gradient of a vector, $\partial A_i / \partial x_k$, is an order-2 tensor.
- The trace A_{ii} of an order-2 tensor A_{ij} is an order-0 tensor (is a scalar).
- A linear relationship between two tensors defines a tensor whose order is the added orders of the original tensors. For instance, $D_i = \varepsilon_{ij} E_j$ implies that ε_{ij} is an order-2 tensor.

5.B Modeling bulk, plate, and surface waves with FEM

5.B.1 Variational formulation for elastic waves

In this section we summarize the main equations of the variational formulation of linear elastic wave propagation in homogeneous elastic and piezoelectric media.

Elastic waves. Let us start from the elastodynamic equation (5.2), written with differential operators

$$-\nabla \cdot T + \rho \frac{\partial^2 \mathbf{u}}{\partial t^2} = \mathbf{f}. \tag{5.90}$$

All vectors and tensors will be represented as P_n finite element functions (using Lagrange elements of degree n) on a mesh Th. They are thus simply continuous polynomial functions, piecewise defined on the triangles composing the mesh. For a basic presentation of finite element methods we refer to Section 3.3. We consider test functions $\mathbf{v}$ that are defined on the same finite element space as the displacement vector $\mathbf{u}$. Projection on the test functions results in

$$-\int_{\Omega} \mathbf{v}^* \cdot (\nabla \cdot T) + \int_{\Omega} \mathbf{v}^* \cdot \rho \frac{\partial^2 \mathbf{u}}{\partial t^2} = \int_{\Omega} \mathbf{v}^* \cdot \mathbf{f}. \tag{5.91}$$

The second and the third integrals can perfectly be computed by a finite element software given the nodal values. We transform the first integral using the Gauss theorem in order to transfer the divergence operator acting on the stress tensor to a gradient acting on the test function vector

$$\int_{\Omega} \nabla \mathbf{v}^* \, T - \int_{\sigma} \mathbf{v}^* \cdot T_n + \int_{\Omega} \mathbf{v}^* \cdot \rho \frac{\partial^2 \mathbf{u}}{\partial t^2} = \int_{\Omega} \mathbf{v}^* \cdot \mathbf{f}. \tag{5.92}$$

Written in component form for tensors, the scalar function $\nabla \mathbf{v}\, T = v_{i,j} T_{ij}$ is the contraction of two order-2 tensors, and vector T_n (the normal traction) has components $T_{ij} n_j$.

Next we can use Hooke's law in contracted form $T_I = c_{IJ} S_J$ with the strain $S(\mathbf{u}) = \nabla \mathbf{u}$. As a reminder, indices I and J run from 1 to 6. The first integral can conveniently be written in contracted notation to yield the variational formulation of the wave equation in elastic media

$$\int_{\Omega} S(\mathbf{v})_I^* c_{IJ} S(\mathbf{u})_J - \int_{\sigma} \mathbf{v}^* \cdot T_n + \int_{\Omega} \mathbf{v}^* \cdot \rho \frac{\partial^2 \mathbf{u}}{\partial t^2} = \int_{\Omega} \mathbf{v}^* \cdot \mathbf{f}. \tag{5.93}$$

In the monochromatic or harmonic case, the variational formulation of the Helmholtz equation is

$$\int_{\Omega} S(\mathbf{v})_I^* c_{IJ} S(\mathbf{u})_J - \int_{\sigma} \mathbf{v}^* \cdot T_n - \omega^2 \int_{\Omega} \mathbf{v}^* \cdot \rho \mathbf{u} = \int_{\Omega} \mathbf{v}^* \cdot \mathbf{f}. \tag{5.94}$$

The boundary integral (second integral) is specified by the boundary conditions. For a Dirichlet boundary condition, $\mathbf{u} = 0$ on σ, the test functions must satisfy the same condition, i.e. $\mathbf{v} = 0$, and the boundary integral vanishes. For a Neumann boundary condition, $T_n = 0$, this is also the case. For an interface between two elastic media, the normal traction is continuous and the boundary integral vanishes as well. When looking for eigenmodes, the force density $\mathbf{f} = 0$.

Elastic waves in piezoelectric media. The piezoelectric case is a generalization over the elastic case. In additional to the elastodynamic equation (5.2), we have to satisfy Poisson's equation (5.30)

$$-\nabla \cdot \mathbf{D} = 0. \tag{5.95}$$

The finite element space as before uses P_n finite element functions, but now the electric potential must be added to the variables; as a result we have a mixed finite element space $(\mathbf{u}, \Phi)$. The test functions are defined in the same space and are denoted $(\mathbf{v}, \Psi)$. The variational formulation is based on the equation

$$-\int_\Omega \mathbf{v}^* \cdot \nabla T + \int_\Omega \mathbf{v}^* \cdot \rho \frac{\partial^2 \mathbf{u}}{\partial t^2} - \int_\Omega \Psi^* \nabla \cdot \mathbf{D} = \int_\Omega \mathbf{v}^* \cdot \mathbf{f}. \tag{5.96}$$

When this equation is satisfied for all test functions $(\mathbf{v}, \Psi)$ then the original elastodynamic equation and Poisson's equation are recovered in a weak sense (i.e. hold for the finite element space). Application of the Gauss theorem to the first and the third integrals lead to

$$\int_\Omega \nabla \mathbf{v}^* \, T - \int_\sigma \mathbf{v}^* \cdot T_n + \int_\Omega \mathbf{v}^* \cdot \rho \frac{\partial^2 \mathbf{u}}{\partial t^2} + \int_\Omega \nabla \Psi^* \cdot \mathbf{D} - \int_\sigma \Psi^* D_n = \int_\Omega \mathbf{v}^* \cdot \mathbf{f}. \tag{5.97}$$

With the contracted notation, we finally have

$$\int_\Omega S(\mathbf{v})_I^* T_I - \int_\sigma \mathbf{v}^* \cdot T_n + \int_\Omega \mathbf{v}^* \cdot \rho \frac{\partial^2 \mathbf{u}}{\partial t^2} + \int_\Omega \nabla \Psi^* \cdot \mathbf{D} - \int_\sigma \Psi D_n = \int_\Omega \mathbf{v}^* \cdot \mathbf{f} \tag{5.98}$$

and the monochromatic or harmonic case,

$$\int_\Omega S(\mathbf{v})_I^* T_I - \int_\sigma \mathbf{v}^* \cdot T_n - \omega^2 \int_\Omega \mathbf{v}^* \cdot \rho \mathbf{u} + \int_\Omega \nabla \Psi^* \cdot \mathbf{D} - \int_\sigma \Psi^* D_n = \int_\Omega \mathbf{v}^* \cdot \mathbf{f}. \tag{5.99}$$

T_I and D_i in these equations are taken from (5.34) and (5.35).

5.B.2 Finite element implementation

We give some hints in the following to explain how we wrote the scripts that were used in this chapter to obtain slowness curves for bulk waves and band structures for plate waves (with FreeFem++ [48]).

Bulk elastic waves. For bulk elastic waves we need only form the 3×3 Christoffel tensor and find its eigenvalues. With FEM, this can be achieved by considering only one triangular element and a P0 vector finite element space $[u_1, u_2, u_3]$. The P0 element has only one node at the barycenter of the triangle; the finite element space that is generated has exactly three degrees of freedom. The direction of propagation is varied by changing a direction vector (n_1, n_2, n_3) and the eigenvalue is the square of the slowness s in this direction. Indeed, the strains in the variational formulation are obtained as

$$\begin{aligned}
S_1(\mathbf{u}) &= -\imath\omega s n_1 u_1\\
S_2(\mathbf{u}) &= -\imath\omega s n_2 u_2\\
S_3(\mathbf{u}) &= -\imath\omega s n_3 u_3\\
S_4(\mathbf{u}) &= -\imath\omega s(n_3 u_2 + n_2 u_3)\\
S_5(\mathbf{u}) &= -\imath\omega s(n_3 u_1 + n_1 u_3)\\
S_6(\mathbf{u}) &= -\imath\omega s(n_2 u_1 + n_1 u_2).
\end{aligned}$$

The strains obtained from the test functions are similar but with the opposite sign because of complex conjugation, and finally a factor ω^2 factors from the FEM matrices, leaving only s^2 in factor of the stiffness matrix. Matrices are built from (5.94).

The piezoelectric case is very similar. We use exactly the same mesh with a single triangle and a P0 finite element. The changes are that the finite element space now has four degrees of freedom $[u_1, u_2, u_3, \Phi]$ and matrices are built from (5.99).

Plate and surface waves. For plate waves, we need two ingredients. The first is that the problem is intrinsically 1D, hence in principle we need only a 1D mesh. With FreeFem++ this is not possible, and we approximate the 1D problem with a rectangular 2D mesh that is only one element wide in the x_1 direction and we apply periodic boundary conditions for the boundaries orthogonal to this axis. In direction x_3 the mesh is exactly h long, with h the thickness of the plate. For propagation in the x_2 direction, as we did for Figure 5.14, we add a variable k_2 component. Following (5.79), the strains are now

$$\begin{aligned}
S_1(\mathbf{u}) &= u_{1,1}\\
S_2(\mathbf{u}) &= -\imath k_2 u_2\\
S_3(\mathbf{u}) &= u_{3,3}\\
S_4(\mathbf{u}) &= -\imath k_2 u_3 + u_{2,3}\\
S_5(\mathbf{u}) &= u_{1,3} + u_{3,1}\\
S_6(\mathbf{u}) &= -\imath k_2 u_1 + u_{2,1}.
\end{aligned}$$

Since we do not know of an eigenvalue equation giving surface waves, it is not possible to obtain a surface wave velocity directly from an eigenvalue problem. We note, however, that it is sufficient to look for the slowest of all plate waves for a large value of k_3 (in practice a value of $k_3 h$ of the order of 10 is sufficient). In practice, all the examples we have shown for surface elastic waves in this chapter were obtained by directly implementing the plane wave formalism of Section 5.7.

6 Phononic crystals for bulk elastic waves

In this chapter we consider artificial crystals for elastic waves in solids, or phononic crystals. We will be mainly concerned with bulk elastic waves in perfectly periodic inhomogeneous media, leaving the discussion of surface and plate wave propagation to Chapter 7. Properties of phononic crystals that arise from periodicity parallel those of sonic crystals. The difference to acoustic waves, however, is that elastic waves are vector waves, i.e. they possess a polarization, meaning that several physical degrees of freedom are now necessary to describe their propagation. In elastic solids, these degrees of freedom are the three displacements in physical space. As a note, this number of degrees of freedom does not decrease if the dimensionality of the structure is reduced. For instance, elastic waves in 2D phononic crystals must still be described with three displacements in space. Popular simplifications such as assuming plane strain or plane stress are generally not grounded. Furthermore, when we consider wave propagation in piezoelectric solids, the number of degrees of freedom is augmented to four, since the electric potential has to be added in the frame of the quasistatic approximation.

Phononic crystals for bulk elastic waves can include either only one or several solid constituents. In the case of heterogeneous phononic crystals containing different materials, Bragg band gaps are generally larger when the contrast between material properties is maximized, as we will illustrate in this chapter. A popular composition is given by heavy and stiff inclusions in a light and soft matrix, such as the steel in epoxy phononic crystal depicted in Figure 6.1 (a). Alternatively, and in contrast with fluids, holes periodically distributed in a solid material provide a phononic crystal for which material properties contrast is replaced by internal boundary conditions at the boundaries of the holes. Such phononic crystals are rather easy to obtain by forming holes in a solid material, as illustrated in Figure 6.1 (b).

6.1 Modeling of phononic crystals

We describe in this section the calculation of the band structure of phononic crystals. Based on the equations for elastic waves presented in Chapter 5, we first discuss the dynamical equations and the relevant boundary conditions.

6.1.1 Dynamical equations

The dynamical equations in phononic crystals are those of elastic waves in elastic or piezoelectric solids that we presented in Chapter 5. A major difference to Chapter 5, however, is that the material constants are now functions of space coordinates.

(a) (b)

Fig. 6.1: Examples of phononic crystals. (a) A three-dimensional phononic crystal of steel spheres in an epoxy matrix is arranged according to an FCC lattice [68] (photograph ©Clément Eustache). (b) A two-dimensional phononic crystal of aluminum in air [50] (copyright 2013 American Institute of Physics) is actually designed to present negative refraction for longitudinal elastic waves; see Chapter 12.

Summary of equations for elastic waves. The basic equations in elastic solids are (5.1), (5.2), and (5.3). We will neglect any internal forces and thus write these equations as

$$T_{ij,j} = \rho \frac{\partial^2 u_i}{\partial t^2}, \tag{6.1}$$

$$T_{ij} = c_{ijkl} S_{kl}, \tag{6.2}$$

$$S_{ij} = \frac{1}{2}\left(u_{i,j} + u_{j,i}\right). \tag{6.3}$$

We have purposely written these equations by fully employing tensorial notations and omitting dependence with space and time variables, but these should not be forgotten. The first equation is Newton's first law, the second equation is the constitutive relation or Hooke's law for elastic solids, and the third equation is the definition of the strain tensor in the (symmetric) linear approximation. These equations are in general not difficult to solve given adequate boundary conditions. In the context of phononic crystals the main difficulty arises from discontinuities of the material constants, e.g. at the interface between different elastic media in the unit cell. As a note, one should refrain from factoring c_{ijkl} and ρ out of the partial derivatives, but also from rewriting equations with their inverses, especially in the case of holey phononic crystals since the material constants are simply zero in holes. Because of the symmetries of the elastic tensor, the stress tensor can be eliminated to obtain an elastodynamic equation involving only the displacements

$$\left(c_{ijkl} u_{k,l}\right)_{,j} = \rho \frac{\partial^2 u_i}{\partial t^2}. \tag{6.4}$$

In the literature the elastodynamic equations are very often written for the specific case of isotropic solids, for instance introducing the Lamé constants or Young's modulus and Poisson's ratio in place of the elastic tensor (which has only two independent elements in the case of isotropic solids, as we discussed in Chapter 5). In doing so, one often wishes to reintroduce divergence and gradient operators as we used in Chapter 4 in our description of pressure acoustic waves in sonic crystals. As a note, this is neither a simplification of the numerical nor of the mathematical expressions, and the tensorial notation is far more efficient for writing codes based on the plane wave expansion (PWE) and the finite element method (FEM).

When looking for Bloch waves, the displacements are taken of the form

$$u_i(t,\mathbf{x}) = \tilde{u}_i(\mathbf{x}) \exp(\imath(\omega t - \mathbf{k}\cdot\mathbf{x})), \tag{6.5}$$

with $\tilde{u}_i$ a periodic function defined in the unit cell of the phononic crystal. The strains and the stresses can be written similarly introducing the periodic functions $\tilde{S}_{ij}$ and $\tilde{T}_{ij}$. The elastodynamic equations then become

$$\tilde{T}_{ij,j} - \imath k_j \tilde{T}_{ij} = -\omega^2 \rho u_i, \tag{6.6}$$

$$\tilde{T}_{ij} = c_{ijkl}\tilde{S}_{kl}, \tag{6.7}$$

$$\tilde{S}_{ij} = \frac{1}{2}\left(\tilde{u}_{i,j} - \imath k_j \tilde{u}_i + \tilde{u}_{j,i} - \imath k_i \tilde{u}_j\right). \tag{6.8}$$

Summary of equations for piezoelectric waves. Let us now write the dynamic equations for piezoelectric waves in a form similar to the equations for elastic waves. We will use the generalized elastic tensor of (5.53), and the generalized displacements and stresses of (5.54) to (5.57) to form the system of equations

$$T^e_{ij,j} = \rho^e_{ik}\frac{\partial^2 u^e_k}{\partial t^2}, \tag{6.9}$$

$$T^e_{ij} = c^e_{ijkl}u^e_{k,l}. \tag{6.10}$$

In these equations, the indices i and k vary from 1 to 4, but the indices j and l only vary from 1 to 3. We thus avoided writing a strain tensor because it would no longer be symmetric. Note that the order-2 tensor ρ^e_{ik} was defined by (5.60). Equation (6.9) contains both Newton's first law and Poisson's equation.

For Bloch waves, the generalized displacements are taken of the form

$$u^e_i(t,\mathbf{x}) = \tilde{u}^e_i(\mathbf{x}) \exp(\imath(\omega t - \mathbf{k}\cdot\mathbf{x})), \tag{6.11}$$

with $\tilde{u}^e_i$ a periodic function defined in the unit cell of the phononic crystal. The generalized stresses are written similarly with the periodic functions $\tilde{T}^e_{ij}$ and the dynamic equations for Bloch waves are

$$\tilde{T}^e_{ij,j} - \imath k_j \tilde{T}^e_{ij} = -\omega^2 \rho^e_{ik}\tilde{u}^e_k, \tag{6.12}$$

$$\tilde{T}^e_{ij} = c_{ijkl}\left(\tilde{u}^e_{k,l} - \imath k_l \tilde{u}^e_k\right). \tag{6.13}$$

The equations for piezoelectric solids may look simpler than those of the subcase of elastic solids, or at least more compact, but the symmetries of the elastic tensor are less apparent and a bit more difficult to use.

Boundary conditions. At the interface between two solids, e.g. at the interface between an inclusion and the matrix, the mechanical boundary conditions are such that the displacements are all continuous but only the stresses normal to the interface are continuous, or

$$[u_i] = 0, \tag{6.14}$$

$$[T_{ij}n_j] = 0, \tag{6.15}$$

with n_j the three components of the unit normal vector $\mathbf{n}$ and where we recall that [...] means the jump at the interface. There are thus six mechanical continuity boundary conditions. The tangential stresses are not specified by these conditions and are generally discontinuous. At the interface between a solid and a vacuum, or air in the limit that we can neglect acoustic wave generation inside it, the displacements are left completely free and the normal stresses vanish identically

$$T_{ij}n_j = 0. \tag{6.16}$$

Electrical boundary conditions must be considered moreover for piezoelectric solids. At the interface between two piezoelectric solids, both the electrical potential and the normal component of the electric displacement are continuous, or

$$[\Phi] = 0, \tag{6.17}$$

$$[D_j n_j] = 0. \tag{6.18}$$

There are thus two electrical continuity boundary conditions. The tangential electric displacements are not specified and are generally discontinuous. As a reminder, $[D_i n_j]$ is the surface charge density appearing at the interface. At the interface between a piezoelectric solid and air or a vacuum, the same boundary conditions still apply and the permittivity of vacuum, ϵ_0, should not be overlooked in general.

As we described in the case of sonic crystals, periodic boundary conditions are needed for the computation of band structures when the solution is defined on a unit cell. In such a case, if the primitive cell is chosen as the unit cell, it is bounded by d pairs of parallel boundaries, with d the number of periodicities. Each pair is such that one boundary (σ_2) is obtained by a translation $\mathbf{a}_m$ of the other (σ_1), with $m = 1, \ldots d$,

$$\tilde{u}_i(\mathbf{x} + \mathbf{a}_m)|_{\sigma_2} = \tilde{u}_i(\mathbf{x})|_{\sigma_1} \tag{6.19}$$

or

$$u_i(\mathbf{x} + \mathbf{a}_m)|_{\sigma_2} = \exp(-\imath\mathbf{k} \cdot \mathbf{a}_m)\; u_i(\mathbf{x})|_{\sigma_1}. \tag{6.20}$$

In this description, the constant $\exp(-\imath\mathbf{k} \cdot \mathbf{a}_m)$ is a phase factor relating both sides of the unit cell. If the Wigner–Seitz cell is chosen as the unit cell, then the number d of pairs of parallel boundaries is larger than the number of periodicities, and the translations $\mathbf{a}_m$ are replaced by some translations $\mathbf{R}$ of the direct lattice. It must be noted that periodic boundary conditions (6.19) and (6.20) are in any case weaker statements of periodicity than the original strong statement $\tilde{u}_i(\mathbf{x}+\mathbf{R}) = \tilde{u}_i(\mathbf{x})$ with $\mathbf{R}$ any lattice vector and $\mathbf{x}$ any position in space. In practice, however, solutions of the wave equation obtained with periodic boundary conditions can be used to pave all space and thus satisfy the strong periodicity statement individually. We have written the periodic boundary conditions for the mechanical displacements only, but similar conditions hold for mechanical stresses, electric potential, and electric displacement, whenever we need them. Note that mechanical stresses and electric displacement are continuous quantities as we cross the external boundary of the unit cell, because of the implied continuity of material constants there. This is in contrast with boundaries internal to the unit cell and separating different materials.

6.1.2 Plane wave expansion (PWE) method

Our presentation of the plane wave expansion (PWE) method follows the one proposed in references [77, 162]. These papers contain the full piezoelectric formulation which for instance will be useful for the discussion of surface waves in Chapter 7. There have been a number of other presentations, of course all equivalent in their domain of application, in the context of isotropic [22], cubic [74], or generally elastic solids [166].

The PWE method relies on a literal application of the Bloch–Floquet theorem. As such it is straightforward and the derivation is easy to follow. We warn the reader, however, that it may not be the best numerical method in many cases and that it is generally outperformed by FEM.

Fourier series of material constants. The starting point for PWE equations are (6.12) and (6.13). We will drop the subscript e in the subsequent equations for simplicity of presentation, but both the elastic and the piezoelectric cases are included in the formulas. The derivation closely follows the one used in Chapter 4 for sonic crystals. Material constants are periodic and we consider their Fourier series expansions

$$\rho_{ik}(\mathbf{x}) = \sum_{m=-\infty}^{+\infty} (\rho_{ik})_m \exp(-\imath \mathbf{G}_m \cdot \mathbf{x}), \tag{6.21}$$

$$c_{ijkl}(\mathbf{x}) = \sum_{m=-\infty}^{+\infty} (c_{ijkl})_m \exp(-\imath \mathbf{G}_m \cdot \mathbf{x}). \tag{6.22}$$

As we discussed in Section 4.1.2, in practice in a computer program we have to limit the summations over m to some finite range, say from $-N$ to $+N$. Furthermore, the material constants defined by the Fourier series are continuous (and infinitely differentiable) in the unit cell, which can only be an approximation when considering inhomogeneous phononic crystal compositions. The convergence limitations summarized by Figure 4.2 also apply here.

PWE equations. From the Bloch theorem, $\tilde{u}_k$ and $\tilde{T}_{ij}$ are periodic in the unit cell and their truncated Fourier series representations are

$$\tilde{u}_k(\mathbf{x}) = \sum_{m=-N}^{+N} (u_k)_m \exp(-\imath \mathbf{G}_m \cdot \mathbf{x}),$$
$$\tilde{T}_{ij}(\mathbf{x}) = \sum_{m=-N}^{+N} (T_{ij})_m \exp(-\imath \mathbf{G}_m \cdot \mathbf{x}).$$

The gradient of the displacements in (6.13) and the divergence of the stress tensor in (6.12) are then

$$\tilde{u}_{k,l} - \imath k_l \tilde{u}_k = -\imath \sum_{m=-N}^{+N} (G_{lm} + k_l)(u_k)_m \exp(-\imath \mathbf{G}_m \cdot \mathbf{x}),$$
$$\tilde{T}_{ij,j} - \imath k_j \tilde{T}_{ij} = -\imath \sum_{m=-N}^{+N} (G_{jm} + k_j)(T_{ij})_m \exp(-\imath \mathbf{G}_m \cdot \mathbf{x}).$$

Defining the vectors of Fourier coefficients with $2N + 1$ elements

$$\mathbf{U}_k = ((u_k)_{-N}, \ldots, (u_k)_0, \ldots, (u_k)_N)^T, \tag{6.23}$$
$$\mathbf{T}_{ij} = ((T_{ij})_{-N}, \ldots, (T_{ij})_0, \ldots, (T_{ij})_N)^T, \tag{6.24}$$

and recalling the definition (4.19) for the diagonal square matrix with $(2N+1)\times(2N+1)$ elements Γ_j, we finally can write

$$\tilde{u}_{k,l} - \imath k_l \tilde{u}_k = -\imath \sum_{m=-N}^{+N} (\Gamma_l \mathbf{U}_k)_m \exp(-\imath \mathbf{G}_m \cdot \mathbf{x}),$$
$$\tilde{T}_{ij,j} - \imath k_j \tilde{T}_{ij} = -\imath \sum_{m=-N}^{+N} (\Gamma_j \mathbf{T}_{ij})_m \exp(-\imath \mathbf{G}_m \cdot \mathbf{x}).$$

$\Gamma_l \mathbf{U}_k$ and $\Gamma_j \mathbf{T}_{ij}$ must be understood as matrix-vector products giving the Fourier coefficients of the first derivatives of the displacements and of the stresses.

Next we define square band matrices with $(2N+1)\times(2N+1)$ elements constructed for the Fourier coefficients of the material constants

$$(\mathcal{R}_{ik})_{mn} = (\rho_{ik})_{m-n}; \quad m, n = -N, \ldots, N, \tag{6.25}$$
$$(\mathcal{C}_{ijkl})_{mn} = (c_{ijkl})_{m-n}; \quad m, n = -N, \ldots, N. \tag{6.26}$$

The PWE versions of (6.12) and (6.13) are the pair of equations

$$\imath \Gamma_j \mathbf{T}_{ij} = \omega^2 \mathcal{R}_{ik} \mathbf{U}_k, \tag{6.27}$$
$$\mathbf{T}_{ij} = -\imath \mathcal{C}_{ijkl} \Gamma_l \mathbf{U}_k. \tag{6.28}$$

By eliminating the stresses, a generalized eigenvalue equation for the displacements is obtained as

$$\mathcal{C}_{ik}\mathbf{U}_k = \omega^2 \mathcal{R}_{ik}\mathbf{U}_k, \tag{6.29}$$

$$\mathcal{C}_{ik} = \Gamma_j \mathcal{C}_{ijkl}\Gamma_l. \tag{6.30}$$

In order to present this mixed tensor-matrix formulation into a pure matrix problem, the following procedure can be followed. Since the indices i and k run from 1 to 3 for elastic solids and from 1 to 4 for piezoelectric solids, $3N \times 3N$ (or $4N \times 4N$) matrices can be constructed with the elementary blocks

$$\mathcal{C}_{mn} = \begin{pmatrix} (\mathcal{C}_{11})_{mn} & \cdots & (\mathcal{C}_{14})_{mn} \\ \vdots & \ddots & \vdots \\ (\mathcal{C}_{41})_{mn} & \cdots & (\mathcal{C}_{44})_{mn} \end{pmatrix}, \tag{6.31}$$

$$\mathcal{R}_{mn} = \begin{pmatrix} (\mathcal{R}_{11})_{mn} & \cdots & (\mathcal{R}_{14})_{mn} \\ \vdots & \ddots & \vdots \\ (\mathcal{R}_{41})_{mn} & \cdots & (\mathcal{R}_{44})_{mn} \end{pmatrix}. \tag{6.32}$$

Further defining a $3N$ (or $4N$) vector of all displacements with the elementary block

$$\mathbf{U}_m = ((u_1)_m \ldots (u_4)_m)^T, \tag{6.33}$$

we can write the generalized eigenvalue problem as

$$\mathcal{C}\mathbf{U} = \omega^2 \mathcal{R}\mathbf{U}. \tag{6.34}$$

Bulk waves are then obtained as the eigensolutions of this equation. More precisely, (6.34) can be solved for ω^2 as a function of $\mathbf{k}$ to obtain the band structure of the phononic crystal for bulk waves. The Bloch wavevector $\mathbf{k}$ is a parameter of the matrices Γ_j and thus matrix $\mathcal{C}$ is a second-order matrix polynomial in $\mathbf{k}$. By letting $\mathbf{k}$ span the first Brillouin zone, band structures for 1D, 2D, and 3D bulk wave propagation in phononic crystals are readily obtained. In addition, the out-of-plane 2D bulk wave propagation problem can be tackled as well [163]. For the latter, assuming periodicity is in the (x_1, x_2) plane, the band structure can be plotted with (k_1, k_2) spanning the 2D Brillouin zone while k_3 can be given any value. Such a procedure allows one to explore phononic band gap guidance along a linear defect in a 2D phononic crystal, for instance [163]; see Chapter 11.

Let us specify a few useful energy formulas for the PWE solutions. The kinetic energy per unit cell of a Bloch wave can be evaluated using

$$E_k = \frac{1}{2}\int_{\Omega} \rho(\mathbf{x})\tilde{u}^*(\mathbf{x})\tilde{u}(\mathbf{x}) = \frac{1}{2}\mathbf{U}_i^\dagger \mathcal{R}_{ik}\mathbf{U}_k. \tag{6.35}$$

This expression can easily be evaluated once the eigenvector $\mathbf{U}$ is known. Similarly, the potential (or elastic) energy per unit cell of a Bloch wave is

$$E_p = \frac{1}{2}\mathbf{U}_i^\dagger \mathcal{C}_{ik}\mathbf{U}_k. \tag{6.36}$$

It is easy to see that the PWE equation (6.34) implies that kinetic and potential energies are equal for each Bloch wave. Furthermore, the polarization contents can be evaluated by computing the mean-square value of each displacement (the repeated index summation rule should not be applied in the second relation)

$$\langle |u_k(\mathbf{x})|^2\rangle_\Omega = \mathbf{U}_k^\dagger \mathbf{U}_k = \sum_{m=-N}^{+N} |(u_k)_m|^2. \tag{6.37}$$

We can then evaluate the polarization contents with the triplet of numbers (p_1, p_2, p_3) with $p_k \propto \langle |u_k(\mathbf{x})|^2\rangle_\Omega$ and imposing $p_1 + p_2 + p_3 = 1$.

Finally, we note that the PWE method has the drawback that the Fourier series expansion of the displacement and the stress fields implicitly makes the solution continuous everywhere inside the unit cell. Whereas the displacements and the normal stresses can be considered continuous at an interface between different solid materials, such is not the case for shear stresses applying along the interface. Also, the boundary condition at the interface of a solid and a fluid cannot in general be satisfied, which makes the PWE method rather unsuitable for solid-fluid problems (its application can result in the appearance of spurious modes in the fluid part).

Band structure convergence. Figure 6.2 displays the phononic band structure computed by PWE for a square-lattice phononic crystal of aluminum cylinders in nickel. The case considered is taken from the seminal publication by Kushwaha et al. [74]. The diameter of the cylinders is very large, $d/a = 0.9772$, i.e. the cylinders are almost in contact. Given the material contrast between nickel and aluminum, complete band gaps would not open for smaller values of d/a. 9×9 harmonics are used in Fourier series expansions in this case and convergence is almost achieved for the frequency range of the plot. As a general rule, improving the convergence of the PWE by increasing the number of Fourier harmonics is very costly in terms of computation time, as the PWE matrices are dense, in opposition to the sparse matrices generated by FEM.

In Figure 6.2 we have presented the band structure as two separate plots, one for in-plane polarized elastic waves and one for out-of-plane polarized elastic waves. In the case of homogeneous and isotropic materials, we have indeed seen in Chapter 5 that bulk elastic waves separate into one longitudinal (L) wave and two orthogonal shear (S) waves. Waves with a pure polarization can thus propagate independently in homogeneous and isotropic materials. In 2D heterogeneous materials composed of isotropic elastic materials, and for propagation in the plane of inhomogeneity (the plane of periodicity for a phononic crystal), only part of this polarization independence remains. Instead of entering the mathematical symmetries of the elastodynamic

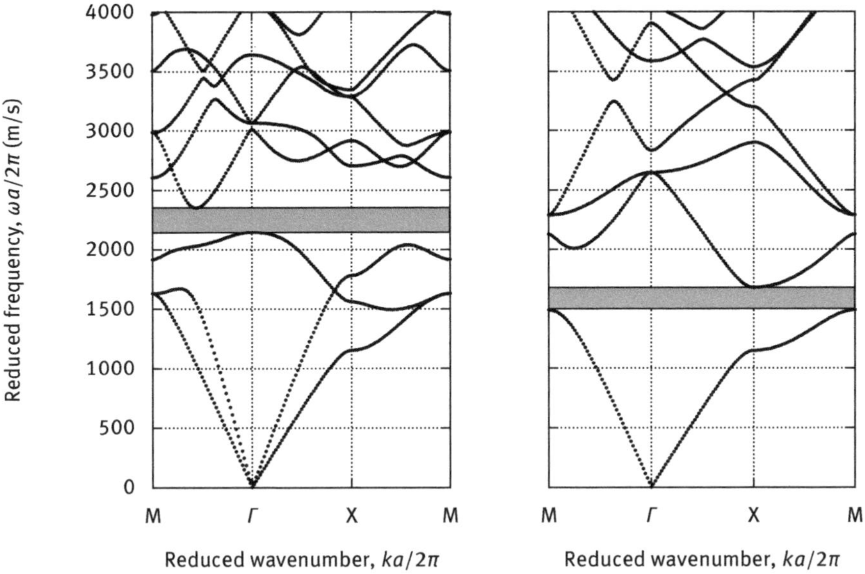

Fig. 6.2: Band structure computed with PWE for a square-lattice phononic crystal of aluminum cylinders in nickel. Geometrical parameters are taken from Figure 2 in Kushwaha et al. [74]. The filling fraction is 75 % ($d/a = 0.9772$). 9×9 harmonics are used in Fourier series expansions. The band structure on the left (on the right, respectively) is for in-plane polarized waves (out-of-plane polarized waves). There are complete band gaps in either case, indicated by the gray regions, but no full band gap.

equation, let us give the following physical explanation. Consider first a pure shear elastic wave with displacements along axis x_3, orthogonal to the periodicity plane (x_1, x_2) (we say out-of-plane polarized). This wave propagates with different velocities inside the different materials, but will not be converted to another polarization when reflected on and transmitted through material interfaces; hence it remains pure shear. Now consider an L or S elastic wave with displacements in the plane (x_1, x_2) (we say in-plane polarized). Whenever this wave meets a material interface, it is partly reflected and partly transmitted to both the in-plane L and S polarizations, because the local normal to the interface spans all directions in the plane of periodicity. The in-plane L and S waves are thus coupled by inhomogeneities. This is the reason why we could separate band structures for out-of-plane S waves and in-plane L and S waves. This separation, however, generally does not resist material anisotropy and does not apply to 3D phononic crystals.

6.1.3 Finite-difference time-domain method (FDTD)

The FDTD method is especially popular for obtaining the transmission of waves through a finite size phononic crystal, giving results in the space and time domain that closely follow possible experiments. It can also be employed to obtain band structures, though the numerical efficiency of the method is quite questionable for this particular problem. We will not discuss the implementation of finite differences in time, as it is similar to the one discussed in Chapter 4 for sonic crystals, but the implementation of finite differences in space is not obvious and deserves some explanation [103].

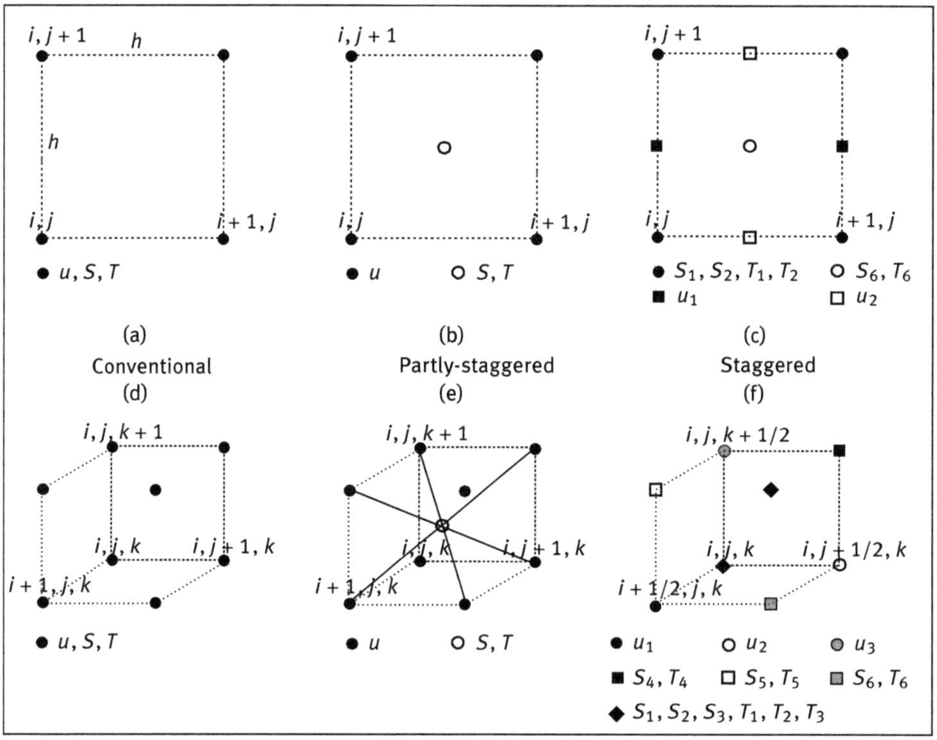

Fig. 6.3: FDTD grids for elastic waves, in 2D and 3D. The first line depicts the 2D (a) conventional, (b) partly-staggered and (c) staggered grids. The second line depicts the 3D (d) conventional, (e) partly-staggered and (f) staggered grids.

The most immediate way to define an interpolating grid is to consider the estimation of physical quantities at the nodes of a regular or conventional grid, as depicted in Figure 6.3 (a). For 2D problems, however, the choice of a staggered grid such as the one depicted in Figure 6.3 (c) is often made. To understand why, let us consider the propagation of in-plane polarized waves in the plane (x_1, x_2) in an elastic solid. All derivatives with respect to x_3 and the u_3 displacement are assumed to vanish explicitly. As a result, only stresses (T_1, T_2, T_6) and strains (S_1, S_2, S_6) (in Voigt notation) appear in the elastodynamic equations. These equations are explicitly

$$\begin{aligned}
\rho\ddot{u}_1 &= T_{1,1} + T_{6,2}, \\
\rho\ddot{u}_2 &= T_{6,1} + T_{2,2}, \\
T_1 &= c_{11}S_1 + c_{12}S_2 + c_{16}S_6, \\
T_2 &= c_{12}S_1 + c_{22}S_2 + c_{26}S_6, \\
T_6 &= c_{16}S_1 + c_{26}S_2 + c_{66}S_6, \\
S_1 &= u_{1,1}, \\
S_2 &= u_{2,2}, \\
S_6 &= u_{1,2} + u_{2,1}.
\end{aligned}$$

Given a grid pitch h, a partial derivative with respect to x_1 is best implemented with a half-pitch shift as for instance in

$$\begin{aligned}
S_1(i,j) &\approx \frac{u_1(i+1/2,j) - u_1(i-1/2,j)}{h}, \\
S_2(i,j) &\approx \frac{u_2(i,j+1/2) - u_2(i,j-1/2)}{h}, \\
S_6(i+1/2,j+1/2) &\approx \frac{u_1(i+1/2,j+1) - u_1(i+\frac{1}{2},j)}{h} \\
&+ \frac{u_2(i+1,j+\frac{1}{2}) - u_2(i,j+1/2)}{h}.
\end{aligned}$$

Inspection of the equations should convince the reader that if (S_1, S_2, T_1, T_2) are estimated at grid nodes, then u_1 must be estimated with a horizontal half-shift, u_2 with a vertical half-shift, and (S_6, T_6) with a diagonal half-shift. But then in Hooke's law the terms $c_{16}S_6$ and $c_{26}S_6$ are not evaluated at the same grid positions as T_1 and T_2, and must be spatially interpolated, thus breaking the coherence of the staggered grid. Similarly, $c_{16}S_1$ and $c_{26}S_2$ are not evaluated at the same grid positions as T_6. Consequently, the staggered grid is particularly useful only if $c_{16} = c_{26} = 0$. Looking at Table 5.2, this happens for isotropic solids and for cubic, hexagonal, tetragonal, orthorhombic, and monoclinic crystals, providing they are oriented according to their natural crystallographic axes. A general rotation of these axes will however in general break the symmetry property indicated by $c_{16} = c_{26} = 0$. In case this symmetry is broken, we can just as well stay with the conventional grid and use spatial interpolation. A slightly better alternative, however, is the partly-staggered grid of Figure 6.3 (b). The partly-staggered grid can be viewed as composed of two interleaved conventional grids, one for displacements and the other for stresses and strains. A formula for the strain S_1 could for instance be

$$S_1(i+\frac{1}{2}, j+\frac{1}{2}) \approx \frac{u_1(i+1,j+1) + u_1(i+1,j) - u_1(i,j+1) - u_1(i,j)}{2h}.$$

Tab. 6.1: Choice of FDTD grids for elastic waves, depending on the formulation of the wave propagation problem.

Formulation	Grid		
	conventional	partly-staggered	staggered
Displacement $\rho\ddot{u}_i = (c_{ijkl}u_{k,l})_{,j}$	×		
Displacement/stress $\rho\ddot{u}_i = T_{ij,j}$ $T_{ij} = c_{ijkl}u_{k,l}$		× (anisotropic)	× (orthotropic)

As a remark, the strains never need to be stored in memory but can be evaluated on the fly from displacements at the time stresses as needed.

The 3D case can be obtained from similar considerations. In the staggered grid shown in Figure 6.3 (f), the longitudinal stresses and strains are initially placed on the same conventional grid. Then the nodes for the three displacements are each half-shifted along one of the three principal directions. The shear stresses and strains are each half-shifted along two of the three principal directions, in accordance with their definition: S_4 and T_4 are for instance half-shifted along x_2 and x_3. With these choices, the dynamical relations $\rho\ddot{u}_i = T_{ij,j}$ are compatible with simple finite differences, as can be seen by inspection. Hooke's law, however, is compatible only if it has the following form

$$\begin{pmatrix} T_1 \\ T_2 \\ T_3 \\ T_4 \\ T_5 \\ T_6 \end{pmatrix} = \begin{pmatrix} c_{11} & c_{12} & c_{13} & \cdot & \cdot & \cdot \\ c_{12} & c_{22} & c_{23} & \cdot & \cdot & \cdot \\ c_{13} & c_{23} & c_{33} & \cdot & \cdot & \cdot \\ \cdot & \cdot & \cdot & c_{44} & \cdot & \cdot \\ \cdot & \cdot & \cdot & \cdot & c_{55} & \cdot \\ \cdot & \cdot & \cdot & \cdot & \cdot & c_{66} \end{pmatrix} \begin{pmatrix} S_1 \\ S_2 \\ S_3 \\ S_4 \\ S_5 \\ S_6 \end{pmatrix} .$$

Looking at Table 5.2 again, this happens for isotropic solids and for cubic, hexagonal, certain tetragonal, and monoclinic crystals, providing they are oriented according to their natural crystallographic axes. A general rotation of these axes, however, will in general break the necessary symmetry property, making the staggered grid less useful. In the general anisotropic case, the partly-staggered grid shown in Figure 6.3 (e) is probably a better option. Table 6.1 summarizes possible choices between the different FDTD grids depending on the way the elastodynamic equations are formulated.

6.1.4 Finite element modeling (FEM)

The finite element method (FEM) has already been set out in Chapter 4 for sonic crystals and the variational formulations for elastic and piezoelectric waves were given in Chapter 5. Here we present how the phononic band structure can be obtained by taking into account periodicity within Bloch's theorem.

FEM for elastic waves. The basic equations for propagation of elastic waves in a solid were given in (6.1) to (6.3). In the context of Bloch's theorem, they further simplify to (6.6) to (6.8), where $\tilde{u}_i$ and $\tilde{T}_{ij}$ become the unknown field distributions instead of the displacements u_i and of the stresses T_{ij}. Each solution is associated with a pair $(\omega, \mathbf{k})$.

The domain of definition of the periodic parts $\tilde{u}_i$ and $\tilde{T}_{ij}$ is naturally a unit cell of the phononic crystal. Finite element meshes for some primitive unit cells for 2D sonic crystals were shown in Figure 4.6 and primitive unit cells for 3D sonic crystals were shown in Figure 4.22. All these meshes can be used for phononic crystals without any changes.

In order to obtain the band structure, we can use the variational formulation of (5.94). There is no source term here and the boundary integral vanishes identically because of the periodic boundary condition. As a result, the variational formulation is

$$\int_{\Omega} S(\mathbf{v})_I^* c_{IJ} S(\mathbf{u})_J = \omega^2 \int_{\Omega} \mathbf{v}^* \cdot \rho \mathbf{u}, \tag{6.38}$$

where the strains should be understood as

$$\begin{aligned}
S_1(\mathbf{u}) &= \frac{\partial \tilde{u}_1}{\partial x_1} - \imath k_1 \tilde{u}_1, \\
S_2(\mathbf{u}) &= \frac{\partial \tilde{u}_2}{\partial x_2} - \imath k_2 \tilde{u}_2, \\
S_3(\mathbf{u}) &= \frac{\partial \tilde{u}_3}{\partial x_3} - \imath k_3 \tilde{u}_3, \\
S_4(\mathbf{u}) &= \frac{\partial \tilde{u}_3}{\partial x_2} + \frac{\partial \tilde{u}_2}{\partial x_3} - \imath(k_3 \tilde{u}_2 + k_2 \tilde{u}_3), \\
S_5(\mathbf{u}) &= \frac{\partial \tilde{u}_3}{\partial x_1} + \frac{\partial \tilde{u}_1}{\partial x_3} - \imath(k_3 \tilde{u}_1 + k_1 \tilde{u}_3), \\
S_6(\mathbf{u}) &= \frac{\partial \tilde{u}_2}{\partial x_1} + \frac{\partial \tilde{u}_1}{\partial x_2} - \imath(k_2 \tilde{u}_1 + k_1 \tilde{u}_2).
\end{aligned}$$

The wavevector $\mathbf{k}$ enters directly inside the variational formulation via the strains, and more precisely inside the stiffness matrix associated with the FEM problem. For 2D problems, it is possible to drop all derivatives with respect to x_3.

The variational equation (6.38) can be used to obtain band structures simply by varying the wavevector in the first Brillouin zone and computing the discrete set of eigenfrequencies. As a result of Hermitian symmetry in case the material parameters are purely real (no material loss), the resulting eigenvalue problem for ω^2 has only real and positive solutions. The FEM formulation just described only requires truly periodic boundary conditions; an alternative derivation with periodic boundary conditions involving complex phase factors can be equivalently used [63]. In the latter case, the wavevector dependence only enters the periodic boundary conditions, but not the FEM matrices. The remarks we made in Chapter 4 about mesh density and convergence apply here as well.

FEM for piezoelectric waves. The basic equations for propagation of piezoelectric waves in a solid were given in (6.9) and (6.10). In the context of Bloch's theorem, they further simplify to (6.12) and (6.13), where $\tilde{u}_i^e$ and $\tilde{T}_{ij}^e$ become the unknown field distributions instead of the generalized displacements u_i^e and the generalized stresses T_{ij}^e.

The piezoelectric variational formulation follows from (5.99), without a source term and with a vanishing boundary integral

$$\int_\Omega S(\mathbf{v})_I^* T_I + \int_\Omega \nabla\Psi^* \cdot \mathbf{D} = \omega^2 \int_\Omega \mathbf{v}^* \cdot \rho\mathbf{u}, \tag{6.39}$$

where we recall (5.34) and (5.35)

$$T_I = c_{IJ} S_J + e_{kI} \Phi_{,k},$$
$$D_i = e_{iJ} S_J - \varepsilon_{ij} \Phi_{,j}.$$

The expressions for the strains were given after (6.38) for elastic solids. We have similarly

$$\Phi_{,i} = \tilde{\Phi}_{,i} - \imath k_i \tilde{\Phi}.$$

Energy and polarization. The FEM variational formulations we have just given can be understood from an energetic point of view by the particular choice that the test function equals the FEM solution. Indeed, putting $\mathbf{v} = \mathbf{u}$ in (6.38) or (6.39) immediately yields the equality of potential (elastic) energy and kinetic energy, or

$$E_p = \frac{1}{2} \int_\Omega S(\mathbf{u})_I^* c_{IJ} S(\mathbf{u})_J = E_k = \frac{1}{2} \omega^2 \int_\Omega \mathbf{u}^* \cdot \rho\mathbf{u}. \tag{6.40}$$

The total energy of a Bloch wave in the unit cell is then simply $E = 2E_k = 2E_p$. As a consequence of the quasistatic approximation of piezoelectricity, there is no stored electromagnetic energy, since $\int_\Omega \nabla\Phi^* \cdot \mathbf{D} = -\int_\Omega \Phi^* \nabla \cdot \mathbf{D} = 0$, which is a weak version of Poisson's equation. The similarity with the PWE equations (6.35) and (6.36) is obvious. As with PWE, the polarization contents can be evaluated by computing the mean-square value of each displacement (the repeated index summation rule should not be applied in the second relation)

$$\langle |u_k(\mathbf{x})|^2 \rangle_\Omega = \int_\Omega |u_k|^2. \tag{6.41}$$

We can then again evaluate the polarization contents with the triplet of numbers (p_1, p_2, p_3) with $p_k \propto \langle |u_k(\mathbf{x})|^2 \rangle_\Omega$ and imposing $p_1 + p_2 + p_3 = 1$.

6.2 2D phononic crystals

In this section, we describe a variety of 2D phononic crystals based on their band structure. We consider solid-solid compositions involving two different materials for the matrix and the inclusions, and solid-void compositions with hollow cylindrical inclusions in a single solid material. When available, band structures are compared with experimental results. Note that we consider only circular cylindrical inclusions here, as they are most common; of course the mathematics in the previous section and the computer programs that were used to obtain band structures can cope with arbitrarily shaped inclusions. We will start with isotropic materials, before moving on to anisotropic elastic materials, and finally piezoelectric materials. All band structures are computed with FEM with the methods set out in Section 6.1.4.

6.2.1 Stiff-in-soft composition

Historically, the concept of the phononic crystal was introduced in the 2D case of a solid inclusion embedded in a solid matrix [74]. The composition imagined in this pioneering work was either nickel in aluminum or aluminum in nickel, two isotropic materials. The complete band gap obtained for a square-lattice phononic crystal was rather small, see Figure 6.2, and was furthermore only presented for pure shear elastic waves polarized out-of-plane (along the axis of the cylinders).

The first experimental demonstration was performed some years later with a quite different composition: steel cylinders in an epoxy matrix [155]. Such a composition is very practical experimentally for macroscopic demonstrations, since metal rods can be easily bought in any length, assembled in a 2D periodic array using perforated plates as holders, and finally filled with liquid epoxy before the latter solidifies. Apart from this practical advantage, the choice of a fast and heavy inclusion ($c_S = 3200\,\text{m/s}$, $c_l = 5940\,\text{m/s}$, and $\rho = 7800\,\text{kg/m}^3$ for steel) inside a slow and light matrix ($c_S = 1140\,\text{m/s}$, $c_l = 2570\,\text{m/s}$, and $\rho = 1142\,\text{kg/m}^3$ for epoxy) leads to wide complete band gaps with reasonable filling fractions.

Figure 6.4 displays the band structure for a hexagonal-lattice phononic crystal of cylindrical steel rods in epoxy, with the dimensions selected by Vasseur et al. [155]: $d/a = 0.6641$ or a filling fraction $F = 0.4$. The band structure is separated in two parts, for in-plane S and L waves on the left, and for out-of-plane S waves on the right. For the range of frequencies shown, there are two complete band gaps for the former waves and three complete band gaps for the latter waves. The first complete band gaps for both polarizations have a rather wide overlap range where the phononic band gap is truly full for all polarizations. If one has a source for waves of a given polarization, however, it is perfectly possible in practice to independently explore band gaps for orthogonal polarizations. The experimental results presented in Figure 6.5 [155] were obtained using longitudinal wave transducers and should thus be compared with the

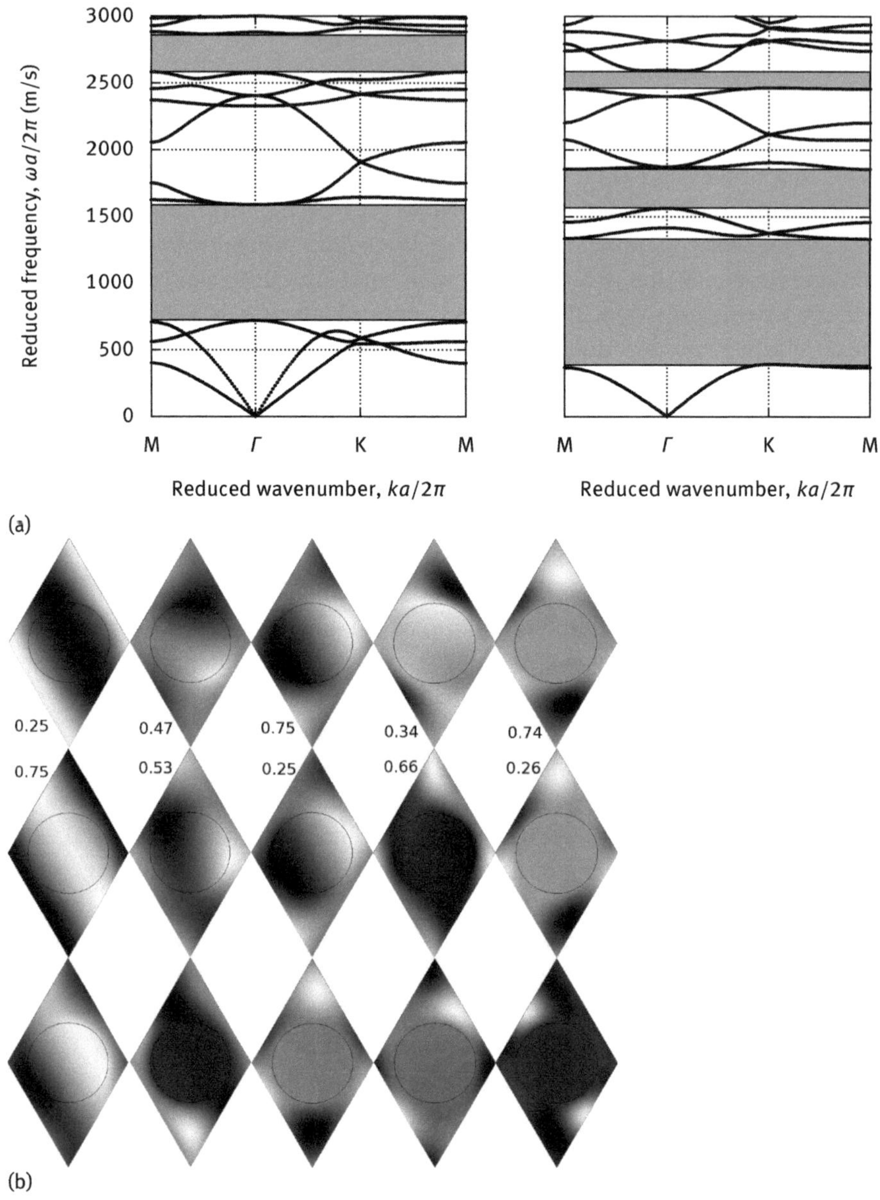

Fig. 6.4: Band structure for a 2D hexagonal-lattice phononic crystal of steel rods in epoxy. The filling fraction is $F = 0.4$ or $d/a = 0.6641$, following the geometry discussed by Vasseur et al. [155]. The band structure is presented on the left for in-plane polarized waves and on the right for out-of-plane polarized waves. The first five Bloch waves are shown in (b) at the M point of the first Brillouin zone for both cases. The first two lines show the distributions of the real parts of u_1 and u_2 for in-plane Bloch waves. The figures indicated in each case are the associated polarization contents p_1 and p_2. The third line shows the distribution of the real part of u_3 for out-of-plane Bloch waves ($p_3 = 1$).

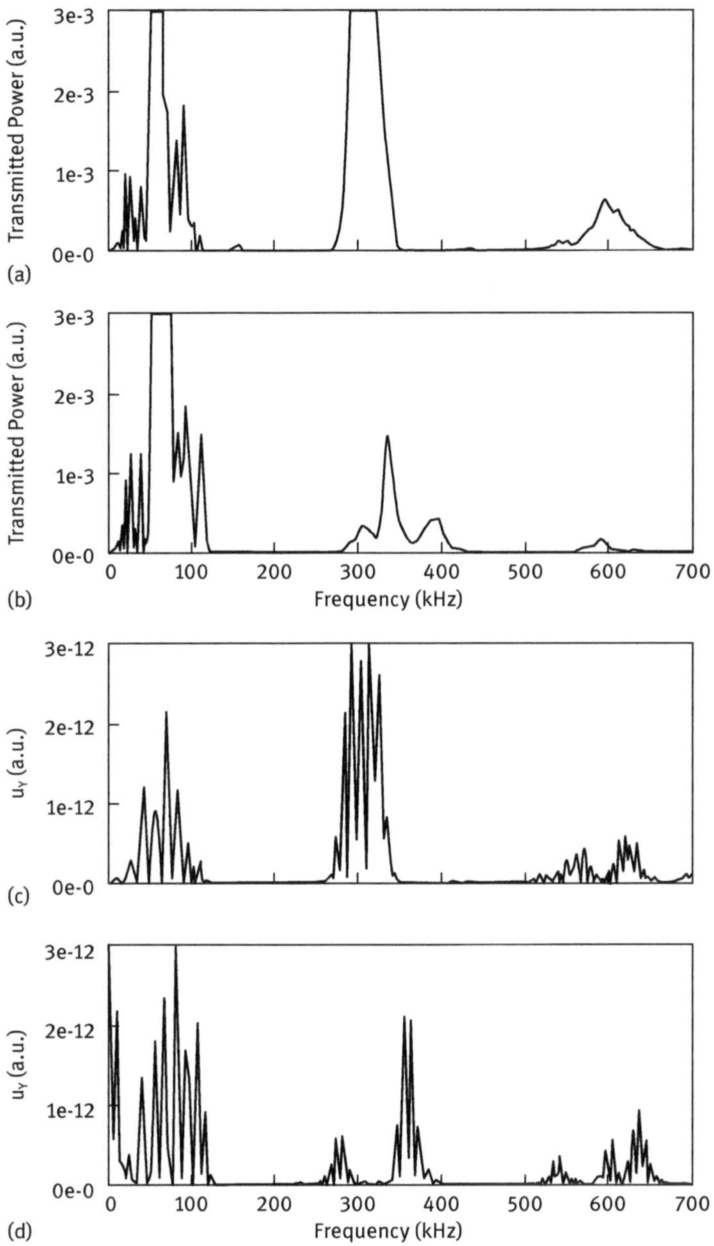

Fig. 6.5: Transmission through a 2D hexagonal-lattice phononic crystal of steel rods in epoxy. The filling fraction is $F = 0.4$ or $d/a = 0.6641$. The measured transmission for longitudinal waves is shown for (a) the ΓK and (b) the ΓM directions. The transmission computed with FDTD and for longitudinal waves is shown for (c) the ΓK and (d) the ΓM directions. u_Y in the graphs is here equivalent to our notation u_1 for the longitudinal displacement (after Vasseur et al. [155], copyright 2001 American Physical Society).

in-plane band structure. Measurements were taken along the two principal symmetry directions, ΓM and ΓK. An absence of transmission is observed in the complete band gap range, as expected, but also at other frequencies where propagating bands exist in principle. This fact can be connected to the existence of deaf bands, similar to the discussion in Chapter 4 for sonic crystals. Instead of identifying deaf bands, an alternative is to compare the experimental transmission with the result of a numerical simulation of transmission through the finite phononic crystal. In Figure 6.5 this was performed thanks to an FDTD computation. The agreement between experiment and simulation is rather good, though absolute values of transmission are not given.

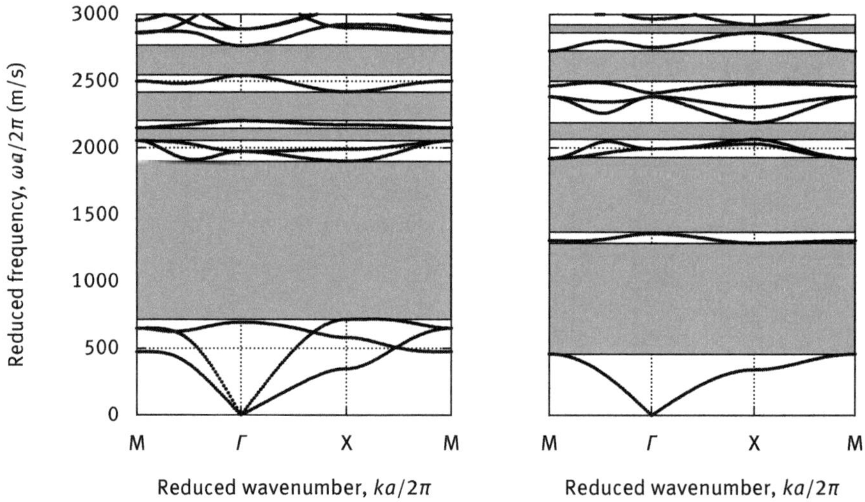

Fig. 6.6: Band structure for a 2D square-lattice phononic crystal of steel rods in epoxy. The filling fraction is $F = 0.54$ or $d/a = 0.83$. The band structure is presented on the left for in-plane polarized waves and on the right for out-of-plane polarized waves.

What if another choice had been made for the lattice type with the same steel rods in epoxy composition? Figure 6.6 displays the band structure for the square lattice for the case $d/a = 0.83$ (filling fraction $F = 0.54$). Again, numerous wide complete band gaps are found for the two separable polarization types. Furthermore, there are several overlap ranges where phononic band gaps are truly complete for all polarizations. With the honeycomb lattice in Figure 6.7, with $d/a = 0.462$ (filling fraction $F = 0.39$), the observations are similar. It is thus no wonder that the stiff-in-soft (or heavy-in-light) composition is popular in the phononic crystal literature.

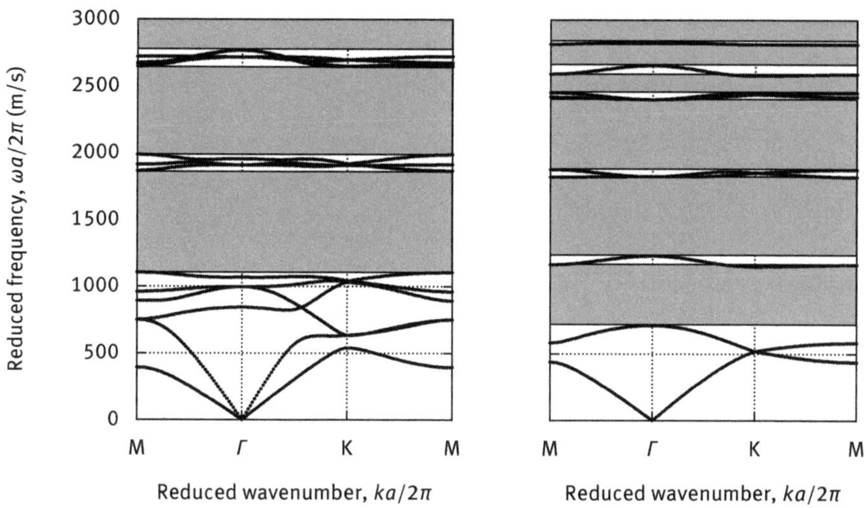

Fig. 6.7: Band structure for a 2D honeycomb-lattice phononic crystal of steel rods in epoxy. The filling fraction is $F = 0.39$ or $d/a = 0.462$. The band structure is presented on the left for in-plane polarized waves and on the right for out-of-plane polarized waves.

6.2.2 Soft-in-stiff composition

Let us now consider the reverse contrast of epoxy rods in a steel matrix, with exactly the same geometrical dimensions as previously.

The hexagonal-lattice phononic crystal in Figure 6.8 does not possess any complete band gap. It does not even possess a directional band gap for out-of-plane S and L waves, and only one directional band gap in the ΓM direction for in-plane S waves.

The square-lattice phononic crystal in Figure 6.9 has one complete band gap for each polarization type, though they do not overlap to form a full band gap.

The honeycomb-lattice phononic crystal in Figure 6.10 has more and slightly wider complete band gaps, with one overlap forming a full band gap. It is clear, however, that the choice of soft (or light) inclusions in a stiff (or heavy) matrix is outperformed by the reverse composition in this case too.

With sonic crystals, we saw in Chapter 4 that the most interesting composition to open wide complete band gaps is the slow-in-fast one, or dilute-in-dense (such as the case of air bubbles in water). In contrast, the case of phononic crystals calls for the converse choice: fast-in-slow or stiff-and-dense in soft-and-dilute. To understand this inversion from the dynamical equations, it is useful to compare the acoustic wave equation and the elastodynamic equation written as a scalar equation for simplicity

$$-\nabla \cdot \left(\frac{1}{\rho_f} \nabla p \right) = \omega^2 \frac{1}{B} p,$$

$$-\nabla \cdot (c \nabla u) = \omega^2 \rho_s u.$$

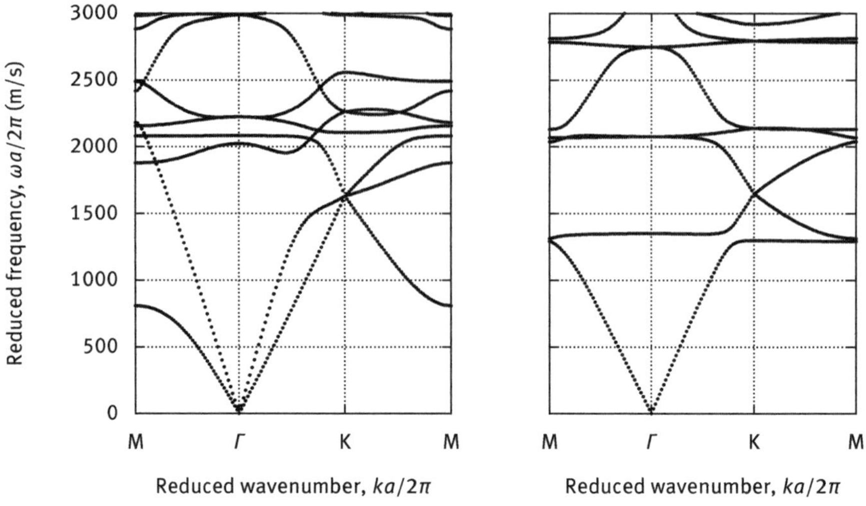

Fig. 6.8: Band structure for a 2D hexagonal-lattice phononic crystal of epoxy rods in steel. The filling fraction is $F = 0.4$ or $d/a = 0.6641$. The band structure is presented on the left for in-plane polarized waves and on the right for out-of-plane polarized waves.

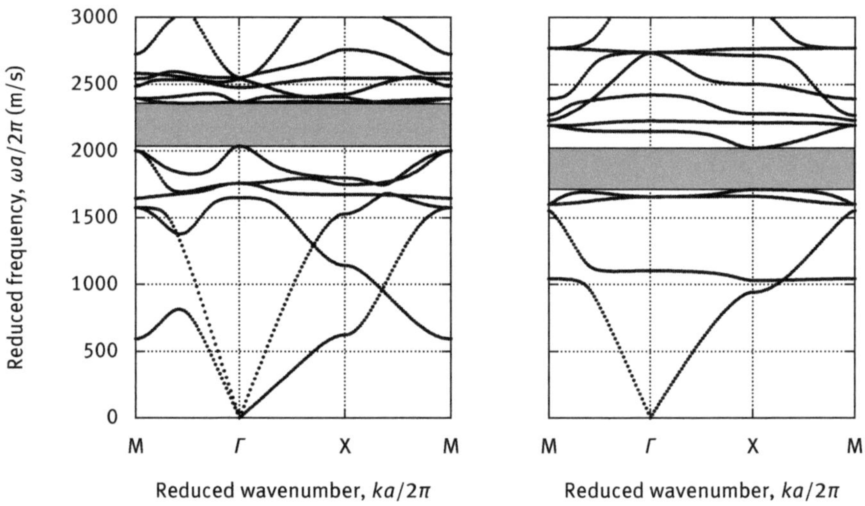

Fig. 6.9: Band structure for a 2D square-lattice phononic crystal of epoxy rods in steel. The filling fraction is $F = 0.54$ or $d/a = 0.83$. The band structure is presented on the left for in-plane polarized waves and on the right for out-of-plane polarized waves.

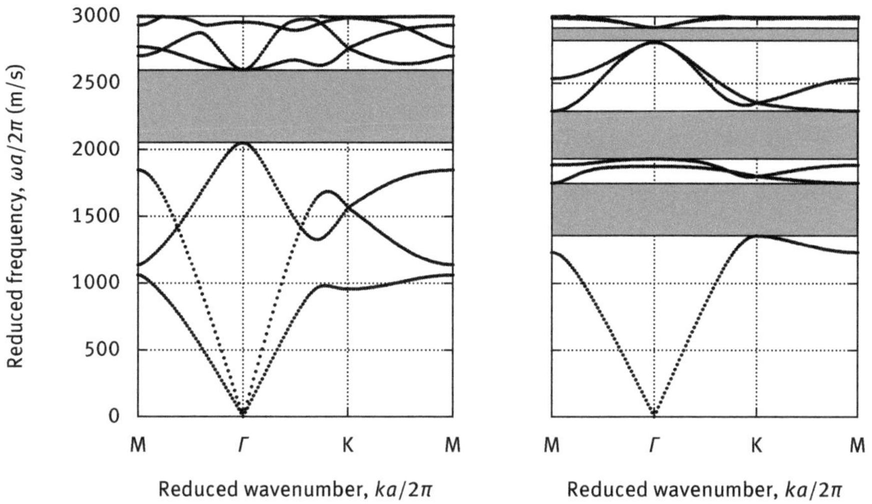

Fig. 6.10: Band structure for a 2D honeycomb-lattice phononic crystal of epoxy rods in steel. The filling fraction is $F = 0.39$ or $d/a = 0.462$. The band structure is presented on the left for in-plane polarized waves and on the right for out-of-plane polarized waves.

It can be seen that the two scalar equations are formally equivalent if we consider $c \equiv \rho_f^{-1}$ and $\rho_s \equiv B^{-1}$. It is hence not surprising that the roles of slow and fast waves are exchanged, since velocities in the fluid (respectively, in the solid) are of the form $\sqrt{B/\rho_f}$ (resp., $\sqrt{c/\rho_s}$).

6.2.3 Solid-void composition

Another possibility that is often considered is the case of phononic crystals with holes in a solid material. This case has an obvious practical value, as macroscopic samples can easily be produced by drilling holes in a piece of metal or ceramic, for instance. Going to small dimensions, there exist many different technological processes to obtain small and deep holes in wafers of glass or crystalline materials such as silicon. In the next chapter, we will encounter practical realizations of this type with lattice constants in the micron range.

The case of holey phononic crystals in steel is considered in Figures 6.11 to 6.13, again for the cases of the hexagonal lattice, the square lattice and the honeycomb lattice. For an isotropic material such as steel and for hollow inclusions, the phononic band structure only depends on two reduced material parameters, for instance c_{11}/ρ and c_{12}/ρ, or equivalently the longitudinal and the shear velocities.

Similar to the epoxy-in-steel composition, there are no complete band gaps for the hexagonal lattice. Compared to Figure 6.8, the band structure in the solid-void

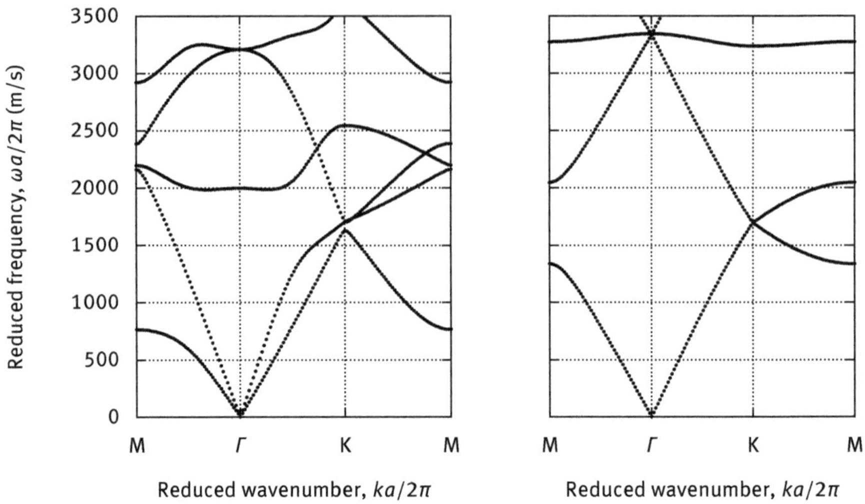

Fig. 6.11: Band structure for a 2D hexagonal-lattice phononic crystal of cylindrical holes in steel. The filling fraction is $F = 0.4$ or $d/a = 0.6641$. The band structure is presented on the left for in-plane polarized waves and on the right for out-of-plane polarized waves.

composition is seen to have fewer bands than the corresponding solid-solid case. This fact reflects the lesser inhomogeneity of holey structures, for which the elastodynamic equation involves only one set of material constants instead of two.

The square-lattice phononic crystal in Figure 6.12 has one complete band gap for each polarization type, in-plane and out-of-plane, with some overlap between them. Compared to the epoxy-in-steel composition in Figure 6.9, the hole-in-steel case has several advantages. It has wider complete band gaps, a less complicated band structure, and notably some isolated bands for in-plane waves that can be used to exhibit the phenomenon of negative refraction, as we will discuss in Chapter 12.

Similar observations can be made for the honeycomb-lattice phononic crystal in Figure 6.13. The symmetry reduction principle in this case operates very efficiently to open complete band gaps by introducing two inclusions instead of only one in the hexagonal lattice.

6.2.4 Crystals containing anisotropic elastic solids

When anisotropic materials are included in the composition of a phononic crystal, it should not be forgotten that the symmetries of the lattice and of the unit cell alone do not fully define those of Bloch waves. Indeed, the total symmetry of the phononic crystal is the combination of three different ingredients: (i) the point symmetry of the lattice defining the periodical repetition of the unit cell, (ii) the symmetry of the unit

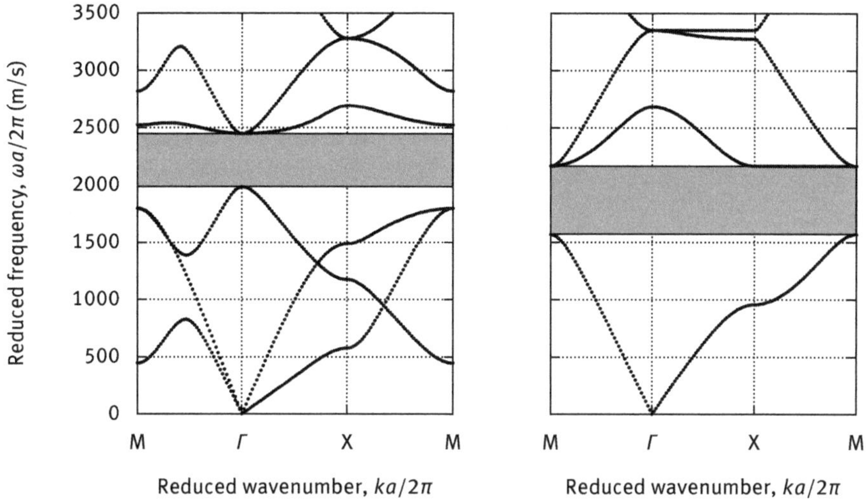

(a)

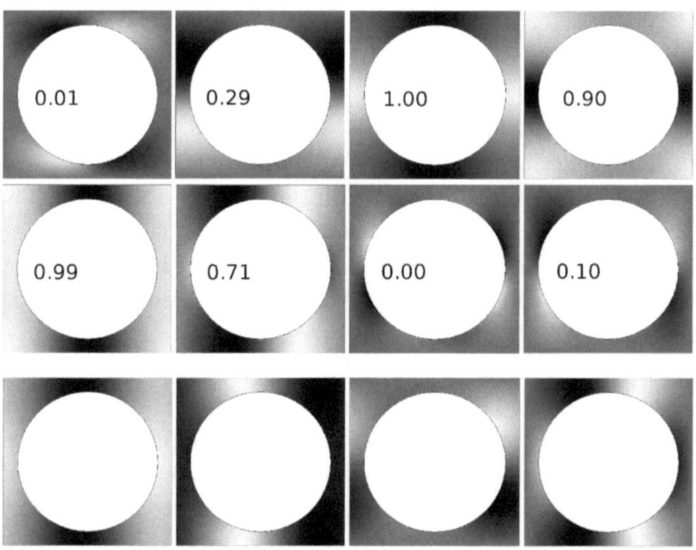

(b)

Fig. 6.12: Band structure for a 2D square-lattice phononic crystal of cylindrical holes in steel. The filling fraction is $F = 0.54$ or $d/a = 0.83$. (a) The band structure is presented on the left for in-plane polarized waves and on the right for out-of-plane polarized waves. (b) The first four Bloch waves are shown in either case at the X point. The first two lines show the distributions of the real parts of u_1 and u_2 for in-plane Bloch waves. The figures indicated in each case are the associated polarization contents p_1 and p_2. The third line shows the distribution of the real part of u_3 for out-of-plane Bloch waves ($p_3 = 1$).

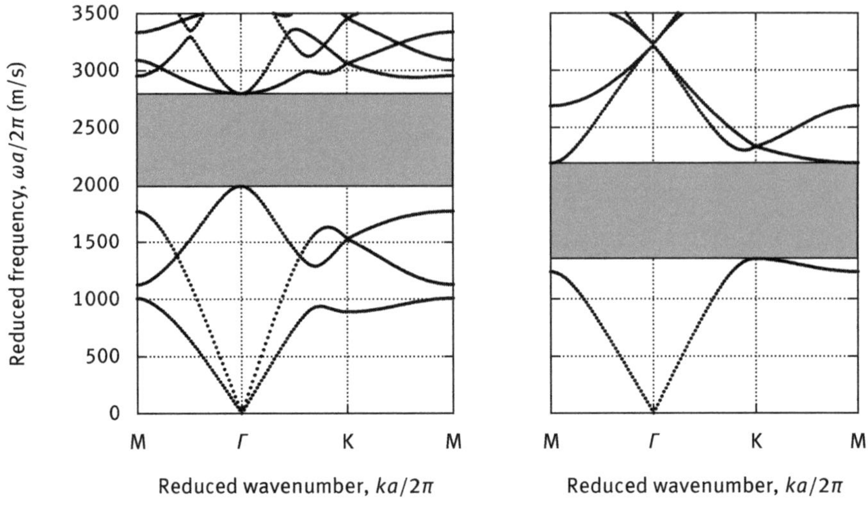

Fig. 6.13: Band structure for a 2D honeycomb-lattice phononic crystal of cylindrical holes in steel. The filling fraction is $F = 0.39$ or $d/a = 0.462$. The band structure is presented on the left for in-plane polarized waves and on the right for out-of-plane polarized waves.

cell itself, which depends on the positions and shapes of the inclusions, and (iii) the crystallographic system of the material constituents. We will continue to consider only circular inclusions, so that they are compatible with all symmetries, and we wish to evaluate the changes occurring because of material anisotropy.

In Figure 6.14 we first consider the case of a phononic crystal of holes in silicon with a square lattice. Two different orientations of silicon are considered, a first one with the crystallographic axes (X, Y, Z) aligned with the reference frame (x_1, x_2, x_3), and a second one where the crystal is rotated by 45° around axis Z. In the IEEE notation, the former orientation is noted (ZX) while the latter is noted (ZXt)/45 [57]. As a cubic crystal, and for the two orientations we consider, silicon preserves the axial symmetry of the 2D square lattice along x_3 and the separation between in-plane and out-of-plane (shear) waves still holds. The 2D irreducible Brillouin is also not changed, as symmetry point Y is still equivalent to symmetry point X, for instance. For in-plane waves, however, rotating silicon around the Z axis changes both the longitudinal and the shear velocities and causes the band structure to transform, as Figure 6.14 (a) clearly evidences. It is even seen that the complete band gap for in-plane waves is significantly reduced when passing from orientation (ZX) to orientation (ZXt)/45. In this case the anisotropy present in the elastic tensor acts against the formation of band gaps. Of course, there can be other orientations of silicon for which the width of the complete band gap would be increased instead of reduced. In the case of out-of-plane waves, the band structure is not changed at all by the rotation around the Z axis. Indeed, this band structure is dictated only by the shear velocity $\sqrt{c_{66}/\rho}$ along axis Z

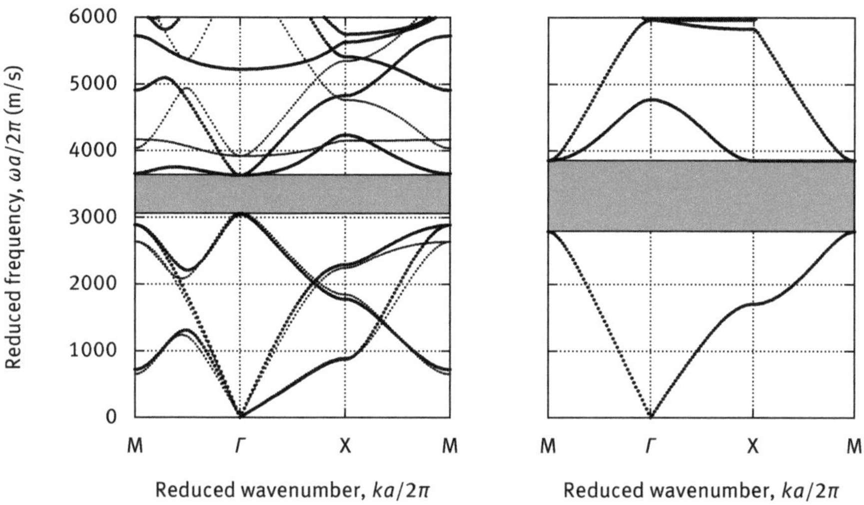

Fig. 6.14: Band structure for a 2D square-lattice phononic crystal of cylindrical holes in silicon. The filling fraction is $F = 0.54$ or $d/a = 0.83$. The band structure is presented on the left for in-plane polarized waves and on the right for out-of-plane polarized waves. In the in-plane band structure, small dots are used for silicon in its natural crystallographic orientation; large dots are used for silicon rotated by 45° in-plane, i.e. the band structure is plotted in the plane (001) but axis x_2 is now midway between directions [100] and [010]. The out-of-plane band structure is unaffected by the rotation.

which is not modified at all in the rotation. In fact, this band structure is identical to the one for holes in steel shown in Figure 6.12 (a), but for a multiplicative factor given by the ratio of shear velocities between silicon and steel.

The most anisotropic crystal that we encountered in Chapter 5 was tellurium dioxide, TiO_2, see Figure 5.5. With this cubic crystal, the shear in-plane velocity can come rather close to the longitudinal velocity for propagation along axis X and can be three times less for propagation at 45° between axes X and Y. This strong anisotropy is seen to efficiently play against the formation of the complete band gap for in-plane waves, as Figure 6.15 shows. The rotation by 45° around axis Z changes the in-plane band structure, but not so as to make the complete band gap open. As with silicon, the band structure for out-of-plane shear waves is unaffected and again simply scaled by the shear velocity $\sqrt{c_{66}/\rho}$.

If we had considered a general anisotropic case, for instance with a more anisotropic crystal or simply by rotating the crystallographic axes by arbitrary angles, then we would have broken the lattice symmetry. The only symmetry of Bloch waves that remains in general is the central symmetry with respect to the Γ point. As a result, for instance, the irreducible Brillouin zone for the square lattice becomes the closed path Γ-X-M-Y-Γ. Moreover, the separation between in-plane and out-of-plane waves is no

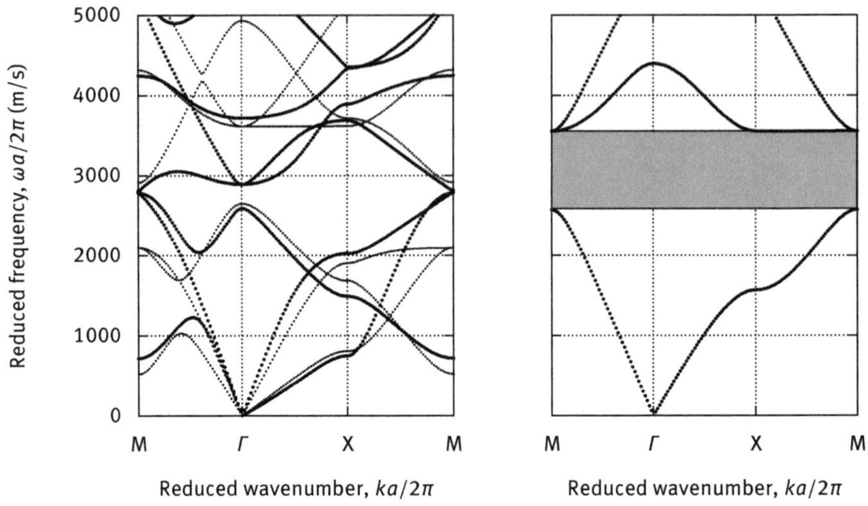

Fig. 6.15: Band structure for a 2D square lattice phononic crystal of cylindrical holes in cubic TiO_2. The filling fraction is $F = 0.54$ or $d/a = 0.83$. The band structure is presented on the left for in-plane polarized waves and on the right for out-of-plane polarized waves. In the in-plane band structure, small dots are used for TiO_2 in its natural crystallographic orientation; large dots are used for silicon rotated by 45° in-plane, i.e. the band structure is plotted in the plane (001) but axis x_2 is now midway between directions [100] and [010]. The out-of-plane band structure is unaffected by the rotation.

longer valid. Consequently, the phononic band structure generally has three different bands starting at the Γ point.

6.2.5 Crystals containing piezoelectric solids

In order to illustrate a more general case of anisotropy, let us consider lithium niobate ($LiNbO_3$) with a square lattice of holes. In Figure 6.16, we plot the band structure for the case where the crystallographic axes (X, Y, Z) are aligned with the reference frame (x_1, x_2, x_3). Both the full case of piezoelectricity and the case where the piezoelectric effect is neglected are plotted for comparison. It can be seen that piezoelectricity tends to make all bands a bit higher in frequency, which can be related to the remark we made in Chapter 5 that the piezoelectrically stiffened elastic constant is generally larger than the purely elastic values. For the (ZX) orientation in Figure 6.16, piezoelectricity plays in favor of the opening of the complete band gap and the effect of piezoelectricity is globally quite strong. Neglecting piezoelectricity, the complete band gap would not even be opened in this case.

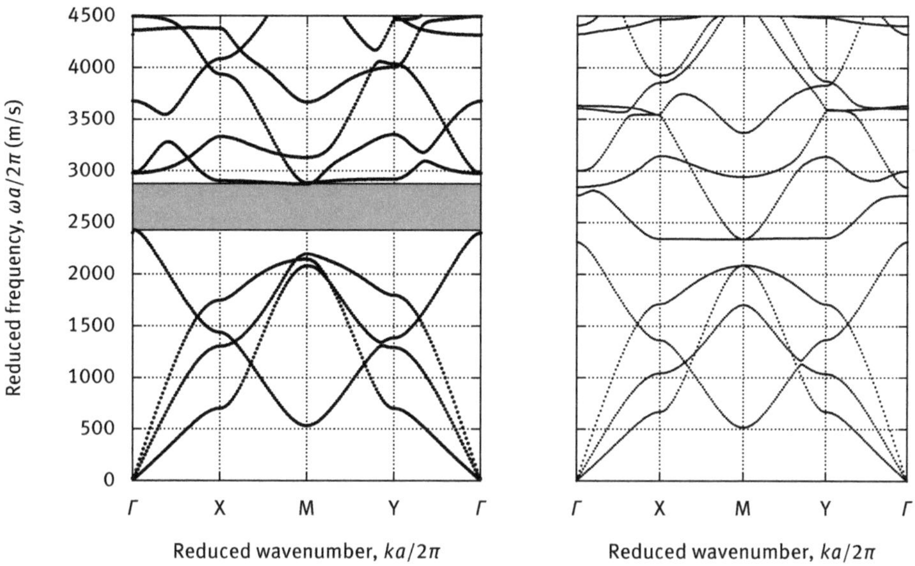

Fig. 6.16: Band structure for a 2D square-lattice phononic crystal of cylindrical holes in lithium niobate. The orientation is (ZX), meaning that the crystallographic axes (X, Y, Z) are aligned with the reference frame (x_1, x_2, x_3). The filling fraction is $F = 0.54$ or $d/a = 0.83$. The left panel is plotted with piezoelectricity taken into account; the right panel is plotted without piezoelectricity (by setting the piezoelectric tensor to zero).

In Figures 6.17 and 6.18 we further plot the band structures for the (YX) and the (XY) orientations, respectively. In the former case, we have $(x_1, x_2, x_3) =$ (X, -Z, Y); in the latter case we have $(x_1, x_2, x_3) =$ (Y, Z, X). With these orientations, the influence of piezoelectricity is much less significant, and though the bands are modified slightly depending on whether it is taken into account or not, the complete band gap remains rather unaffected.

From these examples, it is clear that the influence of piezoelectricity on the phononic band structure is highly dependent on the orientation of the piezoelectric materials and on the direction of propagation. For weak piezoelectrics such as quartz, this influence is generally negligible, but for strong piezoelectrics such as lithium niobate it must absolutely be taken into account. Note that we have not considered any electrical boundary condition at this point, but that these are important in the case of surface and slab piezoelectric phononic crystals, as Chapter 7 details.

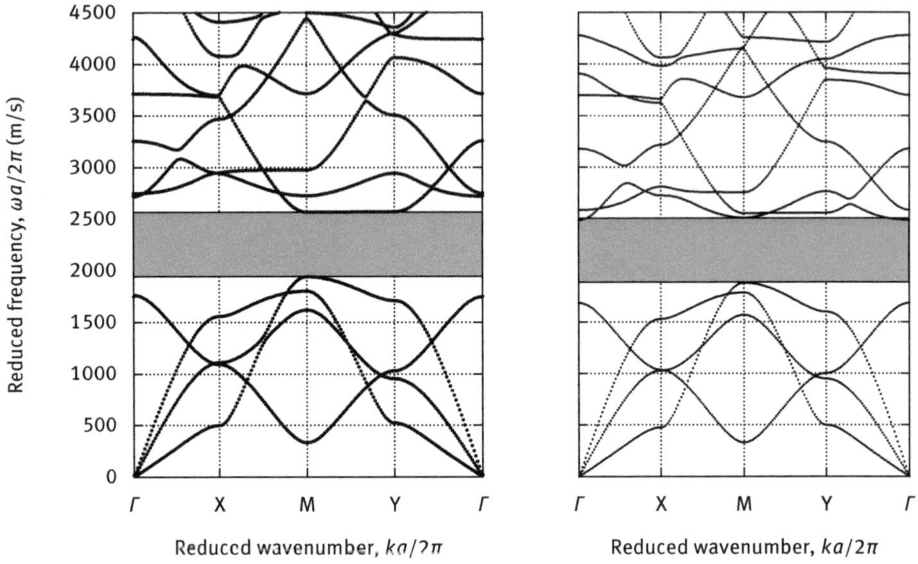

Fig. 6.17: Band structure for a 2D square-lattice phononic crystal of cylindrical holes in lithium niobate. The orientation is (YX), meaning that the crystallographic axes (X, -Z, Y) are aligned with the reference frame (x_1, x_2, x_3). The filling fraction is $F = 0.54$ or $d/a = 0.9$. The left panel is plotted with piezoelectricity taken into account; the right panel is plotted without piezoelectricity (by setting the piezoelectric tensor to zero).

6.3 3D phononic crystals

Surprisingly or not, there are not many examples of three-dimensional phononic crystals in the literature. Furthermore, in most instances compositions are limited to the solid-solid case and to isotropic materials. Obtaining the band structure of 3D phononic crystals can be a lengthy computational task, especially with PWE and to a lesser extent with FEM. If a solid-solid phononic crystal can be described layer by layer and with nonoverlapping scatterers, and providing only isotropic constituents are considered, the LMS method is probably more efficient. 3D solid-void phononic crystals are not obviously easier to fabricate than 3D solid-solid phononic crystals, in contrast to 2D phononic crystals, since it is no longer possible to simply drill or etch holes at the surface of a solid sample. The fabrication of 3D phononic crystals instead relies on precise and controlled assembly or self-assembly of small constituents, or 3D printing. One notable exception is the use of 3D lithography in polymers.

Regarding the composition of 3D phononic crystals, the general conclusions that we obtained with 2D phononic crystals apply. In view of obtaining large complete band gaps, the choice of fast and heavy inclusions in a slow and light matrix is again better than the converse solution, and is also generally better than hollow inclusions. For instance, Sainidou et al. have theoretically considered an FCC lattice phononic crystal of

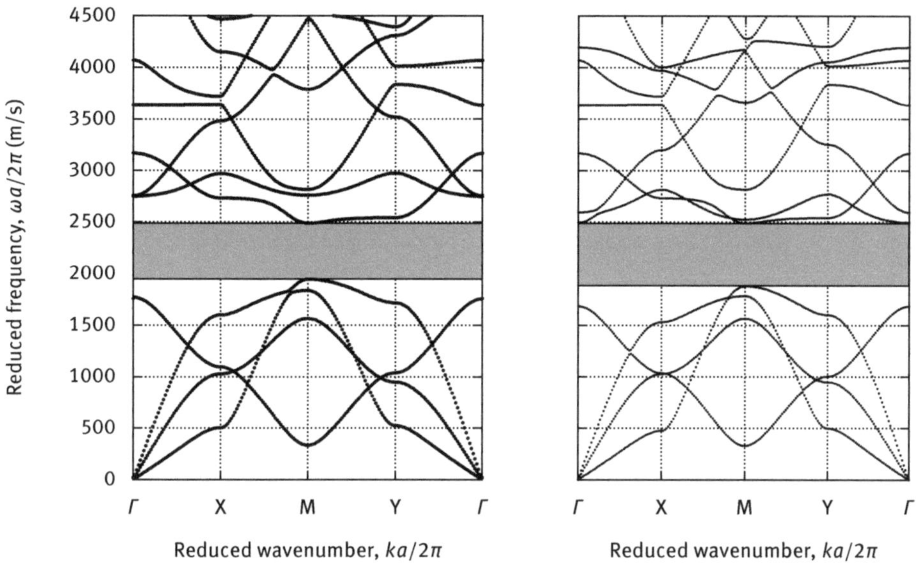

Fig. 6.18: Band structure for a 2D square-lattice phononic crystal of cylindrical holes in lithium niobate. The orientation is (XY), meaning that the crystallographic axes (Y, Z, X) are aligned with the reference frame (x_1, x_2, x_3). The filling fraction is $F = 0.54$ or $d/a = 0.9$. The left panel is plotted with piezoelectricity taken into account; the right panel is plotted without piezoelectricity (by setting the piezoelectric tensor to zero).

steel spheres in a polymer matrix, with a filling fraction $F = 0.184$ ($d/a = 0.445$) [135]. Figure 6.19 shows the phononic band structure obtained with FEM in this case, which is similar to the LMS result obtained in the cited work. A rather wide full band gap is obtained. It can be noted that the first two bands in the ΓX and in the ΓL directions are degenerate. Given that the FCC phononic crystal is composed of isotropic solid constituents, its overall symmetry for Bloch waves is similar to a cubic (natural) crystal and there is, in the considered propagation directions, a natural separation of the three initial Bloch waves into two degenerate bands plus another separate band. The situation is similar to the two shear wave velocities being degenerate in cubic solids for the same directions.

On the experimental side, Khelif et al. considered the case of a close-packed FCC phononic crystal of steel spheres in an epoxy matrix [68]. A close-pack lattice of steel beads was first formed by hand, simply by piling successive layers of beads on a plane surface. The array of beads was subsequently infiltrated with hot liquid epoxy which became solid at ambient temperature. With this technique, only certain Bragg planes of the FCC lattice can be considered at the start. In order to test all required directions of propagation in the crystal several samples have to be manufactured. The phononic band structure computed with FEM is shown on Figure 6.20. There is a very wide full band gap covering at least one frequency octave, or 100% relative bandwidth. The

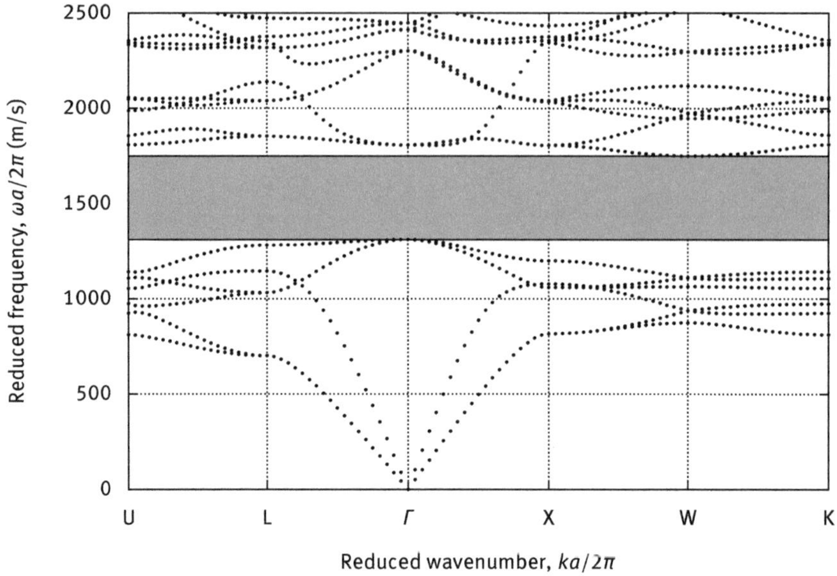

Fig. 6.19: Band structure of an FCC lattice phononic crystal of steel spheres in a polymer matrix. The filling fraction is $F = 0.184$ ($d/a = 0.445$) and material constants are those of Sainidou et al. [135]. The full band gap is indicated by the gray area.

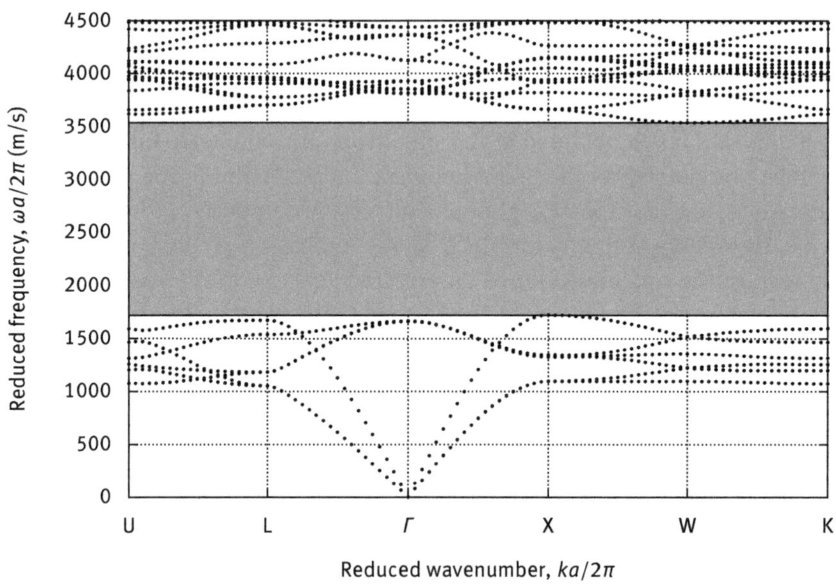

Fig. 6.20: Band structure of a close-packed FCC lattice phononic crystal of steel spheres in an epoxy matrix. The filling fraction is $F = 0.74$ ($d/a = 0.707$) and material constants are those of Khelif et al. [68]. The full band gap is indicated by the gray area.

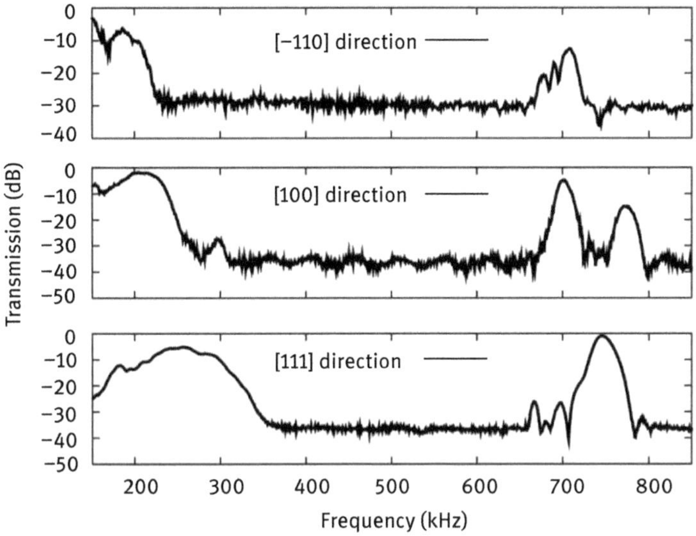

Fig. 6.21: Experimental transmission through a close-packed FCC phononic crystal of steels spheres in epoxy. Experimental transmission power spectra are measured along (a) the [$\bar{1}$10], (b) the [100] and (c) the [111] directions of a phononic crystal composed of four periods of a face-centered cubic array of steel beads in an epoxy matrix. A complete band gap extending at least from 350 to 650 kHz is observed (after Khelif et al. [68], copyright 2010 Institute of Electrical and Electronics Engineers).

existence of this full band gap is confirmed by experiment, as shown in Figure 6.21. Transmissions are performed with acoustic transducers generating and detecting longitudinal bulk waves, placed in direct mechanical contact with the phononic crystal sample.

7 Phononic crystals for surface and plate waves

In Chapter 6 we introduced phononic crystals for bulk elastic waves. In contrast to sonic crystals, phononic crystals are composed of solid materials that can be quite easily engineered. Moreover, their surfaces are naturally available for the integration of elastic wave transducers, particularly in the case that they contain piezoelectric materials. The presence of surfaces leads to a natural combination of the concepts of phononic dispersion and band gaps with elastic surface and plate waves. It has become customary to separate the two cases of phononic crystal slabs – with two plane parallel surfaces enclosing a slab of composite material with in-plane periodicity – and of phononic crystals for surface acoustic waves (SAW) – with only one plane surface terminating a semi-infinite composite substrate with in-plane periodicity.

Examples of phononic crystals for surface and plate waves are given in Figure 7.1. To the best of our knowledge, the first experiments aimed at showing the existence of SAW band gaps were published in 1999 [98]. In this first attempt, cylindrical holes with millimeter-size diameters were drilled in marble plates using rather standard mechanical equipment, as shown in Figure 7.1 (a1). Applications, however, are generally sought for ultrasonic frequencies from 100 MHz upwards. At the micron scale, phononic crystals meet the micro- and nanotechnologies that are used for microelectromechanical systems (MEMS), microelectronics, or photonics applications. The first SAW phononic crystal was realized by etching holes in a silicon wafer as shown in Figure 7.1 (a2) [168]. As can be seen on the schematic drawing, SAW transducers were integrated in close proximity to the phononic crystal itself, resulting in an integrated system that can be easily probed electrically. These transducers were patterned on a thin piezoelectric layer of zinc oxide (ZnO) deposited on the silicon wafer. Subsequently, phononic crystals were directly fabricated by etching holes in a piezoelectric wafer of lithium niobate ($LiNbO_3$), thus avoiding the use of an additional layer for transduction, as shown in Figure 7.1 (a3) [11].

Achieving a phononic crystal slab device implies that some technology is available to release a membrane inside which the crystal can be realized. Such technologies have been continuously developed for MEMS and photonic crystal devices during past decades. Figure 7.1 (b1) shows a phononic crystal slab device realized in a 20 μm-thick silicon membrane, with a honeycomb lattice of holes [106]. Lamb waves transducers have again been integrated thanks to the deposition of a piezoelectric thin layer of aluminum nitride (AlN). An array of holes can also be achieved directly in a piezoelectric membrane, as shown in Figure 7.1 (b2) for the case of AlN [71]. Holes traversing the membrane are not the only solution to defining efficient inclusions in the unit cell of the phononic crystal. They can be replaced by solid inclusions, such as in the case shown in Figure 7.1 (b3) [113]. These different solutions and their expected merits will be discussed in this chapter.

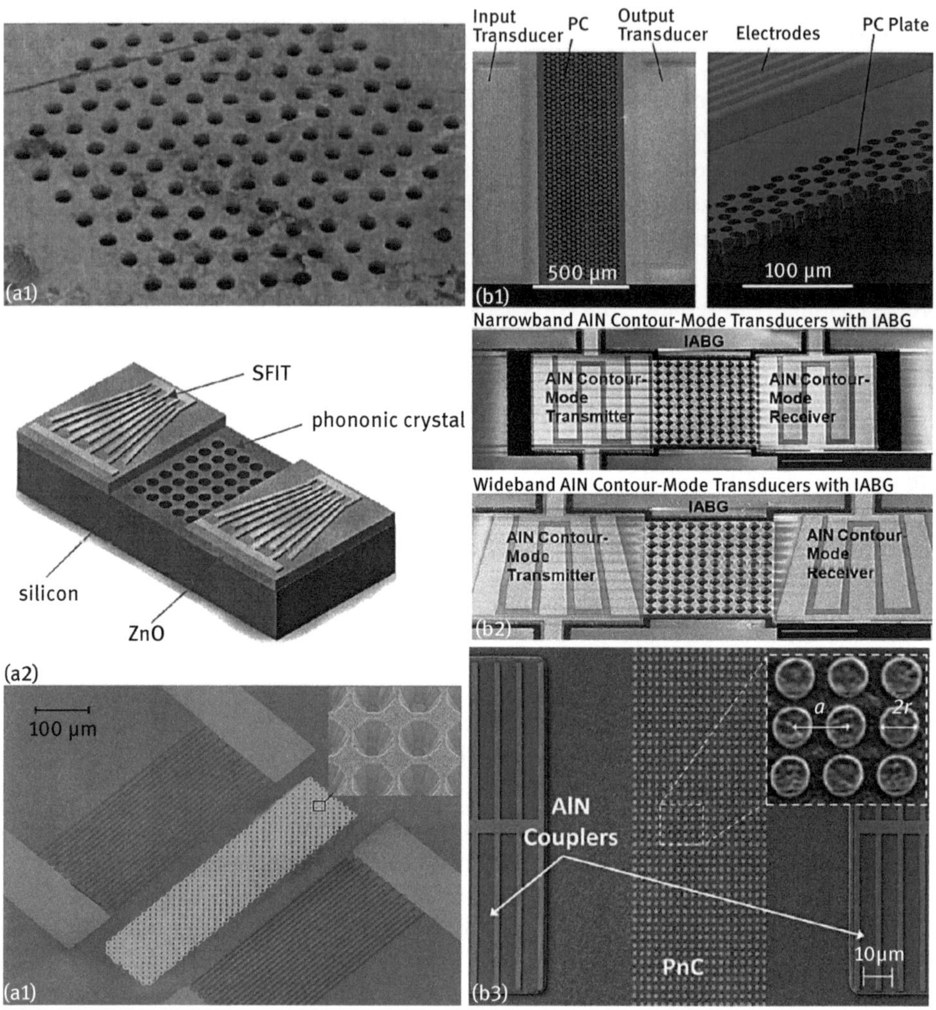

Fig. 7.1: Examples of phononic crystals for surface and plate waves. (a1) A marble stone is drilled with a hexagonal lattice of holes [98] (copyright 1999 American Physical Society). (a2) A silicon wafer with a square lattice of holes is obtained by deep reactive ion etching. Fan-shape SAW transducers (SFIT) are placed on either side of the phononic crystal on top of a thin layer of ZnO and serve as emitters and receivers for SAW [168] (copyright 2005 American Institute of Physics). (a3) A lithium niobate wafer with a square lattice of holes is obtained by reactive ion etching. SAW transducers are patterned directly at the surface [11] (copyright 2006 American Physical Society). (b1) A silicon membrane with a honeycomb lattice of holes and a pair of Lamb wave transducers that are formed on top of a thin layer of AlN [106] (copyright 2008 American Institute of Physics). (b2) A thin membrane of AlN is perforated with a square lattice of holes [71] (copyright 2009 American Institute of Physics). (b3) A solid-solid square-lattice phononic crystal slab is composed of tungsten inclusions in a silica membrane [113].

Phononic crystal slabs historically appeared shortly after phononic crystals for surface elastic waves. From the point of view of obtaining a band structure, the slab case at first appeared a bit more difficult than the SAW case. As we will see in the following, difficulties arise for the slab case mostly with the plane wave expansion method (PWE), but are rather easily circumvented using the finite element method (FEM). In the SAW case, however, obtaining a true band structure for a semi-infinite substrate remains a difficult problem, and there is no clear advantage for either one of the two methods. For these reasons, we will start our analysis with the phononic crystal slab case first and only continue with the SAW case afterward, thus not following historical order. In both cases, we will first discuss the Bloch waves of the phononic crystal and how they are influenced by the presence of the surfaces, before moving on to discuss actual phononic crystals that have been presented in the literature.

7.1 Bloch waves of phononic crystal slabs

In Chapter 6 we presented detailed numerical techniques that can be used to obtain Bloch waves and the band structure of phononic crystals. This presentation did not take into account the possible existence of free surfaces in the unit cell. Here we specify the case of the phononic crystal slab, which is a 2D phononic crystal (i.e. with two periodicities) limited by two plane and parallel surfaces, as depicted in Figure 7.2. The five Bravais lattices for phononic crystal slabs are those of 2D crystals as defined in Chapter 2, i.e. square, hexagonal, rectangular, centered rectangular (rhombic), and oblique. We will also consider the honeycomb lattice, a hexagonal lattice with two inclusions per unit cell. The two plane and parallel surfaces are conventionally taken orthogonal to axis x_3 (or z) and define a membrane with thickness h. This membrane can contain hollow or filled inclusions with diameter d. In this case, there are main geometrical ratios to consider, d/a as in the case of phononic crystals for bulk elastic

(a) (b) (c)

Fig. 7.2: Some possible phononic crystal slab unit cells, shown here for a square lattice. Possible inclusions are (a) holes, (b) filled solid holes or (c) pillars.

waves, and h/a, the thickness to lattice constant ratio. In case the inclusions are pillars of height h_p sitting on the membrane, then the additional geometrical ratio h_p/a must be considered as well, with h_p the height of the pillars. Numerical techniques for phononic crystal slabs differ mainly by the way the surface boundary conditions are taken into account.

7.1.1 Analysis with FEM

The boundary conditions applicable to phononic crystal slabs are of two types. First, periodic boundary conditions must be applied on the lateral sides (parallel to x_3), following formulas (6.19) or (6.20). Thus considering that the mesh is based on the primitive cell of the 2D lattice, there are exactly two pairs of periodical boundary conditions. For the particular case of the hexagonal lattice, choosing the mesh to conform to the Wigner–Seitz cell results in three pairs of periodical boundary conditions. Overall, however, these boundary conditions are basically similar to the 2D phononic crystal case, except they apply on surfaces instead of along edges. Second, free boundary conditions are generally considered for all remaining external faces, implying that the slab is assumed to be in a vacuum and that acoustic radiation to air is neglected. Explicitly taking into account radiation losses would require one to solve the coupled elastic-acoustic problem (see Chapter 8); we will not attempt to solve this difficult problem here.

Next, the elastic variational formulation introduced for phononic crystals for bulk elastic waves applies here unchanged (see Section 6.1.4). In particular, the band structure is obtained by solving either (6.38) (for elastic media) or (6.39) (for piezoelectric media). Examples of 3D FEM meshes suitable for band structure computation are shown in Figure 7.3. They are used for the examples presented in this chapter. One important point pertains to the mesh of the faces on which periodic boundary conditions are applied. Each pair of such faces should have exactly the same mesh so that the vertices of the meshes can be associated in pairs. Hence the periodic condition basically amounts to replacing degrees of freedom from the destination face mesh with degrees of freedom from the origin face mesh.

Holey silicon phononic crystal slab. The silicon phononic crystal slab with holes is attractive because it is compatible with microfabrication techniques that have been developed for both MEMS and photonic crystal slab devices. It has thus attracted a lot of attention from a number of research groups. Let us discuss its phononic band structure for a rather large filling fraction, dictated by the ratio $d/a = 0.9$, and for the "magic" thickness-to-lattice constant ratio $h/a = 0.6$. The reason for this choice comes from gap maps as a function of d/a and h/a that have been published in the literature [105, 107, 118]. It appears from these works that the range of geometrical parameters for which complete band gaps can be found is quite restricted.

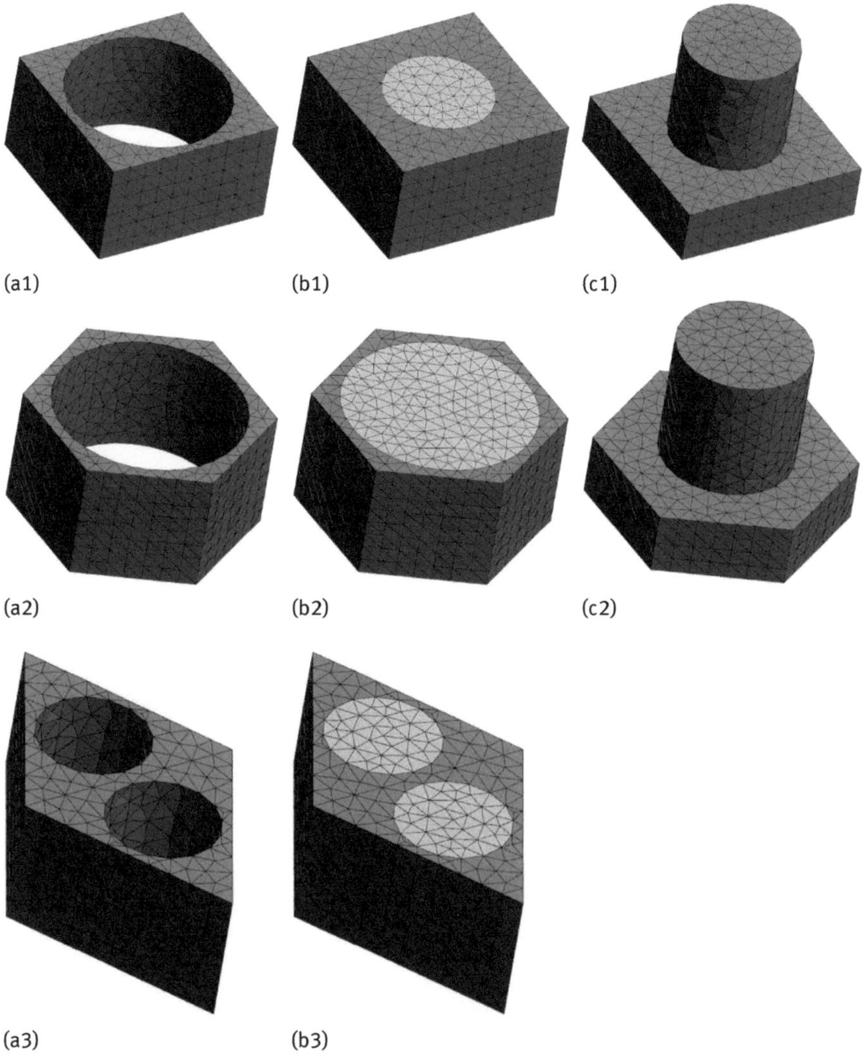

Fig. 7.3: Examples of FEM meshes used to obtain the band structure of phononic crystal slabs. Meshes are shown in rows for (1) the square lattice, (2) the hexagonal lattice and (3) the honeycomb lattice. Columns are for (a) a hollow inclusion, (b) a solid inclusion and (c) a pillar sitting on a membrane.

The case of the square lattice is shown in Figure 7.4. The band structure suggests a rather small full band gap appearing around $\omega a/2\pi = 3000\,\text{m/s}$. Assuming a lattice constant $a = 1\,\mu\text{m}$, which is perfectly feasible technologically, the center frequency would then be 3 GHz. Compared to Figure 6.14, we can see that the full band gap is less wide for the holey silicon phononic crystal slab compared to the 2D holey silicon phononic crystal for bulk waves. Besides, the center frequency is similar. This reduc-

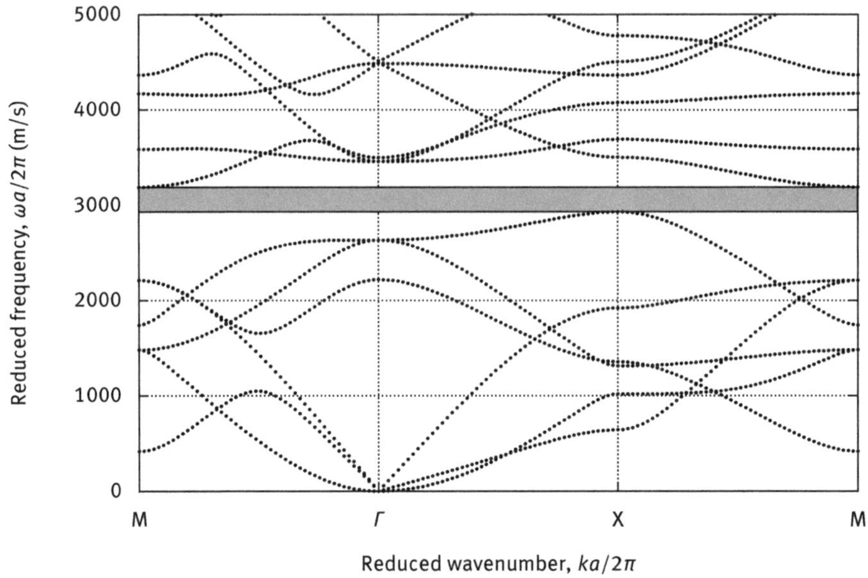

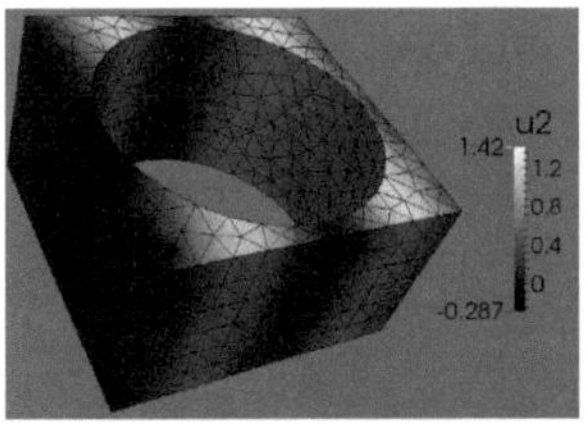

(1)

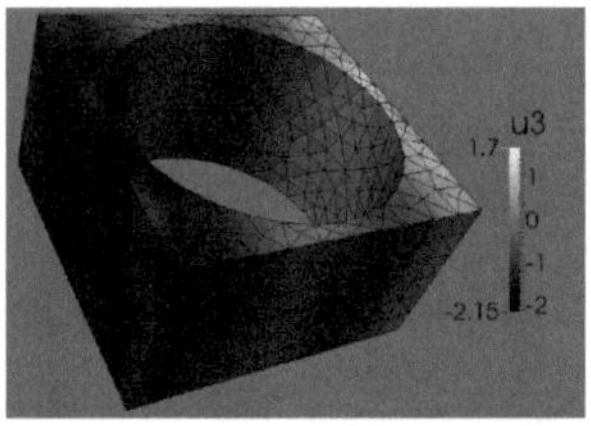

(2)

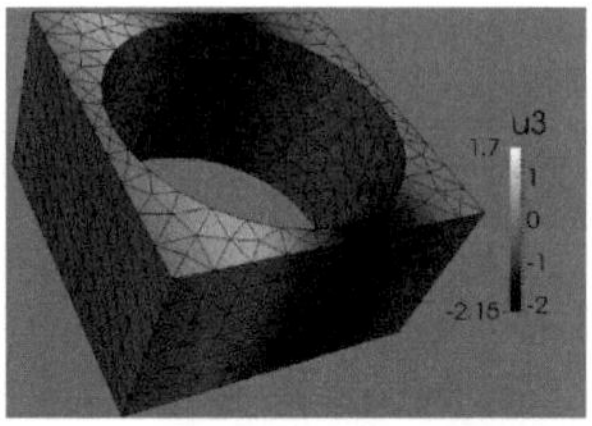

(3)

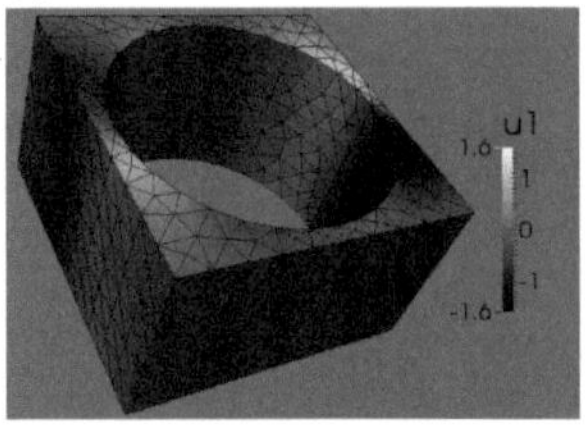

(4)

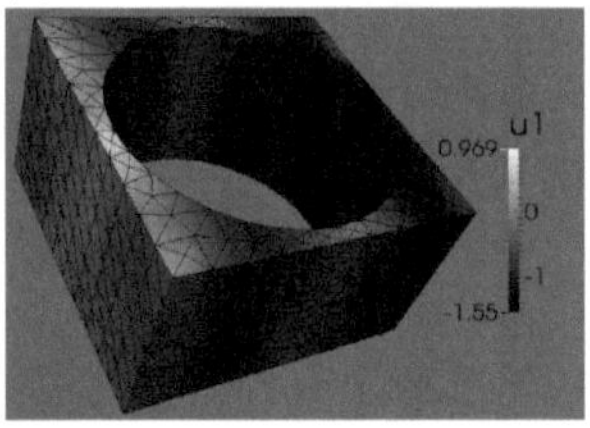

(5)

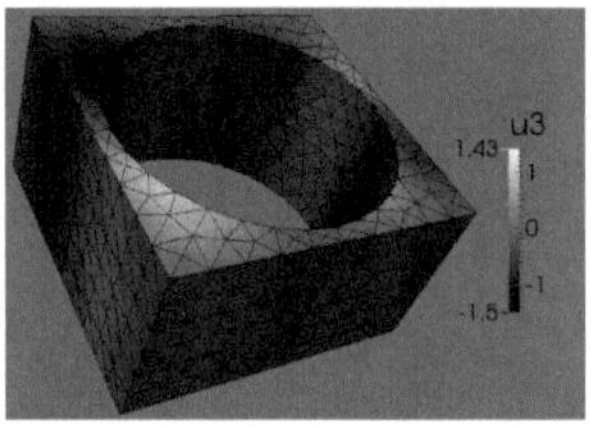

(6)

Fig. 7.4: Band structure for the square-lattice phononic crystal slab composed of cylindrical holes in silicon. The geometrical parameters are $h/a = 0.6$ and $d/a = 0.9$. The gray area in the band structure indicates the full phononic band gap. The first six Bloch waves at the X point of the first Brillouin zone are shown. Only the dominant displacement is shown in each case, but all Bloch waves are mixed modes.

tion in gap width can be understood by observing that there are more bands in the slab case: even in the absence of periodicity, Figure 5.14 shows that there many Lamb waves and pure shear plate waves where there are only three bulk waves in silicon. Let us consider propagation in direction x_1, for definiteness. As in the case of Lamb waves, the top and bottom surfaces couple together the displacements (u_1, u_3) in the sagittal plane. Because of the in-plane 2D periodicity, moreover, the displacements (u_1, u_2) are necessarily coupled. As a result, all three displacements are always coupled and there are no pure waves. This reasoning obviously generalizes to any direction of propagation. The first six Bloch waves at the X point are shown in Figure 7.4. Though we show only the dominant displacement distribution in each case, the other two displacements do not vanish.

We have not attempted to separate Bloch waves by their symmetries in Figure 7.4. There are indeed three planes of symmetry with the unit cell for the square lattice with a hollow circular inclusion. In the ΓX direction of the first Brillouin zone, two of them encompass the direction of the wavevector. Symmetry with respect to the vertical plane (x_1, x_3) could possibly separate Bloch modes between deaf and nondeaf. Yet the deafness property would still be dependent on the polarization contents of the incident plane wave on the entrance of the phononic crystal, which can only be evaluated given all the experimental conditions (i.e. it is not an intrinsic property of a particular Bloch wave). Symmetry with respect to the horizontal midplane (x_1, x_2) would allow us to separate between symmetric and antisymmetric Bloch waves, similarly to the case of plate waves discussed in Section 5.6. Again, this separation can only be useful if the incident wave can be prepared with a definite symmetry with respect to the same plane. However, an interdigital transducer deposited on the top surface of the slab, such as those shown in Figure 7.1, in principle excites Lamb and pure shear waves of both symmetries.

The cases of the silicon phononic crystal slab with cylindrical holes in the hexagonal and the honeycomb lattices are shown in Figures 7.5 and 7.6. As was already noted with the 2D holey silicon phononic crystal in the case of bulk elastic waves, there is no complete band gap with the hexagonal lattice. The symmetry-breaking mechanism introduced by the honeycomb lattice (which is simply a hexagonal lattice with two inclusions rather than only one, as a reminder) acts efficiently to open a full phononic band gap. This is arguably the best choice with cylindrical holey inclusions.

Heavy inclusions in soft membrane. The consideration of phononic crystal slabs was initially spurred by a study of quartz inclusions in an epoxy matrix [63]. From a later perspective, holes in a membrane may seem an easier alternative. It was however known that heavy and stiff inclusions inside a light and soft matrix led to large phononic band gaps, at least in 2D, as we have illustrated in Chapter 6. It was then natural to extend the idea of this system to phononic crystal slabs. Figure 7.7 reproduces the gap map giving the domain of existence of full band gaps as a function of h/a, with a filling fraction fixed to 0.5 ($d/a = 0.8$). It can be seen that this existence is strongly

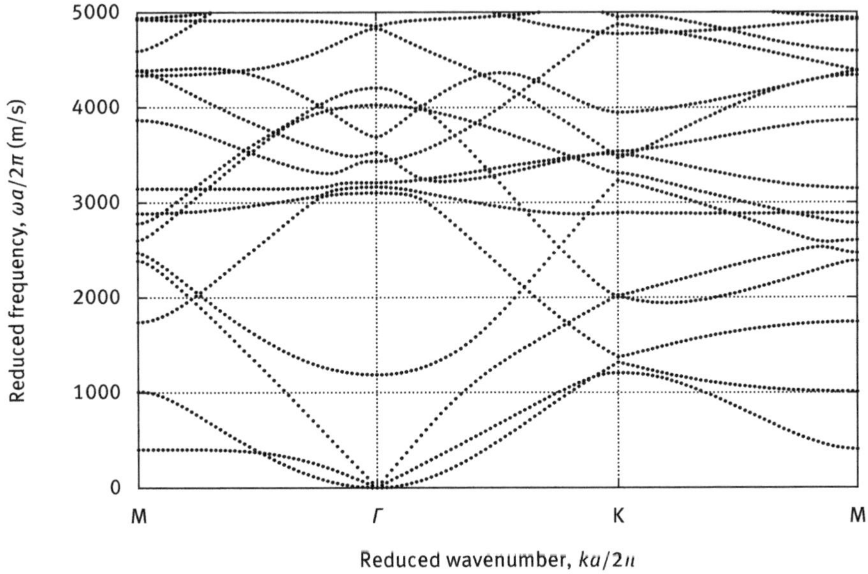

Fig. 7.5: Band structure for the hexagonal-lattice phononic crystal slab composed of cylindrical holes in silicon. The geometrical parameters are $h/a = 0.6$ and $d/a = 0.9$.

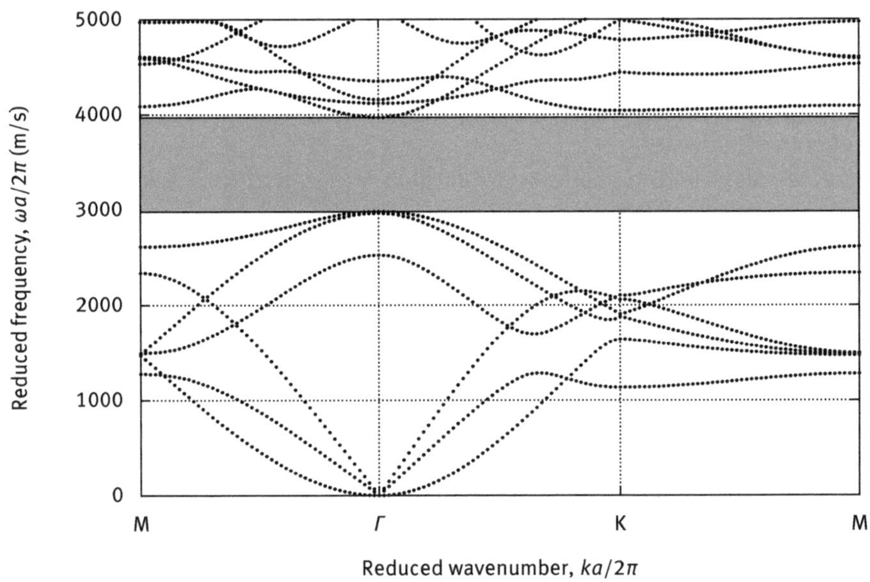

Fig. 7.6: Band structure for the honeycomb-lattice phononic crystal slab composed of cylindrical holes in silicon. The geometrical parameters are $h/a = 0.6$ and $d/a = 0.5$.

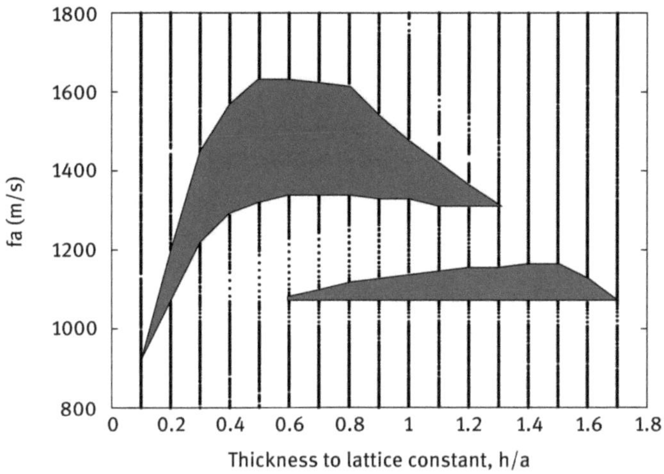

Fig. 7.7: Gap map for the square-lattice phononic crystal slab composed of quartz inclusions in an epoxy matrix. The diagram, reproduced after [63], shows the domain of existence of full band gap frequencies as a function of the thickness to lattice pitch ratio, h/a, with a filling fraction $F = 0.5$. Note that $fa = \omega a/(2\pi)$.

dependent on the thickness of the slab. If h/a is too small, then there are many foldings of the first flexural bands that populate the band structure in the low frequency range. The opening of Bragg band gaps is then made more difficult. If h/a tends to ∞, it might naively be thought that the slab case will tend asymptotically to the 2D infinite case (2D phononic crystal for bulk elastic waves). This is however not the case, as the presence of the top and bottom surfaces induces the appearance of more and more plate waves as h increases. As a result, thick phononic crystal slabs do not possess full Bragg band gaps in general. By continuity, it results that there must be some range for h/a for which full band gaps are maximized. Figure 7.7 indeed indicates that this optimal range is around $h/a = 0.5$ to 0.8 for the particular case considered.

The case of quartz inclusions in an epoxy matrix has not been considered in experiments. Instead, the related case of tungsten (W) inclusions in a silica (SiO_2) matrix was experimented rather extensively [128, 143]. The experiments will be discussed more thoroughly in Section 7.2, but we analyze here the opening of full band gaps in this system. The value of $h/a = 1$ is used in Figures 7.8, 7.9, and 7.10, where the cases of the square lattice, the hexagonal lattice, and the honeycomb lattice are considered in sequence. Tungsten is significantly heavier than silica. Shear and longitudinal velocities, however, are of the same order of magnitude for both materials.

In the square lattice case, in Figure 7.8, a full band gap is seen to appear in a frequency range about twice as low as the full band gap for the holey silicon phononic crystal of Figure 7.4. The difference in center frequency can be attributed to the fact that shear and longitudinal waves are significantly faster in silicon compared to both silica and tungsten.

In the hexagonal lattice case, in Figure 7.9, the full band gap appears in a frequency range similar to the case of the square lattice, but it is a bit wider. We again make an observation that was apparent with 2D phononic crystals for bulk elastic waves: in the case of solid inclusions, the hexagonal lattice is adequate for obtaining full band gaps, while none are found in the case of hollow inclusions. In the honeycomb lattice case, in Figure 7.10, the full band gap is rather small and appears in a frequency range slightly higher than for the square and the hexagonal lattice. In contrast to holey phononic crystal slabs, the honeycomb lattice does not necessarily represent an optimal choice with solid inclusions.

A comparison of solid-hole and solid-solid phononic crystal slabs was conducted by Reinke et al. [128]. These authors considered two choices for the solid matrix, either silicon or silica. For the solid inclusion, they considered tungsten. Their results can be summarized with the diagrams in Figure 7.11, which are shown here only for silicon. They show the optimal choices for the thickness-to-lattice constant ratio, h/a, and for the radius-to-lattice constant ratio, r/a ($r = d/2$). From the diagrams, it is clear that the full band gaps of phononic crystal slabs are always less wide than those of their 2D counterpart, confirming the observation of Khelif et al. [63]. The widest band gaps

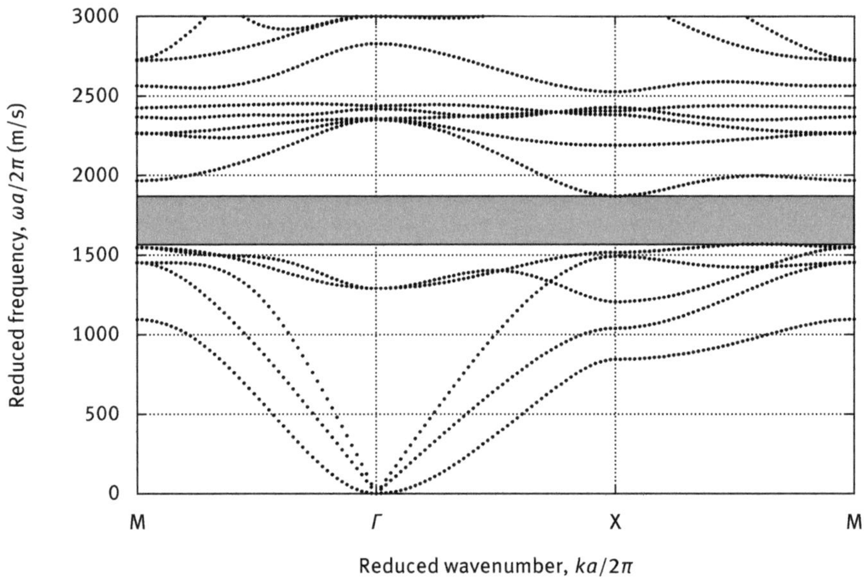

Fig. 7.8: Band structure for the square-lattice phononic crystal slab composed of cylindrical tungsten inclusions in silica. The geometrical parameters are $h/a = 1$ and $d/a = 0.48$. The gray area in the band structure indicates the full phononic band gap. The first six Bloch waves at the X point of the first Brillouin zone are shown. Only the dominant displacement is shown in each case, but all Bloch waves are mixed modes.

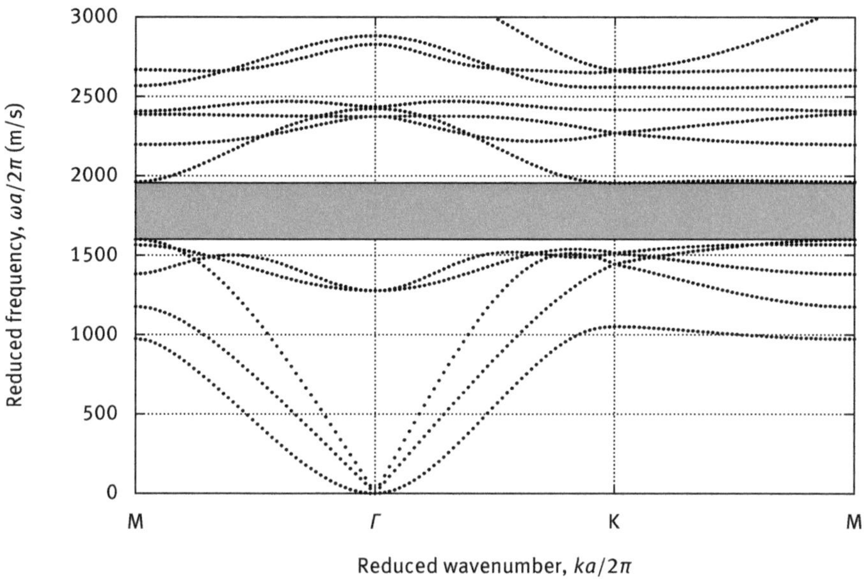

Fig. 7.9: Band structure for the hexagonal-lattice phononic crystal slab composed of cylindrical tungsten inclusions in silica. The geometrical parameters are $h/a = 1$ and $d/a = 0.48$. The gray area in the band structure indicates the full phononic band gap.

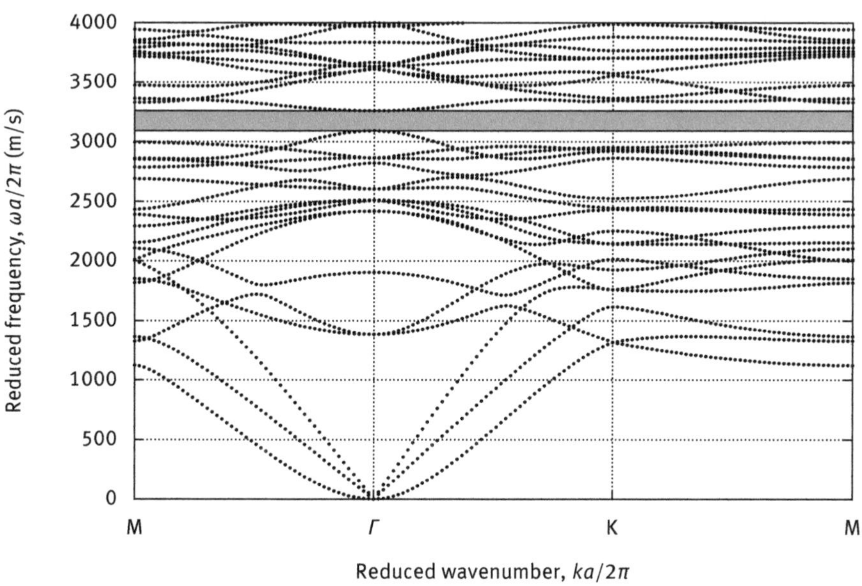

Fig. 7.10: Band structure for the honeycomb-lattice phononic crystal slab composed of cylindrical tungsten inclusions in silica. The geometrical parameters are $h/a = 1$ and $d/a = 0.3$. The gray area in the band structure indicates the full phononic band gap.

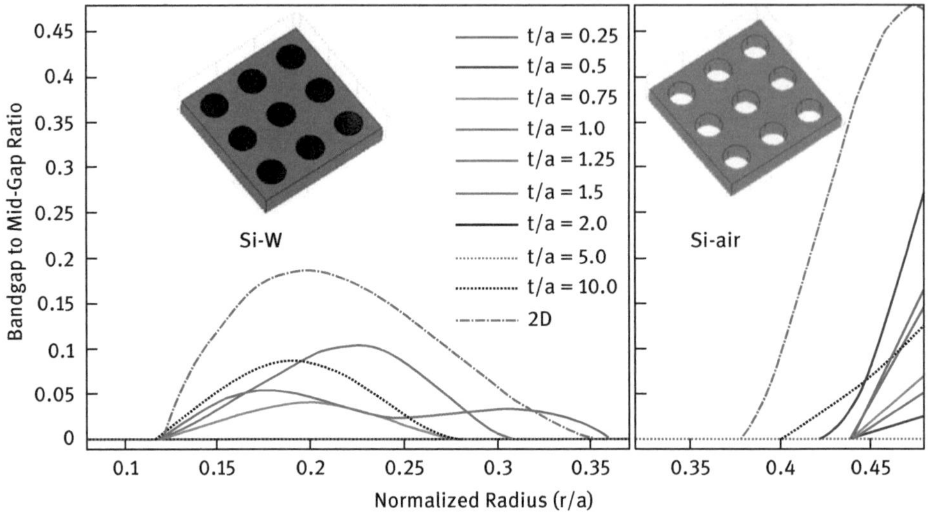

Fig. 7.11: Calculated full phononic band gap versus inclusion radius for a range of slab thicknesses for a phononic crystal formed in a silicon matrix with a square-lattice array of cylindrical air holes (right panel) and tungsten rods (left). The full 2D band gaps are shown by the dotted curve, where the thickness is assumed to be infinite (from Reinke et al. [128], copyright 2011 American Institute of Physics).

that can be obtained are similar for the solid-hole and the solid-solid cases, the main difference is the range of r/a values for which the optimum is attained. The optimal range for r/a is around 0.25 for the solid-solid composition, but approaches 0.5 for the solid-hole composition. As a reminder, for $r/a = 0.5$, the holes are touching. Practically, this extremal value can hardly be met, because the structure will be very fragile or will even collapse. A maximum value of 0.48 would be advisable. For instance, if the lattice constant is $a = 1\,\mu m$, this means that the width of the separations between holes would only be 40 nm. It should be pondered, however, that the technology to obtain filled solid inclusions can be much more complicated than the simple etching of holes in a thin membrane.

7.1.2 Analysis with PWE

The plane wave expansion (PWE) technique is in principle not very well suited to the case of the phononic crystal slab. The PWE is indeed based on Fourier series representations of both the material constants and of the physical wave fields (displacements and stresses). As such, it implicitly assumes that a unit cell is defined and that the final solution is derived from the solution inside the unit cell by periodicity and application of the Bloch theorem. In contrast, the phononic crystal slab has a 3D unit cell, but with two periodic directions (x_1 and x_2) and a spatial closure imposed in the third direction by appropriate boundary conditions, for instance free surfaces. In order to circumvent the problem, a supercell numerical technique can be used. The supercell technique will be described in more depth in Chapter 9, in the context of complex band structures, and in Chapter 11, when we discuss the band structure of phononic waveguides and cavities. We introduce it here in simple terms for our purposes.

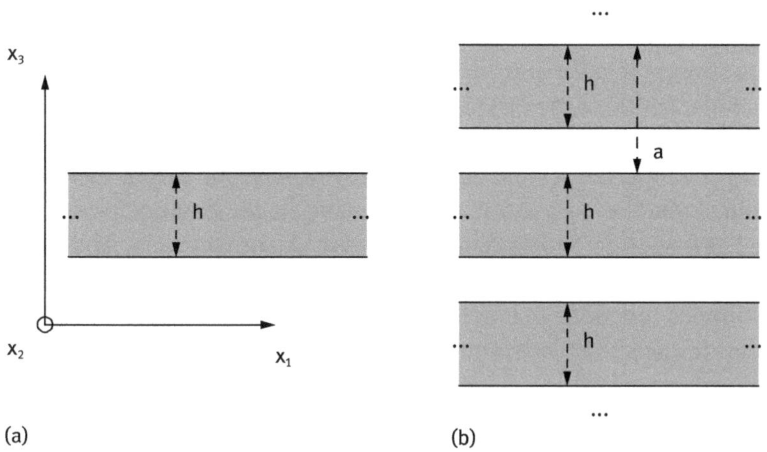

Fig. 7.12: Schematics of the PWE implementation of plate problems using a supercell technique. (a) A solid plate with finite thickness h is supposed to be surrounded by a vacuum. (b) The same plate is artificially repeated periodically in the vertical direction, x_3. The structure is assumed to be periodic in the x_3 direction (with period a) and in the x_1 and x_2 directions. Vacuum thus alternates with the solid material.

The supercell technique for plate problems is illustrated by Figure 7.12. We consider some solid plate with finite thickness h bounded by two plane and parallel surfaces. The surfaces are assumed to be free of mechanical traction. The inside of the plate can be either homogeneous, and we would be led back to the discussion of plate waves in Chapter 5, or 1D or 2D periodic, to account for the phononic crystal slab structure. In the real problem, there are no elastic waves in the vacuum around the plate, and no elastic or kinetic energy there either. Then it is logically safe to consider that the plate can be infinitely repeated periodically in space along the third dimension. The clones of the original plate will not "feel" its elastic waves and reciprocally the original plate will not be influenced by its clones. Since we are able to replace the original problem with an equivalent periodic problem, we are in a position to employ the PWE method with a 3D periodic unit cell and lattice constant a_3 being the sum of h and the additional vacuum thickness separating the plate clones. The mathematical analysis presented in Section 6.1.2 can be applied unchanged. The key question, however, is whether the clones are truly isolated from one another when the Fourier series are truncated in the practical application of the PWE method. To answer this question, we must first examine how a vacuum can be represented by an equivalent elastic solid. The equations of motion in elastic media are

$$T_{ij} = c_{ijkl} \frac{\partial u_k}{\partial x_l},$$
$$\rho \frac{\partial^2 u_j}{\partial t^2} = \frac{\partial T_{ij}}{\partial x_i}.$$

Since there are no stresses in a vacuum, we must set $c_{ijkl} = 0$ in the first equation to impose $T_{ij} = 0$ independently of the displacements. Of course, displacements have a physical meaning in a vacuum only at the interface with a solid. As a consequence, we also have to set $\rho = 0$ in the second equation or otherwise the displacements at the interface would not be free. As a result, the equivalent elastic medium for vacuum should be defined as having zero mass density and zero elastic constants. Following this line, the Fourier series coefficients for the material constants should be able to represent discontinuous functions that are piecewise constant, having constant and nonzero values inside each solid subdomain (i.e. in the matrix and in the solid inclusions) and zero values elsewhere. As we discussed already in Chapter 4, Fourier series define continuous and infinitely differentiable functions that can only approximate step functions. It is then impossible for the PWE method to represent material constants that are exactly zero in the equivalent vacuum domain; they will instead fluctuate continuously around the zero mean value, assuming both positive and negative values. Does this mean that the PWE method will fail with this equivalent elastic medium for vacuum? Actually, it is not the functions $\rho(\mathbf{x})$ and $c_{ijkl}(\mathbf{x})$ of (6.21) and (6.22) that enter the PWE equations (6.34), but their Fourier series coefficients ρ_m and $(c_{ijkl})_m$, so the actual values of the functions are not significant in this discussion. It is found in practice that no numerical instabilities occur, as was for instance demonstrated for SAW on phononic crystals of holes [77], 2D phononic crystals of holes [1], holey phononic crystal slabs [52], or anisotropic and piezoelectric plates [79]. Nevertheless, it is sometimes stated in the literature that the PWE method would be numerically unstable in these cases [95, 151, 156]. As an alternative, these papers propose to replace vacuum with some artificial isotropic elastic medium with a very low mass density (very dilute) but elastic constants chosen such that the shear and longitudinal speeds are very high, precluding the appearance of spurious modes in the lower frequency part of the band structure. For instance, the values $\rho = 10^{-4}$ kg/m^{3}and $c_{11} = c_{44} = 10^6$ N/m^2 are used in reference [156], implying that the shear and longitudinal speeds are 10^5 m/s, significantly larger than in any known solid or fluid material. As a note, the choice $c_{11} = c_{44}$ makes $c_{12} = -c_{11}$, but this does not seem to cause any numerical problem. In practice, it can also be remarked that the Fourier series coefficients of the material constants of the phononic crystal are simply an additive combination of the material constants of the individual constituents, so that the Fourier series coefficients of the matrix dominate over those of the artificial elastic medium. For instance, using the notations from Section 4.1.2, if the unit cell contains two constituents A (e.g. the matrix) and B (e.g. the inclusion), the mass density is

$$\rho(\mathbf{x}) = \rho_A(1 - \alpha(\mathbf{x})) + \rho_B\alpha(\mathbf{x}), \tag{7.1}$$

with $\alpha(\mathbf{x})$ a step function equal to 1 inside B and equal to 0 inside A. The Fourier series coefficients are simply

$$\rho_m = \rho_A(\delta_m - \alpha_m) + \rho_B\alpha_m = \rho_A\delta_m + (\rho_B - \rho_A)\alpha_m. \tag{7.2}$$

Then since $\rho_B \ll \rho_A$,

$$\rho_m \approx \rho_A(\delta_m - \alpha_m). \tag{7.3}$$

Similar relations hold for the elastic constants.

Finally, we remark that air can be considered instead of a vacuum. If the material constants for air are used, then the band structure will appear to be polluted by many flat bands, making it quite difficult to read. These flat bands are however perfectly physical, and reflect the sound modes trapped in air between the solid parts, which, as we remarked in Chapter 4 for sonic crystals, appear as almost perfectly rigid walls to sound waves in air. This will be specifically discussed in Chapter 8.

7.2 Experiments with phononic crystal slabs

7.2.1 Holey phononic crystal slab

Band gaps in silicon phononic crystal slabs were explored by Mohammadi et al. [104, 106, 107]. These authors used silicon-over-insulator (SOI) technology to prepare samples like those shown in Figure 7.1 (b1). The slab thickness is $h = 15$ µm, the $d = 12.8$ µm holes are arranged in a honeycomb lattice with lattice constant $a = 26$ µm. For the measurements shown in Figure 7.13, a total of eight layers of holes was considered and the measurement was taken in the ΓK direction only. Fortunately, this direction alone is sufficient to qualify a full band gap for hexagonal-based lattices, as we discussed in Chapter 4. The transduction of plate waves in an area of the membrane placed outside the phononic crystal was obtained via the deposition of a ZnO thin layer and of the patterning of a interdigital transducer (IDT) on it. A similar IDT is placed on the other side of the phononic crystal in order to receive transmitted plate waves. The transmission as a function of frequency shows a very clear dip in the transmission that is coincident with the computed band structure. Two advantages of silicon are its availability as large-scale wafers suitable for collective micro- and nanofabrication technologies, but also the existence of qualified processes to obtain holes with a large aspect ratio by etching through a mask. Combined with the fact that phonon losses are rather low in silicon at RF frequencies (see Chapter 5), it is not surprising that silicon phononic crystal slabs are the dominant solution for making MEMS phononic resonators or combining with photonic crystal slabs. The thickness $h = 15\,\mu m$in Figure 7.13 can be made much smaller in order to implement phononic band gaps in the GHz frequency range. For operation around 5 GHz, the thickness would be $h \approx 0.5\,\mu m$ and the lattice constant would be $a \approx 1\,\mu m$.

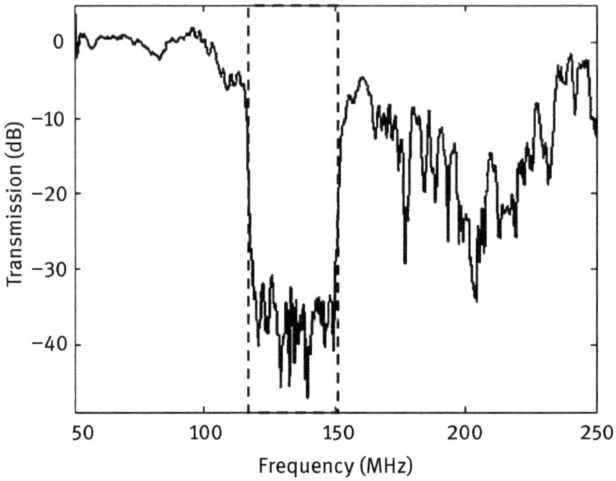

Fig. 7.13: Transmission measured electrically through a honeycomb-lattice phononic crystal of holes in a silicon slab. The sample is similar to the one shown in Figure 7.1 (b1). Geometrical parameters are $a = 26\,\mu m$, $d = 12.8\,\mu m$, and $h = 15\,\mu m$. Transmission is obtained with a pair of interdigital transducers and normalized to the value without the phononic crystal (reproduced from Mohammadi et al. [106], copyright 2008 American Institute of Physics).

The drawback of silicon is that a piezoelectric layer must be added for the electrical transduction of plate waves. An alternative solution is the thermoelastic excitation of plate waves by a focused laser beam, which is known as laser ultrasonics. This technique is not compatible with real-world applications of phononic crystal devices, but still offers the opportunity to explore their properties. At a larger scale, silicon wafers with phononic crystals of holes have been studied by a number of research groups [15, 177]. We shall briefly describe the example reported by Brunet et al. [19] in order to give orders of magnitude. These authors prepared phononic crystal slabs of holes in silicon wafers according to either the square or the rectangular-centered lattice. Dimensions were $a = 1$ mm, $h/a = 0.2$, and $h = 200\,\mu m$. For such dimensions, complete band gaps cannot be obtained; directional band gaps can, however, still be observed. With laser excitation they could excite mostly the lowest antisymmetrical Lamb wave A0. Dispersion measurements along one direction clearly reveal the sequence of Bragg band gaps as the frequency increases.

Phononic crystal slab devices fabricated directly in piezoelectric membranes have been tested successfully as well. Kuo et al. [71] prepared square-lattice samples in aluminum nitride (AlN) as shown in Figure 7.1 (b2). Their design of the unit cell involves plain cylinders connected by tethers. Alternatively, this unit cell can be viewed as a rounded square hole rotated by 45°. The geometrical dimensions were $a = 8.6\,\mu m$, $h = 2\,\mu m$ (AlN thickness), $d = 6.6\,\mu m$, and $w = 1\,\mu m$ (w is the width of the tether connecting plain cylinders). Platinum (Pt) coatings with a thickness of 200 nm were deposited on both sides of the AlN membrane. Transmission measurements indicate the possibility of a phononic band gap in the ΓX direction in the frequency range [185–240] MHz. Contour mode transducers were designed on the membrane for excitation and detection of plate waves.

Gorisse et al. [44] further fabricated phononic crystal slab devices in a composite AlN on SiO_2 membrane. Circular holes were defined by etching according to a square lattice. Geometrical dimensions were $a = 6.6\,\mu m$, $h = 2\,\mu m$ (for AlN) plus $0.5\,\mu m$ (for SiO_2), and $d = 4\,\mu m$. Figure 7.14 shows optical measurements of the vertical (or

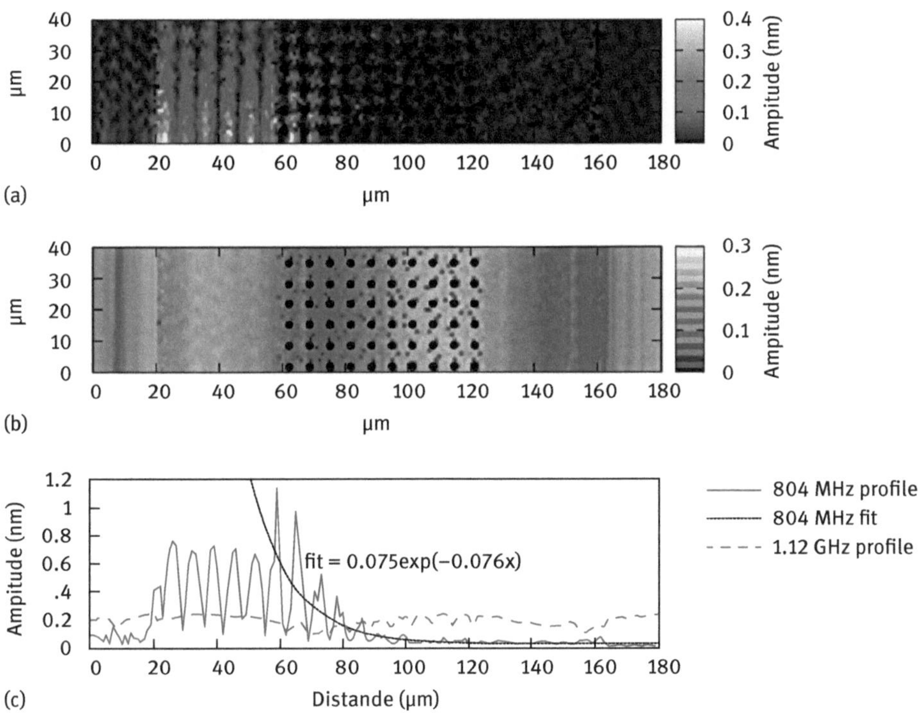

Fig. 7.14: Optical measurement of transmission through a phononic crystal slab of holes in AlN/SiO_2. Measurements are performed inside the band gap at (a) 804 MHz and above it at (b) 1120 MHz. (c) The profile of the amplitude in and above the band gap is integrated along x_2 and fitted against an exponentially decreasing law (adapted from Gorisse et al. [44], copyright 2011 American Institute of Physics).

out-of-plane) surface displacement. These measurements are performed using a scanning heterodyne optical interferometer with 1 µm lateral resolution. Two images are shown, taken at different monochromatic frequencies, first within the phononic band gap (at 804 MHz) and then above it (at 1120 MHz). Within the phononic band gap, a standing wave pattern is formed on the left-hand side, caused by the interference of incoming and reflected plate waves. Incoming waves are generated by an interdigital transducer. Above the band gap, almost unit transmission is observed, as if the array of holes were hardly there at all. Within the band gap, in contrast, waves are strongly attenuated within the phononic crystal, following an exponential decay law. This is a signature of the evanescent Bloch waves of the crystal.

7.2.2 Solid-solid phononic crystal slab

We now turn our attention to phononic crystal slabs with a solid-solid composition, i.e. the inclusions are filled with a solid material different from the solid matrix.

Hsiao et al. [54] demonstrated a full band gap in a simple phononic crystal slab sample made by hand, as depicted in Figure 7.15. The crystal is composed of a single layer of spherical steel beads with a diameter $d = 4$ mm that were first arranged according to a square lattice ($a = 4$ mm). An epoxy matrix was then created by pouring it in the liquid phase and then polishing the sides after it had solidified. Elastic plate waves were excited by a longitudinal wave transducer connected to the plate through a prism with an apex angle of 30°. This way, waves with a nonzero parallel wavenumber are launched inside the phononic crystal slab. Note that the transducer is placed at the location of the first rows of beads, so Bloch waves of the phononic crystal can be directly excited, i.e. there is no modal conversion from the elastic waves of the homogeneous epoxy plate. As the transducer produces short pulses, a wide frequency spectrum was covered in the measurements, similarly to the case of sonic crystals. Detection of the transmitted elastic plate waves was achieved using a laser vibrometer. Two different samples were prepared, intended for measurements of transmission along the ΓX and along the ΓM directions of the first Brillouin zone. A full band gap is found in a frequency range that is in agreement with the computed band structure. The band structure shown in Figure 7.15, to which the transmissions are compared, was obtained with the FEM method and a mesh of the unit cell similar to the one shown in the bottom left corner of Figure 4.22.

At a much smaller scale, Figure 7.16 shows transmission measurements obtained with the square-lattice phononic crystal slab of tungsten inclusions in a silica matrix displayed in Figure 7.1 (b3) [143]. The slab thickness is $h = 1.85$ µm, the lattice constant is $a = 2.5$ µm ($h/a = 0.74$), and the diameter of the inclusions is $d = 1.4$ µm ($d/a = 0.56$). The transmission measurement indicates the existence of a rather wide phononic band gap in the ΓX direction, extending for frequencies below 1 GHz.

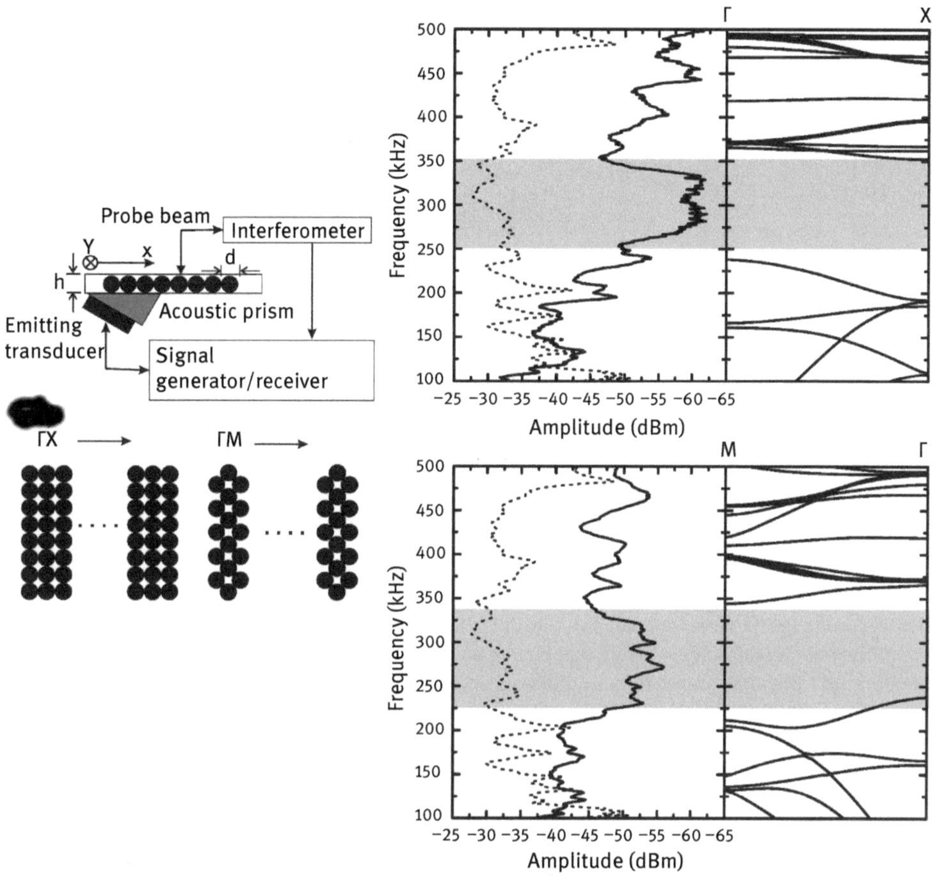

Fig. 7.15: A solid-solid phononic crystal slab of steel beads in epoxy. A sketch of the experiment of Hsiao et al. is shown in the left panel [54]. Plate waves are excited in the phononic crystal slab using an ultrasonic transducer through a prism with an incidence angle of 30°. The phononic crystal slab is formed by exactly one layer of an array of spherical steel beads with a diameter of 4 mm arranged according to a square lattice in an epoxy matrix. Measurements along the *Γ*X and *Γ*M directions of the first Brillouin zone are shown in the right panel. The transmission spectra for the phononic crystal samples (solid line) and for the pure epoxy reference slab (dashed line) are compared. In the band structures, the gray areas indicate the frequency bands where attenuation caused by a band gap is apparent (adapted from Hsiao et al. [54], copyright 2007 American Physical Society).

Before closing this section, we refer the reader to the presentation of phononic crystal slabs of pillars in Section 10.5 to complete the present discussion of phononic crystal slabs. In the case of pillars, the phenomenon of local resonance indeed becomes very important, which justifies our delayed presentation.

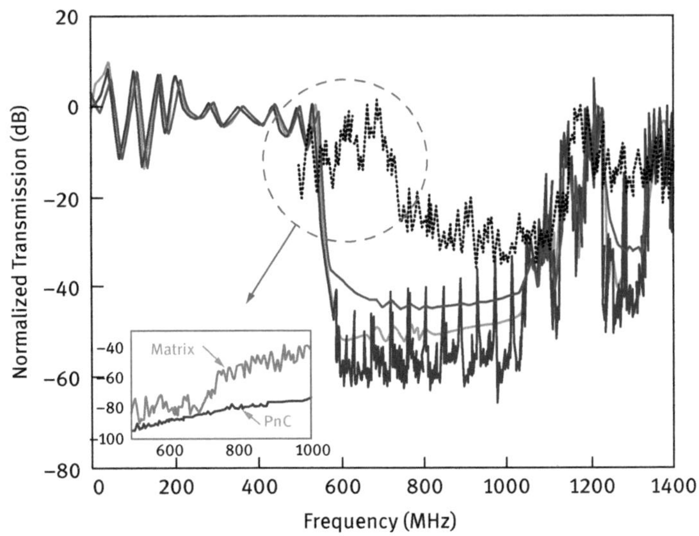

Fig. 7.16: Transmission measurements of a phononic crystal slab composed of tungsten (W) rods in a silicon dioxide (SiO_2) matrix. A representative sample was shown in Figure 7.1 (b3). Experimental results (dotted line) are compared with FDTD results (solid line) for 1.85 μm thick slabs. The different solid lines show the effect of using an increasing number of time steps in the FDTD computation. The inset shows the raw transmission data for the matrix (dotted line) and the actual phononic crystal (solid line), highlighting the untransduced frequency region. The lattice constant is $a = 2.5$ μm and the rod diameter is $d = 1.4$ μm (adapted from Su et al. [143], copyright 2010 American Institute of Physics).

7.3 Surface Bloch waves

In this section, we theoretically examine the conditions of existence of surface elastic Bloch waves. Our considerations include the cases of isotropic, anisotropic elastic, and general piezoelectric materials. We start with the apparently simple problem of a semi-infinite half-space composed of a substrate with a 2D periodic array of cylindrical inclusions that run from the top surface (defined by $x_3 = 0$) down to minus infinity, as depicted in Figure 7.17. For this particular geometry that is invariant along the vertical direction – except for the cut at the top surface – it is useful to introduce an expansion in partial waves, which we will consider in detail for both the PWE and the FEM methods. Whenever the holes have a finite depth, and possibly some shape different from the simple vertical cylinder, such an approach is somewhat less flexible. We will instead present a procedure to obtain the band structure for true surface waves only, limited to the outside of the sound cone. In this process, the possible existence of leaky surface Bloch waves inside the sound cone will, however, be obscured.

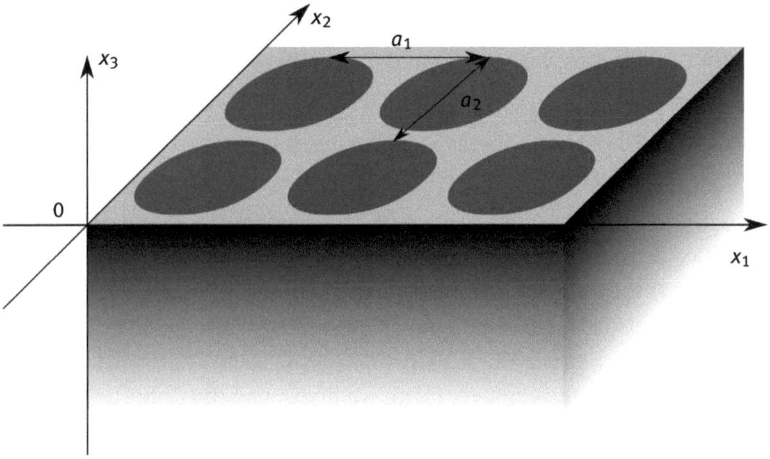

Fig. 7.17: Schematic geometry of elastic wave propagation on the surface of a 2D phononic crystal. A semi-infinite half-space is composed of a substrate with a 2D periodic array of cylindrical inclusions that run from the top surface (defined by $x_3 = 0$) down to $x_3 = -\infty$.

7.3.1 Expansion in partial waves

The 2D phononic crystal is assumed to be periodic in the directions x_1 and x_2 and invariant along direction x_3, except for the presence of a surface at $x_3 = 0$. Surface Bloch waves can then be defined for an angular frequency ω and a wavevector $\mathbf{k}_s = (k_1, k_2, 0)^T$ defined in the plane of the surface. The 2D phononic crystal is not periodic along the x_3 axis, so we can consider a 1D Fourier transform along this particular direction, introducing a component k_3 of the wavevector $\mathbf{k}$. In doing so, we are in fact considering a plane wave spectrum for the wavefield. The wavenumber k_3 can be obtained as a function of the parameters of the Bloch waves, i.e. k_1, k_2 and ω.

PWE analysis. The plane wave expansion formulas for 2D phononic crystals were given in Section 6.1.2. We start again from the equations giving the Bloch wave, following (6.5) or (6.11), as

$$u_i(t, \mathbf{x}) = \tilde{u}_i(\mathbf{x}) \exp(\imath(\omega t - \mathbf{k} \cdot \mathbf{x})),$$
$$T_{ij}(t, \mathbf{x}) = \tilde{T}_{ij}(\mathbf{x}) \exp(\imath(\omega t - \mathbf{k} \cdot \mathbf{x})),$$

with the periodic functions

$$\tilde{u}_i(\mathbf{x}) = \sum_{m=-N}^{+N} (u_i)_m \exp(-\imath \mathbf{G}_m \cdot \mathbf{x}),$$
$$\tilde{T}_{ij}(\mathbf{x}) = \sum_{m=-N}^{+N} (T_{ij})_m \exp(-\imath \mathbf{G}_m \cdot \mathbf{x}).$$

As a reminder, there are $M = 2N + 1$ Fourier harmonics. The PWE equations relating the vectors of Fourier coefficients of displacements and stresses are (6.27) and (6.28),

$$\imath\Gamma_j \mathbf{T}_{ij} = \omega^2 \mathcal{R}_{ik}\mathbf{U}_k,$$
$$\mathbf{T}_{ij} = -\imath\mathcal{C}_{ijkl}\Gamma_l\mathbf{U}_k,$$

where the square band matrices $\mathcal{R}_{ik}$ and $\mathcal{C}_{ijkl}$, with $M \times M$ elements, were defined by (6.25) and (6.26). Indices i and k run from 1 to r, with $r = 3$ for elastic solids and $r = 4$ for piezoelectric solids; they are associated with the polarization of waves. Indices j and l run from 1 to 3; they are associated with partial derivatives taken along directions x_j and x_l.

To put these equations in a pure matrix form, we again use the definitions (6.33) for $\mathbf{U}$ and (6.32) for matrix $\mathcal{R}$ to which we add the vectors of Fourier coefficients of the stresses

$$(\mathbf{T}_j)_m = ((T_{1j})_m \dots (u_{4j})_m)^T \tag{7.4}$$

and the matrices $\mathcal{C}'_{jl}$ with elements

$$(\mathcal{C}'_{jl})_{mn} = \begin{pmatrix} (\mathcal{C}_{1j1l})_{mn} & \dots & (\mathcal{C}_{1j4l})_{mn} \\ \vdots & \ddots & \vdots \\ (\mathcal{C}_{4j1l})_{mn} & \dots & (\mathcal{C}_{4j4l})_{mn} \end{pmatrix}. \tag{7.5}$$

With all these definitions, we can rewrite the PWE equations under the matrix form

$$\imath\Gamma_j\mathbf{T}_j = \omega^2\mathcal{R}\mathbf{U}, \tag{7.6}$$
$$\mathbf{T}_j = -\imath\mathcal{C}'_{jl}\Gamma_l\mathbf{U}. \tag{7.7}$$

It can be remarked that these manipulations were all intended to eliminate indices i and k.

Now we can exploit the nonperiodicity along axis x_3 by writing a generalized eigenvalue problem for the wavenumber k_3. We first remark that $\Gamma_3 = k_3 I_d$. Equation (7.7) is written for $j = 3$ as

$$\imath\mathbf{T}_3 - (\mathcal{C}'_{31}\Gamma_1 + \mathcal{C}'_{32}\Gamma_2)\mathbf{U} = k_3\mathcal{C}'_{33}\mathbf{U}. \tag{7.8}$$

Similarly, we have for $j = 1$ and 2,

$$\imath\mathbf{T}_j = (\mathcal{C}'_{j1}\Gamma_1 + \mathcal{C}'_{j2}\Gamma_2)\mathbf{U} + k_3\mathcal{C}'_{j3}\mathbf{U},$$

so that (7.6) becomes

$$\omega^2\mathcal{R}\mathbf{U} - (\Gamma_1\mathcal{C}'_{11}\Gamma_1 + \Gamma_1\mathcal{C}'_{12}\Gamma_2 + \Gamma_2\mathcal{C}'_{21}\Gamma_1 + \Gamma_2\mathcal{C}'_{22}\Gamma_2)\mathbf{U}$$
$$= \imath k_3\mathbf{T}_3 + k_3(\Gamma_1\mathcal{C}'_{13} + \Gamma_2\mathcal{C}'_{23})\mathbf{U}. \tag{7.9}$$

We next group the displacements and the stresses normal to the surface in the $2r \times M$-component state vector $\mathbf{H} = (\mathbf{U}, \imath\mathbf{T}_3)^T$ and obtain k_3 as the eigenvalue of the equation system

$$\begin{bmatrix} \omega^2 R - B & 0 \\ -C_2 & I_d \end{bmatrix} \mathbf{H} = k_3 \begin{bmatrix} C_1 & I_d \\ D & 0 \end{bmatrix} \mathbf{H}, \tag{7.10}$$

where

$$B = \sum_{j,l=1,2} \Gamma_j \mathfrak{C}'_{jl} \Gamma_l, \quad C_1 = \sum_{j=1,2} \Gamma_j \mathfrak{C}'_{j3}, \quad C_2 = \sum_{l=1,2} \mathfrak{C}'_{3l} \Gamma_l, \quad D = \mathfrak{C}'_{33}. \tag{7.11}$$

The $2r \times M$-component state vector $\mathbf{H} = (\mathbf{U}, \imath\mathbf{T}_3)^T$. Solving this system yields k_3 as the eigenvalue of a generalized eigenvalue problem. The matrices are nonsymmetrical. As a result, the $2r \times M$ eigenvalues k_{3q} and eigenvectors $\mathbf{H}_q$ are complex valued. By grouping in the eigenvectors the polarization components corresponding to the m-th harmonic, we introduce the notation

$$\mathbf{h}_{mq} = \begin{pmatrix} (u_i)_{\mathbf{G}_m q} \\ (T_{i3})_{\mathbf{G}_m q} \end{pmatrix}, \tag{7.12}$$

with $i = 1, \ldots, r$, $m = 1, \ldots, M$, and $q = 1, \ldots, 2rM$. The generalized displacement and normal stress fields are obtained from the superposition with relative amplitudes A_q

$$\mathbf{h}(\mathbf{x}, t) = \sum_{m=1}^{M} \sum_{q=1}^{2rM} A_q \mathbf{h}_{mq} \exp(\jmath(\omega t - (\mathbf{G}_m + \mathbf{k}_q) \cdot \mathbf{x})) \tag{7.13}$$

with $\mathbf{k}_q = (k_1, k_2, k_{3q})^T$. This superposition is a finite approximation to the infinite series. From this partial wave expansion, boundary conditions can be constructed to solve surface and plate problems [77, 162].

FEM analysis. We now show that the previous derivation of a partial wave expansion is in no way specific to the PWE method. The lesson learned is that the functional representation in the (x_1, x_2) plane (a Fourier series expansion for PWE) and along the x_3 axis (a plane-wave like expression) can be separated. We use a 2D mesh similar to one of those shown in Figure 4.6 for the unit cell in the plane. We start again from the Bloch wave expression (6.5) or (6.11),

$$u_i(t, \mathbf{x}) = \tilde{u}_i(\mathbf{x}) \exp(\imath(\omega t - \mathbf{k} \cdot \mathbf{x})),$$
$$T_{ij}(t, \mathbf{x}) = \tilde{T}_{ij}(\mathbf{x}) \exp(\imath(\omega t - \mathbf{k} \cdot \mathbf{x})),$$

and the dynamical equations for Bloch waves

$$T_{ij,j} = -\omega^2 \rho_{ik} u_k,$$
$$T_{ij} = c_{ijkl} u_{k,l}.$$

The periodic functions $\tilde{u}_i(\mathbf{x})$ and $\tilde{T}_{i3}(\mathbf{x})$ are approximated by a 2D FEM expansion and are the unknowns of the formulation. Again we select only the stress components that correspond to traction on the top surface, T_{i3}. The stresses T_{i1} and T_{i2} can be expressed as a function of the gradient of u_i from Hooke's law. We introduce test functions according to

$$u_i^t(t,\mathbf{x}) = \tilde{u}_i^t(\mathbf{x})\exp(\imath(\omega t - \mathbf{k}\cdot\mathbf{x})),$$
$$T_{i3}^t(t,\mathbf{x}) = \tilde{T}_{i3}^t(\mathbf{x})\exp(\imath(\omega t - \mathbf{k}\cdot\mathbf{x})).$$

The variational formulation is written as

$$\int_\Omega u_i^{t*}(\omega^2\rho_{ik}u_k + T_{ij,j}) + \int_\Omega T_{i3}^{t*}(T_{i3} - c_{i3kl}u_{k,l}) = 0. \tag{7.14}$$

This is a mixed FEM formulation combining displacements and stresses. Next we use the Gauss theorem to transform the integral involving the divergence of the stress tensor

$$\omega^2\int_\Omega u_i^{t*}\rho_{ik}u_k - \int_\Omega u_{i,j}^{t*}T_{ij} + \int_\Omega T_{i3}^{t*}T_{i3} - \int_\Omega T_{i3}^{t*}c_{i3kl}u_{k,l} = 0, \tag{7.15}$$

where we have dropped the boundary integral because it vanishes for periodic boundary conditions as we explained before. The four integrals in this expression should now be expressed using the periodic test and unknown functions. The first and the third integrals are simply

$$\omega^2\int_\Omega u_i^{t*}\rho_{ik}u_k = \omega^2\int_\Omega \tilde{u}_i^{t*}\rho_{ik}\tilde{u}_k,$$
$$\int_\Omega T_{i3}^{t*}T_{i3} = \int_\Omega \tilde{T}_{i3}^{t*}\tilde{T}_{i3}.$$

The fourth integral is

$$\int_\Omega T_{i3}^{t*}c_{i3kl}u_{k,l} = -\imath k_3\int_\Omega \tilde{T}_{i3}^{t*}c_{i3k3}\tilde{u}_k + \sum_{l=1}^{2}\int_\Omega \tilde{T}_{i3}^{t*}c_{i3kl}\tilde{u}_{k,l}.$$

The second integral is

$$\int_\Omega u_{i,j}^{t*}T_{ij} = \int_\Omega (u_{i,1}^{t*}T_{i1} + u_{i,2}^{t*}T_{i2} + u_{i,3}^{t*}T_{i3})$$
$$= \sum_{j,l=1}^{2}\int_\Omega \tilde{u}_{i,j}^{t*}c_{ijkl}\tilde{u}_{k,l} + \imath k_3\sum_{j=1}^{2}\int_\Omega \tilde{u}_{i,j}^{t*}c_{ijk3}\tilde{u}_k + \imath k_3 u_i^{t*}T_{i3}.$$

Gathering all these contributions to the variational formulation, we arrive at the equation

$$\omega^2 \int_\Omega \tilde{u}_i^{t*} \rho_{ik} \tilde{u}_k - \sum_{j,l=1}^{2} \int_\Omega \tilde{u}_{i,j}^{t*} c_{ijkl} \tilde{u}_{k,l} - \sum_{l=1}^{2} \int_\Omega \tilde{T}_{i3}^{t*} c_{i3kl} \tilde{u}_{k,l} + \int_\Omega \tilde{T}_{i3}^{t*} \tilde{T}_{i3}$$
$$= \imath k_3 \left(\sum_{j=1}^{2} \int_\Omega \tilde{u}_{i,j}^{t*} c_{ijk3} \tilde{u}_k + u_i^{t*} T_{i3} - \int_\Omega \tilde{T}_{i3}^{t*} c_{i3k3} \tilde{u}_k \right). \quad (7.16)$$

This equation can be directly compared to the PWE generalized eigenvalue equation (7.10); in order, the integrals of the first line correspond to matrices R, B, C_2, and I_d, and the integrals of the second line correspond to matrices C_1, I_d, and D. When the variational equation (7.16) is expressed by the FEM software as matrices operating on the degrees of freedom $\mathbf{h} = (\tilde{u}_i, \tilde{T}_{i3})$, $i = 1 \ldots r$, taken at the nodes of the mesh, then we again obtain a generalized eigenvalue problem with $\imath k_3$ as the eigenvalue. Compared to PWE matrices, FEM matrices are sparse, which means that the computation of eigenvalues is much faster for a given matrix size (a given number of degrees of freedom). The partial wave expansion can be written as

$$\mathbf{h}(\mathbf{x}, t) = \sum_{q=1}^{Q} A_q \tilde{\mathbf{h}}_q(\mathbf{x}) \exp(\jmath(\omega t - \mathbf{k}_q \cdot \mathbf{x})), \quad (7.17)$$

with Q the number of eigenvalues $\mathbf{k}_q = (k_1, k_2, k_{3q})$ and eigenvectors $\tilde{\mathbf{h}}_q(\mathbf{x})$ that are retained. This number may be chosen equal to the number of degrees of freedom, in order to obtain square matrices as with the PWE, but this is not compulsory.

Selection rule for partial waves. Since we consider a semi-infinite (though composite) substrate, only physically valid partial waves must be included in the normal mode expansion (7.13) or (7.17). Due to the tensor symmetries of the material constants, and for real-valued k_1 and k_2, the partial waves belong in pairs to an ensemble of $Q/2 = rM$ slowness curves. Hence, for each partial wave, using a criterion based on the sign of the component of the time-averaged Poynting vector that is normal to the surface [117] (in the case of a propagating partial wave) or on the sign of the imaginary part of k_{3q} (in the case of an evanescent partial wave), an unambiguous modal selection can be performed. The selection rule for partial waves entering or decaying in the substrate ($x_3 \leq 0$) is

$$\text{Select partial wave } q \text{ if } \begin{cases} P_{3q} < 0 & \text{if } \operatorname{Im}(k_{3q}) = 0, \\ \operatorname{Im}(k_{3q}) > 0 & \text{otherwise.} \end{cases} \quad (7.18)$$

From (5.9), the Poynting vector component can be evaluated as

$$P_{3q} = \frac{1}{2S(\Omega)} \int_\Omega \operatorname{Re}(\imath\omega \tilde{T}_{i3q} \tilde{u}_{iq}^*) \quad (7.19)$$

in the FEM case, and

$$P_{3q} = \frac{1}{2S(\Omega)} \mathrm{Re}\left(\imath\omega \sum_{m=1}^{M} (T_{i3})_{\mathbf{G}_m q} (u_i)^*_{\mathbf{G}_m q} \right) \tag{7.20}$$

in the PWE case, where we recall that $S(\Omega)$ is the surface of the 2D unit cell. We are then left with exactly $Q/2$ partial waves describing waves in the interior of the substrate. The eigenvectors $\tilde{\mathbf{H}}_q$ can then be restricted to the selected $Q/2$ partial waves.

7.3.2 Surface boundary conditions

The boundary conditions for surface modes apply at the $x_3 = 0$ surface. The mechanical boundary conditions require the nullity of stress components normal to the surface at any point within the unit cell. With PWE, because of the orthogonality of Fourier exponentials, this leads to

$$\sum_{q=1}^{Q/2} A_q (T_{i3})_{\mathbf{G}_m q} = 0, \quad i = 1\dots3, \ m = 1\dots M, \tag{7.21}$$

for a total of $3M$ equations. In the derivation of boundary conditions, the orthogonality of the $\exp(\jmath \mathbf{G}_m \cdot \mathbf{r})$ harmonic functions over one period of the surface is used [162]. From the electrical point of view, the free and shorted boundary conditions are considered. The free boundary condition is that the component of the electric displacement normal to the surface is continuous, resulting in

$$\sum_{q=1}^{Q/2} A_q \left[(D_3)_{\mathbf{G}_m q} - \imath \epsilon_0 k_{3q} \phi_{\mathbf{G}_m q} \right] = 0, \quad m = 1\dots M, \tag{7.22}$$

with ϵ_0 the permittivity of vacuum. The shorted boundary condition considers that the electric potential at the surface vanishes, for instance because of the presence of a thin, perfectly conducting metallic layer, yielding

$$\sum_{q=1}^{Q/2} A_q \phi_{\mathbf{G}_m q} = 0, \quad m = 1\dots M. \tag{7.23}$$

Both electrical boundary conditions result in a total of M conditions.

We have written the above boundary conditions from the PWE partial wave expansion. With FEM, the formulas are similar but functional values at the nodes of the mesh replace the Fourier coefficients. Equation (7.21) becomes

$$\sum_{q=1}^{Q/2} A_q \tilde{T}_{i3q}(\mathbf{x}_m) = 0, \quad i = 1\dots3, \ m = 1\dots M \tag{7.24}$$

where $\mathbf{x}_m$ denotes the coordinates of node m and M is now the total number of nodes. Similarly, (7.22) and (7.23) become

$$\sum_{q=1}^{Q/2} A_q \left[\tilde{D}_{3q}(\mathbf{x}_m) - \imath\epsilon_0 k_{3q}\tilde{\phi}_q(\mathbf{x}_m)\right] = 0, \quad m = 1 \ldots M \tag{7.25}$$

$$\sum_{q=1}^{Q/2} A_q \tilde{\phi}_q(\mathbf{x}_m) = 0, \quad m = 1 \ldots M. \tag{7.26}$$

Summarizing, for elastic surface phononic crystals, we have $3M$ boundary conditions given by (7.21) (alternatively, (7.24)). For piezoelectric surface phononic crystals, we have in addition M boundary conditions given by (7.22) or (7.23) (alternatively, (7.25) or (7.26)). There are thus $Q/2 = rM$ boundary conditions in either case which can be cast in the linear form

$$\mathcal{M}_{pq} A_q = 0, \tag{7.27}$$

with $\mathcal{M}$ a square matrix of size $Q/2 \times Q/2$.

7.3.3 Semi-infinite surface phononic crystals

Let us now describe how the expansion in partial waves, the selection rule and boundary conditions can be combined together to obtain a solution to the semi-infinite surface phononic crystal problem. The surface boundary conditions summarized by (7.27) are not indicative enough to obtain a definite solution. Instead, they allow us to check whether for some given values of the wave parameters (ω, k_1, k_2) some combination of the partial waves can satisfy the surface boundary conditions. As a consequence, we are unable to write an eigenvalue problem to which the surface wave would explicitly be an eigensolution. This situation is the consequence of the fact that partial waves are not themselves solutions to the wave equation with the required surface boundary conditions, but only solutions to the wave equation without these boundary conditions (i.e. they are complex-valued bulk solutions). As we noted in Chapter 5, this was already the case for surface waves on homogeneous semi-infinite media. The way around this mathematical inconvenience is again to write down boundary condition determinants; surface wave solutions will then be obtained for the zeros of the determinants [77, 149, 166]. We will write these determinants with the PWE notations in the following, but the expressions for the FEM notation should be obvious from them.

For elastic solids, the surface boundary conditions is given by (7.21). This yields a system of $3M$ linear equations in the $3M$ unknown amplitudes A_q. The trivial solution $A_q = 0$ must obviously not be retained, since the wave would then have zero energy. Then the following determinant must vanish

$$\Delta(\omega, k_1, k_2) = \left|(T_{i3})_{\mathbf{G}_m q}\right| = 0, \tag{7.28}$$

where it is be understood that q is the index for columns while lines are written for $i = 1 \ldots 3$ and $m = 1 \ldots M$. In practice, a diagram similar to a band structure can be obtained by scanning the possible values of (ω, k_1, k_2) and looking at the value of the determinant; if it is close to zero then we are close to a possible surface wave.

Piezoelectric surface phononic crystal can be treated as follows. In the case of a shorted surface, the surface boundary conditions (7.21) and (7.23) must be satisfied simultaneously. This yields a system of $4M$ linear equations in the $4M$ unknown amplitudes A_q, which has a nontrivial solution only if the following determinant vanishes,

$$\Delta_s(\omega, k_1, k_2) = \begin{vmatrix} (T_{i3})_{\mathbf{G}_m q} \\ \phi_{\mathbf{G}_m q} \end{vmatrix} = 0. \tag{7.29}$$

Thus the SAW solutions on a shorted surface can be identified by locating the zeros of Δ_s. Similarly, in the case of a free surface, the SAW solutions can be identified by locating the zeros of the determinant

$$\Delta_f(\omega, k_1, k_2) = \begin{vmatrix} (T_{i3})_{\mathbf{G}_m q} \\ (D_3)_{\mathbf{G}_m q} - \imath \epsilon_0 k_{3q} \phi_{\mathbf{G}_m q} \end{vmatrix} = 0. \tag{7.30}$$

As in the case of a homogeneous substrate, it can be expected that a simultaneous zero of both determinants Δ_s and Δ_f is the signature of a nonpiezoelectrically coupled surface mode. Indeed, in this case, the surface mode is insensitive to the electrical boundary conditions. Conversely, in the case where there exists a (small) frequency shift between two zeros of the determinants, this shift can be used to obtain an estimate of the piezoelectric coupling factor K^2 of the surface mode, according to the usual formula $K^2 = 2(\omega_f - \omega_s)/(\omega_f + \omega_s)$.

Examples of boundary condition determinants for a free and a shorted surface are shown in Figure 7.18. The phononic crystal considered is constituted of circular cylindrical holes in a Y-cut lithium niobate semi-infinite substrate [77]. The variations of the free and shorted boundary condition determinants are plotted as a function of frequency for the X and the M points of the first irreducible Brillouin zone. The occurrence of surface modes is indicated by zeros (or sharp minima in the case of leaky modes). As a general rule, it can be observed that there exist many surface modes in the phononic crystal, both below and above the full band gap for bulk waves, but not within it. This multiplicity of surface modes was not mentioned in the case of the solid-solid phononic crystals of references [149, 150, 166], but is clearly apparent in previous studies of surface modes in superlattices [31, 32], which can be viewed as one-dimensional phononic crystals. The contributions of bulk waves can also be seen in the form of discontinuities of the first derivative of the determinants.

A surface Green's dyadic, generalizing the surface Green's dyadic or Green's function of the homogeneous piezoelectric substrate, can further be obtained by eliminating the partial waves amplitudes from (7.13). At the surface, we take the scalar product over one period of the surface of (7.13), restricted to the rM selected partial waves, with

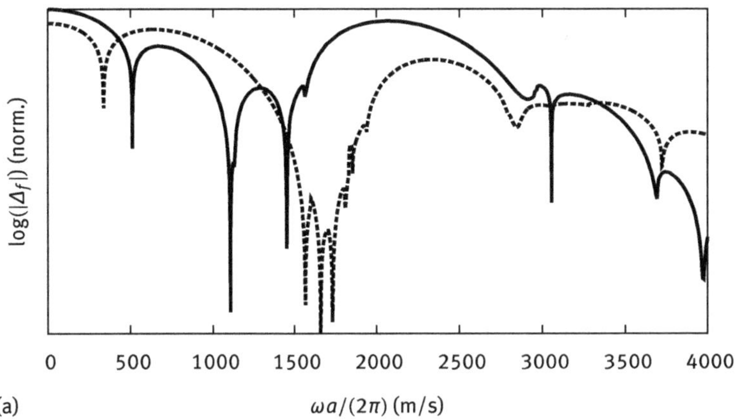

(a)

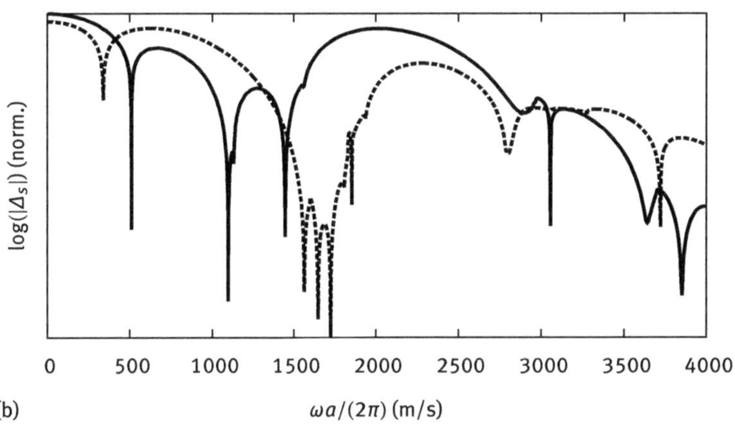

(b)

Fig. 7.18: Surface boundary condition determinants as a function of reduced frequency. Free (a) and shorted (b) boundary condition determinants are plotted for the X (solid line) and the M (dotted line) points of the first Brillouin zone, for a Y-cut square-lattice lithium niobate surface phononic crystal with $d/a = 0.9$ (after reference [77]).

the M harmonic functions $\exp(\jmath \mathbf{G}_m \cdot \mathbf{r})$, or

$$\frac{1}{S(\Omega)} \int \mathbf{h}(\mathbf{x}, t) \exp(\jmath \mathbf{G}_m \cdot \mathbf{x}) d\mathbf{r} = \sum_{q=1}^{rN} A_q \tilde{\mathbf{h}}_{mq} \exp(\jmath(\omega t - k_1 x_1 - k_2 x_2))$$
$$= \tilde{\mathbf{h}}_m \exp(\jmath(\omega t - k_1 x_1 - k_2 x_2)).$$

The second equation is for the coefficients of the Fourier series expansion of $\mathbf{h}(\mathbf{x}, t)$. Identifying terms in the series, we have

$$\tilde{\mathbf{h}}_m = \sum_{q=1}^{rN} \tilde{\mathbf{h}}_{mq} A_q.$$

Written in matrix form, this equation is

$$\begin{pmatrix} (u_i)_{\mathbf{G}_m} \\ \phi_{\mathbf{G}_m} \\ (T_{3j})_{\mathbf{G}_m} \\ (D_3)_{\mathbf{G}_m} \end{pmatrix} = \begin{pmatrix} u_{i\mathbf{G}_m q} \\ \phi_{\mathbf{G}_m q} \\ (T_{3j})_{\mathbf{G}_m q} \\ D_{3\mathbf{G}_m q} \end{pmatrix} (A_q) .$$

There are twice as many rows as columns in the matrix since we have dropped half of the partial waves. Eliminating the amplitudes A_q, we then have the vector-matrix relation between the harmonics of the generalized displacements and generalized stresses

$$\begin{pmatrix} (u_i)_{\mathbf{G}_m} \\ \phi_{\mathbf{G}_m} \end{pmatrix} = G \times \begin{pmatrix} (T_{3j})_{\mathbf{G}_m} \\ (D_3)_{\mathbf{G}_m} \end{pmatrix} , \tag{7.31}$$

with the Green's dyadic

$$G = \begin{pmatrix} u_{i\mathbf{G}_m q} \\ \phi_{\mathbf{G}_m q} \end{pmatrix} \times \begin{pmatrix} (T_{3j})_{\mathbf{G}_m q} \\ D_{3\mathbf{G}_m q} \end{pmatrix}^{-1} . \tag{7.32}$$

The Green's dyadic is an $rM \times rM$ square matrix and relates the generalized displacements to the generalized stresses. This is a direct generalization of the Green's dyadic of a homogeneous elastic or piezoelectric semi-infinite substrate.

The concept of an effective permittivity for surface acoustic waves on a homogeneous substrate is very useful [59]. Such an effective permittivity is a scalar function relating the normal electric displacement to the potential of a plane wave solution with a given wavevector and frequency, in the case of a mechanically-free surface. In the case treated here, there are M Fourier harmonics which have to be considered for every wavevector and frequency couple. Then an effective permittivity matrix ϵ_{eff} can be defined by relating the normal electric displacement harmonics to the potential harmonics, i.e. $[(D_3)_{\mathbf{G}_m}] = \epsilon_{\text{eff}} k [\phi_{\mathbf{G}_m}]$, with $k = \sqrt{k_1^2 + k_2^2}$. Using the boundary condition (7.22), we obtain at once that the effective permittivity matrix is proportional to the lower right $M \times M$ submatrix of the inverse Green's dyadic.

Before closing this section, we define two useful scalar functions for locating surface modes. First, we remark that in the case of a homogeneous substrate, the effective permittivity is proportional to the function $b(\omega, k_1, k_2) = \Delta_f / \Delta_s$. This property is no longer valid for a piezocomposite material as considered here. However, this function still gives direct information on piezoelectrically coupled surface modes, since its poles indicate surface modes on the shorted surface while its zeros indicate surface modes on a free surface. Second, we can define the variation of the total density of surface states according to the formula [31]

$$\Delta n(\omega, k_1, k_2) = \frac{1}{\pi} \text{Im} \left\{ \frac{\partial}{\partial \omega} \log |G| \right\} . \tag{7.33}$$

This function is zero when there are no surface modes and presents poles where there are. Figure 7.19 shows the band structure for surface modes plotted along the path

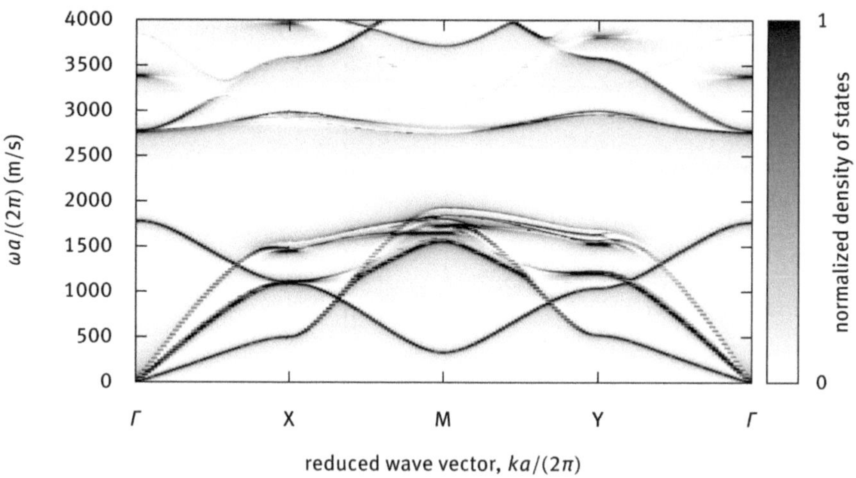

Fig. 7.19: Total density of surface states plotted along the irreducible Brillouin zone. The plot is along the path Γ-X-M-Y-Γ around the irreducible Brillouin zone, for a Y-cut square-lattice lithium niobate surface phononic crystal with $d/a = 0.9$. The total density of surface states is maximal for surface Bloch waves. Maxima, appearing in black, thus outline the band structure for surface Bloch waves (after reference [77]).

Γ-X-M-Y-Γ using the total density of states, for the same phononic crystal as in Figure 7.18 [77]. This band structure clearly has similarities with the band structure of bulk waves propagating in the plane of the surface. The surface modes branches define a full band gap that is coincident with the one for in-plane propagating bulk waves. This result is not obvious since the rN partial waves defining a surface mode include evanescent waves as well as bulk waves propagating obliquely in the phononic crystal, with possibly any direction and polarization state. We observe that surface modes branches often exist just below bulk branches, as in the case of the homogeneous substrate, but not exclusively.

7.3.4 Finite-depth surface phononic crystals

As we will exemplify in the next section, actual phononic crystals for surface elastic waves do not involve a fully perforated semi-infinite half-space, but rather a phononic crystal layer with a certain depth sitting on a homogeneous substrate. This situation is depicted in Figure 7.20. In practice, if the depth of the phononic crystal layer is sufficient, it is intuitively expected that the phononic properties will tend towards those of the ideal infinite-depth phononic crystal. Conversely, if the depth of the phononic crystal layer is too small, it is likely that the phononic band structure will differ strongly from the infinite-depth case and in particular that full band gaps can be lost.

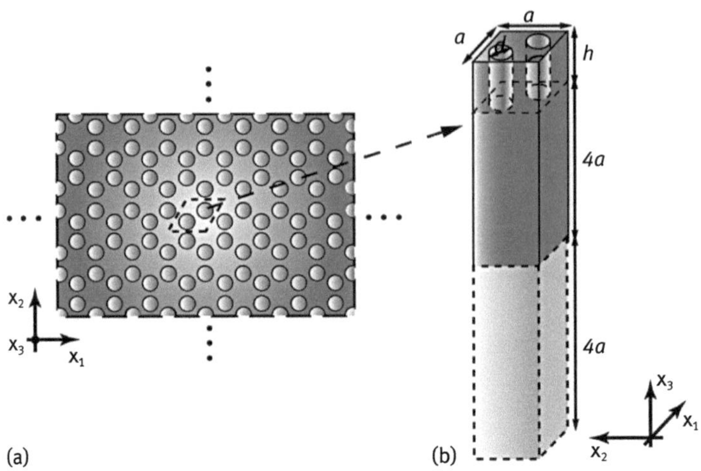

Fig. 7.20: The finite-depth surface phononic crystal problem. (a) A 2D piezoelectric phononic crystal composed of an air hole array arranged according to a honeycomb lattice is viewed from the top. (b) The unit cell is represented in 3D where the dark gray and the light gray areas denote the $LiNbO_3$ layer and the artificial layer, respectively. The depth of the holes is limited to h and the phononic crystal layer sits on a semi-infinite homogeneous substrate that is here artificially limited to a depth of $4a + 4a$ (after Yudistira et al. [175], copyright 2012 American Institute of Physics).

We have seen in Chapter 5 that there are three bulk elastic waves propagating in any direction inside a homogeneous solid. Consider a direction of propagation given by azimuth angle ψ in the surface plane (x_1, x_2). Bulk elastic waves in the substrate underneath the phononic crystal layer are specified by azimuth angle ψ and a polar angle θ measured in the sagittal plane. Let us concentrate on the slowest of these bulk elastic waves. For every value of ψ, the maximum value $s(\psi)$ for the bulk slowness is obtained for θ such that Poynting's vector satisfies $P_3 = 0$.[1] This particular solution is called a surface-skimming bulk wave as its energy propagation direction is parallel to the surface. In the space $(\omega, \mathbf{k})$ of dispersion relations, $(\omega, s(\psi)\omega)$ defines the surface of a cone that limits the domain of existence of bulk elastic waves to the interior of the cone. This cone is known as the sound cone [11, 62]. Every point $(\omega, \mathbf{k})$ that lies outside the sound cone potentially represents the dispersion of a surface guided wave; every point that lies inside the sound cone potentially represents the dispersion of a leaky surface guided wave that can couple to some bulk elastic wave.

1 The group velocity is orthogonal to the slowness surface and identifies with the energy velocity in homogeneous media. The energy velocity is proportional to Poynting's vector by definition. Then any maximum of the slowness $s(\psi, \theta)$ as a function of θ is a stationary point, i.e. satisfies $\frac{\partial s(\psi,\theta)}{\partial \theta} = 0$, which in turn implies that $P_3 = 0$.

Numerically, the finite-thickness surface phononic crystal problem can be solved with FEM providing an artificial way of mimicking the semi-infinite substrate is implemented. Khelif et al. used a certain depth for the substrate complemented with a perfectly matched layer (PML) and obtained band structures limited to surface guided Bloch waves, i.e. valid only under the sound cone [62]. PMLs for elastic waves can be introduced for monochromatic waves much as presented in Section 3.3.5 for pressure acoustic waves, following a technique introduced originally for surface elastic waves in interdigital transducers [30]. Assouar et al. used some large finite thickness plus a clamped bottom surface ($u_i = 0$) and found that the band structures limited to surface guided Bloch waves could be obtained faithfully thanks to the guaranteed exponential decay under the sound cone [7]. Yudistira et al. used the variant depicted in Figure 7.20, with a homogeneous region of finite thickness $4a$ sitting on an artificial homogeneous region also of thickness $4a$ but with high bulk velocities [175]. The purpose of the artificial homogeneous region is to force a faster exponential decay, hence diminishing the meshed substrate depth. All these different numerical techniques potentially have problems in the lower part of the band structure, because the wavelength is very long there (the wavenumber k is small), and even if exponential decay is guaranteed under the sound cone this decay can become very slow and the bottom boundary condition might induce reflected waves.

Figure 7.21 shows both the effect of a finite phononic crystal layer and the effect of a conical shape for the hole [175]. All band structures in the figure are for Y-cut lithium niobate and for a honeycomb lattice of circular cylinders or cones with $d/a = 0.5$ at the top surface. The band structures in Figure 7.21 (a)–(c) are plotted for $h/a = 0.7$, while those in Figure 7.21 (d)–(f) are plotted for $h/a = 0.9$; they are plotted for cylindrical holes. They show that the full band gap for surface guided waves is dependent on the thickness of the phononic crystal layer. Furthermore, the sound cone is quite anisotropic, as we know from the slowness curves for Y-cut lithium niobate in Figure 5.18. As a result, the end points for the bands that are considered as truly guided by the surface depend on the direction in the first Brillouin zone and the irreducible Brillouin zone extends to a larger area than the usual Γ-K-M-Γ triangle. The gap map in Figure 7.21 (g) shows that the full band gap for surface guided waves exists for a rather large range of values for h/a. As with phononic crystal slabs, the full band gap closes as the thickness of the phononic crystal layer is increased; it must be stressed however that the semi-infinite band structure in Figure 7.19 is not obtained as an asymptotic limit as h goes to infinity. This is because for large values of h there are many bands for Bloch waves that “see” the interface between the phononic crystal layer and the bottom substrate but which would be ruled out by the radiation condition at infinity in the mathematical asymptotic process. Finally, in the case of conical holes in 7.21 (h), we can see that the full band gap is actually resilient to the value of the cone angle. This is good news for practical realizations, especially in materials that are difficult to etch or mill, such as lithium niobate. Practical examples are given in the following section.

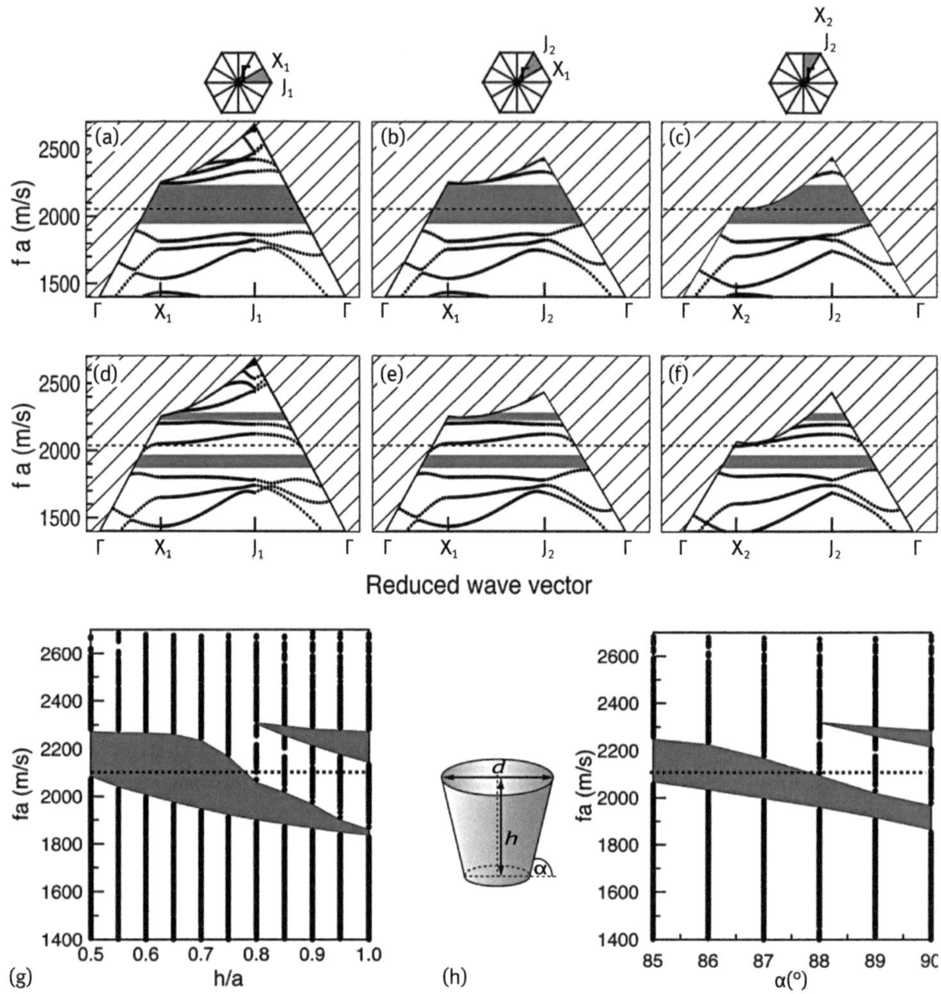

Fig. 7.21: Phononic band structure for surface guided waves for finite-depth honeycomb-lattice phononic crystals. The substrate is Y-cut lithium niobate. Calculated SAW band structures for the case: $h = 0.7a$, $d = 0.5a$ (a)–(c), and $h = 0.9a$, $d = 0.5a$ (d)–(f), obtained for each area inside the irreducible Brillouin zone as indicated by the insets. The black dots, the thick lines, and the patterned areas correspond to the localized SAW modes, the sound line, and the substrate or leaky modes, respectively. The shaded gray areas indicate SAW band gaps. (g) Calculated SAW gap map for the case when the depth h is varied from $5a$ to $1a$ while the diameter d is fixed at $d = 0.5a$. (h) Calculated SAW gap map for the case when the shape of the hole is changing from conical to cylindrical by varying angle a from 85° to 90°, with $h = 0.9a$ and $d = 0.5a$. The inset illustrates the shape of the conical hole. Note that the symmetry points denoted J (respectively X) are denoted K (resp. M) in this book (adapted from Yudistira et al. [175], copyright 2012 American Institute of Physics).

7.4 Surface phononic crystals

In Figure 7.1 we presented three examples of surface phononic crystals. The silicon structure in Figure 7.1 (a2) and the lithium niobate structure in Figure 7.1 (a3) are the most extensively discussed structures in the literature and we discuss them in succession. Further examples can be found in Chapter 10 (surface phononic crystals of pillars) and in Chapter 11 (surface phononic crystals used as mirrors and to define defect waveguides).

Silicon surface phononic crystal. The silicon phononic crystal device in Figure 7.1 (a2) was prepared by etching a silicon wafer through a photoresist mask [168]. The parameters of the square-lattice array of cylindrical holes are $a = 10\,\mu m$ and $d = 7\,\mu m$. The etch depth is quoted to be at least 80 µm, so that the samples practically approach the semi-infinite case. A zinc oxide layer is added underneath the interdigital transducers but does not cover the phononic crystal area. There are six rows of holes in the measurements which are based on an electrical transmission measurement, with one interdigital transducer operating as a source of SAW while the other operates as a receiver. The experimental transmission in Figure 7.22 shows a clear dip in the transmission that can be attributed to a Bragg phononic band gap in the direction ΓX. Note that for the filling fraction that is considered, only a directional band gap can be opened.

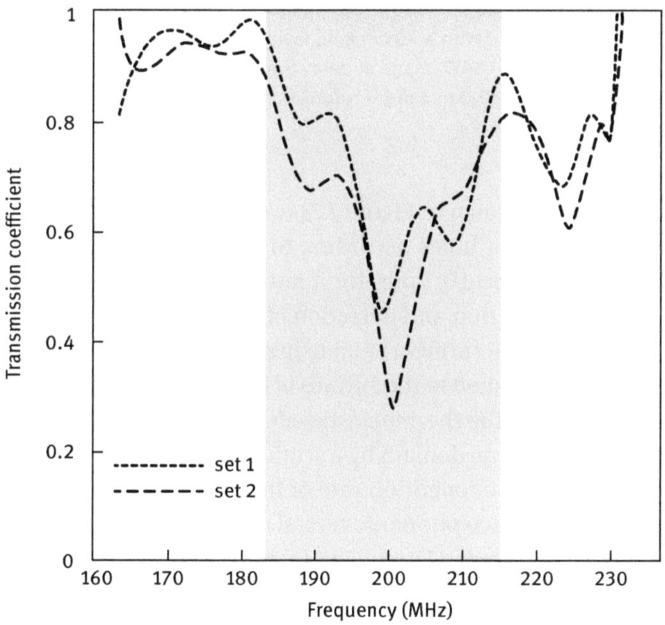

Fig. 7.22: Experimental transmission of a surface phononic crystal in silicon. (After Wu et al. [168], copyright 2005 American Institute of Physics).

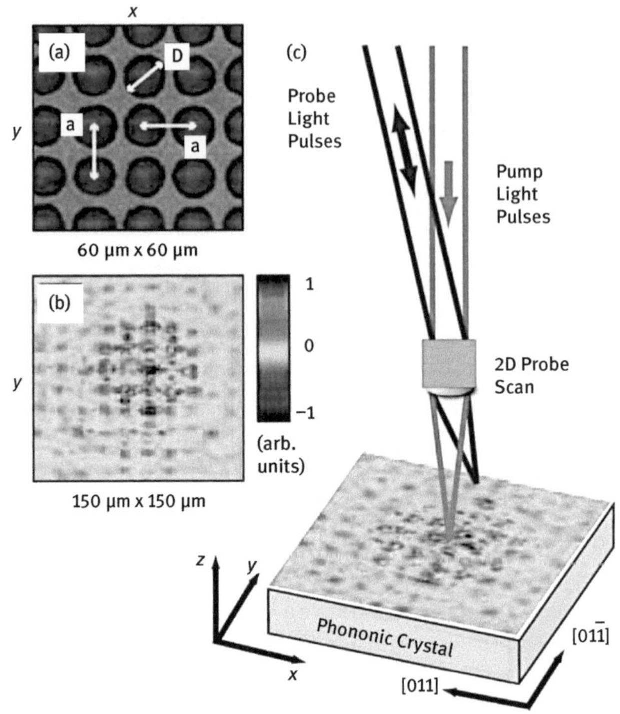

Fig. 7.23: Optical pump-probe experiments with a silicon surface phononic crystal. (a) Optical micrograph of the Si phononic crystal coated with a 40 nm gold layer. The hole diameter is $d = 12\,\mu\text{m}$ with lattice constant $a = 15\,\mu\text{m}$. (b) SAW image at delay time $\tau = 7.4$ ns. (c) Outline of experimental set-up with SAW image at $\tau = 10.3$ ns (after Profunser et al. [123], copyright 2009 American Physical Society).

The surface phononic crystal sample shown in Figure 7.23 was also obtained by etching a silicon wafer to define an array of holes according to the square lattice [123]. The etching depth was determined to be 10.5 µm, for a lattice constant $a = 15\,\mu\text{m}$ and a hole diameter $d = 12\,\mu\text{m}$. Generation and detection of surface elastic waves is performed all-optically in a pump-probe configuration using femtosecond ultrashort laser pulses. A thin layer of gold is deposited at the surface of the sample and provides absorption of incident light pulses. Via the thermoelastic effect, surface elastic waves are generated and are observed in the time domain by a scanning probe optical beam. Frames can be acquired every 2.9 ns, the repetition rate of the femtosecond laser, so that propagation of SAW inside the surface phononic crystal can be observed. Frames obtained at two different times are depicted in Figure 7.23. Overall, pump-probe experiments allow for generating elastic waves on nonpiezoelectric substrate and to observe wave propagation in the space-and-time domain. Another measurement example of the same kind is given in Figure 11.13 in the case of a phononic crystal waveguide for surface elastic waves.

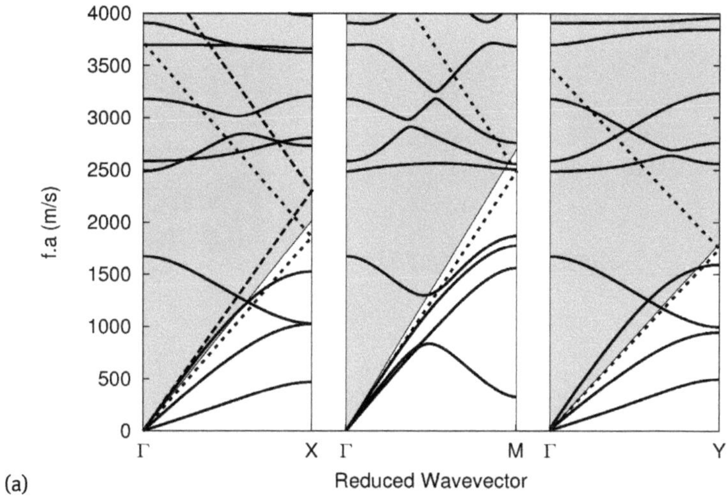

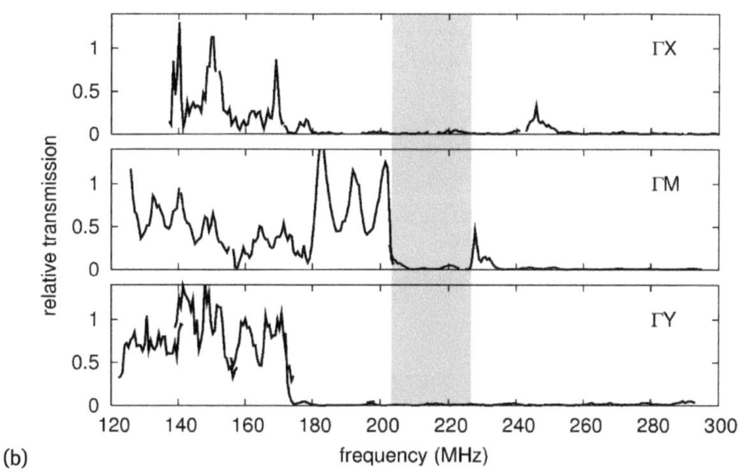

Fig. 7.24: Theoretical and experimental full phononic band gap for a square-lattice Y-cut lithium niobate surface phononic crystal. (a) Theoretical band diagrams for bulk waves propagating in the plane of a square lattice void/lithium-niobate phononic crystal with a 64 % filling fraction ($d/a = 0.9$) along the ΓX, ΓM and ΓY directions of the first Brillouin zone. In the gray regions, above the sound lines, there is coupling of surface modes with the radiation modes of the substrate. The dispersion relations on a free surface are also indicated for the Rayleigh (short dashed line) and the leaky (long dashed line) surface waves. The theoretical full band gap extends from 190 to 250 MHz when the lattice parameter a equals 10 μm. (b) Relative transmission, defined as the ratio of the transmittance with and without a phononic crystal. The experimental full band gap extends from 203 to 226 MHz (after Benchabane et al. [11], copyright 2006 American Physical Society).

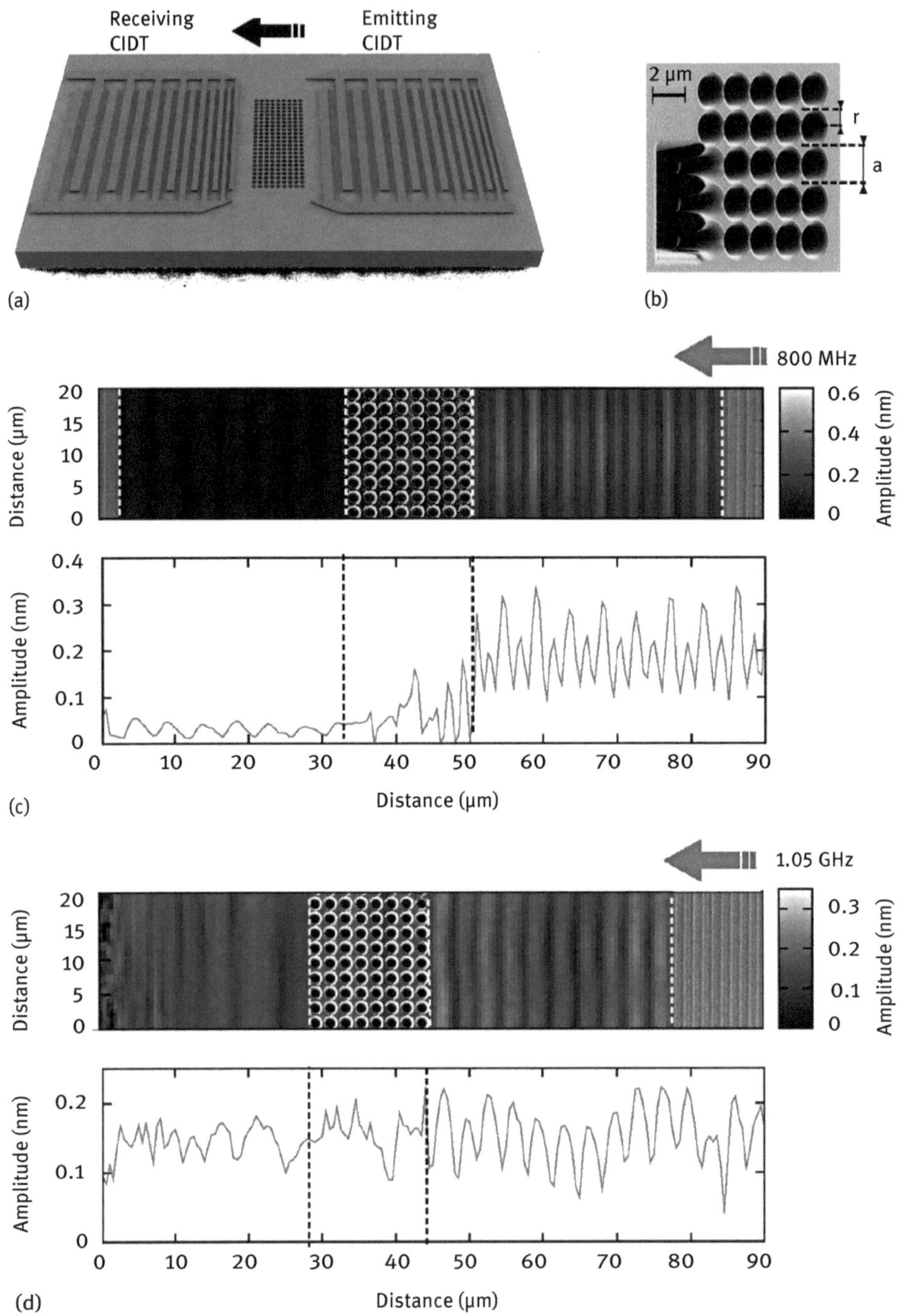
Receiving CIDT
Emitting CIDT
(a)
2 µm
r
a
(b)
800 MHz
Distance (µm)
Amplitude (nm)
0.6
0.4
0.2
0
20
15
10
5
0
0.4
0.3
0.2
0.1
0
10
20
30
40
50
60
70
80
90
Distance (µm)
(c)
1.05 GHz
0.3
0.2
0.1
Distance (µm)
Amplitude (nm)
(d)

Lithium niobate surface phononic crystal. The observation of a full band gap in a surface phononic crystal was reported by Benchabane et al. [11]. The experiment was based on a series of phononic crystal samples obtained by reactive ion etching of Y-cut lithium niobate wafers through a photoresist mask, as shown in Figure 7.1 (a3). The lattice constant $a = 10\,\mu m$ and the hole diameter $d = 9\,\mu m$ were chosen to obtain a full phononic band gap for a semi-infinite surface phononic crystal. The experimentally determined etching depth was estimated to be larger than 10 μm and the conical hole shape had an angle $\alpha \approx 85°$. Compared to the theoretical [190–250] MHz band gap range, the experimental results summarized in Figure 7.24 indicated a [203–226] MHz band gap range. These observations were later confirmed by optical interferometer measurements [70].

It has to be noted that electrical transmission measurements, even though they reveal the existence of band gaps, do not give further information on the spatial distribution of the displacement field over the surface of the phononic crystal. Figure 7.25 shows heterodyne optical interferometer measurements of the out-of-plane displacement u_3 at different operating frequencies. The surface phononic crystal sample is a square-lattice array of conical holes in X-cut lithium niobate [12]. For a frequency of 800 MHz, inside the full band gap for surface elastic waves, the waves incident from the right are almost totally reflected on the phononic crystal. A clear decay of displacement u_3 is observed inside the phononic crystal and the eight rows of holes are sufficient to obtain a small transmission of the left side. For a frequency of 1050 MHz, above the full band gap and inside the sound cone, transmission through the phononic crystal is mostly observed, accompanied with some reflection. This observation means that even though radiation to the bulk waves of the semi-infinite substrate is certainly possible for frequencies falling inside the sound cone, the necessary modal conversion may not always be efficient, thus preserving at least partly the propagation of elastic waves at the surface. It is not unlikely, though yet unproven, that leaky surface elastic waves with moderate propagation loss can be present in phononic crystals, such as is the case for instance for some homogeneous piezoelectric substrates we listed in Table 5.5.

◂ **Fig. 7.25:** Heterodyne optical interferometer measurement of a square-lattice X-cut lithium niobate surface phononic crystal. (a) The 2D phononic crystal made of air holes in an X-cut lithium niobate matrix is surrounded by a pair of chirped interdigital transducers. A scanning electron microscope image of the square lattice crystal is shown in (b). The period *a* of the crystal is around 2.2 μm, for a hole diameter *d* of 2 μm. Measured elastic wave fields propagating through the phononic crystals are shown for an excitation frequency (c) within (880 MHz) and (d) above (1.04 GHz) the band gap. In each case, the emitting transducers is at the right-hand side and a scanning electron microscope image of the actual crystal and transducers is overlayed to the field map. The amplitude data are averaged in the y-direction and the resulting line profiles along the wave propagation direction are displayed (after Benchabane et al. [12]).

Part III: **Wave phenomena in phononic crystals**

Si les signes vous faschent, ô quant vous fascheront les choses signifiées.

François Rabelais (1483–1553)
Le Tiers Livre, chapitre 20

8 Coupling of acoustic and elastic waves in phononic crystals

The case of sonic or phononic crystals containing both solid and fluid materials has been considered quite often in the literature, especially in relation to experiments, for instance for the steel rods or spheres in water or air cases. The reason is that macroscopic realizations are rather easy with these material systems. Experiments involving measurements from the fluid side are also made easy by the existence of loudspeakers, for sound in air, or ultrasonic transducers, for acoustic waves in water. As is apparent from the equations for the solid and the fluid cases that we introduced in the previous chapters, it is not possible to use just one set of equations with coefficients assuming different values in different regions of space; the equations are different in different regions of space. The rigorous way to solve this problem is generally to consider what is often termed fluid-structure interaction, a concept that contains the coupling of acoustic and elastic waves as a subcase. Our treatment in this chapter is based on a domain decomposition method that is general enough to include most practical cases. We will introduce the method first on the simple example of an interface separating a solid and a fluid domain, before moving on to specific sonic and phononic crystal examples.

8.1 Coupling of acoustic and elastic waves

As an introduction, we consider the simplest problem of two connected domains, of which one is a solid and the second is a fluid, as depicted in Figure 8.1. If it were not for the common interface σ between the two domains, elastic waves in the solid and acoustic waves in the fluid would be completely disconnected. It should then not come as a surprise that the introduction of adequate boundary conditions at the interface is all that is required to obtain a solution.

What are these interface boundary conditions and can the methods we have discussed in the previous chapters cope with them? First, if the displacement field has been computed inside the solid and is thus known beforehand, we know that displacements, velocities, and accelerations are continuous at the interface. Then (4.1) and the use of the divergence theorem from vector calculus leads to a boundary condition relating the pressure created in the fluid with the normal acceleration of the solid boundary. Second, if we assume that the pressure distribution is known, then this pressure is exerted by the fluid on the solid boundary. As normal traction is continuous, the stress inside the solid and exactly at the boundary is related to the pressure exerted. We thus have a pair of boundary conditions that relate wavefields in the solid and in the fluid when going from one domain to the other. Such boundary conditions can be employed naturally with FDTD and FEM methods, since they are based on a grid or

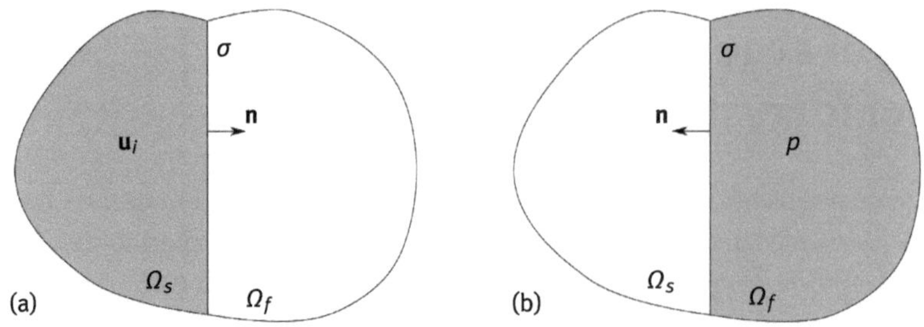

Fig. 8.1: Schematic representation of the coupling of elastic waves in a solid domain, Ω_s, with the acoustic waves in a fluid domain, Ω_f. Both domains are a decomposition of the total domain $\Omega = \Omega_s \cup \Omega_f$ and they are separated by boundary $\sigma = \Omega_s \cap \Omega_f$. (a) If the displacement field $\mathbf{u}_i$ is known inside Ω_s, then the normal acceleration at the interface is also known and becomes a boundary condition on σ for the fluid domain. (b) Reciprocally, if the pressure field p is known inside Ω_f, then the normal traction on interface σ is known and becomes a boundary condition for the solid domain.

on a mesh, respectively. In contrast, PWE assumes continuous and periodic fields, a condition that is usually too strong for crystals involving both fluid and solid domains.

Let us express the interface boundary conditions. First, at interface σ the normal acceleration is continuous across both domains and is related to the normal derivative of the pressure p in the fluid, as we noted above, so that,

$$\omega^2 u_n = \frac{1}{\rho_f} \frac{\partial p}{\partial n}, \tag{8.1}$$

with u_n the normal component of the displacement at the interface and ρ_f the mass density of the fluid. Mathematically, this equation is obtained from (4.1) by integration over a closed contour around a specific point of the interface, then using the divergence theorem and then letting the width of the contour go to zero. Second, the normal traction must be continuous across the interface, i.e.

$$T_{ij} n_j = -p n_i, \tag{8.2}$$

with T_{ij} the stress tensor inside the solid and n_i the components of the unit vector normal to boundary σ. It must be stressed that the normal unit vector $\mathbf{n}$ is oriented oppositely when seen from within the fluid or from within the solid; the normal is orthogonal to boundary σ and points outward from the considered domain.

A variational formulation of elastic-acoustic interaction can then be formulated as follows. We consider time harmonic waves with angular frequency ω and we recall the dynamical equations of motion for convenience. The scalar equation of motion for acoustic waves in the fluid can then be written for the pressure only as

$$\frac{\omega^2}{B} p + \nabla \cdot \left(\frac{1}{\rho_f} \nabla p \right) = 0, \tag{8.3}$$

with B the adiabatic compressibility for zero heat conduction of the fluid (implying that we neglect thermal phenomena in the model). Motion in a solid is further described by Hooke's law

$$T_{ij} = c_{ijkl}\frac{\partial u_k}{\partial x_l}, \tag{8.4}$$

and by Newton's second law

$$\omega^2 \rho_s u_i + T_{ij,j} = 0. \tag{8.5}$$

Here, c_{ijkl} is the rigidity tensor of the solid and ρ_s is its mass density.

A variational formulation can be derived using the basic equations (8.3)–(8.5), and the coupling boundary conditions (8.1, 8.2) at the fluid-solid interface. We multiply (8.3) by test function p' living in the same finite element space as p and (8.5) by test function u_i' ($i = 1, 2, 3$) living in the same finite element space as u_i, and integrate independently over the fluid and the solid domains. The terms containing a divergence are transformed by applying Green's formula and incorporating coupling boundary conditions at the fluid-solid interface, according to

$$\int_{\Omega_f} p'\nabla.\left(\frac{1}{\rho_f}\nabla p\right) = \omega^2 \int_{\sigma} p' u_n - \int_{\Omega_f} \nabla p' \cdot \left(\frac{1}{\rho_f}\nabla p\right) \tag{8.6}$$

and

$$\int_{\Omega_s} u_i' T_{ij,j} = -\int_{\sigma} u_n' p - \int_{\Omega_s} u_{i,j}' T_{ij}, \tag{8.7}$$

with $u_n = u_i n_i$ and $u_n' = u_i' n_i$. In the last two expressions, it should be noted that the fluid-solid boundary σ is oriented differently, since the unit normal n_i is exterior to domain Ω_f in (8.6) but exterior to domain Ω_s in (8.7). Upon rearranging terms, a variational formulation unifying elements in space and on the boundary is obtained as

$$\omega^2\left(\int_{\Omega_f} p'\frac{1}{B}p + \int_{\sigma} p' u_n\right) = \int_{\Omega_f} p'_{,i}\frac{1}{\rho_f}p_{,i}, \tag{8.8}$$

$$\omega^2 \int_{\Omega_s} u_i' \rho_s u_i = \int_{\sigma} p u_n' + \int_{\Omega_s} u_{i,j}' c_{ijkl} u_{k,l}, \tag{8.9}$$

This formulation is valid generally and leads to a generalized eigenvalue problem for ω^2. In case of periodicity, the inclusion of the wavevector dependence follows the application of the Bloch–Floquet theorem. Specifically, it is sufficient [82] to make the following replacements in the pair of equations above

$$p_{,i} = \tilde{p}_{,i} - \jmath k_i \overline{p}, \qquad p'_{,i} = \tilde{p}'_{,i} + \jmath k_i \tilde{q}, \tag{8.10}$$

$$u_{k,l} = \tilde{u}_{k,l} - \jmath k_l \tilde{u}_k, \quad u_{i,j}' = \tilde{u}_{i,j}' + \jmath k_j \tilde{u}_i'. \tag{8.11}$$

Finally, the pair of equations (8.8) and (8.9) can be cast in the form of a generalized eigenvalue problem $\omega(k)$

$$A(k)\mathbf{y} = \omega^2 B\mathbf{y}. \tag{8.12}$$

where vector **y** is composed of the values of p and u_i, $i = 1, 2, 3$ at the nodes of the meshes of the domains Ω_f and Ω_s. The block matrices $A(k)$ and B can be decomposed as

$$A(k) = \begin{bmatrix} A_{11}(k) & 0 \\ A_{21} & A_{22}(k) \end{bmatrix}, \tag{8.13}$$

$$B = \begin{bmatrix} B_{11} & B_{12} \\ 0 & B_{22} \end{bmatrix}. \tag{8.14}$$

The square block matrices $A_{11}(k)$ and B_{11} ($A_{22}(k)$ and B_{22}, respectively) define the uncoupled acoustic (elastic, resp.) problem. The rectangular block matrices A_{21} and B_{12} are the coupling terms of the equation, as the acoustic and elastic problems would be uncoupled if they were to vanish. Coupling thus occurs solely from the boundary conditions (8.1) and (8.2). Solving the eigenvalue problem (8.12) as a function of the wavevector k, dispersion diagrams are readily obtained as $\omega(k)$. The eigenvectors **y** further give the modal shape of each eigenvector, if needed.

Let us come back to the example of the interface between a solid and a fluid media as depicted in Figure 8.2. We want to numerically obtain the velocity c_{ss} of the Scholte–Stoneley wave (SSW) propagating at the interface and compare it to an exact analytical formula [17]. The SSW is evanescent in both the solid and the fluid domains

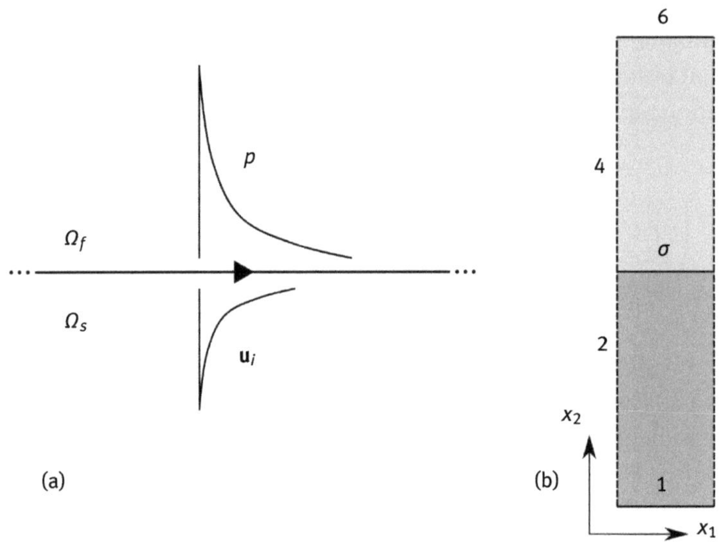

Fig. 8.2: The Scholte–Stoneley wave problem. (a) The Scholte–Stoneley wave propagates at the interface between a solid and a fluid semi-infinite media, and is evanescent on both sides. It is described by a pressure distribution in the fluid, p, and up to three displacements in the solid, u_1, u_2, and u_3. (b) The FEM mesh used to obtain a band structure for the composite solid-fluid plate is artificially truncated vertically compared to the original doubly semi-infinite problem. Boundaries 1 and 6 are left free; boundaries (2,3) and (4,5) are periodic pairs.

and somehow generalizes the Rayleigh wave. Indeed, the Scholte–Stoneley wave dispersion relation can be deduced from the acousto-elastic equations with assumption of absence of tangential stress, continuity of pressure and normal stress, and continuity of normal displacements at the interface. The dispersion relation is given in implicit form as a real root s of

$$4\sqrt{1-s}\sqrt{1-qs}-(s-2)^2=(\rho s^2)\sqrt{1-sq}/\sqrt{1-sr}, \tag{8.15}$$

where

$$\begin{aligned} \rho &= \rho_f/\rho_s, \\ q &= (c_s/c_l)^2, \\ r &= (c_s/c_f)^2, \\ s &= (c_{ss}/c_s)^2. \end{aligned}$$

In these expressions, ρ_f and ρ_s are the mass densities of the fluid and of the solid, c_s and c_l are the shear and longitudinal sound velocities in the solid, c_f is the sound velocity in the fluid, and c_{ss} is the velocity of the Scholte–Stoneley wave.

Although we could try to numerically model the propagation of waves at the interface between two semi-infinite media, we consider for practical reasons finite heights for both the fluid and the solid, as depicted in Figure 8.2 (b). In doing so, we assume that the wave decays exponentially sufficiently fast from both sides of the interface so that it never feels the external surfaces. As we discuss later, this is generally true only for the part of the dispersion diagram that lies under the sound cone. The sound line, in this particular case, is defined by the slowest of all bulk waves. We should thus distinguish the case when the slowest bulk wave is on the fluid side or on the solid side. For this purpose, following Moiseyenko et al. [111], three different combinations of materials are considered: mercury/Plexiglas, water/PVC, and water/silicon. The velocities of bulk acoustic and elastic waves for these materials are given in Table 8.1. First, for the couples mercury/Plexiglas and water/PVC, the shear velocity is smaller than the sound velocity in the fluid. As a result, there is a significant difference between c_f and c_{ss}, and the SSW velocity can be easily extracted by experiment. Second, for the couple water/silicon, the shear velocity is larger than the sound velocity in water. When this is the case, it is usually found that c_{ss} is very close to c_f and their experimental separation is somewhat more difficult.

The FEM domain in Figure 8.2 (b) is two-dimensional. Periodic boundary conditions are applied on both pairs of lateral boundaries, in view of obtaining a solution with no variation along axis x_1, thus mimicking a linear wavefront. The vertical coordinate x_2 registers the spatial distribution of the wavefield, which should tend to an exponentially decreasing function for the Scholte–Stoneley wave. Finally, k_3 is varied while fixing $k_1 = k_2 = 0$, in order to obtain frequency as a function of wavenumber from (8.12). Results for the different fluid/solid couples in Table 8.1 are presented in Figure 8.3.

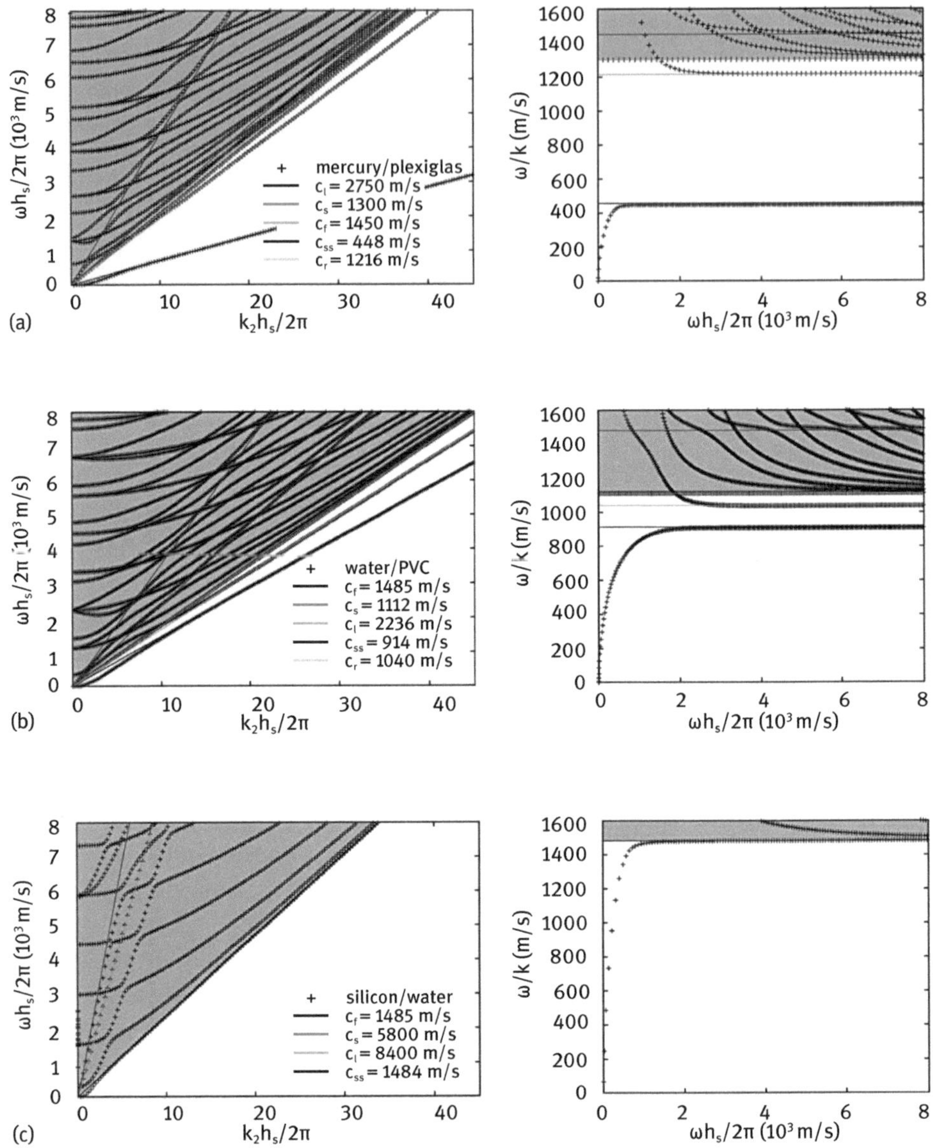

Fig. 8.3: Dispersion relation for acoustic-elastic waves guided along the planar interface between a solid and a fluid. The band structure – or the $\omega(k_z)$ dispersion relation – is shown on the left and the phase velocity ω/k_z is shown on the right as a function of frequency. h_s is the solid thickness. The computation is performed with FEM. The cases of (a) mercury/Plexiglass, (b) water/PVC and (c) water/silicon are presented (after Moiseyenko et al. [111]).

Tab. 8.1: Scholte–Stoneley velocity for some fluid/solid couples. The Scholte–Stoneley velocity, c_{SS}, as obtained from Brekhovskih's dispersion relation [17] for a planar interface [111].

Material	c_{ss} m/s	c_f m/s	c_s m/s	c_l m/s	ρ_f kg/m^3	ρ_s kg/m^3
Mercury/Plexiglas	448	1450	1300	2750	13500	1190
Water/PVC	914	1485	1112	2236	1000	1560
Water/silicon	1484	1485	5800	8400	1000	2330

The radiative region, or sound cone, is depicted in gray in the simulated dispersion diagrams presented in Figure 8.3. It is limited by the sound line, whose slope is given by the smallest of c_s and c_f (since $c_s < c_l$ for any solid material). The dispersion curves are rigorously those of waves propagating inside a composite fluid-solid plate. Outside the sound cone, however, these waves are evanescent both in the fluid and in the solid. The dispersion curve of any wave propagating along the interface must then appear outside the sound cone. In the long wavenumber limit, in addition, the Scholte–Stoneley wave must appear as the asymptotic limit of some dispersion curve. When the product of frequency times plate thickness $\omega h_s/2\pi$ is small, waves outside the sound cone are only weakly evanescent and their velocity is influenced by the surfaces terminating the computational domain. As a result, the bands that eventually converge to the Scholte–Stoneley wave as the frequency increases first emerge as flexural waves at low frequencies. This is not a limitation in practice, since the convergence rate can be made faster by simply increasing the plate thickness.

The cases of mercury/Plexiglas in Figure 8.3 (a) and of water/PVC in Figure 8.3 (b) correspond to $c_s < c_f < c_l$, i.e. the acoustic velocity in the fluid is in-between the shear and the longitudinal elastic velocities in the solid. Under this setting, and with the condition of zero surface traction considered numerically for boundary 1, the Rayleigh surface wave at the free solid surface appears in the nonradiative region, together with the Scholte–Stoneley wave. The Rayleigh wave velocity is easily found from the dispersion relation (8.15) by setting $\rho_f = 0$, i.e. the Scholte–Stoneley dispersion relation resolves into the Rayleigh dispersion relation when the fluid density vanishes. This particular band is spurious for the problem we are considering and should be disregarded in the analysis. In the case of water/silicon, $c_f < c_s < c_l$, the Rayleigh wave appears inside the sound cone and thus only the Scholte–Stoneley wave is expected in the nonradiative region. In summary, for the three cases considered in Figure 8.3, there are only one or two dispersion bands appearing under the sound cone, in accordance with the discussion above. As k goes to infinity, the phase velocity ω/k of these bands approaches asymptotic limit corresponding to the SSW or to the Rayleigh wave, when the latter is present. The agreement between the SSW velocity evaluated by FEM and according to (8.15) is excellent.

8.2 Sonic crystal of solid inclusions in a fluid

In this section, we consider the case of a sonic crystal of solid inclusions immersed in a fluid, usually water, and we investigate the changes in the band structure of 2D and 3D crystals. As we argued in Chapter 4, in the case of a sonic crystal immersed in air, and providing we look at wave propagation in the fluid, it is generally justified to consider the solid as perfectly rigid (no elastic waves in the solid) and hence to replace it with the adequate boundary condition. For a sonic crystal immersed in water, the analysis in the previous section should give us the intuition that the coupling of acoustic and elastic waves will change the picture whenever the contrast between material constants is not too high.

8.2.1 Solid rods in water

In Chapter 4 we presented the crystal of steel rods in water as if it were a mere sonic crystal; the band structures that we discussed were then obtained by replacing steel by a equivalent fluid material supporting only longitudinal acoustic waves. Given the theory developed in the previous section, we can now compute band structures accounting for the coupling of acoustic waves in water and elastic waves in steel. The most salient question is whether complete band gaps will still be present and if they were to be deeply changed. Let us start with the case of the square lattice crystal considered experimentally by Khelif et al. [65] (see Figure 4.13). The band structure considering acoustic-elastic coupling is plotted in Figure 8.4. It looks quite close to the purely acoustic band structure shown in Figure 4.14, apart from some small differences. The bands are generally slightly displaced toward lower frequencies, and in effect the complete band gap occurs for a range of frequency that is slightly lower. The width of the complete band gap, however, is not really changed. Looking more closely at the measurements, the band structure in Figure 8.4 appears to more closely match the gap entrances and exits at the main symmetry points. All in all, the consideration of acoustic-elastic coupling leads to a better agreement between theory and experiment, but does not radically change the picture.

The cases of the hexagonal lattice and its honeycomb variant were considered already in Figures 4.14 and 4.16 for the experiments of Hsiao et al. [53]. The corresponding band structures obtained under the purely acoustic approximation were plotted in Figures 4.15 and 4.17. They can now be compared to the acoustic-elastic coupling result displayed in Figures 8.5 and 8.6. In the hexagonal lattice case, it is again found that the complete band gap is displaced to lower frequencies, by an amount similar to the square lattice case, while the full band gap width is not significantly affected. In the honeycomb case, strikingly, there is almost no change when acoustic-elastic coupling is taken into account.

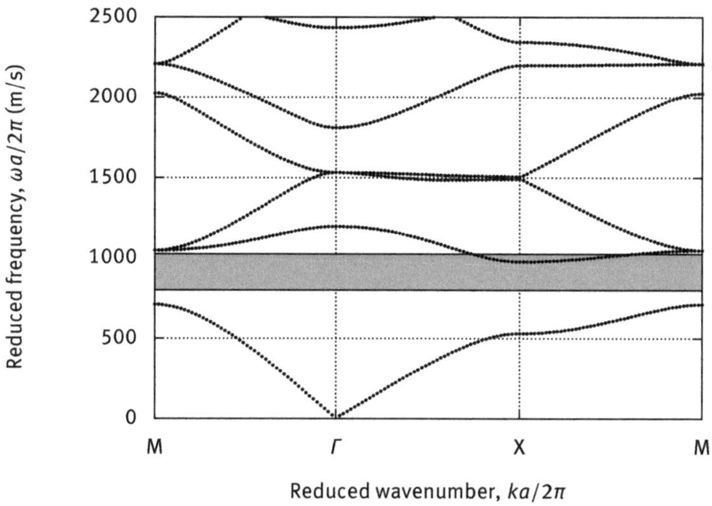

Fig. 8.4: Two-dimensional square-lattice sonic crystal of steel rods in water. The diameter of the cylinders normalized to the lattice constant is $d/a = 0.833$, as in Figure 4.13. The FEM computation includes the coupling of acoustic waves in water with elastic waves inside the steel rods. The gray area marks for comparison the full band gap that was predicted with the pressure wave model in Figure 4.13.

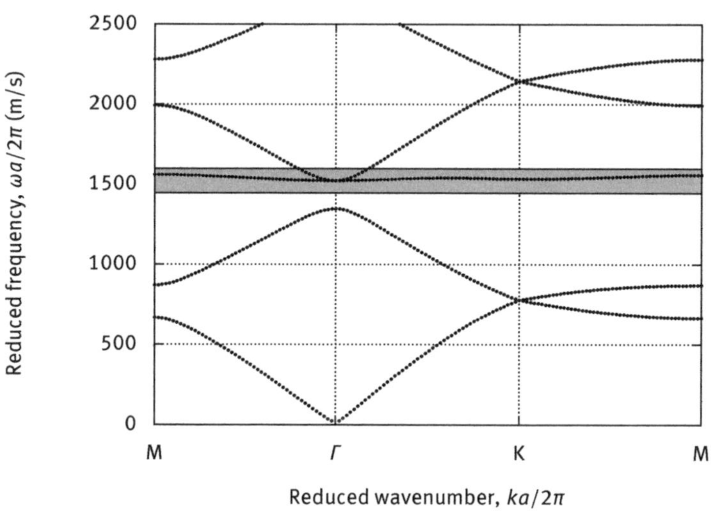

Fig. 8.5: Two-dimensional hexagonal-lattice sonic crystal of steel rods in water. The diameter of the cylinders normalized to the lattice constant is $d/a = 0.8$, as in Figure 4.15. The FEM computation includes the coupling of acoustic waves in water with elastic waves inside the steel rods. The gray area marks for comparison the full band gap that was predicted with the pressure wave model in Figure 4.15.

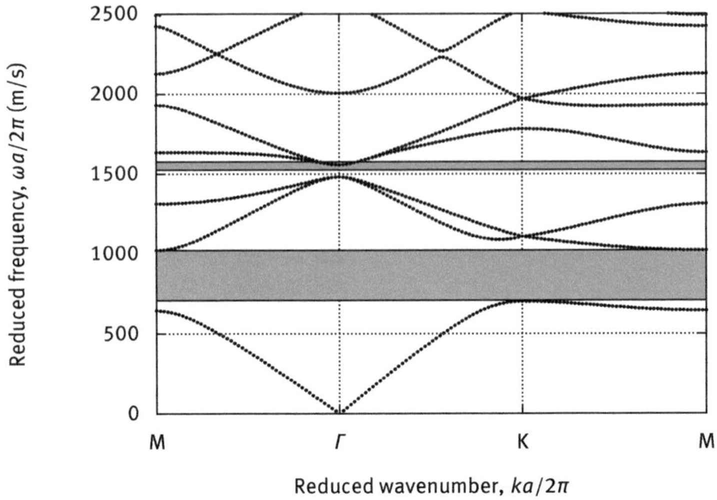

Fig. 8.6: Two dimensional honeycomb-lattice sonic crystal of steel rods in water. The diameter of the cylinders normalized to the lattice constant is $d/a = 0.46$, as in Figure 4.17. The FEM computation includes the coupling of acoustic waves in water with elastic waves inside the steel rods. The gray areas mark for comparison the full band gaps that were predicted with the pressure wave model in Figure 4.17.

8.2.2 Nylon rods in water

It should not be thought that acoustic-elastic coupling only has a small effect on the band structure of fluid-solid sonic crystals. The effect is in fact quite important when the material contrast is less pronounced, so that elastic energy stored inside the solid becomes significant compared to acoustic energy stored in the fluid. As an example, we consider in Figure 8.7 a square-lattice crystal of nylon rods in water. Such a crystal can be easily fabricated from a fishing line [27]. We consider exactly the same geometrical parameters as before for the water-steel square-lattice sonic crystal. The purely acoustic band structure is seen to be very different from the band structure considering acoustic-elastic coupling. Because of the reduced material constant contrast, there is no complete band gap in either case. There are, however, more bands appearing in the coupled case. These bands arise from in-plane polarized elastic waves that would be trapped in the nylon rods if water were not present; the presence of water creates the possibility for them to radiate away. As the coupling of elastic waves in the rods and acoustic waves in water is significant, the two kinds of waves hybridize together to form strongly dispersive bands. Pure shear waves polarized along x_3 do not couple at all with pressure waves in water in the two-dimensional case, and thus cannot be excited by a source of waves placed in water; they form flat bands in the band structure as they are completely confined within the rods and are thus unable to

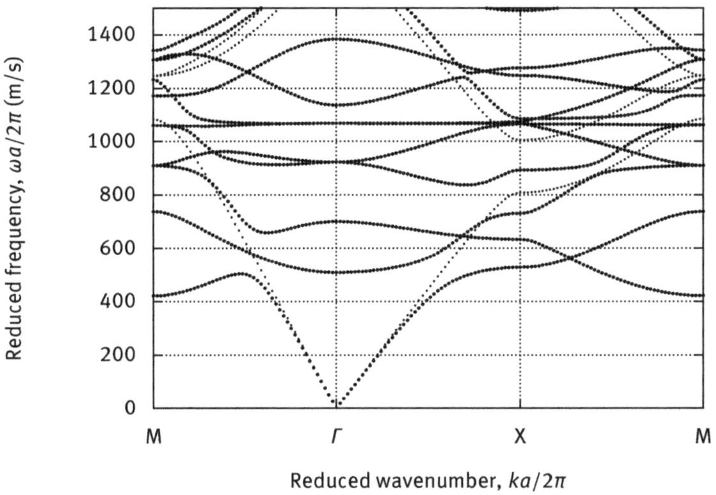

Fig. 8.7: Two-dimensional square-lattice sonic crystal of nylon rods in water. The diameter of the cylinders normalized to the lattice constant is $d/a = 0.8$. The FEM computation shown with large dots includes the coupling of acoustic waves in water with elastic waves inside the nylon rods. For comparison, the FEM computation for pressure waves is shown with small dots.

form a propagating band. One such flat band is observed for instance in Figure 8.7 for $\omega a/2\pi = 1070\,\text{m/s}$.

8.3 Fluid-filled inclusions in 2D phononic crystals

8.3.1 Air holes in 2D phononic crystals

In Chapter 6, we treated phononic crystals with holey inclusions as if they were in a vacuum. Experiments, of course, are generally performed in air and it might seem possible that acoustic waves can be generated inside the holes and maybe form noticeable resonances in measured transmissions. As an example, let us reconsider the square-lattice 2D phononic crystal of holes in silicon in Figure 6.14. Compared to the previous section, the role of the matrix and of the inclusions are exchanged. If there is a hybridization, it will be between the elastic wave continuum and the acoustic vibrations confined within the air holes. It can indeed be observed in Figure 8.8 that many flat bands populate the band structure when acoustic-elastic coupling is taken into account. These flat bands seemingly go through the dispersive bands for elastic waves in the silicon phononic crystal without apparent interaction (compare with the band structure in Figure 6.14; only the lowest frequency range is plotted here). In fact, there is a very faint coupling between both type of waves, but the resolution used to plot the band structure is not sufficient to reveal it. In transmission experiments with wave

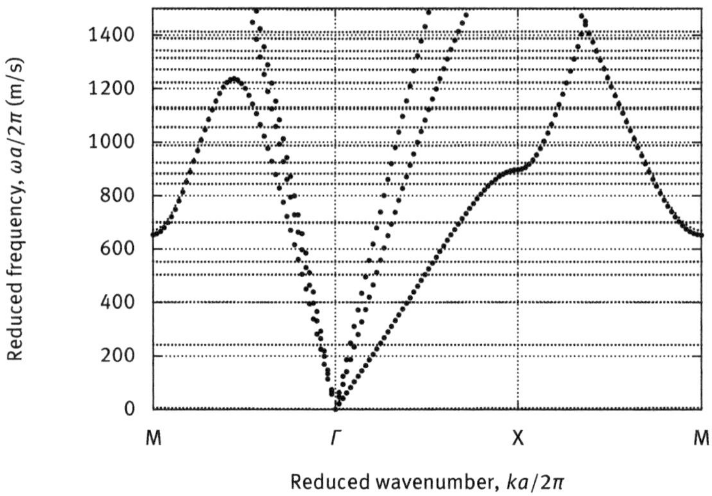

Fig. 8.8: Two-dimensional square-lattice phononic crystal of cylindrical air holes in silicon. The diameter of the cylinders normalized to the lattice constant is $d/a = 0.8$. The FEM computation shown with small dots includes the coupling of acoustic waves in air with elastic waves in silicon. The result of the FEM computation with free boundary conditions at the holes is overlayed with large dots: these bands are coincident with an equal number of bands of the acoustic-elastic computation (compare with Figure 6.14). The many flat bands are produced by resonant modes of the air cylinders.

generation and detection performed in the solid matrix, we would need a very sharp frequency resolution to notice any sign of acoustic resonances, of the order of 10^{-6} in relative terms; their appearance should furthermore not be expected out of short pulse measurements with a limited time window. In this very high contrast case, coupling of acoustic and elastic waves can be safely ignored in practice.

8.3.2 Liquid-filled inclusions as sensors

What if the holey inclusions are filled with a fluid denser than air? This case was considered by Lucklum and Li in view of sensing applications [92]. The underlying idea is the combination of microfluidics with phononic crystal technology. By adding a dense fluid like water inside the holes of a solid-void 2D phononic crystal, we should expect the hybridization of the elastic wave continuum with acoustic vibrations in the holes, as before. The degree of hybridization is also dependent on the material constants contrast.

To be more precise, we consider the same geometrical parameters as Lucklum and Li [92]. The square-lattice 2D phononic crystal is supposed to be made either of tungsten

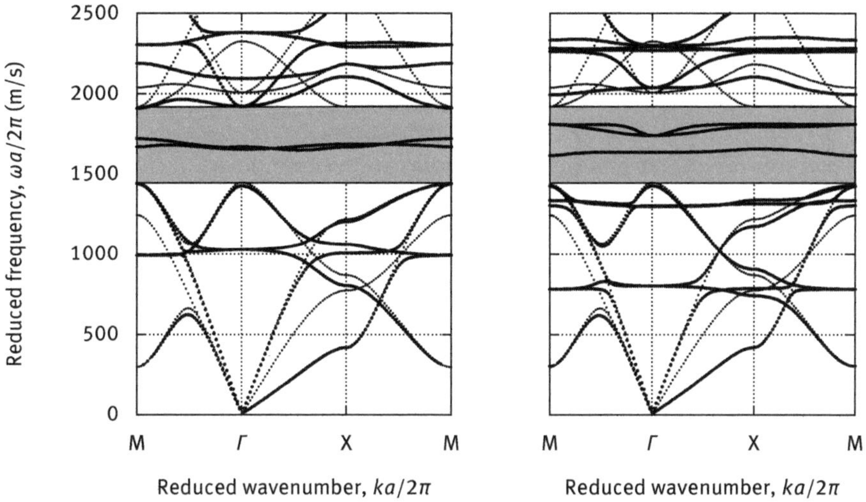

Fig. 8.9: Two-dimensional square-lattice phononic crystal of holes in tungsten, filled with a fluid. The diameter of the cylinders normalized to the lattice constant is $d/a = 0.874$. The FEM computation shown with small dots is for the hollow phononic crystal, while the large dots indicate that the coupling of acoustic waves with elastic waves is taken into account. The fluid is (a) water or (b) Propanol.

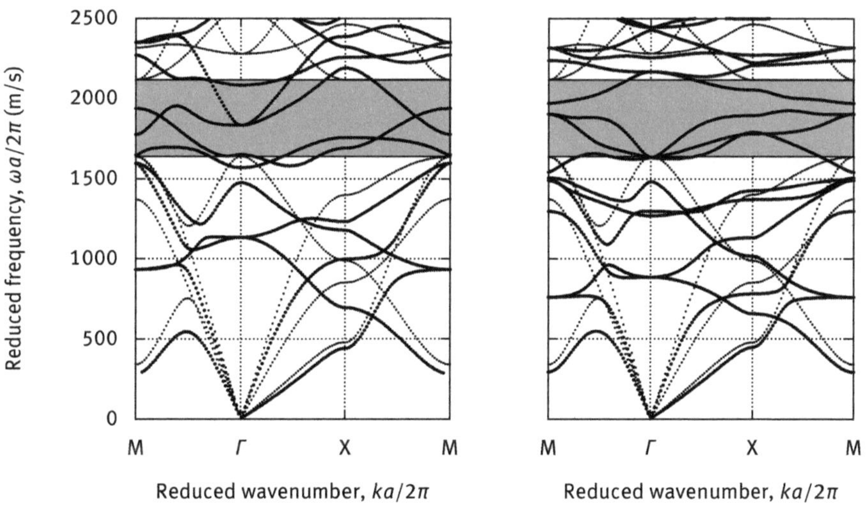

Fig. 8.10: Two-dimensional square-lattice phononic crystal of holes in aluminum, filled with a fluid. The diameter of the cylinders normalized to the lattice constant is $d/a = 0.874$. The FEM computation shown with small dots is for the hollow phononic crystal, while the large dots indicate that the coupling of acoustic waves with elastic waves is taken into account. The fluid is (a) water or (b) Propanol.

or of aluminum and has a filling fraction of 0.6 ($d/a = 0.874$). Both metals actually have comparable elastic wave velocities, but tungsten is significantly denser than aluminum ($\rho = 19\,300\,\mathrm{kg/m^3}$ instead of $\rho = 2\,710\,\mathrm{kg/m^3}$). Two choices are further considered for the fluid, either water or Propanol ($\rho = 786\,\mathrm{kg/m^3}$ and $c_l = 1170\,\mathrm{m/s}$). The phononic band structure for the case of a tungsten matrix is shown in Figure 8.9. The band structures for hollow and fluid-filled holes are superimposed. It can be seen that the complete phononic band gap for hollow holes is globally conserved. Hybridization of acoustic vibrations with the elastic waves of the continuum remains rather weak and manifests through the appearance of reasonably flat bands. For water, for instance, the first resonance appears around $\omega a/2\pi = 1\,000\,\mathrm{m/s}$, below the complete band gap, but the second and the third resonances appear inside it, around $\omega a/2\pi = 1\,700\,\mathrm{m/s}$. A similar observation can be made with Propanol-filled holes, but for a total of three resonances inside the complete band gap. The significance of this finding is that the fluid in the holes can be resonantly excited by incident elastic waves of adequate frequencies. A Fabry–Perot type resonance can for instance be observed in the transmission, without the definition of any defect inside the phononic crystal [92].

What happens if the material contrast between elastic matrix and fluid inclusion is decreased? To illustrate the effect, in Figure 8.10 we plot the case of the same phononic crystal but considering aluminum instead of tungsten for the matrix. For both water and Propanol, the frequency centers of the hybridized fluid resonances are the same as before, but the dispersion of the corresponding bands is significantly increased. As a result, the complete band gap is almost washed out and Fabry–Perot type resonances are no longer to be expected.

8.4 Corrugated surfaces and plates

We now turn to the corrugated solid-fluid interface situation depicted in Figure 8.11 (a) and consider the propagation of interface elastic waves generalizing the Scholte–Stoneley wave. For numerical experiments we consider the interface between silicon and water, following Moiseyenko et al. [111]. This choice of materials is justified from the availability of MEMS technologies for etching deep corrugations in silicon wafers. The general trends that we describe below, however, are expected to remain valid for other material combinations. In this case, propagation is considered along the x_1 axis, and we plot dispersion diagrams given the angular frequency ω as a function of the k_1 component of the wavevector. The computational unit cell is of the form displayed in Figure 8.11 (b), with a the grating period, d the corrugation width, and h the corrugation depth. For definiteness, $d/a = 0.5$ is chosen and h/a is varied. Periodic boundary conditions are imposed on boundary pairs: for pressure on boundaries 4 and 5, and for displacements on boundaries 2 and 3. In addition, boundaries 1 and 6 are left free (Neumann boundary condition), as in the planar interface case.

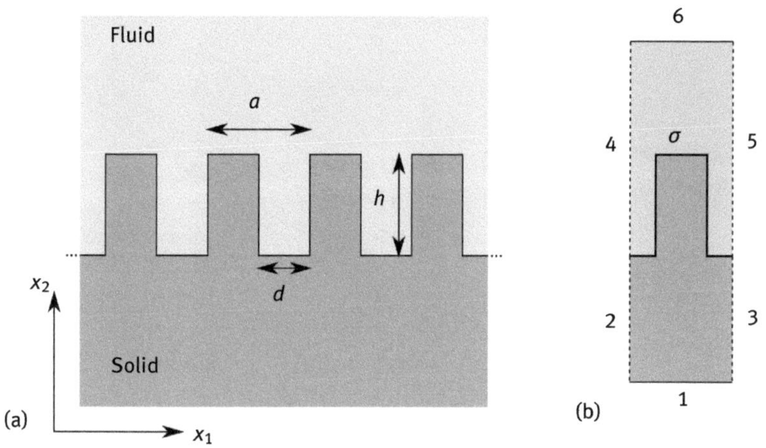

Fig. 8.11: One-dimensional solid grating immersed in water. (a) The one-dimensional grating is periodic along axis x_1. (b) The FEM mesh used to obtain a band structure for the composite solid-fluid plate is artificially truncated vertically compared to the original doubly semi-infinite problem. Boundaries 1 and 6 are left free; boundaries (2,3) and (4,5) are periodic pairs including a Bloch wavenumber k_1.

Figure 8.12 shows dispersion diagrams for six different values of h/a, ranging from 0 to 2.0. The interesting part of these diagrams is the nonradiative region appearing below the sound line in water. It can be seen that increasing the corrugation depth results in two main phenomena. First, new bands appear in the nonradiative region as the corrugation depth is increased, corresponding to the existence of more and more interface waves, in contrast to the existence of only the Scholte–Stoneley wave for the planar interface. Second, the bands in the nonradiative region become flatter as the depth increases, indicating that the group velocity becomes very small. For the lowest band and starting from $h/a = 1$, the group velocity is even slightly negative for intermediate values of $k_1 a/(2\pi)$.

Even though each interface wave clearly lies on a distinct dispersion curve, the detailed polarization field can undergo strong transformations as the corrugation depth increases. The modal field distributions shown in Figure 8.13 are computed for $h/a = 2.5$. The interface wave on the lowest dispersion curve, labeled A, is a strongly evanescent quasi-longitudinal surface wave giving rise to almost no pressure variation in water, as Figure 8.13 (a) shows. Interface waves B and C illustrated in Figure 8.13 (b) and Figure 8.13 (c) have coupled displacements in the sagittal plane and create a pressure variation that is confined inside the corrugation. Interface wave F in Figure 8.13 (f) is also of a similar type, with the pressure field extending deeply in open water. Interface wave D displayed in Figure 8.13 (d) is a pure shear horizontal wave that induces no pressure variation in water. Such a wave can clearly only be excited from the solid. Interface wave E is mostly vertically shear polarized inside the solid ridges and is ac-

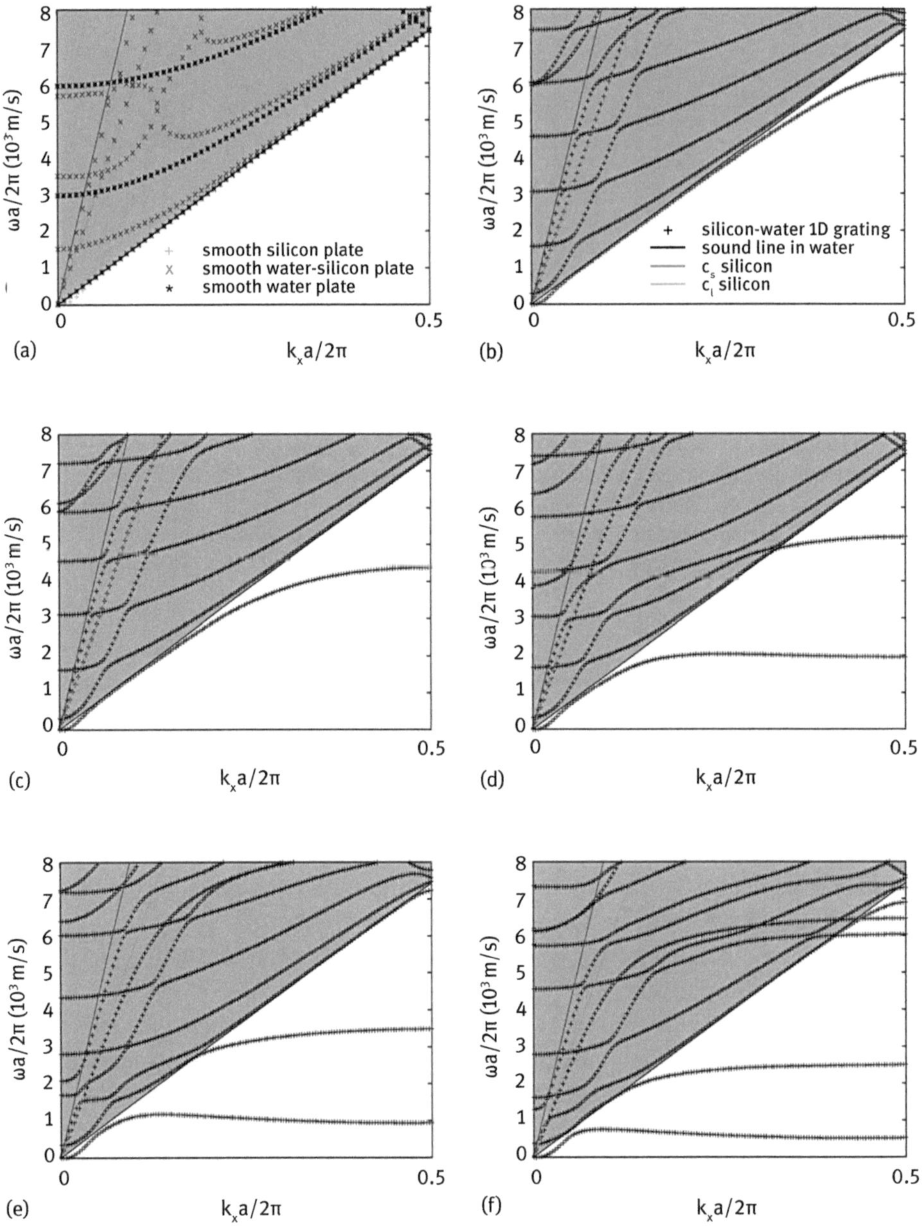

Fig. 8.12: Band structures $\omega(k_x)$ for guided waves obtained by FEM for a corrugated interface of rectangular profile with normalized corrugation depth (a) $h/a = 0$, (b) 0.25, (c) 0.5, (d) 1, (e) 1.5, and (f) 2. a is the corrugation pitch. The example is given for a corrugated silicon plate immersed in water. In the case of the planar interface (a), the band structures for a free silicon plate and for a free water plate are also shown for comparison [111].

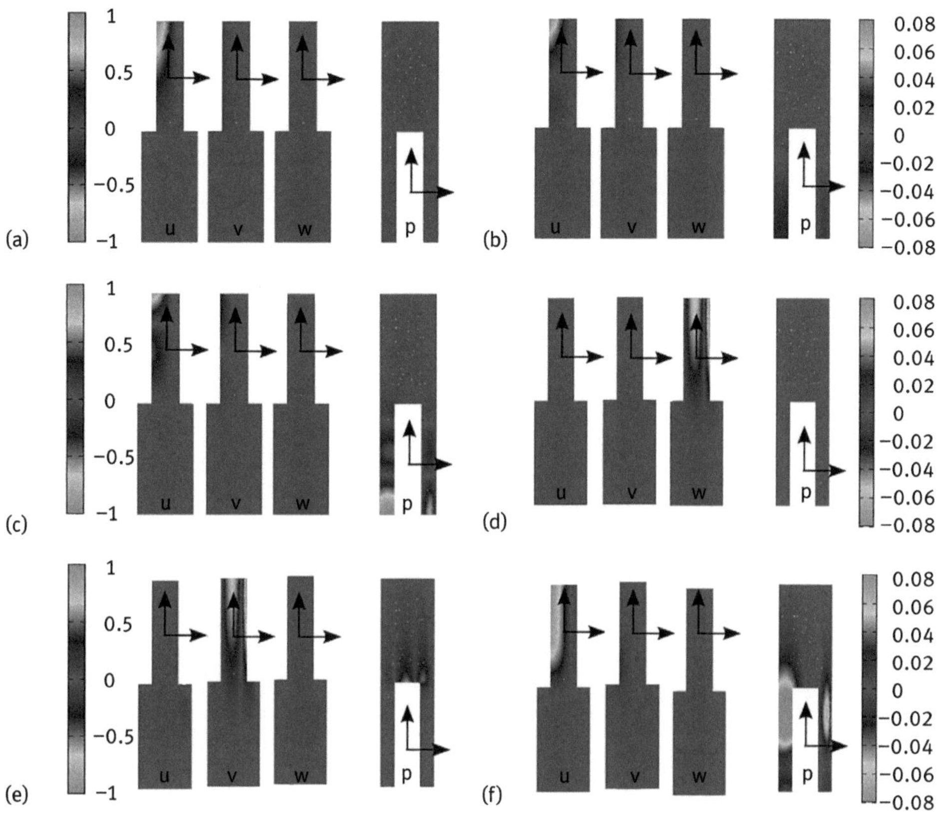

Fig. 8.13: Normalized displacements and pressure for eigenmodes A2, B, C, D, E and F in Figure 8.12. The corrugation depth $h/a = 2.5$. Normalization is performed so that the maximum total displacement is 1 nm; correspondingly, pressure is in units of 10^8 Pa [111].

companied with an evanescent pressure field extending well into open water. The extension of the pressure field into open water for the different interface waves depends directly on the distance from the dispersion curve to the sound line, as suggested by the square-root dependence in the dispersion relation for the Scholte–Stoneley wave.

9 Evanescent Bloch waves

9.1 Evanescent waves and Bloch's theorem

In this chapter, we discuss evanescent waves in sonic and phononic crystals. We met evanescent waves very early in Chapter 2, when we discussed some simple one-dimensional problems. We actually needed them to describe complex solutions of the dispersion relation of homogeneous media, waves guided between two mirrors, Bloch waves of periodic chains of masses connected by springs and of sinusoidal gratings, and waves in systems of resonators connected to waveguides. We met evanescent waves again in the description of surface guided waves in Chapters 5, 7, and 8. We have not yet, however, made the full connection with Bloch's theorem and we have so far presented band structures limited to propagating Bloch waves. This connection is the subject of the present chapter. Before detailing how evanescent Bloch waves can be obtained and how complex band structures can be understood, we start by a discussion around three basic questions.

Why do we need to describe evanescent waves in sonic and phononic crystals? The most relevant answer to this question is that we need evanescent Bloch waves in order to describe the inside of frequency band gaps. Actually, it is not enough to say that there are no propagating waves inside a band gap, we must also say what is there and what Bloch waves have transformed into. In addition to the situations mentioned above, evanescent waves are also involved in the description of all phenomena where the near-field properties matter. For instance, the scattering of a monochromatic plane wave on a small object, if described with an angular spectrum of waves approach, must include a summation over all possible plane waves satisfying the dispersion relation in the incident medium, implying the consideration of a continuum of complex-valued wavevectors [16].

There is another classical situation with a strong relation to sonic and phononic crystals: diffraction gratings. Let us consider the one-dimensional diffraction grating depicted in Figure 9.1 (a) [78]. An incident plane wave with angular frequency ω and wavevector $\mathbf{k}$ has an amplitude proportional to $\exp(-\imath(k_1x_1 + k_2x_2))$. The field diffracted by the grating is composed of a superposition of plane waves:

$$u(x_1, x_2) = \sum_{n=-\infty}^{\infty} a_n \exp\left(-\imath(k_{1n}x_1 + k_{2n}x_2)\right), \tag{9.1}$$

$$k_{2n} = k_2 + n\frac{2\pi}{a}, \quad k_{1n} = \pm\sqrt{\frac{\omega^2}{c^2} - k_{2n}^2}, \tag{9.2}$$

where c is the velocity for the particular kind of wave considered. These expressions derive from the phase matching principle (for the expression of k_{2n}) and from the dispersion relation in the homogeneous medium. In (9.1), every Fourier harmonic has

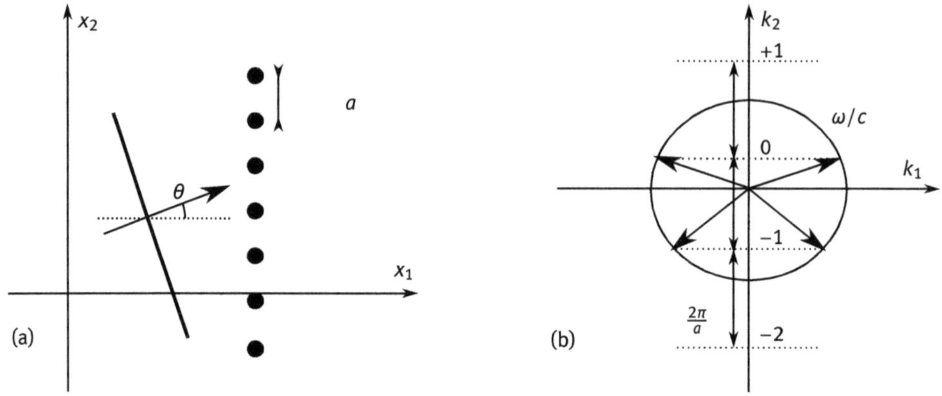

Fig. 9.1: A one-dimensional diffraction grating. (a) A monochromatic plane wave is incident on a periodic array of objects placed in an isotropic homogeneous medium. (b) Graphical construction of propagating and evanescent diffracted waves at a given angular frequency ω and angle of incidence θ. All wavevectors falling out of the dispersion relation – a circle of radius ω/c – are evanescent; this happens whenever $|k_{2n}| > \omega/c$. Wavevectors falling on the circle are propagating; in the example shown, only orders of diffraction $n = 0$ and $n = -1$ are propagating. Orders of diffraction with $n \geq 1$ or $n \leq -2$ are evanescent.

amplitude a_n and the wavevector component k_{1n} is either a real or a purely imaginary complex number, depending on the sign of the expression under the square root in (9.2). Note that k_{2n} remains real valued. Figure 9.1 (b) shows a graphical construction to judge which diffraction orders are propagating and which are evanescent.

At a given frequency ω, evanescent waves are characterized by the fact that their wavevector, $\mathbf{k}(\omega)$, is complex valued. In the simple grating model above, each diffracted wave with $n \neq 0$ becomes evanescent under the cut-off frequency $\omega_n = |k_{2n}|c$ but is propagating above it. Now, since the incident wave satisfies the dispersion relation, we have $k_1 = (\omega/c)\sin\theta$ and $k_2 = (\omega/c)\cos\theta$ if θ is the angle of incidence. The expression for the cut-off frequencies becomes

$$\omega_n = \begin{cases} \frac{n}{1-\sin\theta}\frac{2\pi c}{a} & \text{if } n \geq 0, \\ \frac{|n|}{1+\sin\theta}\frac{2\pi c}{a} & \text{if } n < 0. \end{cases} \tag{9.3}$$

The complex band structure for diffracted waves on transmission (in the direction of positive x_1) is

$$k_{1n}(\omega) = \begin{cases} \sqrt{\frac{\omega^2}{c^2} - (\frac{\omega}{c}\sin\theta + n\frac{2\pi}{a})^2} & \text{if } \omega \geq \omega_n, \\ -\imath\sqrt{(\frac{\omega}{c}\sin\theta + n\frac{2\pi}{a})^2 - \frac{\omega^2}{c^2}} & \text{otherwise.} \end{cases} \tag{9.4}$$

Note that for reflected diffracted waves it is sufficient to use the same expressions with reversed signs. The complex band structure (9.4) is plotted in Figure 9.2 for both normal incidence and oblique incidence. Note that diffraction orders $\pm n$ are degenerate

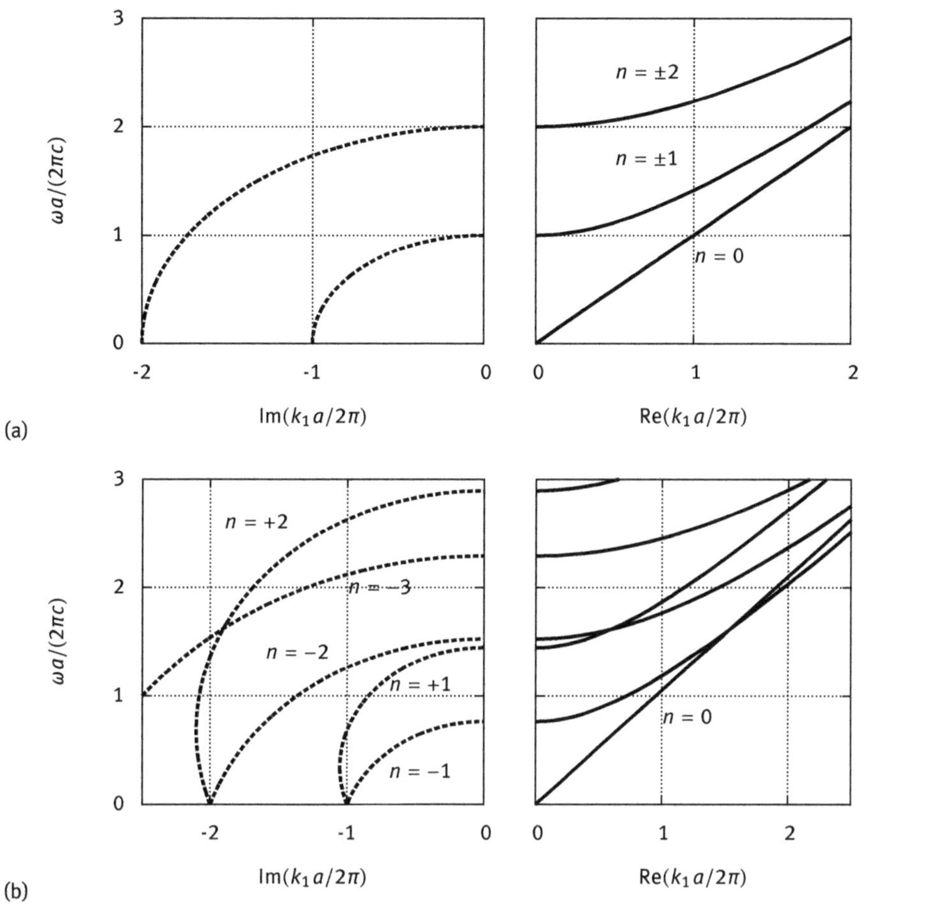

Fig. 9.2: Complex band structure of the one-dimensional diffraction grating. The complex band structure is plotted from the analytical expression (9.4) for the complex wavenumber k_1 as a function of angular frequency ω and of diffraction order n for (a) normal incidence ($\theta = 0$) and for (b) oblique incidence ($\theta = \pi/10$).

in the complex band structure for normal incidence, but that the degeneracy is lifted for oblique incidence. Whatever the angle of incidence, the n-th complex band starts at the zero frequency with the imaginary wavenumber

$$k_{1n}(\omega = 0) = -\imath|n|\frac{2\pi}{a}.$$

Should we select complex k or complex ω? In the example of the one-dimensional grating, we have been naturally driven toward a formulation where we found $k_1(\omega)$, i.e. we obtained the complex wavenumber in the direction of propagation as a function of a real frequency. Anticipating the following derivation, we might say the spatial inhomogeneity was the reason for us taking this path. This was also the case with the linear chain of masses connected by springs and with the array of resonators grafted along a waveguide in Chapter 2. For waves propagating in spatially periodic regions of space, it is intuitive that the complex $k(\omega)$ approach makes sense for monochromatic waves in the sense that it measures the spatial decay of waves upon propagation.

Now it is equally true that complex frequencies are very useful to describe the attenuation of vibrations in time. The quality factor of a resonance, Q, measures the number of oscillations a resonator can support before the vibration energy has decreased by a factor $e^{-2\pi} \approx 0.2\%$. It can be used to define a complex frequency $\omega_0 = \omega_r(1 + \frac{\imath}{2Q})$, with $\omega_r = \frac{2\pi}{T}$ the angular resonant frequency and T the period of oscillation. Indeed, after a time $t = QT$, the vibration energy is proportional to

$$\exp(-2\,\mathrm{Im}(\omega_0)t) = \exp\left(-\frac{\omega_r}{Q} \times QT\right) = \exp(-2\pi).$$

Looking at the equation systems we have solved in the previous chapters for both sonic and phononic crystals, Bloch waves are generally found as the solutions of the linear problem

$$M(\omega, \mathbf{k})\mathbf{y} = 0, \tag{9.5}$$

with $M(\omega, \mathbf{k}) = K(\mathbf{k}) - \omega^2 M$ and $\mathbf{y}$ a vector of coefficients describing the Bloch wave in a particular function basis (e.g. an exponential basis for PWE or a piecewise continuous polynomial basis for FEM). The dispersion relation can then generally be written as the secular equation

$$\det M(\omega, \mathbf{k}) = 0. \tag{9.6}$$

As an algebraic equation, the dispersion relation has in general complex-valued $(\omega, \mathbf{k})$ solutions. From a purely mathematical point of view, there is no reason to prefer real over complex values for either the frequency or the wavevector. What matters is actually the physical meaning we can associate to them.

Coming back to the wave propagation problem, we can try to formally look for a Green's function that would give the elementary response to an elementary excitation. Writing the Green's function as a time and space Fourier transform

$$g(t, \mathbf{x}) = \frac{1}{(2\pi)^4} \int_{-\infty}^{+\infty} \mathrm{d}\omega \int_{-\infty}^{+\infty} \mathrm{d}\mathbf{k}\bar{g}(\omega, \mathbf{k}) \exp(\imath\omega t) \exp(-\imath\mathbf{k} \cdot \mathbf{x}), \tag{9.7}$$

the Fourier transform is formally

$$\bar{g}(\omega, \mathbf{k}) = M(\omega, \mathbf{k})^{-1}\bar{\delta}(\omega, \mathbf{k}), \tag{9.8}$$

with $\delta(t, \mathbf{x})$ denoting the elementary excitation. Note that the variables of integration ω and $\mathbf{k}$ in the integrals are real valued for the purpose of defining the Fourier transforms. We will assume that the order of the integrals can be interchanged. The spectral representation of the Green's function, $\bar{g}(\omega, \mathbf{k})$, is well defined everywhere except where the dispersion relation (9.6) is satisfied. At such $(\omega, \mathbf{k})$ points, it is singular.

Let us first assume that we can write the singular part of the spectral Green's function as

$$\bar{g}(\omega, \mathbf{k}) \approx \frac{g(\mathbf{k})}{\omega - \omega_p(\mathbf{k})}, \tag{9.9}$$

with $\omega_p(\mathbf{k})$ a complex-valued band in the band structure. We could for instance use the classical band structure computation with possibly complex-valued material constants to obtain this function. Then the singular part of the Green's function would be

$$g(t, \mathbf{x}) \approx H(t)\ \mathrm{Re}\left(\int \mathrm{d}\mathbf{k} g(\mathbf{k}) \exp(\imath(\omega_p(\mathbf{k})t - \mathbf{k}\cdot\mathbf{x}))\right). \tag{9.10}$$

This function would describe the attenuation in time of a causal wavepacket. It should be recognized, however, that this description is not very well suited to including the effects of spatial inhomogeneity, including periodicity. More precisely, inhomogeneity induces a strong dispersive character that is not captured by writing in (9.9) that the pole amplitude $g(\mathbf{k})$ is independent of frequency. Such a formulation is well suited for homogeneous space and for transient problems.

Alternatively, we can write the singular part of the spectral Green's function as

$$\bar{g}(\omega, \mathbf{k}) \approx \frac{g(\omega)}{\mathbf{k} - \mathbf{k}_p(\omega)} \tag{9.11}$$

with $\mathbf{k}_p(\omega)$ a band in the complex band structure. This choice basically amounts to reverting the roles of frequency and wavenumber in band structures. We argue that $\mathbf{k}_p(\omega)$ can be defined easily even in the case of spatial inhomogeneity and periodicity, because all that is required is to solve the monochromatic Helmholtz equation under appropriate boundary conditions. The singular part of the Green's function would then be

$$g(t, \mathbf{x}) \approx \int \mathrm{d}\omega g(\omega) \exp(\imath(\omega t - \mathbf{k}(\omega)\cdot\mathbf{x})). \tag{9.12}$$

In the context of sonic and phononic crystals, complex $\omega(\mathbf{k})$ approaches have been used to describe the effect of loss on attenuation [56, 97]. Complex $\mathbf{k}(\omega)$ have also been used to describe viscoelastic loss, or more generally frequency-dependent loss [109, 124]. The latter complex band structure approach is more general in that it also describes evanescent Bloch waves in relation to periodicity, both in the direction of propagation (Bragg band gaps) and in the transverse direction (orders of diffraction). As a result, we will systematically develop it in the rest of this chapter.

Do we have any demonstration that Bloch's theorem extends to complex wavenumbers? It is clear that the derivation of Bloch's theorem that we presented in Chapter 2 requires real wavenumbers because spatial periodicity is strictly enforced by the translation operators. Still, we have provided several examples where complex wavenumbers have introduced themselves naturally, without invoking Bloch's theorem: one-dimensional Bragg band gaps in linear chains of masses connected by springs, one-dimensional local resonances of resonators grafted on a waveguide, or diffraction on gratings. It appears that the mathematical expression of the Bloch theorem (2.28) applies to a larger class of physical problems than the derivation would seem to imply. But still, this is not a demonstration.

9.2 Evanescent Bloch waves of sonic crystals

A complex band structure is the mapping $k(\omega)$ of the wavenumber in the direction of propagation as a function of angular frequency. A general method for obtaining an eigenvalue problem in the form $k(\omega)$ instead of the classical $\omega(k)$ eigenvalue problem is to come back to the first-order differential equations for strain and stress (for phononic crystals) or for pressure and velocity (for sonic crystals) [78]. Since we seek to obtain the wavenumber in the direction of propagation, it will be found that in the unknown physical fields, displacements must be complemented with stresses in the propagation direction or that pressure must be complemented with velocity in the propagation direction. Though the method was historically first introduced for phononic crystals, we start here with the simpler case of sonic crystals.

9.2.1 Analysis via the plane wave expansion

The PWE version of the sonic complex band structure (for pressure waves) was introduced by Romero–Garcia et al. [129, 130] closely following the original phononic derivation for elastic waves [78]. We use the notations from Chapter 4 to set out the method. We start from the PWE equations written for pressure and velocity (4.22) and (4.23),

$$\mathcal{A}\Gamma_i \mathbf{P} = \omega \mathbf{V}_i, \tag{9.13}$$

$$\omega \mathcal{B} \mathbf{P} = \Gamma_i \mathbf{V}_i, \tag{9.14}$$

where we recall that the matrices $\mathcal{A}$ and $\mathcal{B}$ are formed from the Fourier series of the material constants, and that $\mathbf{V}_i$ and $\mathbf{P}$ are vectors composed with the Fourier series coefficients of velocity and pressure. We consider a certain direction of propagation along the unit vector $\boldsymbol{\alpha}$, which can be a direction in reciprocal space.[1] We also add a possible fixed translation in reciprocal space $\mathbf{k}_0$, so that the wavevector is written

$$\mathbf{k} = k\boldsymbol{\alpha} + \mathbf{k}_0. \tag{9.15}$$

1 Since we are looking for the possible values of the wavenumber k at given angular frequency ω, the wavevector is not constrained to the first Brillouin zone. Still, periodicity is enforced in the equations.

The scalar k is the wavenumber we plan to play the role of the eigenvalue. Differential matrices Γ_i then have coefficients

$$(\Gamma_i)_{mn} = \delta_{m-n}(k_i + G_{im}) = \delta_{m-n}(k_{i0} + G_{im}) + \delta_{m-n}k\alpha_i. \tag{9.16}$$

The basic idea of the following derivation is to isolate terms with the eigenvalue k from those that do not contain it. We denote Γ_{i0} the matrix with elements $\delta_{m-n}(k_{i0} + G_{im})$. This matrix includes both periodicity and the fixed translation in reciprocal space. Thus

$$\Gamma_i = \Gamma_{i0} + k\alpha_i I_d. \tag{9.17}$$

Replacing in (9.13) and (9.14), we have

$$\begin{aligned}\mathcal{A}\Gamma_{i0}\mathbf{P} + k\mathcal{A}\alpha_i\mathbf{P} &= \omega\mathbf{V}_i,\\ \omega^2\mathcal{B}\mathbf{P} = \Gamma_{i0}\omega\mathbf{V}_i + k\omega\alpha_i\mathbf{V}_i &= \Gamma_{i0}\mathcal{A}\Gamma_{i0}\mathbf{P} + k\Gamma_{i0}\mathcal{A}\alpha_i\mathbf{P} + k\omega\alpha_i\mathbf{V}_i.\end{aligned}$$

Looking at the second equation, the natural eigenvector appears to be $\mathbf{H} = (\mathbf{P}, \omega\alpha_i\mathbf{V}_i)^T$, with the second term having the units and meaning of the acceleration in the direction of propagation. Left multiplying the first equation with α_i and summing over i, we obtain the equation system

$$\begin{pmatrix} \omega^2\mathcal{B} - B & 0 \\ -C_2 & I_d \end{pmatrix}\mathbf{H} = k\begin{pmatrix} C_1 & I_d \\ D & 0 \end{pmatrix}\mathbf{H}, \tag{9.18}$$

with

$$B = \Gamma_{i0}\mathcal{A}\Gamma_{i0}, \quad C_1 = \Gamma_{i0}\mathcal{A}\alpha_i, \quad C_2 = \alpha_i\mathcal{A}\Gamma_{i0}, \quad D = \alpha_i\mathcal{A}\alpha_i = \mathcal{A}. \tag{9.19}$$

We have hence obtained a generalized eigenvalue problem for the wavenumber k, as planned. Note the very close similarity of (9.18) and (9.19) with the PWE expression for surface elastic waves (7.10) and (7.11). In fact, the surface problem included the normal stress components, orthogonal to the directions of periodicity, while we consider here the acceleration in the direction of propagation which is a direction of periodicity.

The generalized eigenvalue problem (9.18) has no particular symmetries and hence in general has complex-valued solutions. In the absence of loss, the material constants are purely real and so are matrices $\mathcal{B}$ and $\mathcal{A}$, and the matrices of the eigenvalue problem. In this case, complex eigenvalues appear in complex-conjugate pairs, with the implication that evanescent Bloch waves always come in pairs of equally decreasing and increasing waves with the same phase velocity (the same real part of the wavenumber and the opposite sign of the imaginary part). It can also be remarked that the classical PWE band structure is contained in the top left block of (9.18). Indeed, putting $k = 0$ we obtain $(\omega^2\mathcal{B} - \Gamma_{i0}\mathcal{A}\Gamma_{i0})\mathbf{P} = 0$, which is just another way of writing (4.24).

9.2.2 Finite element modeling

We now derive the FEM version of the complex band structure of sonic crystals. To obtain a variational formulation of this problem, we again seek to obtain an eigenvalue equation for the wavenumber k while ω is a parameter. The derivation is in fact similar to the PWE derivation. In the following, $\nabla\bar{p}$ must be understood as a shorthand for $\nabla\bar{p} - \imath\mathbf{k}_0\bar{p}$, where the translation in reciprocal space $\mathbf{k}_0$ was introduced in (9.15). We rewrite (4.6) and (4.7) as

$$-\frac{1}{\rho}\nabla\bar{p} + \imath k\boldsymbol{\alpha}\frac{1}{\rho}\bar{p} = \imath\omega\bar{\mathbf{v}}, \tag{9.20}$$

$$\omega^2\frac{1}{B}\bar{p} = \nabla\cdot(\imath\omega\bar{\mathbf{v}}) - \imath k\boldsymbol{\alpha}\cdot(\imath\omega\bar{\mathbf{v}}). \tag{9.21}$$

Next we define the normal acceleration (or acceleration along the propagation direction) as

$$\phi = \imath\omega\boldsymbol{\alpha}\cdot\bar{\mathbf{v}}, \tag{9.22}$$

and we transform the two equations above to the two scalar equations

$$-\boldsymbol{\alpha}\cdot\frac{1}{\rho}\nabla\bar{p} + \imath k\frac{1}{\rho}\bar{p} = \phi, \tag{9.23}$$

$$\omega^2\frac{1}{B}\bar{p} = -\nabla\cdot\left(\frac{1}{\rho}\nabla\bar{p}\right) + \imath k(\nabla\cdot\boldsymbol{\alpha})\frac{1}{\rho}\bar{p} - \imath k\phi. \tag{9.24}$$

The scalar differential operator $\boldsymbol{\alpha}\cdot\nabla = \nabla\cdot\boldsymbol{\alpha} = \alpha_i\frac{\partial}{\partial x_i}$. Rearranging the terms we get

$$\phi + \frac{1}{\rho}(\boldsymbol{\alpha}\cdot\nabla)\bar{p} = \imath k\frac{1}{\rho}\bar{p}, \tag{9.25}$$

$$\omega^2\frac{1}{B}\bar{p} + \nabla\cdot\left(\frac{1}{\rho}\nabla\bar{p}\right) = \imath k(\boldsymbol{\alpha}\cdot\nabla)\frac{1}{\rho}\bar{p} - \imath k\phi. \tag{9.26}$$

A variational form of these equations is obtained by considering a vector of two test functions $(\phi', \bar{p}')$ living in the same finite element space as $(\phi, \bar{p})$ and by taking the scalar product[2] of (9.25) with ϕ' and (9.26) with $\bar{p}'$. We obtain

$$\int_\Omega d\mathbf{x}A(\phi', \bar{p}'; \phi, \bar{p}) = (\imath k)\int_\Omega d\mathbf{x}B(\phi', \bar{p}'; \phi, \bar{p}), \forall(\phi', p'), \tag{9.27}$$

with

$$A(\phi', \bar{p}'; \phi, \bar{p}) = \phi'^*\phi + \phi'^*\frac{1}{\rho}(\boldsymbol{\alpha}\cdot\nabla\bar{p}) + \omega^2\bar{p}'^*\frac{1}{B}\bar{p} - (\nabla\bar{p}')^\dagger\frac{1}{\rho}\nabla\bar{p}, \tag{9.28}$$

$$B(\phi', \bar{p}'; \phi, \bar{p}) = \phi'^*\frac{1}{\rho}\bar{p} - (\boldsymbol{\alpha}\cdot\nabla\bar{p}')^*\frac{1}{\rho}\bar{p} - \bar{p}'^*\phi. \tag{9.29}$$

2 The scalar product is Hermitian here, as the physical fields are complex. The Floquet phase factors $\exp(-\imath\mathbf{k}_0\cdot\mathbf{x})$ for ϕ and $\bar{p}$, and $\exp(\imath\mathbf{k}_0\cdot\mathbf{x})$ for ϕ'^* and $\bar{p}'^*$ are omitted here for simplicity, but they must be taken into account for the full derivation of the variational formulation; see Section 4.1.5.

The matrix form of this variational problem obtained by FEM again assumes the form of a generalized eigenvalue problem (but now for $\imath k$), with nonsymmetrical matrices. As a consequence, the eigenvalues are in general complex. In case B and ρ are real-valued functions, eigenvalues come out as complex conjugate pairs, as we observed before for the PWE equations.

9.2.3 Complex band structure

In order to illustrate the complex band structure of sonic crystals, we consider the case of a square-lattice crystal of steel rods in water. Of course, steel is represented in this context by the equivalent fluid material supporting only pressure (longitudinal) waves. We also include a discussion of the impact of symmetry, and in particular of deaf bands, on the complex band structures, by considering both circular and half-circular rods. Numerical results are obtained with the FEM model for sonic crystals. Comparison with the classical band structure can be made by comparing with Figures 4.20 and 4.21. The diameter to lattice constant ratio is fixed to $d/a = 0.8$ in the following.

Figure 9.3 presents the complex band structure for a 2D sonic crystal of circular steel rods in water, obtained for the X direction of the first Brillouin zone, i.e. for $\mathbf{k}_0 = 0$ and $\boldsymbol{\alpha} = (1, 0, 0)^T$. Two types of presentation are shown. The first presentation uses two side-by-side panels giving angular frequency versus the imaginary and the real part of the wavenumber. It should be clear that these are scatter plots that were obtained by scanning ω and obtaining the eigenvalues k for each value. The panels are limited artificially to $0 \leq \text{Re}(k) < \pi/a$ and $\text{Im}(k) \leq 0$, but eigenvalues cover a wider range. The choice made is justified by the following facts.

- The real part respects periodicity and thus can always be folded back to the first Brillouin zone.
- Eigenvalues come in complex-conjugate pairs, so that we show only those with a negative imaginary part, corresponding to evanescent waves decreasing in the direction of propagation. There is an equal number of evanescent waves with a positive imaginary part of k, that are increasing in the direction of propagation and of course decreasing in the reverse direction.
- Under the exchange $\boldsymbol{\alpha} \longrightarrow -\boldsymbol{\alpha}$ (for $\mathbf{k}_0 = 0$), the eigenvalue problem is unchanged apart from $k \longrightarrow -k$ as can be checked directly from (9.19) for instance.

The second presentation of the complex band structure is on a cube for which ω varies along the vertical direction while the complex plane $(\text{Re}(k), \text{Im}(k))$ is placed horizontally. With this presentation, the classical band structure would be located on the rear face of the cube. Any evanescent Bloch waves thus appear away from this rear face. Purely imaginary wavevectors are located on the lateral faces, while general complex wavevectors appear inside the cube.

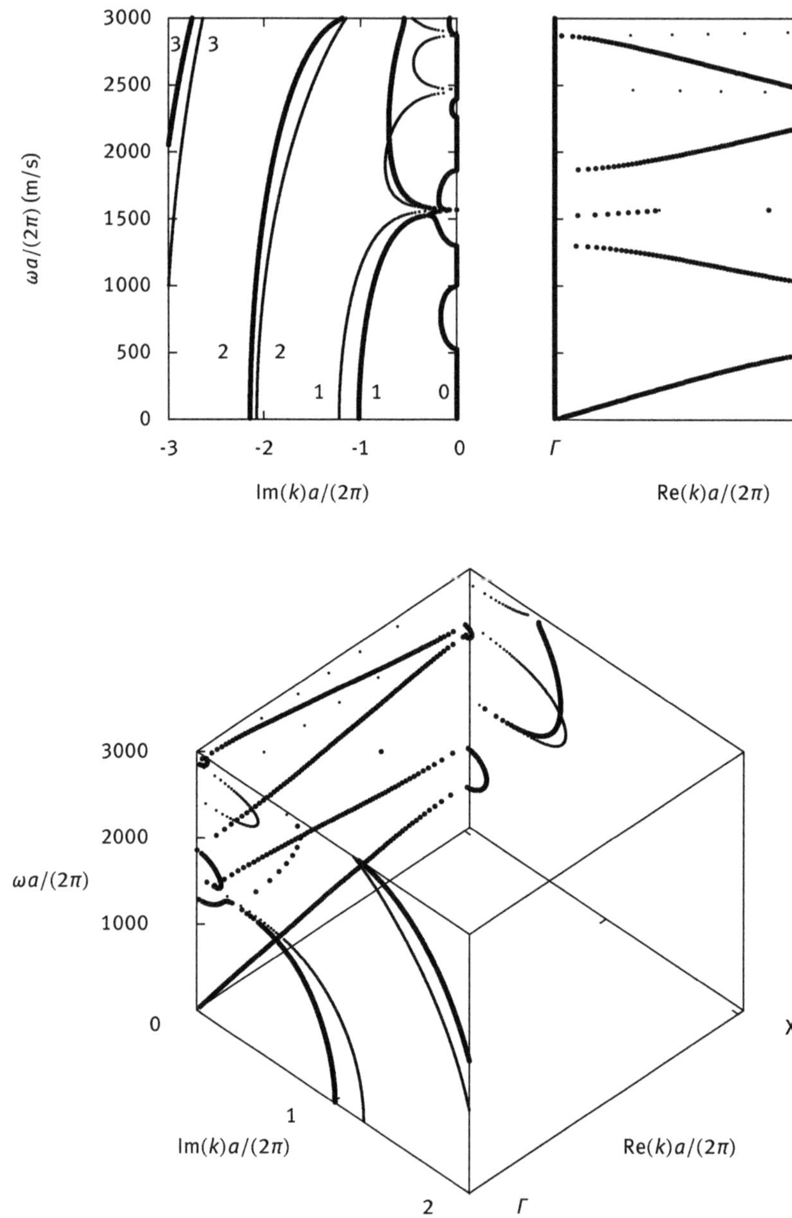

Fig. 9.3: Complex band structure for a square-lattice 2D sonic crystal of circular steel rods in water with $d/a = 0.8$, shown along direction X of the first Brillouin zone. Antisymmetric (deaf) bands are shown with small dots, while symmetric (nondeaf) bands are shown with large dots.

In Figure 9.3, it can be observed that the physical origin of every band can be followed from the zero frequency upward. The 0-th order Bloch wave is the only one that is propagating for low frequencies. The band formed by this wave rises from the Γ point to the X point, where it enters a Bragg band gap. An evanescent arc circle continuously joins the lower branch to its folding above. Three further foldings of this band can be observed for increasing ω, at Γ, X and Γ again. Higher-order Bloch waves are first purely evanescent ($\mathrm{Re}(k) = 0$) and converge continuously toward their frequency cut-off, which is different for each Bloch wave. Above this frequency cut-off, they also form a succession of propagating and evanescent branches. The symmetry character of each Bloch wave, either symmetric or antisymmetric, is indicated in Figure 9.3. This information is useful to analyze experimental results, including the influence of deaf bands on the transmission, as we discussed in Chapter 4. Note that the bands for the first-order Bloch waves ± 1 both cross the band for the zeroth-order Bloch wave in the middle of the second Bragg band gap of the latter, close to the condition $\omega/c = 2\pi/a$ which is the onset condition for diffraction for the square lattice in the X direction. The problem of diffraction of external waves on sonic and phononic crystals will be discussed specifically in Chapter 12.

Figure 9.4 presents the complex band structure of the 2D sonic crystal of circular steel rods in water, obtained for the M direction of the first Brillouin zone, i.e. for $\mathbf{k}_0 = 0$ and $\boldsymbol{\alpha} = (1/\sqrt{2}, 1/\sqrt{2}, 0)^T$. Compared to the X direction in Figure 9.3, there are mostly strong similarities regarding essential features but two main differences. A first difference is that the higher-order Bloch waves start at low frequency at either the Γ or the M point, depending on their parity: Γ point for even n and M point for odd n. A second difference is that the antisymmetric (AS) first order Bloch wave crosses the band for the zeroth-order Bloch wave at the first Bragg band gap of the latter, while the symmetric (S) first order Bloch wave does so at the third Bragg band gap. The first crossing is close to the condition $\omega/c = \sqrt{2}\pi/a$ which is the onset condition for diffraction for the square lattice in the M direction (see Chapter 12).

Let us now consider the case of half-circular steel rods. As we discussed in Chapter 4, the existence of symmetrical and antisymmetrical Bloch waves depends on the combination of the lattice symmetry and of the unit cell symmetry (material anisotropy does not play a part for pressure waves). Figure 9.5 shows the complex band structure for the Y direction of the first Brillouin zone, i.e. for $\mathbf{k}_0 = 0$ and $\boldsymbol{\alpha} = (0, 1, 0)^T$. Since the Y direction remains a symmetry direction, it is not surprising that the result looks quite like the X direction result for circular rods in Figure 9.3 (the X and Y directions were equivalent in the latter case). The complex band structures are obviously not the same, but general observations remain the same.

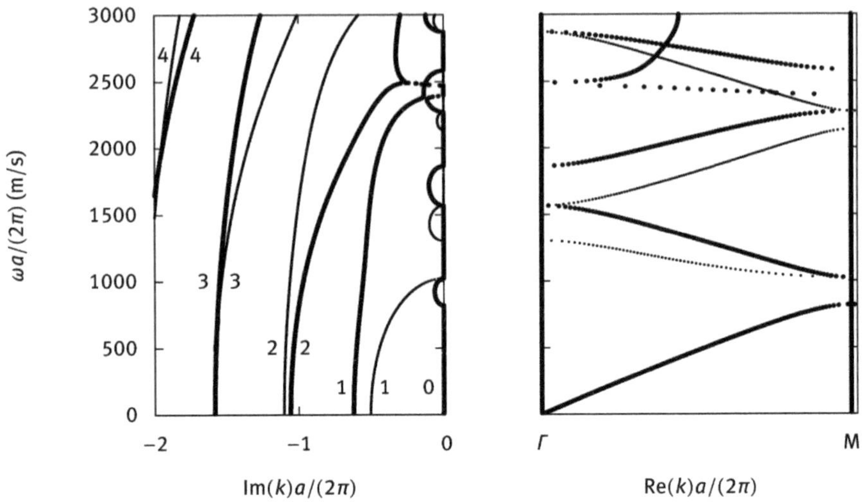

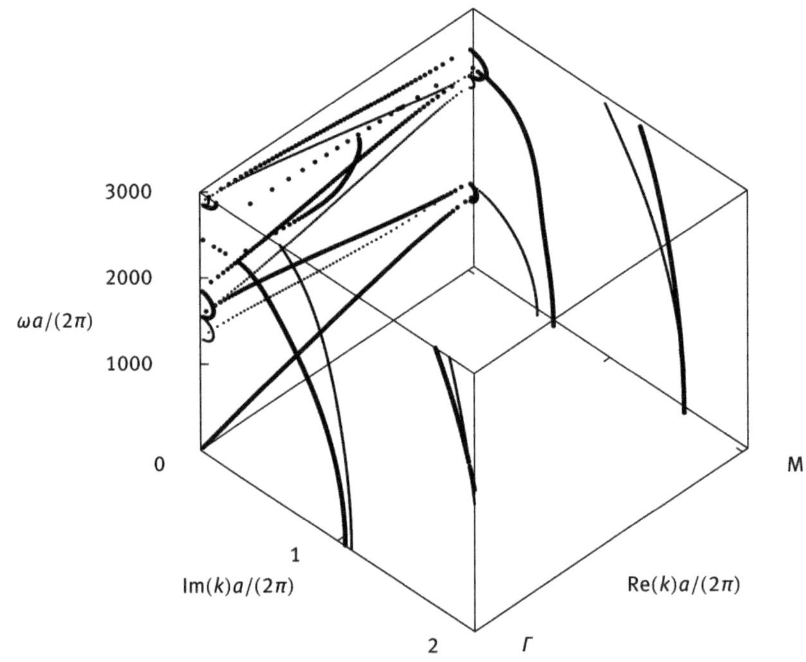

Fig. 9.4: Complex band structure for a square-lattice 2D sonic crystal of circular steel rods in water with $d/a = 0.8$, shown along direction M of the first Brillouin zone. Antisymmetric (deaf) bands are shown with small dots, while symmetric (nondeaf) bands are shown with large dots.

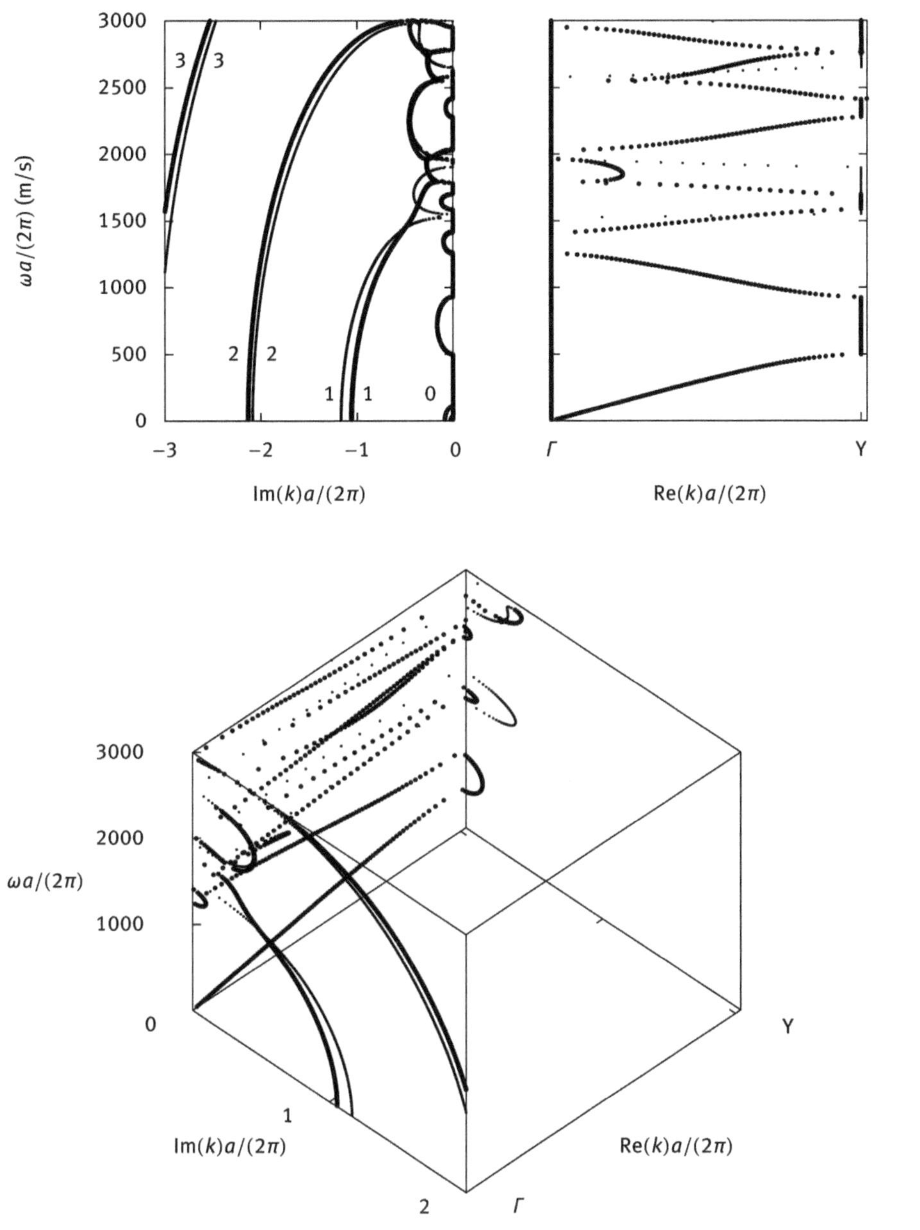

Fig. 9.5: Complex band structure for a square-lattice 2D sonic crystal of half-circular steel rods in water with $d/a = 0.8$, shown along direction Y of the first Brillouin zone. Antisymmetric (deaf) bands are shown with small dots, while symmetric (nondeaf) bands are shown with large dots.

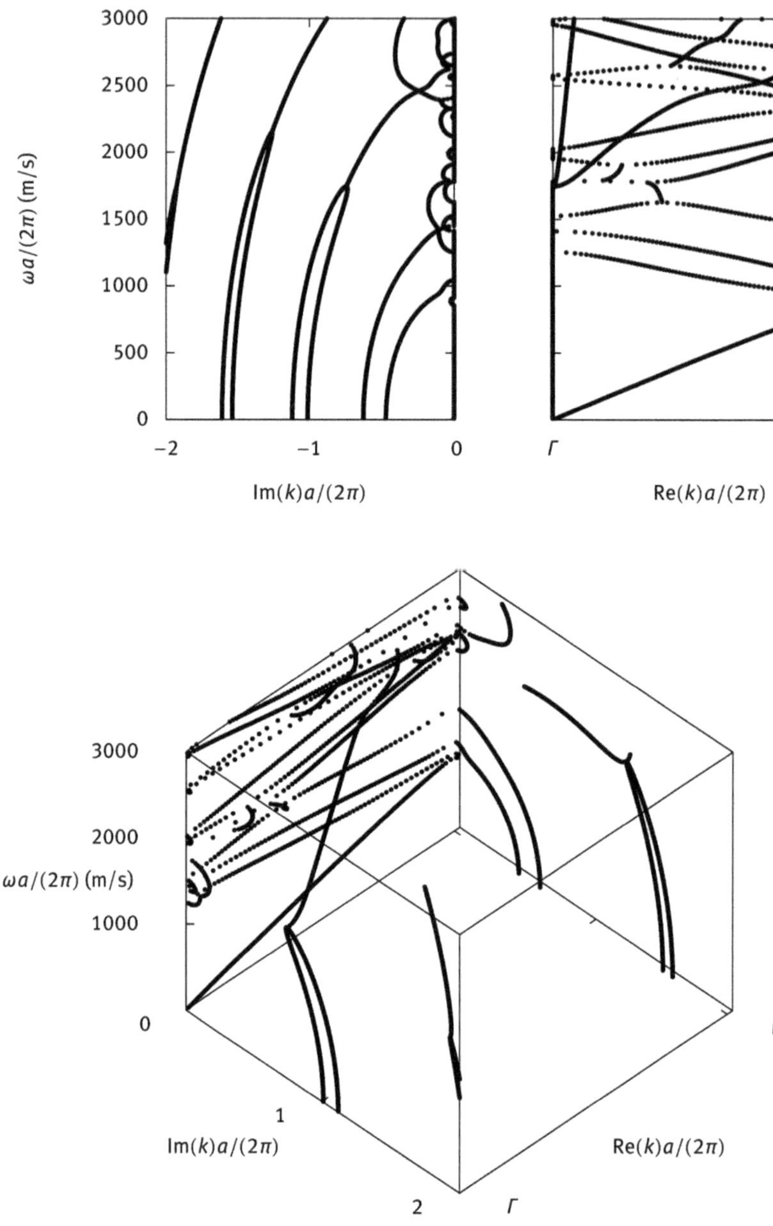

Fig. 9.6: Complex band structure for a square-lattice 2D sonic crystal of half-circular steel rods in water with $d/a = 0.8$, shown along direction M of the first Brillouin zone.

In the M direction considered in Figure 9.6, however, the situation is changed radically. This direction is no longer a symmetry direction of the sonic crystal and there is no separation between S and AS Bloch waves. As a result, there are more interactions between the different bands and the complex band structure is more difficult to read. Note the appearance of several inhomogeneous bands, with the wavenumber generally complex and not simply either real or imaginary.

9.3 Evanescent Bloch waves of phononic crystals

In this section, we derive the complex band structure for elastic waves in phononic crystals. The mathematics are merely extensions of the sonic crystal case, apart from the addition of anisotropy and of piezoelectricity, where needed.

9.3.1 Analysis via the plane wave expansion

The notations for the PWE method for elastic waves were given in Chapter 6. The basic PWE equations are (6.27) and (6.28),

$$\imath \Gamma_j \mathbf{T}_{ij} = \omega^2 \mathcal{R}_{ik} \mathbf{U}_k, \tag{9.30}$$

$$\mathbf{T}_{ij} = -\imath \mathcal{C}_{ijkl} \Gamma_l \mathbf{U}_k. \tag{9.31}$$

As a reminder, indices j and l run from 1 to 3 since they account for partial derivatives with respect to the coordinates of space. Indices i and k run from 1 to r, with $r = 3$ for elastic solids and $r = 4$ for piezoelectric solids, since they are associated with the polarization of the elastic waves. The main mathematical trick is again the decomposition (9.15) for the wavevector and the expression (9.17) for matrices Γ_j. As a result of their introduction,

$$\imath \Gamma_{j0} \mathbf{T}_{ij} + \imath k \alpha_j \mathbf{T}_{ij} = \omega^2 \mathcal{R}_{ik} \mathbf{U}_k,$$

$$\mathbf{T}_{ij} = -\imath \mathcal{C}_{ijkl} \Gamma_{l0} \mathbf{U}_k - \imath k \mathcal{C}_{ijkl} \alpha_l \mathbf{U}_k.$$

The natural eigenvector associated with eigenvalue k is $\mathbf{H} = (\mathbf{U}, \imath \mathbf{T}_\alpha)^T = (\mathbf{U}_k, \imath \alpha_j \mathbf{T}_{ij})^T$ and has $2r \times M$ components. The second term in the eigenvector is proportional to the stresses in the direction of propagation. Inserting the second equation in the first (but only for the first occurrence of $\mathbf{T}_{ij}$), we have

$$(\Gamma_{j0} \mathcal{C}_{ijkl} \Gamma_{l0}) \mathbf{U}_k + k (\Gamma_{j0} \mathcal{C}_{ijkl} \alpha_l) \mathbf{U}_k + k (\imath \alpha_j \mathbf{T}_{ij}) = \omega^2 \mathcal{R}_{ik} \mathbf{U}_k.$$

Multiplying the second equation with α_j and summing over index j, we get

$$(\imath \alpha_j \mathbf{T}_{ij}) = (\alpha_j \mathcal{C}_{ijkl} \Gamma_{l0}) \mathbf{U}_k + k (\alpha_j \mathcal{C}_{ijkl} \alpha_l) \mathbf{U}_k.$$

Gathering up the results, we can form a generalized eigenvalue problem similar to (9.18)

$$\begin{pmatrix} \omega^2\mathcal{R} - B & 0 \\ -C_2 & I_d \end{pmatrix} \mathbf{H} = k \begin{pmatrix} C_1 & I_d \\ D & 0 \end{pmatrix} \mathbf{H}, \tag{9.32}$$

with the matrices B, C_1, C_2, and D formed from $r \times r$ blocks of size $M \times M$ given by

$$B_{ik} = \Gamma_{j0}\mathfrak{C}_{ijkl}\Gamma_{l0}, \quad (C_1)_{ik} = \Gamma_{j0}\mathfrak{C}_{ijkl}\alpha_l,$$
$$(C_2)_{ik} = \alpha_j\mathfrak{C}_{ijkl}\Gamma_{l0}, \quad D_{ik} = \alpha_j\mathfrak{C}_{ijkl}\alpha_l,$$

using the rules expressed by (6.31) and (6.32) [78].

As a whole, except for the size of matrices that are larger by a factor r^2 (the number of degrees of freedom for the polarization of elastic waves), the general eigenvalue problem for phononic crystals is very similar to the one for sonic crystals. The properties of evanescent Bloch waves and of complex band structures we mentioned before are alike in both case.

9.3.2 Analysis via the finite element method

The FEM version of the complex band structure of phononic crystals is derived as follows. Assuming monochromaticity and the Bloch–Floquet theorem, we can write for the periodic part of the solution

$$\bar{T}_{ij} = c_{ijkl}(\bar{u}_{k,l} - \imath k\alpha_l\bar{u}_k), \tag{9.33}$$
$$\bar{T}_{ij,j} - \imath k\alpha_j\bar{T}_{ij} = -\rho\omega^2\bar{u}_i, \tag{9.34}$$

where it is implicitly assumed that the following replacements are being made

$$\bar{u}_{k,l} \longleftarrow \bar{u}_{k,l} - \imath\mathbf{k}_{0l}\bar{u}_k,$$
$$\bar{T}_{ij,j} \longleftarrow \bar{T}_{ij,j} - \imath\mathbf{k}_{0j}\bar{T}_{ij},$$

to account for the possible existence of a shift in reciprocal space $\mathbf{k}_0$.

At this point, it appears that the unknown physical fields that will give us a generalized eigenvalue problem for k are $\bar{u}_k$ and $\tau_i = \alpha_j\bar{T}_{ij}$. $\boldsymbol{\tau}$ is the vector of stresses in the direction of propagation. Inserting (9.33) into (9.34) we get

$$\rho\omega^2\bar{u}_i + c_{ijkl}\bar{u}_{k,jl} = \imath k c_{ijkl}\alpha_l\bar{u}_{k,j} + \imath k\tau_i. \tag{9.35}$$

Furthermore, multiplying (9.33) with α_j we get

$$\tau_i - c_{ijkl}\alpha_j\bar{u}_{k,l} = -\imath k c_{ijkl}\alpha_j\alpha_l\bar{u}_k. \tag{9.36}$$

A variational form is obtained by considering a vector of $2r$ test functions (u_i', τ_i') living in the same finite element space as (u_i, τ_i), $i = 1, \ldots, r$. More precisely, we sum the scalar product of (9.35) with u'_i and the scalar product of (9.36) with τ'_i to obtain

$$\int_{\Omega} d\mathbf{r} A(\boldsymbol{\tau}', \mathbf{u}'; \boldsymbol{\tau}, \mathbf{u}) = (\imath k) \int_{\Omega} d\mathbf{r} B(\boldsymbol{\tau}', \mathbf{u}'; \boldsymbol{\tau}, \mathbf{u}), \forall(\boldsymbol{\tau}', \mathbf{u}'), \tag{9.37}$$

with

$$A(\boldsymbol{\tau}', \mathbf{u}'; \boldsymbol{\tau}, \mathbf{u}) = (\tau')_i^* \tau_i - c_{ijkl}\alpha_j(\tau')_i^* \bar{u}_{k,l} + \rho\omega^2(\bar{u}')_i^* \bar{u}_i - c_{ijkl}(\bar{u}')_{i,j}^* \bar{u}_{k,l}, \tag{9.38}$$

$$B(\boldsymbol{\tau}', \mathbf{u}'; \boldsymbol{\tau}, \mathbf{u}) = -c_{ijkl}\alpha_j\alpha_l(\tau')_i^* \bar{u}_k - c_{ijkl}(\bar{u}')_{i,j}^* \alpha_l \bar{u}_k + (\bar{u}')_i^* \tau_i. \tag{9.39}$$

We define the symbols

$$C_{Ik}^{(1)} = c_{Ikl}\alpha_l, I = (ij), \tag{9.40}$$

$$C_{iJ}^{(2)} = c_{ijJ}\alpha_j, J = (kl), \tag{9.41}$$

$$D_{ik} = c_{ijkl}\alpha_j\alpha_l, \tag{9.42}$$

where we recall that I and J are indices running from 1 to 6 that replace the pairs of symmetric indices (ij) and (kl), respectively. The variational form can be rewritten

$$A(\boldsymbol{\tau}', \mathbf{u}'; \boldsymbol{\tau}, \mathbf{u}) = (\tau')_i^* \tau_i - C_{iJ}^{(2)}(\tau')_i^* S_J(\bar{u}) + \rho\omega^2(\bar{u}')_i^* \bar{u}_i - c_{IJ} S_I^*(\bar{u}') S_J(\bar{u}), \tag{9.43}$$

$$B(\boldsymbol{\tau}', \mathbf{u}'; \boldsymbol{\tau}, \mathbf{u}) = -D_{ik}(\tau')_i^* \bar{u}_k - C_{Ik}^{(1)} S_I^*(\bar{u}')\bar{u}_k + (\bar{u}')_i^* \tau_i. \tag{9.44}$$

With these notations, the resemblance with the PWE equations (9.32) becomes clear. In particular, once discretized on a given finite element space, the variational formulation yields a generalized eigenvalue problem of the same type as the PWE problem. Once again, the advantages of the FEM implementation with respect to the PWE implementation are that sparse matrices result and that the geometry can be approached more closely with a mesh than using a fixed array of Fourier harmonics.

9.3.3 Complex band structure

In Figures 9.7 and 9.8 we present examples of complex band structures for the case of a 2D phononic crystal of circular holes in silicon. A classical phononic band structure for this system and in the X direction was presented in Figure 6.14. Silicon is considered to be oriented in its natural crystallographic axes. As a result, out-of-plane shear waves separate from in-plane polarized longitudinal and shear waves, as we discussed in Chapter 5. Figure 9.7 displays the complex band structures, in both cases in the X direction. As compared to scalar pressure waves in Section 9.2, the complexity of the out-of-plane shear complex band structure is similar and the observations we made earlier still apply.

Figure 9.9 displays the Bloch waves for points of particular interest labeled A, B, C and D in Figure 9.7 (a), for propagation in the X direction. Points A, B and C all belong

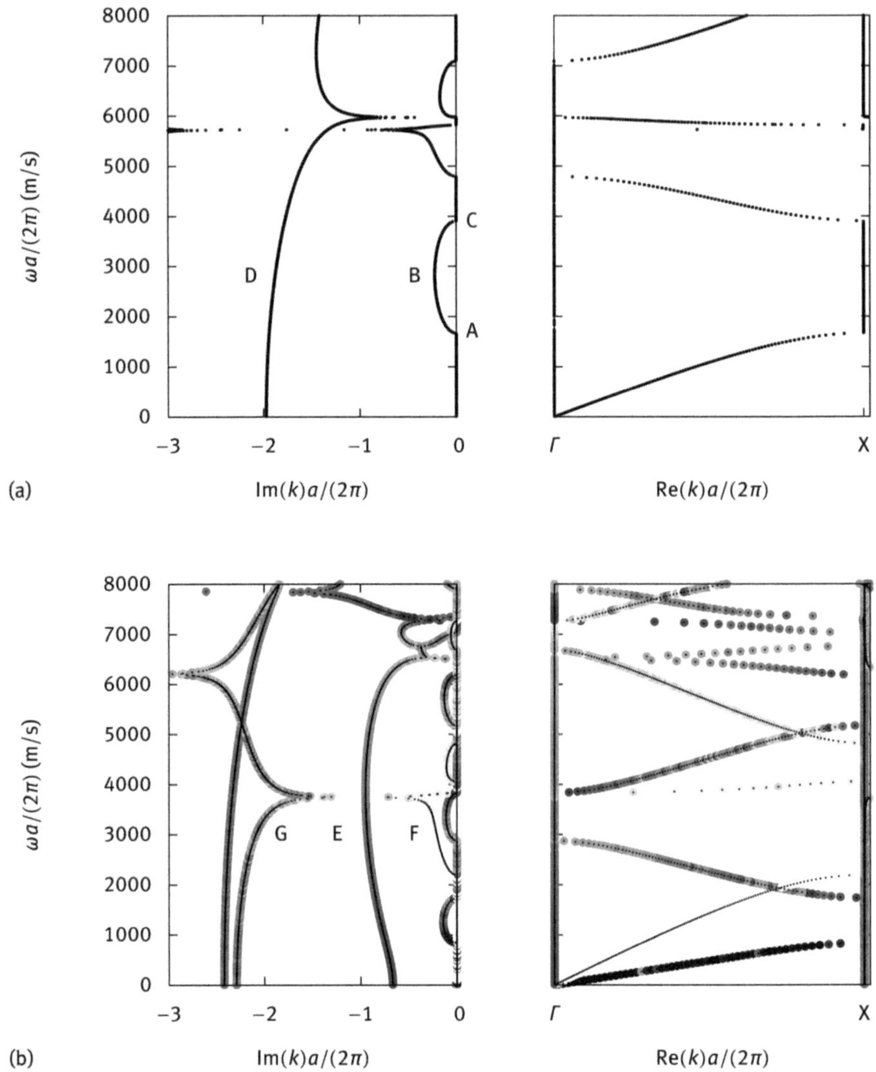

Fig. 9.7: Complex band structure for a square-lattice two-dimensional phononic crystal of holes in silicon. The ΓX direction is considered with $d/a = 0.85$. Band structures are (a) for pure shear Bloch waves (polarized along x_3) and (b) for Bloch waves polarized in-plane (in the (x_1, x_2) plane). Eigensolutions to the $k(\omega)$ problem are plotted with dots. In (b), circles are overlayed at each eigensolution to indicate the polarization content of the Bloch wave. The gray scale varies from black for pure shear to white for pure longitudinal.

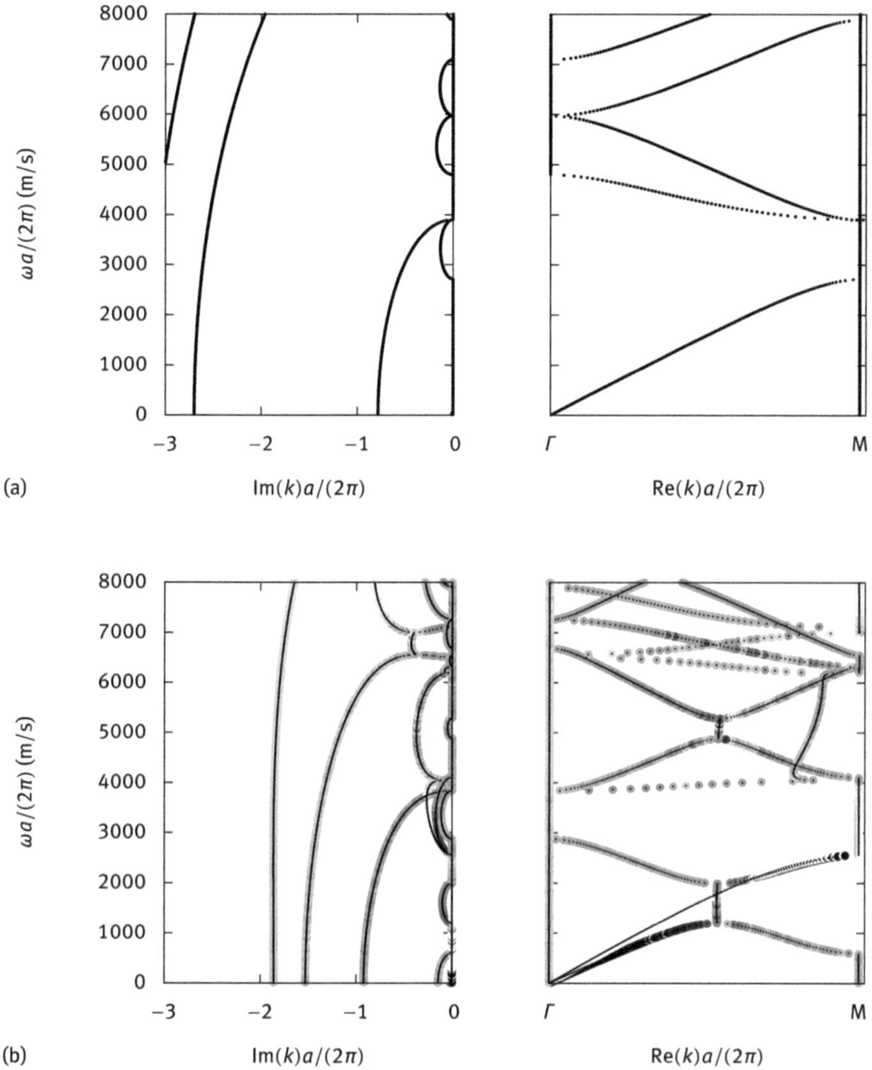

Fig. 9.8: Complex band structure for a square-lattice two-dimensional phononic crystal of holes in silicon. The *ΓM* direction is considered with $d/a = 0.85$. Band structures are (a) for pure shear Bloch waves (polarized along x_3) and (b) for Bloch waves polarized in-plane (in the (x_1, x_2) plane). Eigensolutions to the $k(\omega)$ problem are plotted with dots. In (b), circles are overlayed at each eigensolution to indicate the polarization content of the Bloch wave. The gray scale varies from black for pure shear to white for pure longitudinal.

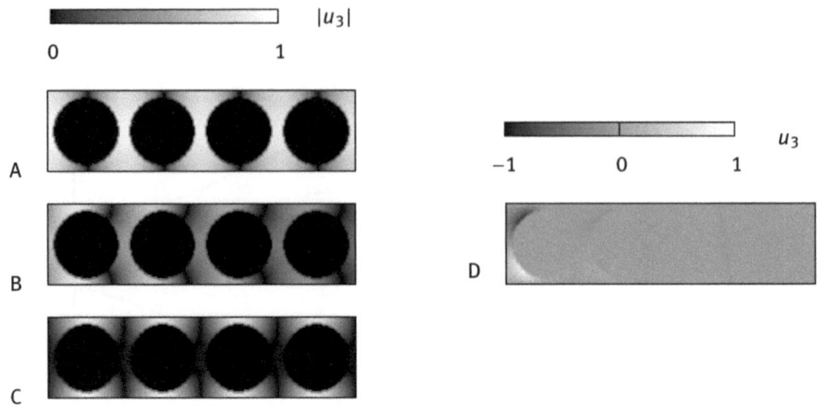

Fig. 9.9: Evolution of the polarization across the lower band gap for some pure shear Bloch waves. Points A, B and C are defined in Figure 9.7 (a). The modulus of the displacement u_3 is displayed for points A, B and C, and the real part of u_3 is displayed for point D. Points A, B and C are for the entrance, the middle, and the exit of the lower band gap, respectively. Point D is for an evanescent Bloch wave at the same frequency as point B.

to the zeroth Bragg order of diffraction. Point A is for the entrance of the first Bragg band gap and point C is for its exit. It can be seen that the distribution of displacement u_3 changes radically between both points and actually these two Bloch waves are orthogonal in the sense of the scalar product defined by integrating over the unit cell. At the center of the band gap, at point B, the distribution of displacement is somewhat in-between that at points A and C. The decrease in the amplitude as the Bloch wave travels to the right can be noticed. The Bloch wave at point D belongs to the first-order Bragg order of diffraction. This is an antisymmetric wave with respect to the direction of propagation, as the distribution of displacement reveals.

The complex band structure for in-plane polarized longitudinal and shear waves is shown in Figure 9.7 (b). There are more bands than in the out-of-plane shear waves case and the interaction between the two polarization components (u_1 and u_2) changes with frequency for each system of bands. Information about the amount of each polarization (p_1, p_2) has been added to the graph with gray circles changing continuously from black for pure shear waves to white for pure longitudinal waves. Figure 9.10 displays the evanescent Bloch waves for points of particular interest labeled E, F and G in Figure 9.7 (b). Bloch wave F is mostly longitudinal (u_1 dominates over u_2). Its symmetry character is a bit more intricate compared to the case of scalar waves, since this wave is symmetric with respect to the propagation direction for u_1 but antisymmetric for u_2. Bloch waves E and G are mostly shear. Bloch wave E (G, respectively) is symmetric (antisymmetric, resp.) with respect to the propagation direction for u_2 but antisymmetric for u_1 (symmetric, resp.).

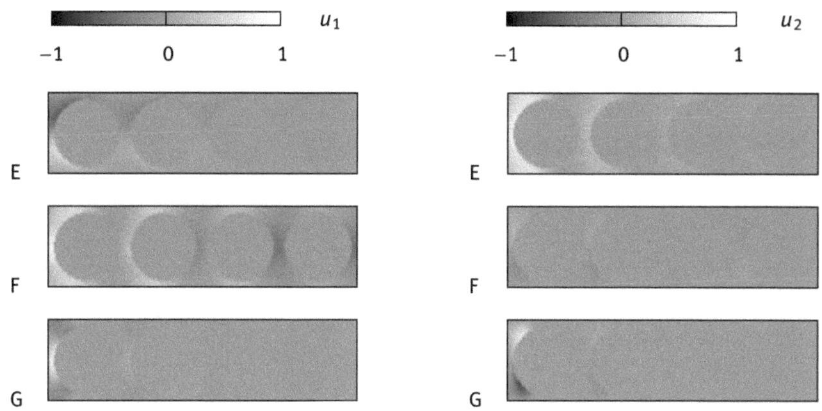

Fig. 9.10: Evolution of the polarization across the lower band gap for some in-plane polarized Bloch waves. Points E, F and G are defined in Figure 9.7 (b). The real parts of the displacements u_1 and u_2 are displayed.

Figure 9.8 displays the complex band structures for the M direction. Their features are comparable to those of the complex band structures for the X direction in Figure 9.7. One noticeable difference, however, is the presence for in-plane polarized waves of inhomogeneous (or generally complex) evanescent bands. For instance, there is an inhomogeneous band connecting the first and the second mostly-shear real bands in Figure 9.7 (b), for reduced frequencies $\omega a/(2\pi)$ roughly between 1000 and 2000 m/s.

9.3.4 Viscoelastic losses

In Section 5.2 we introduced a simple model for the viscoelastic losses of solids. In homogeneous media, the addition of an imaginary part of the elastic tensor proportional to frequency explains well the frequently observed ω^2-dependence of plane wave attenuation on propagation. In phononic crystals, viscoelastic loss has additional consequences linked to periodicity [109].

Figure 9.11 displays the complex band structure for a 2D square-lattice phononic crystal of circular holes in silicon, as we have already considered before. The graph is limited to pure shear waves, for simplicity. Compared to Figure 9.7 (a), the representation extends over the first Brillouin zone and slightly beyond for the real part of the wavenumber, and both positive and negative imaginary parts are shown as well. The purpose is to indicate the physical meaning of each particular complex band, and especially to indicate if real-valued Bloch waves are propagating to the right (R) or to the left (L). The bands holding the dispersion of the first evanescent Bloch waves are also indicated as either decreasing to the right (ER) or to the left (EL). Globally, this

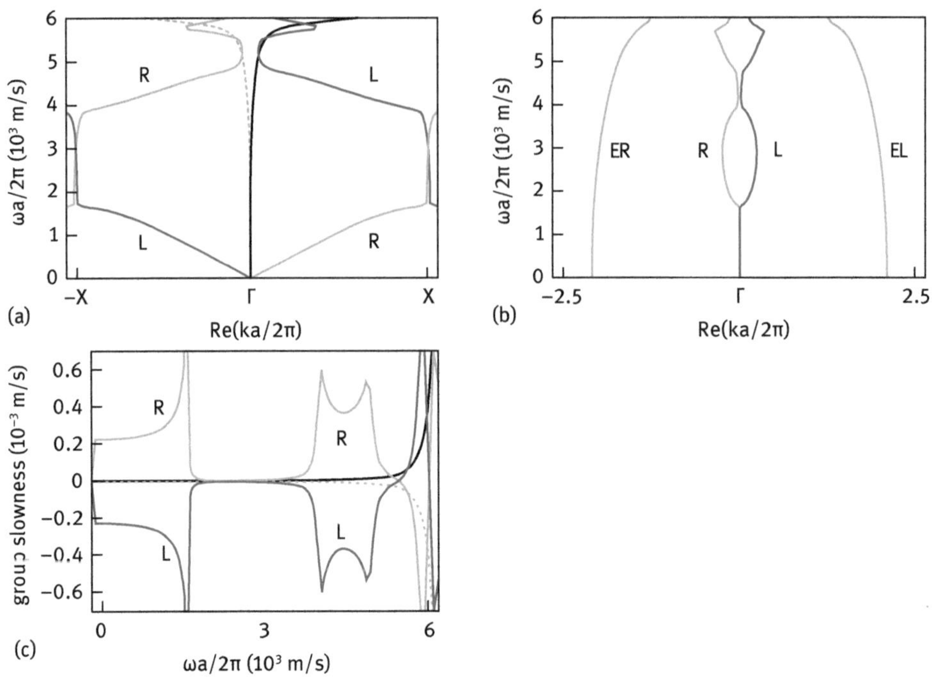

Fig. 9.11: Complex band structure computed with PWE for a lossless square-lattice phononic crystal of silicon with circular holes. The filling fraction is 57 %. In the complex band structure, the reduced frequency is presented as a function of (a) the real part and (b) the imaginary part of the wave number. (c) The group slowness obtained by the eigenvector method is shown as a function of the reduced frequency.

representation is intended to highlight the continuity of each system of bands in the complex band structure $k(\omega)$, a property that is absent in the classical band structure $\omega(k)$. Figure 9.11 (c) shows the group slowness, $\mathrm{Re}(\frac{\partial k}{\partial \omega})$, computed from the solution of the eigenvalue problem using a method that is discussed in Chapter 12. We recall that group slowness is the inverse of group velocity. For the propagating Bloch waves labeled R and L, it goes to $\pm\infty$ at the entrance and at the exit of the Bragg band gap (meaning that the group velocity is zero at these frequencies).

Viscoelastic losses are then added according to (5.23):

$$c'_{ijkl} = c_{ijkl} + \imath\omega\eta_{ijkl},$$

where η_{ijkl} is the phonon viscosity tensor, with the same symmetries as the elastic tensor. The complex band structure is calculated with a complex-valued elastic tensor exactly as with a real-valued elastic tensor. Indeed, the matrices in (9.32) or (9.43) were already frequency dependent. In practice, the solution to the generalized eigenvalue problem remains similar to the lossless case, except that all Bloch waves are now evanescent, since all wavenumbers are complex valued.

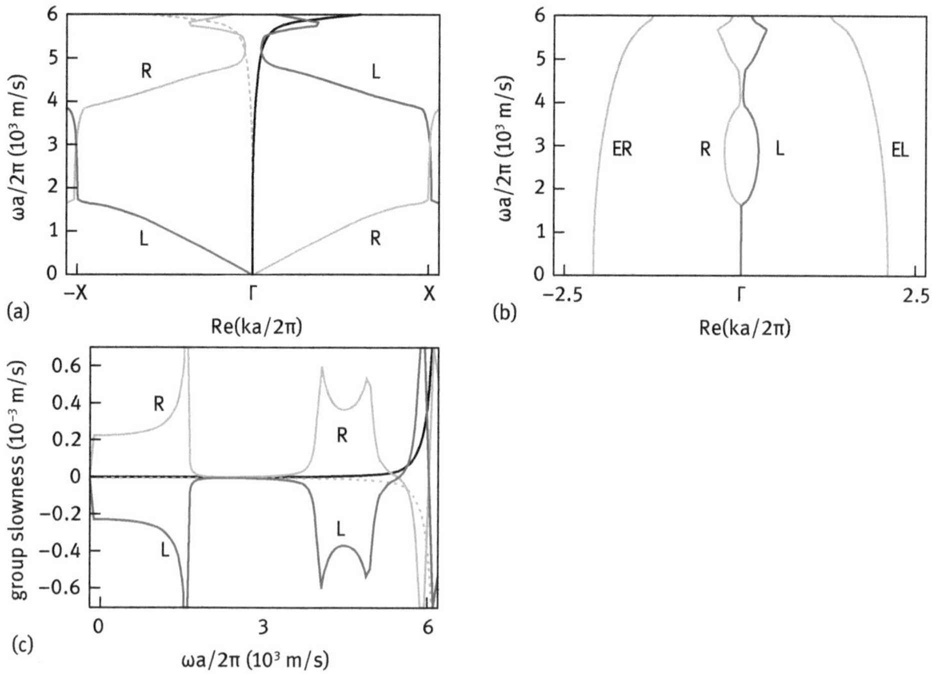

Fig. 9.12: Complex band structure computed with PWE for a lossy square-lattice phononic crystal of silicon with circular holes. The filling fraction is 57 %. In the complex band structure, the reduced frequency is presented as a function of (a) the real part and (b) the imaginary part of the wave number. (c) The group slowness obtained by the eigenvector method is shown as a function of the reduced frequency.

As an example, Figure 9.12 shows the transformation of the complex band structure of Figure 9.11 when viscoelasticity is introduced. Since only pure shear waves are considered in this example, only the value of η_{44} is relevant. Table 9.1 lists the numerical values that are used here. The arbitrary value $\eta_{44} = 0.08$ Pa.s is purposely exaggerated by approximately a factor 100 compared to the actual known value for silicon in Table 5.3. This choice is made so that the modifications of the complex band structure become clearly visible. The lattice constant is fixed to $a = 1\ \mu$m so that the frequency range in Figure 9.12 is actually from 0 to 6 GHz. The following observations can be made.

- Initially propagating Bloch waves R and L are differently affected by losses in different frequency ranges. Outside band gaps, the ω^2-dependence of the imaginary part of the wavenumber applies similarly to bulk elastic waves of homogeneous media. Within a band gap, the imaginary part of the wavenumber is almost unchanged from the lossless case, because it was already large in magnitude, but the real part acquires a distinctive dispersion. Instead of degenerate vertical real bands, the R and the L Bloch waves have clearly distinct and opposite variations.

Tab. 9.1: Material constants considered for pure shear elastic wave propagation in silicon. These values are used to obtain Figures 9.11 and 9.12. The shear viscosity value is arbitrarily chosen (it is not the intrinsic shear viscosity of bulk silicon).

Material	c_{44} (GPa)	ρ (kg/m^3)	η_{44} (Pa · s)
Lossless silicon	79.6	2330	—
Arbitrary viscous silicon	79.6	2330	0.08

These variations imply that the group slowness becomes smaller around the band gap edges: the bands appear rounded instead of forming sharp corners. Consequently, the group velocity is no longer zero but has a lower bound dictated by the amount of loss [80]. Exact formulas for this lower bound can be obtained [80].

– The initially evanescent Bloch waves ER and EL are not strongly affected by losses. While the imaginary part of the wavenumber is not very significantly modified, the real part acquires a ω^2-dependence that becomes clearly apparent in the upper range of frequencies of Figure 9.12 (a).

9.4 Supercells and defect modes

A sonic or a phononic crystal is defined by some primitive cell that is repeated periodically over space with lattice vectors $(\mathbf{a}_1, \mathbf{a}_2, \mathbf{a}_3)$. A supercell is constructed from the primitive cell by repeating it $l \times m \times n$ times along the three lattice vectors. The supercell is itself periodic with lattice vectors $(l\mathbf{a}_1, m\mathbf{a}_2, n\mathbf{a}_3)$. Its volume is lmn times larger than the volume of the original primitive cell. Reciprocally, the reciprocal primitive cell has lattice vectors $(\mathbf{b}_1/l, \mathbf{b}_2/m, \mathbf{b}_3/n)$. Its volume is lmn times smaller than the volume of the original reciprocal primitive cell. These considerations are depicted in Figure 9.13 for the two-dimensional square-lattice case and for $l = 3$ and $m = 2$.

The band structure of the supercell of course has a direct relation to the band structure of the original primitive cell, since the supercell contains exactly the same information as the original primitive cell. Indeed, the band structure for the supercell contains lmn more bands than the original band structure. Denoting $\mathbf{k}'$ a wavevector of a Bloch wave of the supercell, bands are obtained from wavevectors $\mathbf{k}$ of the original band structure given by

$$\mathbf{k} = \mathbf{k}' + \frac{p}{l}\mathbf{b}_1 + \frac{q}{m}\mathbf{b}_2 + \frac{r}{n}\mathbf{b}_3, \tag{9.45}$$

with the integers $p = 1 \ldots l - 1$, $q = 1 \ldots m - 1$, and $r = 1 \ldots n - 1$. It can be easily checked that all the contents of the original reciprocal primitive cell are recovered for these relations as $\mathbf{k}'$ varies in the reciprocal primitive cell associated with the supercell.

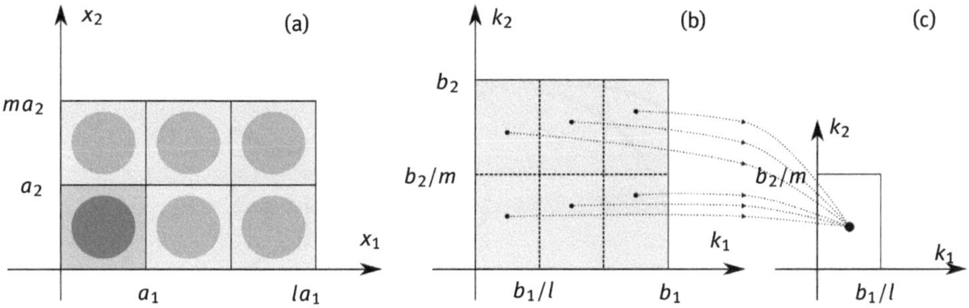

Fig. 9.13: Definition of a supercell and of its reciprocal primitive cell in two dimensions. (a) The supercell is defined in direct space by considering a cluster of $l \times m$ primitive cells. The supercell has lattice vectors $l\mathbf{a}_1$ and $m\mathbf{a}_2$. In the drawing, $l = 3$ and $m = 2$. (b) The reciprocal primitive cell is consequently partitioned into $l \times m$ equal parts. (c) The reciprocal unit cell associated with the supercell is constituted by merging the latter $l \times m$ parts. The resulting cell has reciprocal lattice vectors $\mathbf{b}_1/l$ and $\mathbf{b}_2/m$.

Since all bands for the supercell originate from some path in the reciprocal primitive cell, the full or complete band gaps are preserved. Directional band gaps, however, are not preserved. As an illustration, let us consider in Figure 9.14 the complex band structure for a 2D square-lattice sonic crystal of steel rods in water, plotted for both the ΓX and the ΓM directions. The complete band gap is limited from below by the ΓM direction and from above by the ΓX direction. Now let us consider the ΓX direction for a 6×1 supercell, as depicted in Figure 9.15 (a). From (9.45), with $l = 1$ and $m = 6$, we expect the complex band structure to contain a combination of bands $(k_1, 0)$, $(k_1, b/6)$, $(k_1, 2b/6)$, $(k_1, 3b/6)$, $(k_1, 4b/6)$, and $(k_1, 5b/6)$ from the original problem. Given the equivalence of directions $(k_1, b/6)$ and $(k_1, 5b/6)$, and of directions $(k_1, 2b/6)$ and $(k_1, 4b/6)$, we are left with four distinct origins for the complex bands in Figure 9.15 (a). This is indeed what is obtained numerically.

Of course, considering a purely periodic supercell is rather useless, as the system size is made larger and the computation time much longer just to obtain a combination that can be deduced from the original band structure. The actual usefulness of supercells is the introduction of defects in the otherwise perfectly periodic crystal. As an example, a defect can be introduced in the middle of the previous 6×1 supercell, as depicted in Figure 9.15 (b). As the supercell is repeated periodically in the vertical direction, the actual separation of the central defect from its nearest copies is exactly six rows of the crystal. For frequencies within the complete band gap, there are only evanescent Bloch waves in the perfect crystal and hence the periodic defects are only coupled evanescently; this coupling becomes very small as the number of separation rows is made larger and eventually the defects can be considered as isolated. In the region of the defect, acoustic or elastic waves can be allowed to propagate, and hence form waveguides, or to become spatially localized, and hence form cavities. These mechanisms are specifically discussed in Chapter 11. For now, let us only comment

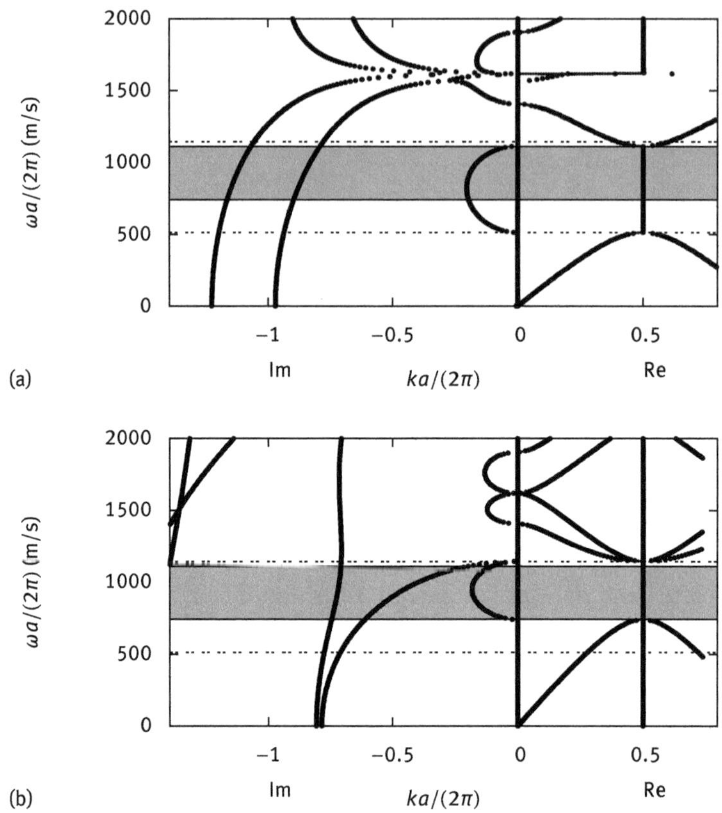

Fig. 9.14: Complex band structure for a 2D square-lattice sonic crystal of steel cylindrical rods in water. The diameter to lattice constant ratio is $d/a = 0.9$. The complex band structure is shown for (a) the ΓX direction and (b) the ΓM direction. The complete band gap is indicated by the gray rectangular area. This figure is provided for comparison with the supercell complex band structure in Figure 9.15.

on the complex band structure in Figure 9.15 (b). Additional bands appear inside the complete band gaps. They actually give the dispersion of waves guided inside the defect and propagating along axis x_1. It can further be seen that two systems of guided waves are formed, each with both propagating and evanescent branches. Indeed, the waveguide is periodic along axis x_1 and thus it should not be surprising that 1D band gaps can be introduced for waves guided by a sonic or phononic band gap effect. In general, waveguides formed this way are multimode and the bandwidth inside which guidance is obtained is smaller than the complete band gap range.

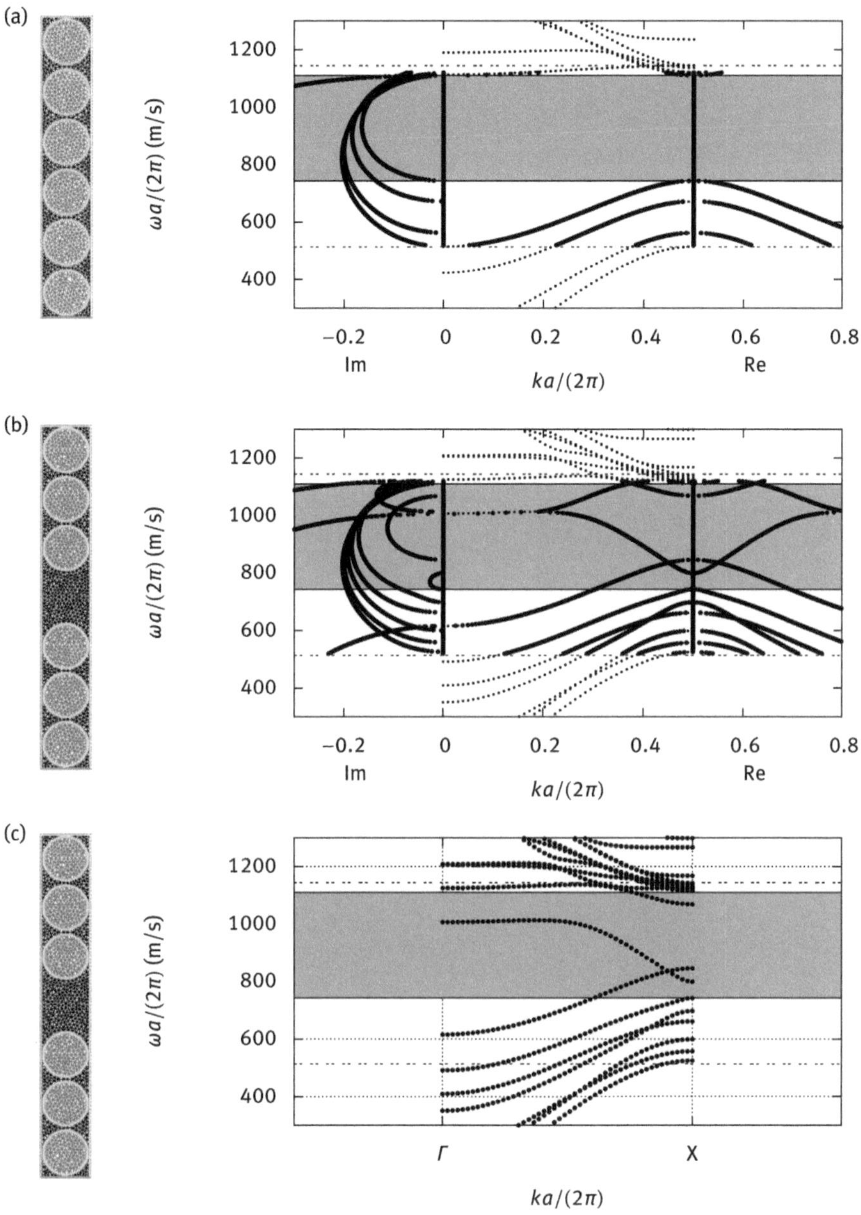

Fig. 9.15: Complex band structure for a 2D supercell of a square-lattice sonic crystal of steel cylindrical rods in water. (a) The 6×1 supercell for the periodic crystal introduces a superposition of four distinct band systems (see text). (b) A central defect is introduced to define the core of a sonic crystal waveguide. Complex bands appear in the complete band gap range. (c) The classical band structure corresponding to (b) shows only the propagating guided Bloch waves.

10 Locally-resonant crystals

We have introduced the concept of local resonance in Chapter 2 through the example of finite-length tubes grafted on a waveguide. Sonic or phononic crystals containing internal structures in their unit cell that present resonances are known as locally-resonant crystals. As resonance frequencies are not dictated by periodicity as in the case of Bragg interference, they have attracted a lot of attention for applications at low frequencies. Indeed, locally-resonant (LR) crystals can present band gaps, or rather transmission zeros, for frequencies such that the wavelength in an effective equivalent medium would be much larger than the lattice constant. In the present chapter we explore various implementations of the concept of local resonance. We start however in the first section with a physical discussion of its origin and of its relation to Fano resonance.

10.1 Local resonance and Fano resonance

The concept of locally-resonant crystals was introduced in 2000 by Liu et al. [89]. Their original structure is shown in Figure 10.1. A simple cubic lattice of coated spheres is placed in a matrix composed of epoxy. The spheres are made of lead and are coated with a layer of silicon rubber. The mass density of lead is very large, so that the lead spheres can be thought of as heavy rigid solids. Silicon rubber, in contrast, is a very soft material having small elastic constants. The silicon rubber layer can thus be imagined to form an elastic spring with rather small rigidity, so that it can be easily deform as the central sphere is moving around. Intuitively, low-frequency elastic waves propagating in the epoxy matrix will be relatively insensitive to the exact internal details of the unit cell of the crystal[1] except at certain frequencies where internal resonances of the unit cell can be excited. What are those resonances that appear in the low frequency range? Let us first imagine that the central lead sphere is free to move in space as a rigid body, as if it were levitating. There are at least six independent natural modes of motion of the sphere requiring no energy: three elementary rotations and three elementary translations in physical space. The frequencies of these six natural rigid body modes are initially zero. Second, we now consider the same motions for the rigid sphere coated with the soft rubber layer, which is itself fixed to the epoxy matrix. Some energy will now be required to set the sphere into rotation or translation, with the result that the resonant frequencies will move away from zero. Since rota-

1 More precisely, low-frequency elastic waves have a very long wavelength compared to the lattice constant, a regime where mathematical homogenization can be performed to obtain effective material constants for wave propagation. Hence, the effective mass density and effective elastic constants will be some mean value of the constants of the three different constituents of the unit cell.

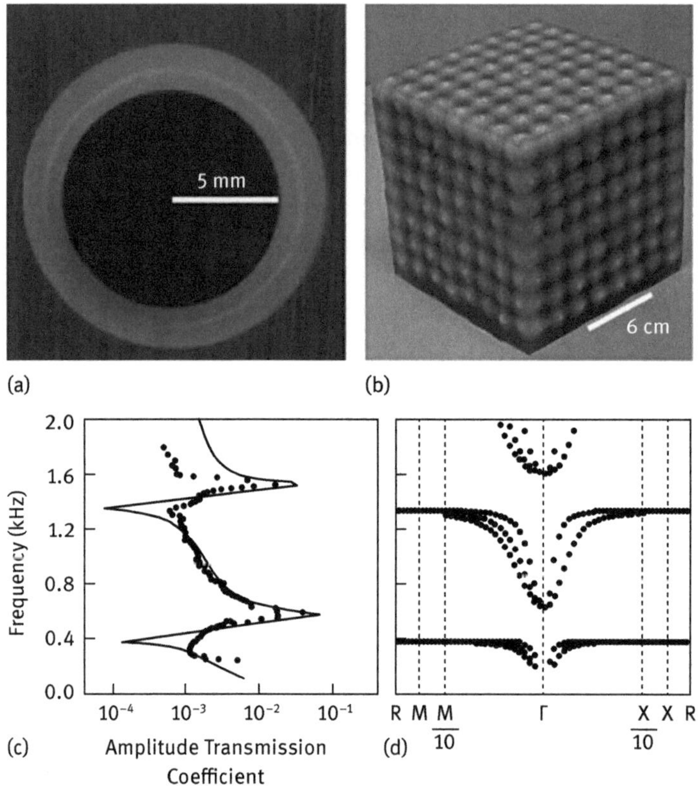

Fig. 10.1: A locally-resonant phononic crystal. (a) Cross section of a coated lead sphere that forms the basic structure unit (b) for an $8 \times 8 \times 8$ crystal. (c) Calculated (solid line) and measured (circles) amplitude transmission coefficient along the [100] direction are plotted as a function of frequency. The calculation is for a four-layer slab of simple cubic arrangement of coated spheres, periodic parallel to the slab. The observed transmission characteristics correspond well with the calculated band structure (d), from 200 to 2000 Hz, of a simple cubic structure of coated spheres. Three modes (two transverse and one longitudinal) are distinguishable in the [110] direction, to the left of the Γ point. The two transverse modes are degenerate along the [100] direction, to the right of the Γ point. Note the expanded scale near the Γ point (after Liu et al. [89], copyright 2000 American Association for the Advancement of Science).

tion involves mostly shear strain in the silicon rubber layer, while translations involve some combination of compressional and shear strains, locally-resonant frequencies corresponding to rotation will appear at frequencies lower than those corresponding to translations.[2]

2 In Chapter 2 we explicitly related the locally-resonant frequency shift for tubes grafted on a waveguide to the excitation of evanescent waves in the waveguide, see (2.61).

The band structure in Figure 10.1 (d) shows peculiar and specific patterns. Two band gaps are introduced in the low frequency range, around 0.5 and 1.5kHz. Given the lattice constant $a = 15.5$ mm, these band gaps appear much lower in frequency than would be expected in the case of Bragg band gaps, which we have seen many different examples of in the previous chapters. One striking detail is also that the band gaps are opened at the X point of the first Brillouin zone (or at any other high symmetry point different to the Γ point) and closed at the Γ point. Hence, bands seemingly jump from one high symmetry point to another, whereas we observed in Chapter 9 that the evanescent Bloch waves connecting the entrance to the exit of Bragg band gaps originated from one single high-symmetry point at a time. The complex band structure of a locally-resonant sonic crystal is discussed explicitly in Section 10.2.

The physical origins of the Bragg band gaps and of the locally-resonant band gaps are actually different. We have already outlined a few examples in Chapter 2: Bragg band gaps were derived for the bilinear chain (see the discussion around Figure 2.14) and for 1D gratings (see Figure 2.16); locally-resonant gaps were described for a periodic array of resonators grafted on a waveguide (see Figure 2.21). A physical picture can be drawn as shown in Figure 10.2. The left panel schematically indicates the dispersion of different elements in a band structure (k, ω). A wave propagating in an intrinsically nondispersive medium is represented with a straight line whose slope is a velocity c. A resonant element without any connection with the propagation medium can be represented with a horizontal line, with the meaning that it has a precise resonant frequency and that it is perfectly confined in space: the spatial Fourier transform extends over all possible values of the wavenumber k. One-dimensional periodicity is represented by the edges of the first Brillouin zone, as vertical lines for $k = \pm\pi/a$. The right panels depicts the hybridization that occurs when the resonant element interacts with waves in the propagation medium but also the folding of bands caused by periodicity. Hybridization describes the avoided crossing that occurs in the frequency

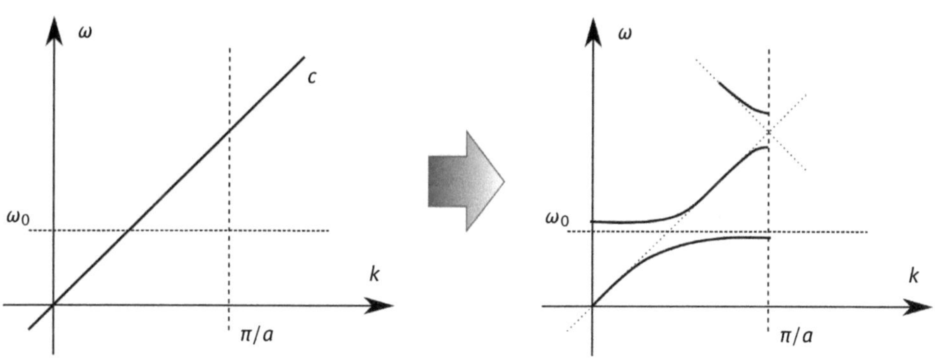

Fig. 10.2: Schematic representation of the formation of Bragg band gaps and of locally-resonant band gaps.

region where the inclined and the horizontal line should have met. If the propagating wave and the vibration excited in the resonator can couple, they combine to form a continuous pair of bands defining a frequency band gap around the resonance. It must be stressed that this band gap is in essence independent of periodicity; hence if the coupling does not vanish for any direction of propagation, the locally-resonant band gap is always a complete band gap. In contrast, Bragg band gaps occur because of periodicity around foldings at high-symmetry points of the Brillouin zone.

Figure 10.3 shows numerical simulation results obtained by Goffaux et al. [42]. The two-dimensional structure considered is similar to the three dimensional one in Figure 10.1: gold cylinders are coated with a rubber polymer and embedded in an epoxy background. The rubber polymer is assumed to have small longitudinal and shear velocities, or equivalently very small elastic constants. Under these conditions, low frequency local resonances appear and introduce band gaps for frequencies around 1 kHz for a lattice constant $a = 15$ mm. As a comparison, Figure 10.3 (b) shows the band structure of the same phononic crystal but without the rubber coating: a complete Bragg band gap opens around 50 kHz. Hence, in this case it can be said that the locally-resonant band gap is 50 times lower than the Bragg band gap, or that the lattice constant that is required to introduce a locally-resonant band gap around 1 kHz is 50 times smaller than the one required to introduce a Bragg band gap, with obvious advantages in terms of compacity of the structure. The computation of the transmission through a locally-resonant phononic crystal with either three or six layers shown in Figure 10.3 (c), however, shows that the extinction that can be expected is not as strong as is usually expected with Bragg band gaps. We will come back to this problem later in this chapter.

One striking feature of the locally-resonant transmission that was pointed out by Goffaux et al. [42] is that it is asymmetrical with respect to the central dip. Such an asymmetric transmission is actually reminiscent of the Fano resonance, a phenomenon first described in atomic systems (see, e.g. [99] for a review). As a remark, the Fano resonance designates a particular profile of the transmission that generalizes over the Lorentzian profile, but it should not be interpreted as meaning that the local resonance is merely a classical counterpart of a quantum mechanical concept. The Fano line shape is generally interpreted as arising from the interference of two traveling waves that have followed different "paths"; in our case a part of the incident wave is transmitted without interacting with the resonator while another part interacts with it and can be stored for some time before it is radiated away by the vibrating resonator; the part of the radiated wave that converts back into the incident propagation mode can clearly interfere with the directly transmitted part of the incident wave. The Fano line shape has the following form [99]

$$\sigma(\epsilon, q) = \frac{(\epsilon + q)^2}{1 + \epsilon^2}, \tag{10.1}$$

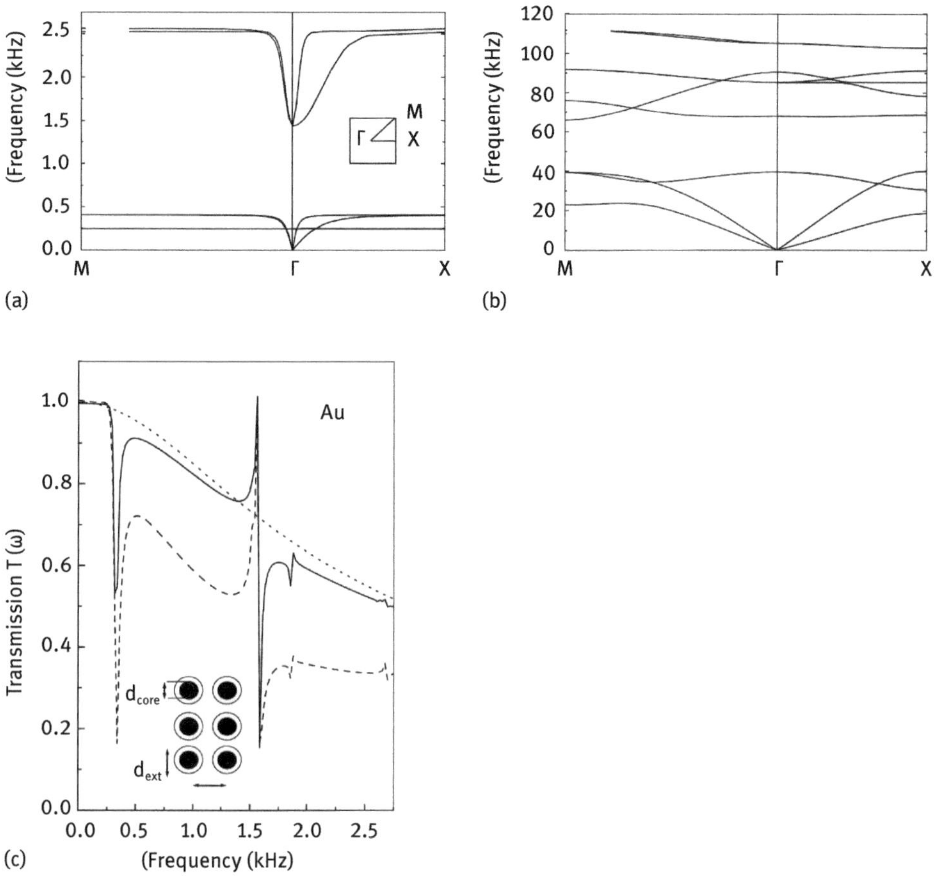

Fig. 10.3: A 2D locally-resonant phononic crystal. (a) In-plane elastic modes of a square lattice of Au cylinder coated with rubber polymer in an epoxy background. (b) In-plane modes for a similar structure made with uncoated cylinders. The parameters used in the calculation for the coating are $\rho = 1300\,\text{kg/m}^3$, $c_l = 23\,\text{m/s}$, $c_t = 6\,\text{m/s}$. The ones of the background are $\rho = 1180\,\text{kg/m}^3$, $c_l = 2535\,\text{m/s}$, $c_t = 1157\,\text{m/s}$. (c) The continuous (dashed) line represents the FDTD transmission amplitude across a slab of 3(6) rows of coated cylinders ($d = 14\,\text{mm}$) arranged on a square lattice ($a = 15\,\text{mm}$). The cylinder inner core is gold ($d = 10\,\text{mm}$). The dotted line defines the transmission for a similar structure made of uncoated cylinders (after Goffaux et al. [42], copyright 2002 American Physical Society).

where ϵ is a reduced energy (dimensionless) and q is the Fano parameter. Figure 10.4 depicts the Fano profile for some values of the parameter q; more precisely, the function $\sigma(\epsilon, q)/(1 + q^2)$ is shown. The two transmission dips in Figure 10.3, around 0.3 and 1.5 kHz, can both be fitted against the Fano profile giving slightly negative values of q [42].

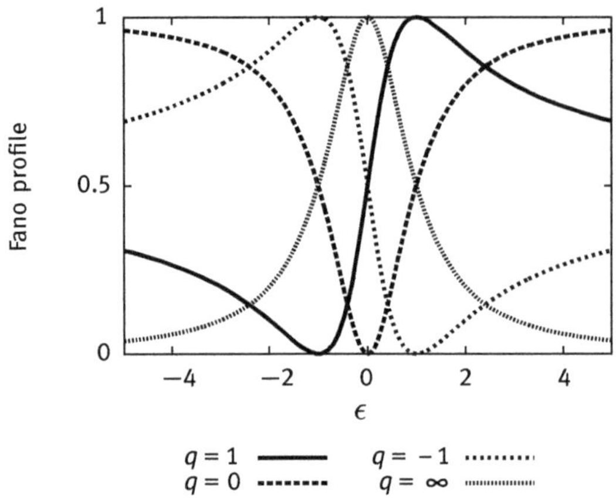

Fig. 10.4: Fano line shape. The plot shows function $\sigma(\epsilon, q)/(1 + q^2)$ with $\sigma(\epsilon, q)$ defined by (10.1) as a function of the reduced energy ϵ, for q considered as a parameter. Note that by tuning the value of q the line shape changes from Lorentzian ($q = \infty$), to a transmission dip ($q = 0$), and to an asymmetric transmission profile ($0 < |q| < \infty$).

10.2 1D arrays of resonators grafted on waveguides

In Chapter 2 we presented the example of local resonances introduced by resonators grafted along a waveguide. In the case of Helmholtz resonators, the local resonances were interpreted as an acoustic metamaterial with a negative effective modulus [39]. As depicted in Figure 10.5 (a), a Helmholtz resonator is composed of a cavity filled with a fluid, or bottle, connected through a neck to the surroundings. With V the volume of the cavity, S the surface of the neck and L its length, the resonance frequency is classically given by the formula

$$\omega_0 = c\sqrt{\frac{S}{VL}}, \tag{10.2}$$

with c the celerity of sound in the fluid. The Helmholtz resonance is usually said to be different to an acoustic resonance, in the sense that it is analogous to a mass-spring resonance, while the acoustic modes of the cavity appear at much higher frequencies. A completely closed bottle, however, has a fundamental acoustic mode at the zero frequency (the pressure distribution of the eigenmode is simply $p = 1$ everywhere for a hard-wall boundary condition). The Helmholtz resonance can thus also be viewed as this fundamental acoustic mode of the bottle that is shifted in frequency through coupling via the neck. With the dimensions of Figure 10.5, the Helmholtz resonance frequency in water is 29.8 kHz from (10.2). The resonant frequency is observed experimentally around 33 kHz with a periodic array of identical resonators. A transmission dip in frequency with a certain degree of asymmetry is found. The interpretation that the (dynamic) effective modulus is negative in a given frequency range means that

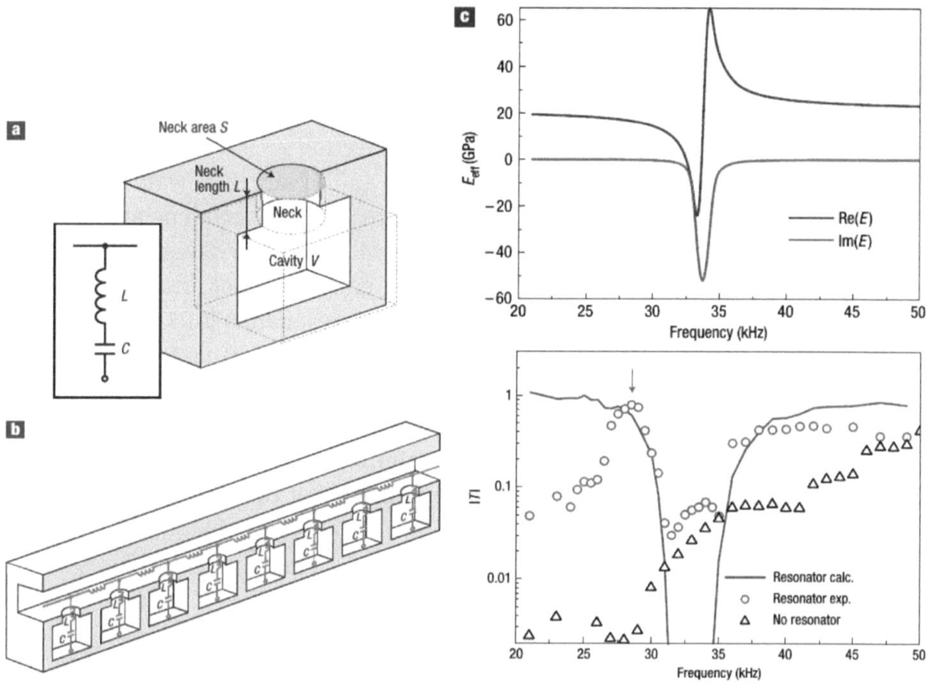

Fig. 10.5: A new class of ultrasonic metamaterials consisting of arrays of subwavelength Helmholtz resonators. (a) A schematic cross-sectional view of a Helmholtz resonator is depicted. The sample is made of aluminum, consisting of a rectangular cavity of $3.14 \times 4 \times 5$ mm, and a cylindrical neck 1 mm long and 1 mm in diameter. The cavity and neck are filled with water and are connected at the same side to a square water duct with a 4×4 mm opening. The resonators are placed in a periodicity of 9.2 mm. The inset illustrates the analogy between a Helmholtz resonator and an inductor-capacitor circuit, showing the fluidic inductance due to the neck, and the acoustic capacitance due to the cavity. (b) A periodical daisy-chained Helmholtz resonator gives rise to a complex acoustic admittance in the fluidic network. (c) The calculated effective bulk modulus in the above one-dimensional subwavelength Helmholtz resonators is shown. As the structures are considerably smaller than the wavelength at the frequency range, the medium can be homogenized with an effective modulus. (d) The measured and calculated transmission (amplitude ratio) are plotted as a function of frequency between upstream and downstream detectors. The arrow indicates a Fano-like asymmetric peak, attributed to the existence of nonresonant paths in the channel. The deviation from theoretical prediction at the absorption band is probably due to the sensitivity of the hydrophones in the experiment (after Fang et al. [39], copyright 2006 Nature Publishing Group).

for this range the variation of pressure has the same sign as the local volume variation (see (3.5); this sign would be the opposite for a usual compressible fluid). At this point, it may be useful to remark that we have already encountered negative effective permittivities for the surface of piezoelectric solids in Section 5.7. The negativity then arose from the coupling of the electromagnetic waves of the surface with propagating but much slower surface elastic waves. The negativity of certain effective parameters in physical situations involving traveling waves interacting with a resonant object is indeed not uncommon.

Figures 10.6 and 10.7 report the observations of Wang et al. for simple cylindrical resonators grafted on a cylindrical waveguide [161]. This is actually a direct experimental implementation of the model presented in Section 2.3.3. The waveguide is a 2 meter long PVC tube with an inner radius of 50 mm and a 2 mm thick wall. This tube is a single-mode waveguide for pressure acoustic waves in air for frequencies between 0 and 2 009 Hz. Resonators in the form of PVC tubes with a smaller diameter and closed at their extremity are grafted onto the waveguide, either alone or arranged in an array with a period of 80 or 250 mm. The radius of the resonators is 25 mm, and their length is either 240 mm for long resonators or 40 mm for short resonators. Such cylindrical resonators have a series of natural resonances whose frequencies are directly related to the tube length. As we noted above, the fundamental mode for a completely closed tube appears at the zero frequency; the same tube forms a Helmholtz resonator when connected to the waveguide. It can be noticed that the transmissions in Fig-

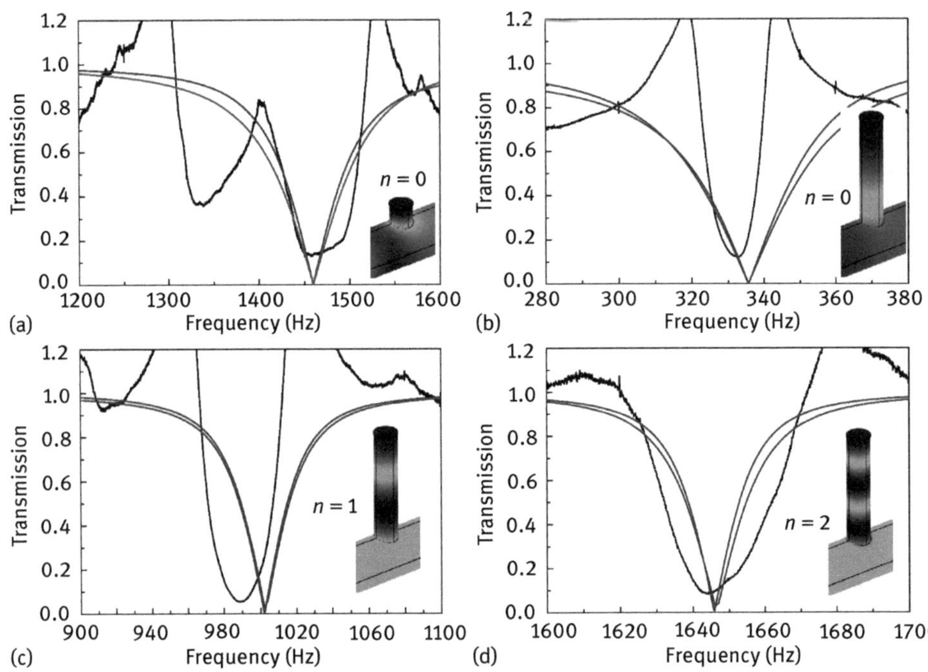

Fig. 10.6: Transmission through single resonators grafted on an air tube forming a waveguide for sound. Transmission of a single (a) short and (b)–(d) long resonator grafted onto the waveguide. Panels (b)–(d) show the transmission at frequencies related to the 0th, 1st and 2nd acoustic resonance of the single long resonator, respectively. The insets show the vibration modes at the resonant frequency obtained from an FEM simulation. The black and the gray lines represent the experimental and simulated results, respectively (after Wang et al. [161]).

ure 10.6 vary with different resonators and different resonant frequencies of the same resonator. Transmission dips clearly appear around 1 460 Hz for the short resonator, and around 336, 1 000 and 1 646 Hz for the long resonator. Simulated pressure distributions at the frequency of the dips are shown as insets. Clearly, the pressure field is localized inside the resonators. The modal distribution of these local resonances look very much like the n-th ($n = 0, 1, 2$) order natural vibration modes of the resonator. The natural frequencies of the isolated resonator with two closed ends and length l are given by $\Omega n/(2\pi) = nc/(2l)$, and they are clearly different from the resonant frequencies obtained from transmission dips. In general, the phenomenon of transmission canceling appears at a resonant frequency that is different from the natural frequency. This property was explained in Section 2.3.3 by the excitation of evanescent guided waves attached to the grafting point.

Figure 10.7 shows the results for periodic arrays of five resonators, with the lattice constant a = 250 mm, and their comparison with complex band structures obtained with the theory in Chapter 9. LR band gaps appear around the resonant frequencies previously found for the single resonators. They are comparatively broader than the transmission dips for the single resonator, in accordance with the complex band structures computed with FEM using the method described in Chapter 9. In the short resonator case, an additional Bragg band gap appears around 1 372 Hz, just below the LR band gap. The attenuation inside the LR band gap is found to be larger than that inside the Bragg band gap. All these features can be clearly explained by the complex band structures. The smallest imaginary part of the complex wavenumber for the LR band gap is larger than that of the Bragg band gap. As a result, a stronger attenuation is expected in the LR band gap as compared to the Bragg one. The complex band structure has a different shape within the Bragg and the LR band gaps. In the Bragg case, the band gap opens at the Γ point of the first Brillouin zone and the complex imaginary band connecting the entrance and the exit of the band gap is continuous and very smooth. In contrast, the entrance and the exit of the LR band gap occur at opposite symmetry points of the Brillouin zone and the complex imaginary band connecting them is composed of two crossing bands forming a cusp. Note the resemblance with the simple model of the complex band structure for locally-resonant crystals in Figure 2.21. In the long resonator case, LR band gaps appear around the three different resonant frequencies under the cut-off frequency for the second guided mode of the waveguide. Bragg band gaps are not clearly observed in this case due to their small attenuation. The complex band structure around the LR band gaps also has the features discussed above for the short resonator. It can also be remarked that inside the experimental LR band gaps, the transmission seems to be a combination of a series of dips. This may be due to the interference of guided waves trapped between the resonators and creating a super-modal structure of the series of five resonators [83].

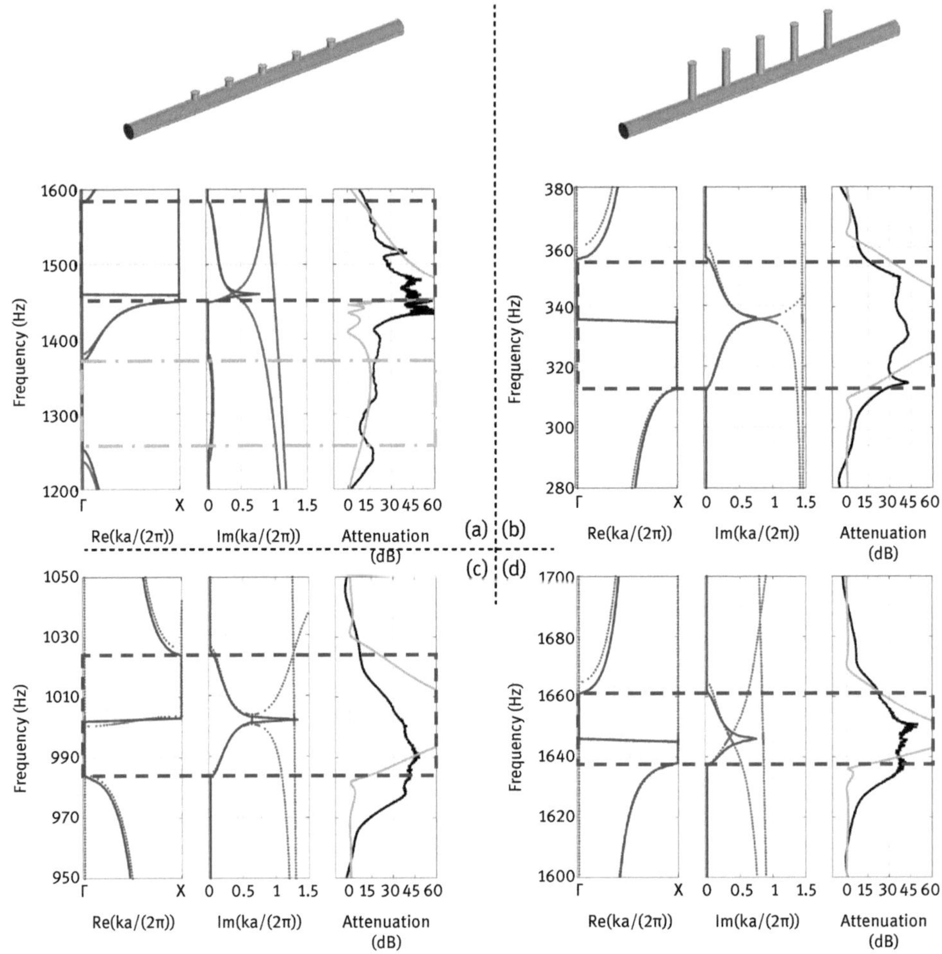

Fig. 10.7: Transmission through arrays of five resonators grafted on an air tube forming a waveguide for sound. Complex band structures (real part on the left and imaginary part in the middle of each panel) and transmission properties (on the right of each panel) are plotted for 1D phononic crystals with a periodic array of (a) short and (b)–(d) long resonators (lattice constant a = 250 mm). The black and gray lines in the complex band structures represent the theoretical and the numerical results, respectively. The black and gray lines in the transmission spectra represent the experimental and numerical results, respectively. The dashed lines mark the LR band gap predicted by the theoretical model. The dashed-dotted lines in panel (a) mark the Bragg band gap predicted by the theoretical model (after Wang et al. [161]).

10.3 Locally-resonant sonic crystals

As an example of a two-dimensional locally-resonant sonic crystal, let us consider the experiment proposed by Lemoult et al. [83] and depicted in Figure 10.8. They studied a periodic ensemble of 7×7 soda cans arranged on a square lattice of period 6.6 cm, in a closed-packed configuration (i.e. the cans are touching). As a further example of Helmholtz resonators, the soda cans possess a resonance frequency of 420 Hz. This frequency would correspond to a wavelength $\lambda = 0.8$ m in air. Around the array, eight speakers were positioned 1 m away, so that only propagating waves reach the cans. The speakers are connected to a multichannel sound card. Pressure can be measured at any location above the soda cans with a small microphone mounted on a 3D moving stage. The time domain Green's functions between the eight speakers and the top of the 49 cans are first acquired, by using a 10 ms pulse centered around 400 Hz. A typical recorded sound form is plotted in Figure 10.8 (b). The Fourier transform of the measurement in Figure 10.8 (c) presents a series of resonance peaks spanning roughly from 340 Hz up to the resonance frequency of a single soda can, 420 Hz. Each peak can in principle be associated with a periodic mode of the 7×7 array and possesses spatial variations ranging from the total dimension of the array down to the size of a single can. As argued in reference [83], because of the finite size of the array, exactly 49 modes with discrete frequencies and wavevectors exist, as a result of the coupling of the res-

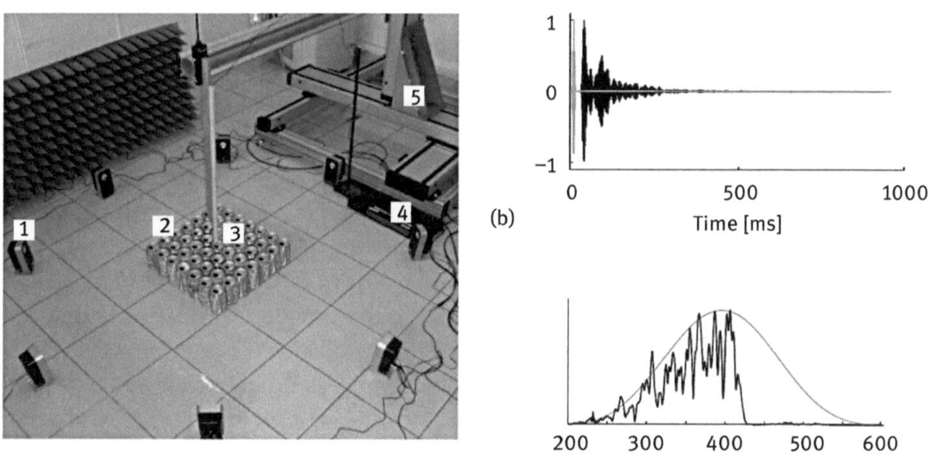

(a) (b) (c)

Fig. 10.8: A square-lattice periodic array of 7×7 soda cans in close-packed position. (a) The array is surrounded with eight audio speakers placed in its far field. A small microphone can be scanned at any position just above the array to detect the pressure distribution. A typical recorded sound form (b) and its Fourier transform (c) indicate a pass-band extending below the resonance frequency of the cans, 420 Hz, and a locally-resonant band gap extending above it (after Lemoult et al. [83], copyright 2011 American Physical Society).

onators. Coupling between adjacent cans occurs via the air surrounding them, rather than through the mechanical contacts at their sides. As a result, the local resonances of the cans hybridize with the air continuum, creating super-modes similar to a surface wave polariton. Above the frequency of 420 Hz, the spectrum in Figure 10.8 (c) drops to zero, a signature of the locally-resonant band gap introduced by the square array.

In Chapter 4 we presented the case of air bubbles in water as the prototype for obtaining ultrawide sonic band gaps. We noted that resonant modes of the air bubbles could be identified in the band structures of Figure 4.23 and 4.24. This naturally leads to the idea that the wide sonic band gaps in crystals of air bubbles are due to the hybridization of waves propagating in water with the resonant modes of the air bubbles. The exact mechanism of band formation is however very difficult to read on the band structures, because only very flat bands are actually obtained. One important question is whether the apparent feature of locally-resonant band gaps, that hybridization causes a jump from one high-symmetry point to another, is generally true? This is the situation that we depicted in Figure 10.2 (b) and that can be clearly observed in Figures 10.1 (d), 10.3 (a), and 10.7. The counterexample provided by Ao and Chan actually proves that this is not always the case [5]. They considered a square-lattice 2D sonic crystal of cylindrical inclusions of a hypothetical heavy and slow fluid ($\rho = 20\,000\,\text{kg/m}^3$, $c_l = 100\,\text{m/s}$) embedded in water ($\rho = 1000\,\text{kg/m}^3$, $c_l = 1490\,\text{m/s}$), with the cylinder diameter given by $d/a = 0.5$. Such a material as that considered for the inclusion can of course hardly exist. The complex band structure for this crystal is shown in Figure 10.9 (a). It displays a series of locally-resonant band gaps in the low frequency range, as we should expect. Inside the lowest band gap, there is a clear cusp formed in the imaginary part of the complex band structure, but the gap is opened and closed at the X point of the first Brillouin zone. There actually seems to be a phenomenon of interference with several locally-resonant modes at once, complicating the interpretation of the complex band structure (and complicating the interpretation of the classical band structure even more). The formation of the cusp in the imaginary part is also seen to result from the crossing of two complex bands, rather than from a single complex band. Even more surprising is the case when the inclusions have material properties $\rho = 1000\,\text{kg/m}^3$ (i.e. exactly as water) and $c_l = 50\,\text{m/s}$, in Figure 10.9 (b). In this case, there is no formation of cusps and the locally-resonant band gaps might easily be confused with Bragg-type band gaps: they open and close at the same high-symmetry point and they give rise to imaginary branches very similar to those we have described in Chapter 9, where all band gaps were of the Bragg type.

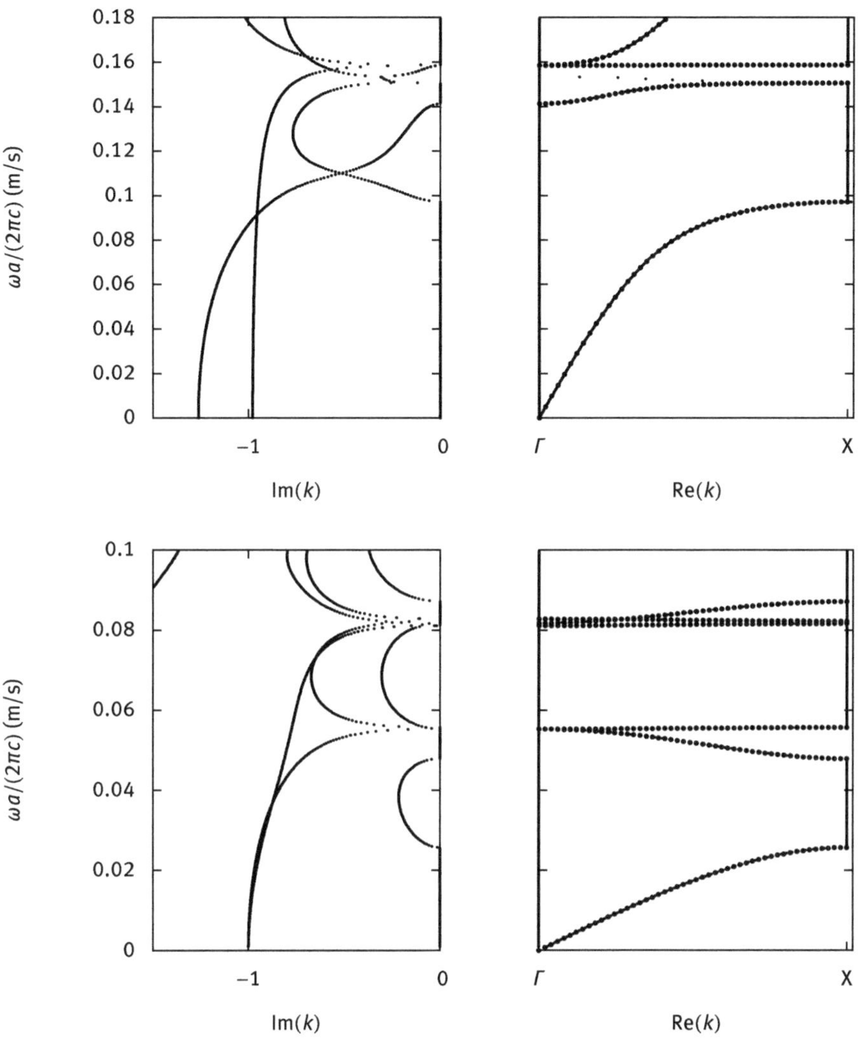

Fig. 10.9: Complex band structure for 2D locally-resonant sonic crystal. (a) Case of a square-lattice crystal of cylindrical inclusions of a hypothetical heavy and slow fluid ($\rho = 20\,000\,\text{kg/m}^3$, $c_l = 100\,\text{m/s}$) embedded in water ($\rho = 1000\,\text{kg/m}^3$, $c_l = 1490\,\text{m/s}$), with $d/a = 0.5$. (b) Case of the same crystal but with the inclusions being defined with $\rho = 1000\,\text{kg/m}^3$, $c_l = 50\,\text{m/s}$. These values reproduce those employed by Ao and Chan [5]. The complex band structures were computed here using the finite element method set out in Chapter 9, instead of the layer multiple scattering method in the cited reference. Frequencies are normalized using the celerity c of longitudinal waves in water. The classical and the complex band structures are plotted with large and small dots, respectively.

10.4 Locally-resonant phononic crystals

We introduced this chapter with the foundational example of the 3D crystal of rubber-coated lead spheres in an epoxy matrix [89], followed by its 2D version [42]. In contrast to locally-resonant sonic crystals, locally-resonant phononic crystals are composed of solid constituents and they do not necessarily need a further support or frame to hold them together. In the 2D case, however, the band structure in Figure 10.3 was for in-plane waves only, implying that out-of-plane waves were not similarly affected by the local resonances. For isotropic solid constituents, we saw in Chapter 5 that in-plane and out-of-plane waves separate completely in 2D phononic crystals. This property of course remains true in the presence of local resonances: only localized vibrations and propagating waves of compatible, i.e. nonorthogonal, polarizations can interfere and hybridize to create locally-resonant band gaps.

Different structures have been proposed as locally-resonant phononic crystals; their variety actually seems very large, as there are many different ways to introduce an internal resonant element within the unit cell. We review some of them in the following, for the purpose of illustration. Important practical properties that can make a difference between them are their subwavelength character, or maybe more interestingly the smallest volume that is necessary to induce a resonance at a given frequency, the relative width of the band gaps that are introduced, and the overall attenuation that can be obtained through a given number of layers. It remains that locally-resonant band gaps are usually quite narrow, except for the exceptional case of the sonic crystal of air bubbles in water that we discussed in Chapter 4.

Ho et al. used unit cells composed of a rigid plastic frame containing a central heavy mass embedded in silicon rubber [51]. Similar to the ternary crystals, the silicon rubber matrix acts as a soft spring from the heavy mass, and the rigid frame replaces the epoxy container. Combining unit cells with different central masses, Ho et al. could experimentally show an enlargement of the attenuation bandwidth as a result of the superposition of the transmission dips introduced by different resonators. The same structure but limited to two-dimensions was explored numerically by Goffaux et al. [43]. They found that the presence of the rigid plastic frame was indeed needed in order to form Fano resonances in the transmission. They showed, however, that simply considering the phononic crystal formed by the heavy rigid spheres in a matrix of silicon rubber led to Bragg band gaps showing more efficient attenuation in almost the same frequency range. This conclusion is however not general and is connected to the very low shear and longitudinal velocities considered for silicon rubber (namely, $\rho = 1300\,\text{kg/m}^3$, $c_l = 22.86\,\text{m/s}$, and $c_t = 5.54\,\text{m/s}$).

An interesting question is whether it is necessary to consider a ternary composition (three different materials) in order to induce locally-resonant band gaps, or if binary compositions are sufficient. G. Wang et al. answered this question by theoretically showing that a 2D phononic crystal of silicon rubber rods in an epoxy matrix can indeed achieve the required property [158]. In a certain sense, this composition is

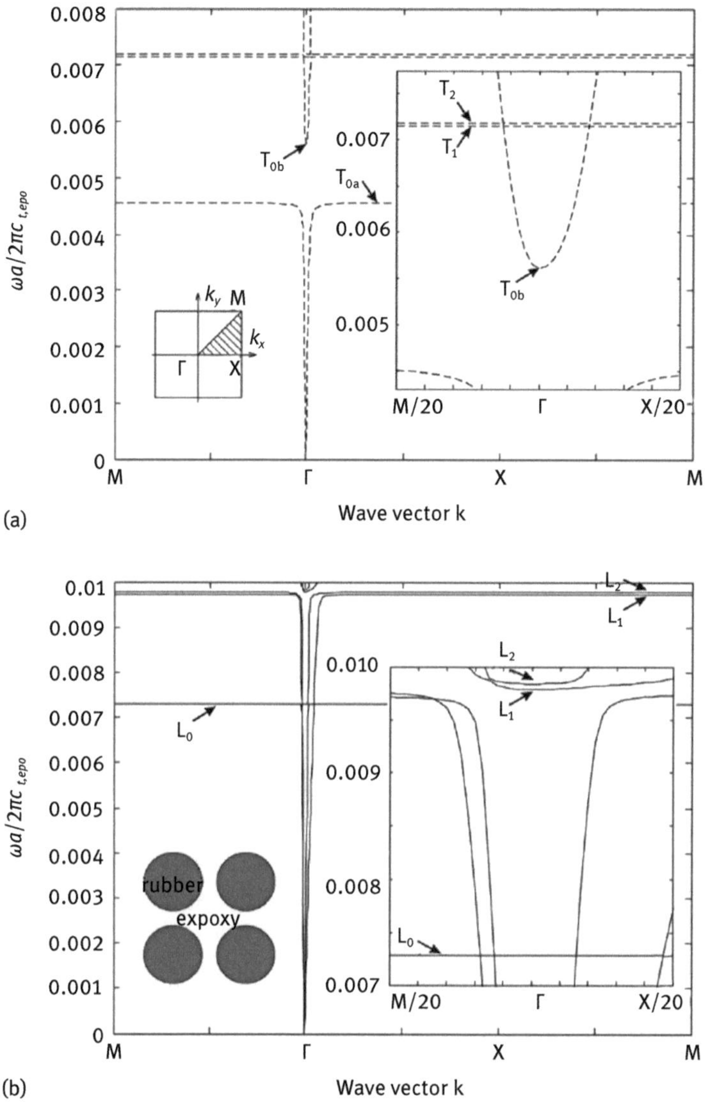

Fig. 10.10: A 2D square-lattice phononic crystal of silicon rubber rods in epoxy, with filling fraction $F = 0.503$. Band structures are shown for (a) out-of-plane modes and (b) in-plane modes in binary phononic crystals composed of the square lattice of soft rubber cylinders in epoxy. The insets show the corresponding zooms. The lattice displacement u_3 is plotted for selected out-of-plane modes at points (c) T0a, (d) T0b, (e) T1 and (f) T2. Selected in-plane modes (g) L0, (h) L1 and (i) L2 are plotted with arrows whose direction and length represent the direction and amplitude of the displacement vectors, respectively (after G. Wang et al. [158], copyright 2004 American Physical Society).

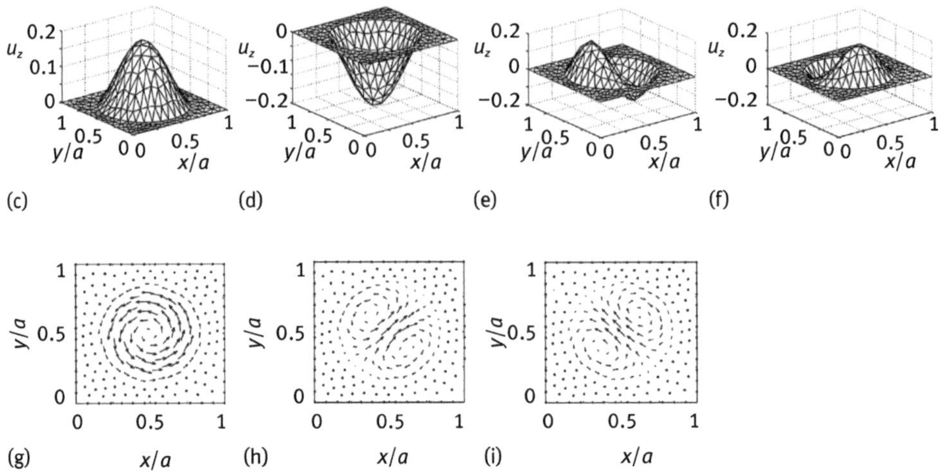

Fig. 10.10 (continued)

analogous to the sonic crystal of air bubbles in water: a very soft solid material with very slow shear and longitudinal velocities as compared to the matrix velocities seems to be the key. Figure 10.10 shows the dispersion of the phononic crystal in the low frequency range. The band structure for out-of-plane waves, in Figure 10.10 (a), shows a locally-resonant band gap that is opened by the interaction of propagating waves with resonant mode T0. This pure shear mode has maximum displacement u_z at the center of the rods and vanishing displacements at the boundary with epoxy. Plotted at the entrance of the band gap (T0a) and at the exit (T0b), the hybridized Bloch wave is mostly localized inside the rod. With little perturbation, two pure shear modes, T1 and T2, pass horizontally through the band structure without opening any gap. They thus do not hybridize with propagating Bloch waves at all. These modes should be degenerated and orthogonal, but appear at slightly different frequencies in the calculation because of the absence of symmetry of the mesh, probably. The band structure for in-plane waves, in Figure 10.10 (b), shows a comparatively small locally-resonant band gap that is opened by the interaction of propagating waves with resonant modes L1 and L2. These modes are probably originally degenerated, but they hybridize slightly differently with propagating waves depending on the direction and modulus of the wavevector. Torsional mode L0 passes horizontally through the band structure without opening any gap.

10.5 Phononic crystal slab with pillars

Locally-resonant phononic crystals can be achieved quite practically by using the third dimension of space in a two-dimensional crystal. For instance, a periodic array of pillars can be formed on top of a membrane, as depicted in Figure 10.11 [119]. The membrane mechanically supports the whole structure and at the same time provides for vertical confinement of elastic waves. The pillars provide natural mechanical resonators because of their added mass and of their bonding to the membrane that acts like a distributed spring. The natural modes that can be expected in the low frequency

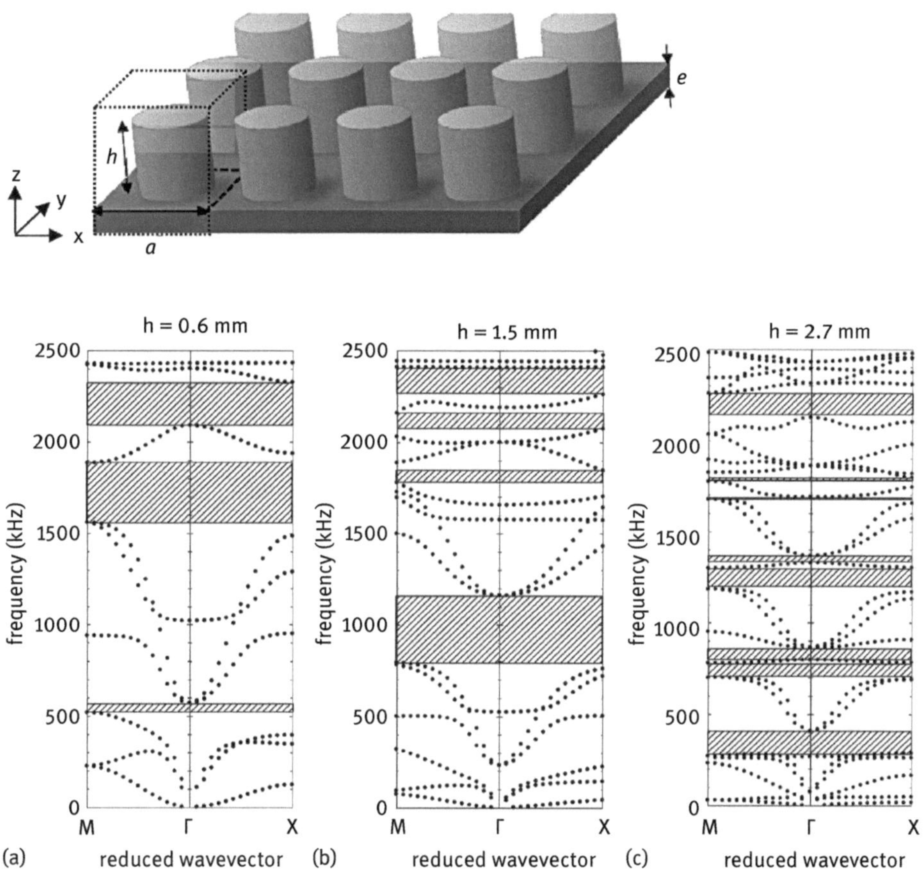

Fig. 10.11: Locally-resonant phononic crystal slab of pillars. A square-lattice array of finite-height cylinders (height h) is deposited on a thin membrane (thickness e). The lattice constant is a. The dashed cube represents one unit cell of the crystal. The band structures are plotted for different values of the relative height of the pillars: (a) $h/a = 0.6$, (b) $h/a = 1.5$, and (c) $h/a = 2.7$. The thickness of the membrane is $e/a = 0.2$ and the filling fraction is 56.4% ($d/a = 0.84$) (after Pennec et al. [119], copyright 2008 American Physical Society).

range are typically clamped at the interface with the membrane and free at the other end of the pillar, providing a rich choice of possible flexural, torsional, and compressional motions for the pillar. At the same time, if the membrane is chosen to be rather thin, the phononic crystal slab will support mostly flexural motion.

Pennec et al. studied the evolutions of locally-resonant band gaps as a function of the height of the pillars, in the case of a rather thin supporting membrane [119]. Examples of band structures are given in Figure 10.11 for three different values of the ratio h/a of the pillar height over the lattice constant. The case considered is a square-lattice phononic crystal slab of steel cylinders on a silicon membrane. It can be seen that a succession of locally-resonant band gaps is introduced in the low frequency range, whose number depends on the height of the pillar: as the height of the pillar is increased, the resonant frequencies move downward in frequency.

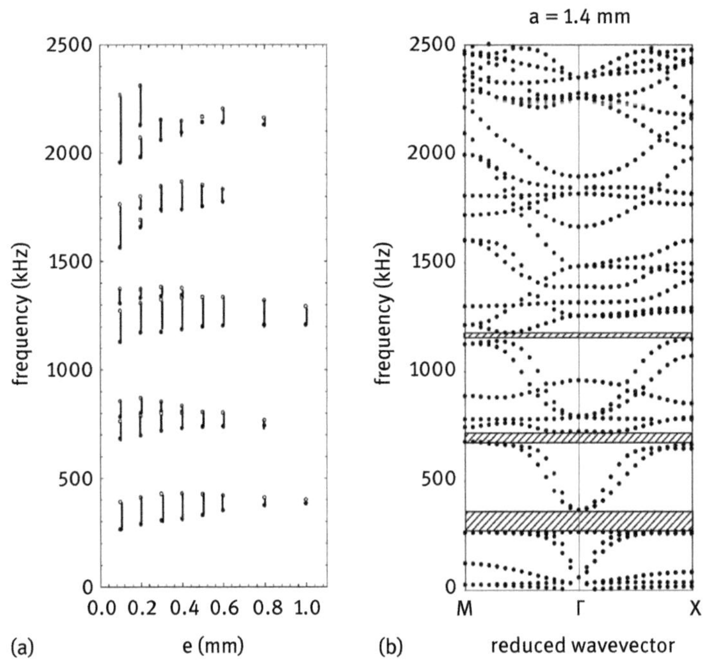

Fig. 10.12: Locally-resonant phononic crystal slab of pillars. The structure is the same as in Figure 10.11. Panel (a) shows the evolution of the locally-resonant band gaps as the thickness of the supporting membrane is increased from $e/a = 0.1$ to 1.0, for $a = 1$ mm. The height of the pillars is $h/a = 2.7$ and the filling fraction is 56.4% ($d/a = 0.84$). The band structure in panel (b) is obtained for a smaller filling fraction of 38% ($d/a = 0.6$) with $a = 1.4$ mm. The height of the pillars $h/a = 1.93$ and for the membrane thickness $e/a = 0.142$ (after Pennec et al. [119], copyright 2008 American Physical Society).

The effect of the membrane thickness was further studied by Pennec et al. [119]. Figure 10.12 (a) displays the evolution of the locally-resonant band gaps as a function of the thickness of the supporting membrane, for a given height of the pillar. The result essentially shows that the width of the band gaps is inversely proportional to the thickness of the membrane, and that they gradually decrease in extent and eventually vanish as the thickness is increased. As in the case of the phononic crystal slabs of empty or filled holes, there is thus an optimal membrane thickness for which the hybridization of flexural waves in the membrane and of the modes of the pillars is maximized. The filling fraction of the phononic crystal also plays a significant role, as Figure 10.12 (b) illustrates. As a rule of the thumb, providing the height of the pillar is significantly larger than its diameter, changing the diameter will not affect thickness-resonant modes but will slightly affect flexural modes.

At almost the same time as Pennec et al., Wu et al. studied a similar phononic crystal slab of aluminum cylinders on a thin aluminum plate and provided experimental evidence of the induced local resonances and locally-resonant band gaps [167]. Figure 10.13 shows both the experimental arrangement, the measurements and their comparison to the theoretical band structure, obtained with FEM. The predicted band gaps can be identified in the measurements as dips in the transmission, but even more strikingly, the local resonances of the pillars appear very clearly as transmission peaks, suggesting that elastic energy transport in the phononic crystal slab is enhanced at those frequencies. As a possible explanation, enhanced transmission may be connected to the low group velocity for frequencies around the band edges, meaning that the generated elastic waves can escape only very slowly in time from the source location and thus increase the detection rate.

Another variation over the concept of the phononic crystal slab of pillars was provided by Oudich et al. [116]. They considered pillars made of silicon rubber on top of a thin aluminum plate, as depicted in Figure 10.14. Owing to the very low shear and low longitudinal velocities of silicon rubber, determined to be respectively $c_l = 975$ m/s and $c_s = 23.5$ m/s as compared to $c_l = 6302$ m/s and $c_s = 3236$ m/s for aluminum, the resonance frequencies of the pillars are expected to appear in a much lower frequency range for the same height of the pillar. Of course, this observation is made without consideration of elastic propagation losses in silicon rubber, which can effectively reduce the Q-factor of the resonances. As Figure 10.14 shows, the lowest locally-resonant band gap with the silicon rubber pillars is obtained around $\omega a/(2\pi) = 22$ m/s, about 60 times lower than with the aluminum pillars of Figure 10.13 ($\omega a/(2\pi) \approx 1200$ m/s).

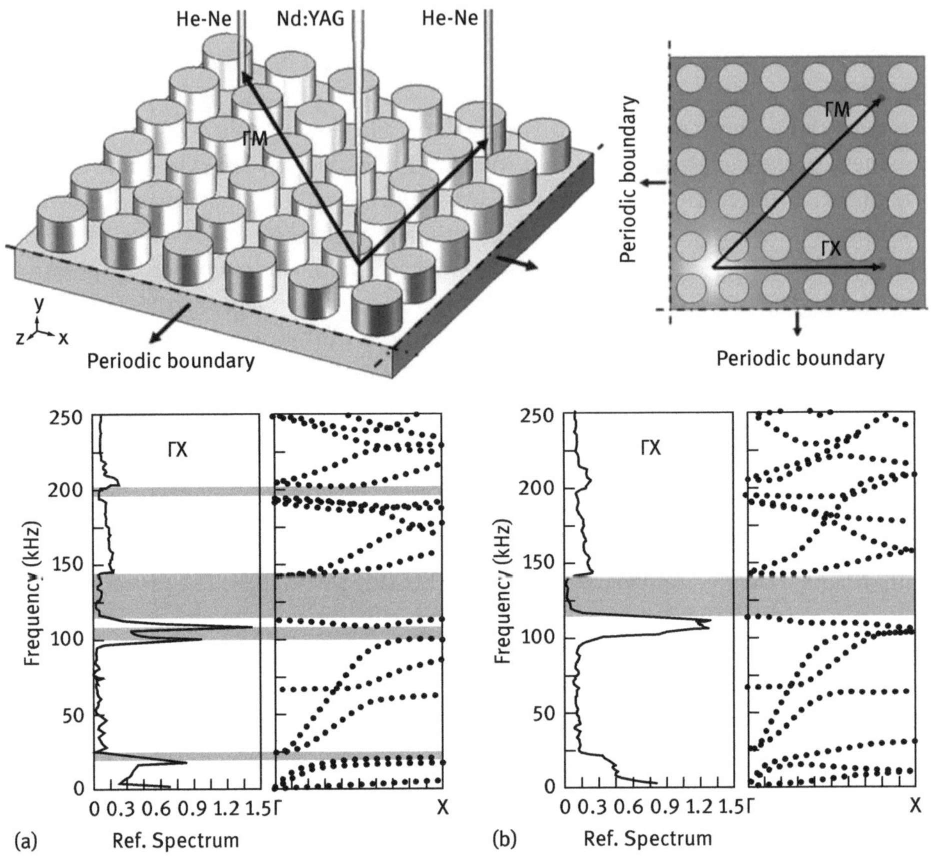

Fig. 10.13: Locally-resonant phononic crystal slab of aluminum pillars on an aluminum plate. A square-lattice array of 16 × 10 finite-height cylinders (height $h/a = 1$) is deposited on a thin membrane (thickness $e/a = 0.1$). The lattice constant is $a = 10$ mm. The diameter of the cylindrical pillars is $d/a = 0.7$. An Nd:YAG (yttrium ion garnet) pulsed laser is used to generate broadband elastic waves in the phononic crystal, while an optical interferometer measures the normal component of the displacement at any point over the surface. Measurements are performed along the two principal symmetry directions (a) *Γ*X and (b) *Γ*M and are compared to the theoretical band structure. Source and receiver are four rows apart in both directions (after Wu et al. [167], copyright 2008 American Institute of Physics).

10.6 Surface phononic crystal of pillars

As a final example of a phononic crystal presenting local resonances, in this section we describe the phononic crystal of pillars on a surface introduced by Khelif et al. [62]. From a technological point of view, this system is very similar to the phononic crystal slab of pillars on a membrane. A major difference, however, is that 4coupling of the resonating pillars to the propagating continuum waves does not occur via flexural vi-

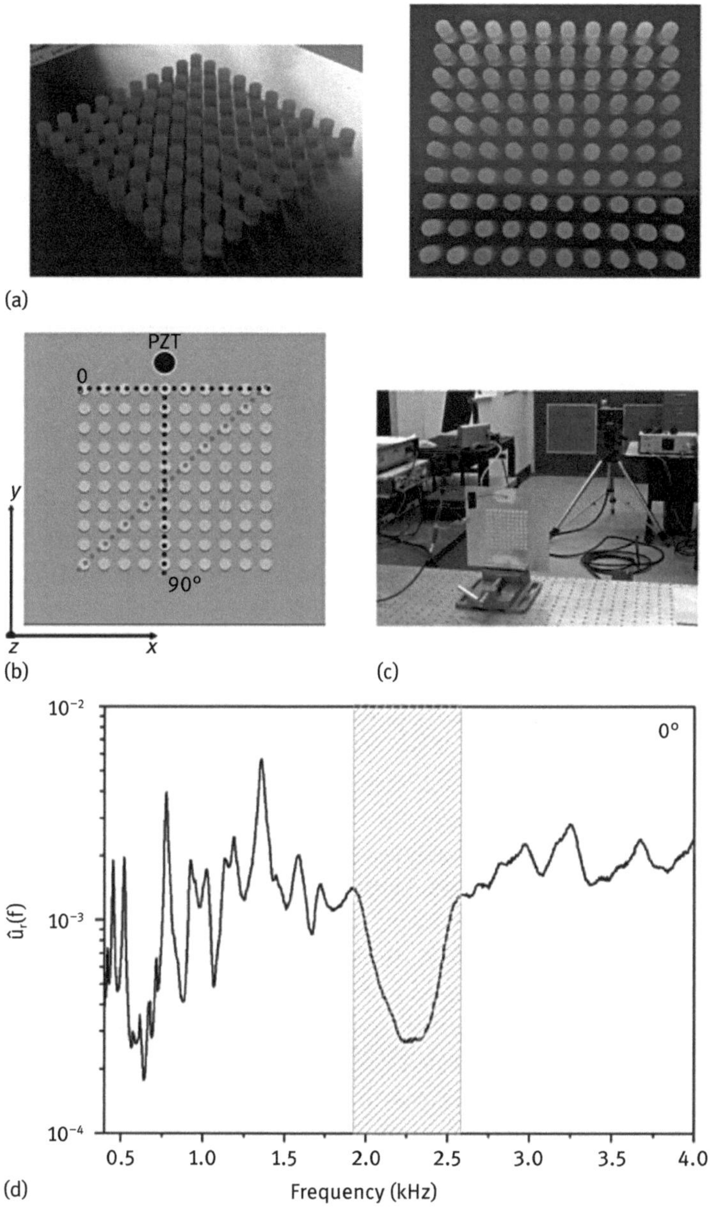

Fig. 10.14: Locally-resonant phononic crystal slab of silicon rubber pillars on an aluminum plate. (a) A square-lattice array of 10 × 10 finite-height cylinders (height $h/a = 0.5$) is deposited on a plate (thickness $e/a = 0.05$). The lattice constant is $a = 10$ mm. The diameter of the cylindrical pillars is $d/a = 0.6$. (b) Elastic waves are excited by a PZT transducer attached to the plate. (c) A laser vibrometer is used to measure the normal component of the displacements at any position over the plate. (d) A locally-resonant band gap is observed at frequencies around 2.2 kHz (after Oudich et al. [116], copyright 2011 American Physical Society).

brations but via the evanescent surface waves of a semi-infinite substrate. As a result, there are far fewer bands in the low frequency range before hybridization, but the possibility of radiation loss to the substrate must be taken into account. As an example, let us consider the phononic crystal of nickel pillars on a lithium niobate substrate in Figure 10.15 [2]. The nickel pillars shown have a radius $r = 3.2\,\mu m$ and a height $h = 4.7\,\mu m$. They are arranged according to a square lattice with a pitch $a = 10\,\mu m$. The experimental set-up used for investigating the propagation of surface phonons involves two chirped interdigital transducers to generate and detect them. The band structure is computed using the finite element method set out in Chapter 7 in the case of surface phononic crystals of holes. The gray region indicates the radiative region, or sound cone, of the substrate. From this band structure, there appear to be two complete band gaps for surface guided waves.[3] The first complete band gap opens from a frequency of about 80 MHz; its exit, however, is not so clearly defined as it appears to be mostly dictated by the sound line rather than by a band folding at a high-symmetry point of the first Brillouin zone. Similarly, the second complete band gap opens from a frequency of about 150 MHz and closes around 180 MHz, again with reference to the sound line. Leaky surface elastic waves inside the sound cone are missing from such a band structure. As a consequence, it is a priori difficult to figure out if transmission through the phononic crystal is strongly affected or not by the possible existence of these leaky surface elastic waves.

The experimental results are shown in Figure 10.16. Two different frequency ranges are identified, around one of each of the two complete band gaps. The first complete band gap is opened by a local resonance in the pillars. The experimental transmittance through the periodic array of pillars is shown in the left panel, from 70 MHz to 100 MHz, for the three highest symmetry directions of the first Brillouin zone. This transmission is obtained from the electrical measurements, by normalizing the transmission with the phononic crystal by the direct transmission on the surface. The transmission dip around 78 MHz is the signature of the local resonance. The averaged cross section of the displacement of the surface, as measured using heterodyne optical interferometry for a frequency of 70 MHz (point A) and the X propagation direction, essentially shows a transmission of the incident surface waves. Then for a frequency of 78 MHz (point B), the same measurement shows the decaying storage of elastic energy along the pillar array. An optical scan of a single pillar of the second row reveals the modal distribution of the surface displacement at the frequency of the transmission dip. The finite element computation of the vertical displacement of the pillar confirms that the observed modal distribution is that of the first local resonance of the pillar. It is striking that the transmission along the row of

3 As in Chapter 7, by surface guided waves we mean all Bloch waves for the surface problem that appear under the sound cone in the band structure. Having a phase velocity lower than any of the bulk elastic waves in the substrate, they are necessarily evanescent along the direction x_3 normal to the surface.

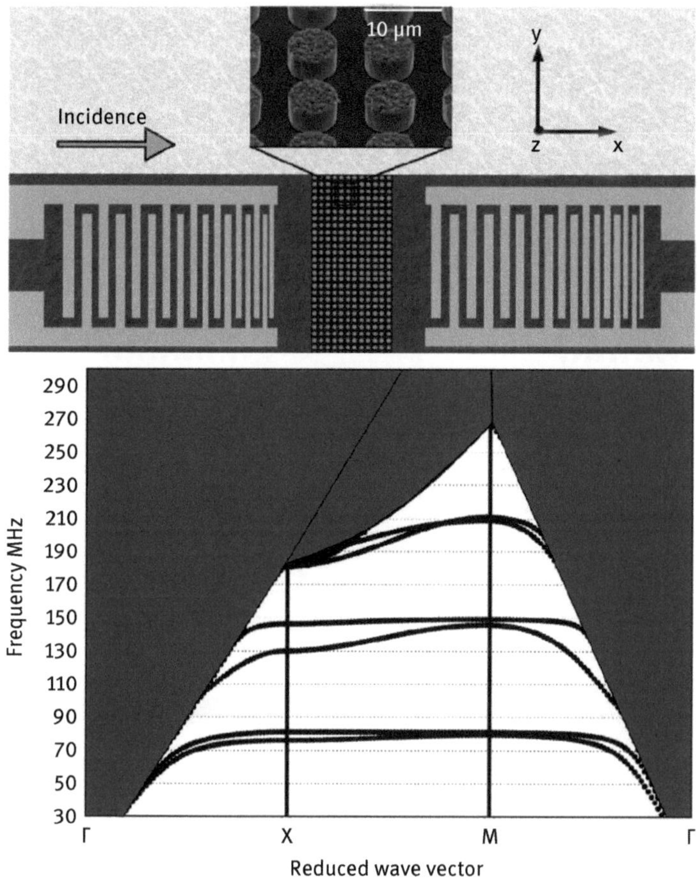

Fig. 10.15: A phononic crystal of pillars on a surface. Schematic of the experimental set-up used for investigating the propagation of surface phonons in a periodic array of pillars on a semi-infinite substrate. The nickel pillars shown as an inset have a radius of 3.2 µm and a height of 4.7 µm. They are arranged according to a square lattice with a pitch of 10 µm. Two chirped interdigital transducers generate and detect surface phonons thanks to the piezoelectric properties of the lithium niobate substrate. The band structure is computed using a finite element method. The gray region indicates the radiative region, or sound cone, of the substrate [2].

pillars decreases almost linearly, rather than exponentially, as we would expect for a Bragg band gap. At resonance, the vibrating pillars drag energy from the incoming waves in the substrate. Because the pillar is connected to the surface, it must later radiate the energy that it has stored, and this radiation can be to any allowed mode of the periodic array, including elastic waves propagating along the surface but also inside the bulk of the substrate. Though in principle in synchronicity with the incoming wave, the radiated waves are emitted along the row with a variable phase and have no reason to interfere constructively or destructively, in contrast to Bragg inter-

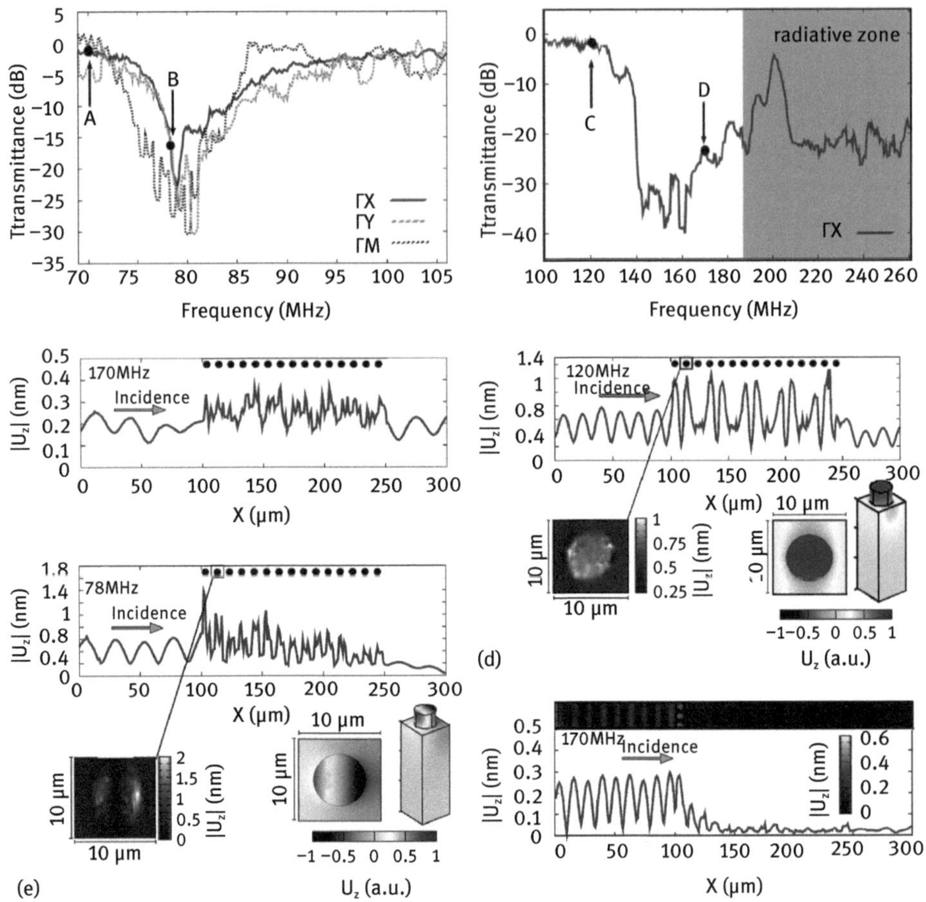

Fig. 10.16: Experimental results for the phononic crystal of pillars on a surface of Figure 10.15. The left panel shows the electrical and optical transmissions for the frequency range [70–100] MHz. The transmission dip around 78 MHz is the signature of a complete band gap for surface elastic waves. Out-of-plane displacements are measured optically for the frequencies 70 MHz (point A) and 78 MHz (point B). An optical scan of a single pillar of the second row reveals the modal distribution of the surface displacement at the latter frequency. It is consistent with a finite element computation of the vertical displacement of the pillar. The right panel shows similar information for the frequency range [100–250] MHz (after Achaoui et al. [2], copyright 2011 American Physical Society).

ference. This difference can be checked by observing the transmission around the second complete band gap, which is due to Bragg interference, in the right panel of Figure 10.16. The frequency band gap for surface elastic waves starts at 140 MHz before which frequency the transmission was close to unity. The averaged cross section of the displacement of the surface, as measured using heterodyne optical interferometry for a frequency of 120 MHz (point C), shows a combination of mostly transmission and some reflection of the incident surface waves. An optical scan of a single pillar of the second row reveals the modal distribution of the surface displacement at this frequency, which is mostly like a piston mode. The finite element computation of the vertical displacement of the pillar is in agreement with the observed modal distribution of the pillar. The averaged cross section of the displacement of the surface for a frequency of 170 MHz (point D) suggests that the incident surface waves are almost totally reflected and shows a strong exponential decay in the periodic array. The optical scan displayed as an inset illustrates how this reflection distributes spatially.

One important test to decide whether a band gap is provided by local resonance or by Bragg interference is the dependence of the band gap on the symmetry of the lattice. Actually, local resonance is essentially dictated by the coupling of an intrinsic mode of the resonator with propagating waves of the matrix (or continuum). When the resonators are placed in a periodic array, this leads to the formation of hybridized bands, providing all resonators are identical and placed in a similar environment. This idea was experimentally checked by Achaoui et al. with pillars essentially similar to those

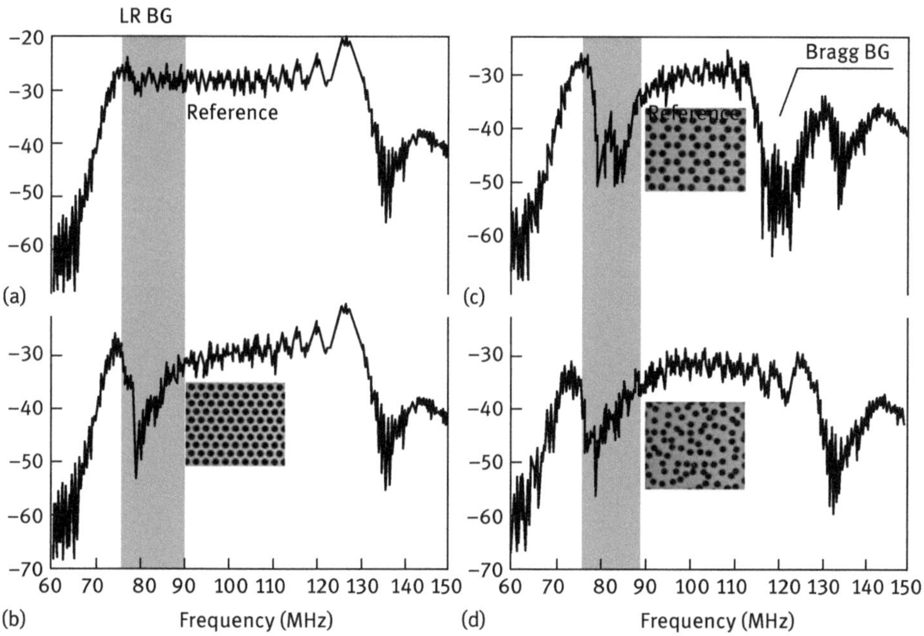

Fig. 10.17: Influence of lattice symmetry on local-resonances of a phononic crystal of pillars on a surface. Measured electrical transmission S_{12} as a function of frequency for (a) a pair of chirped interdigital transducers (IDTs) operating in a delay line configuration without PC, hence serving as a reference, (b) with a triangular lattice PC, (c) with a honeycomb lattice PC and (d) with a randomly distributed pillar array. The gray region represents the common position of the locally resonant band gap. Optical microscope images of each pillar array are shown as insets (after Achaoui et al. [3], copyright 2013 American Institute of Physics).

presented in Figures 10.15 and 10.16 [3]. The three phononic crystals of pillars on a surface in Figure 10.17 are arranged according to a hexagonal, a honeycomb, and a random lattice. Electrical measurements show that transmission dips appear for all three lattices around 78 MHz, at the same resonant frequency that was observed with the square lattice. There are some further slight differences, however, as a double transmission dip seems to be formed with the honeycomb lattice, which is not apparent with either the square or the hexagonal lattice. Maybe the most striking result is that the transmission dip resists a rather large degree of randomness in the spatial distribution of the pillars. Subjected to a similar degree of disorder, the Bragg band gap would have been long lost.

11 Mirrors, waveguides, and cavities

In this chapter, we examine the functionalities of phononic and sonic crystals that result from the existence of a complete phononic band gap. At the end of the discussion of evanescent Bloch waves and of the complex band structure in Chapter 9 we introduced the concept of the supercell. This technique is employed here to explore what happens when a defect is introduced in an otherwise perfectly periodic crystal.

11.1 Phononic crystal functions

In the previous chapters we implicitly classified sonic and phononic crystals according to their geometry and their number of periodicities. In addition, in this chapter we will introduce a series of basic functionalities based on the contents of the supercell.

Classification of artificial crystal problems by geometry. We start by summarizing the crystal geometries we have explored so far. In case the phononic crystal is unbounded, there are no boundary conditions required to terminate the infinite domain of definition. We speak of the propagation of bulk elastic or acoustic waves. In a physical three-dimensional space, we can consider from one to three periodicities and it is thus customary to refer to 1D, 2D, and 3D phononic crystals. When the phononic crystal fills a semi-infinite domain terminated by a given surface, there are additional waves that can be guided along the surface that we term surface waves. In the case where there can be radiation to the remaining half space, we speak of interface waves. Plate (or slab) phononic crystals are limited by two parallel planes and thus two surface or interface boundary conditions. Rather than attempting a lengthy description, in Table 11.1 we depict the different possibilities that arise from this discussion. We have deliberately assumed that the crystal extends homogeneously to infinity in any direction along which it is not periodic. As a result, the sum of the number of periodicities and of the number of homogeneous directions always equals three. Note that there are even more complicated cases that can be considered (e.g. finite phononic crystals with arbitrary enclosing surfaces), so that our tentative classification is clearly not exhaustive.

Classification by problem type. The previous classification by geometry insists upon the physical properties of phononic crystals, and especially upon the types of waves that they can support. A second classification can be performed based on the phononic crystal structure, as attempted in Table 11.2. Perfectly periodic phononic crystals can be thought of ideally as extending to infinity, in which case any solution of the wave equation can be written as a superposition of Bloch waves. The dispersion relation of all Bloch waves forms the band structure which can in general be obtained

Tab. 11.1: Classification of phononic crystals according to their geometry. Three consecutive dots (···) indicate a direction of periodic repetition of the unit cell, while an arrow (→) means the structure extends homogeneously to infinity.

Problem type	Surface boundary conditions	Number of periodicities		
		1	2	3
Bulk	0	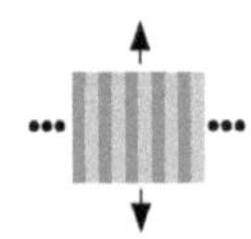	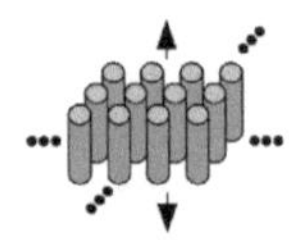	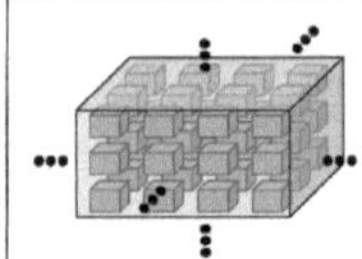
Surface	1	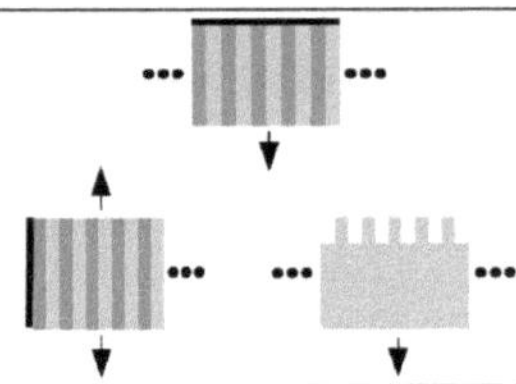	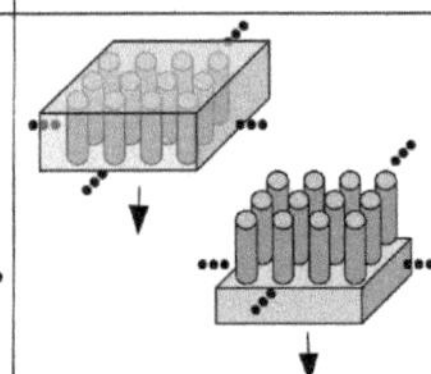	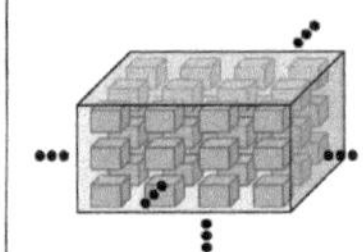
Plate (or slab)	2	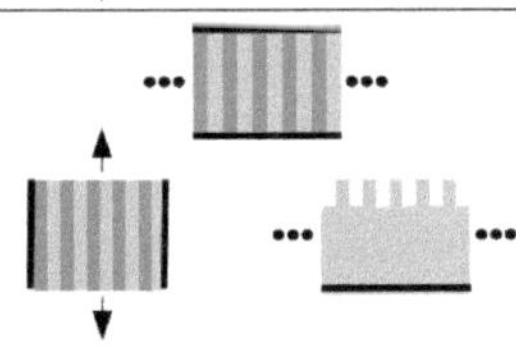	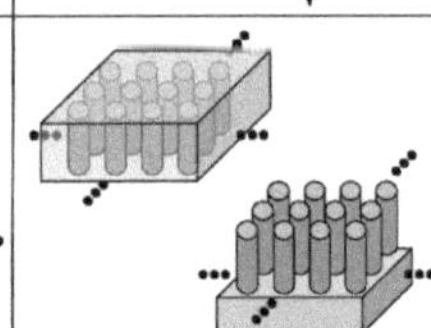	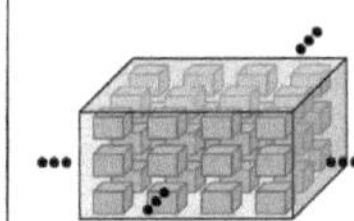

Tab. 11.2: Basic computational problems for perfectly periodic and defect-based phononic crystals. A typical supercell is indicated in each case.

Structure		Dispersion (infinite structure)	Scattering problem (finite structure)
Crystal	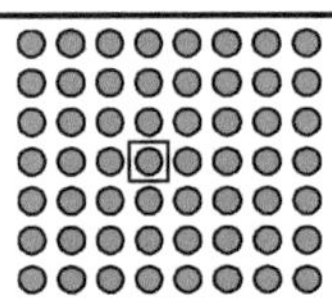	Band structure, $\omega(k)$ complex band structure, $k(\omega)$	Transmission, reflection and diffraction of an incident plane wave
Waveguide	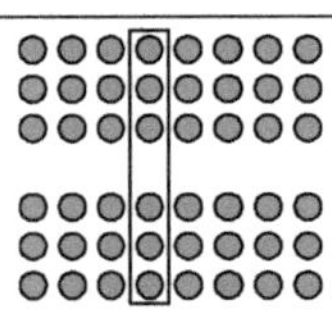	Guided modes, with supercell	Waveguide transmission
Cavity	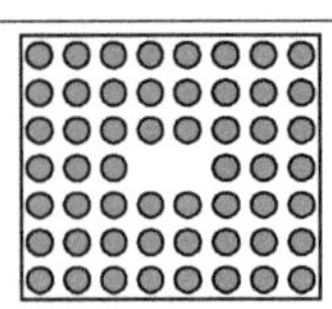	Confined modes, with supercell	Cavity transmission

by solving a generalized eigenvalue problem, possibly supplemented with surface or interface boundary conditions. Note that Bloch waves constitute a complete basis (in the monochromatic sense) only if all complex $k(\omega)$ solutions are considered [78].

If a perfectly periodic phononic crystal of finite size is considered, for instance for comparison with experiments, then the scattering of an incident plane wave to all orders of diffraction is an important basic problem. This situation will be considered specifically in Chapter 12.

Phononic crystal cavities and waveguides can be formed by locally breaking the periodicity of the phononic crystal structure. Defect-based phononic structures are generally mostly interesting for frequencies that fall inside a complete phononic band gap, in which case perfect guidance or confinement can be achieved. Otherwise, phononic crystal waveguides and cavities are leaky. By defining a supercell that artificially restores periodicity, waveguide and cavity modes can be estimated, and their band structures can be obtained. In practice, the supercell should extend at least a few periods away from the defect it encloses. Modal computations then give physically meaningful results when only evanescent Bloch waves of the elementary phononic crystal exist, i.e. inside a complete band gap. Furthermore, the number of phononic crystal rows must be sufficient so that the Bloch wave with the least imaginary part of the wavevector can be considered negligible on the boundary of the supercell. Finite-size defect-based phononic crystal structures are in general considered for comparison with experiments, with the accent on obtaining the transmission of the waveguide or the resonance properties of the cavity.

11.2 Mirrors

Let us consider the application of sonic and phononic crystals as mirrors. This is probably the first idea that comes to mind: since there are no propagating waves inside a complete band gap, then a semi-infinite crystal must be a perfect mirror. This idea is perfectly correct as long as the crystal is infinitely thick or at least has a large number of successive periodic rows. A finite-thickness crystal, however, will present a non-vanishing transmission. But how much transmission exactly? With locally-resonant band gaps, the transmission coefficient might not drop as much as one would like when the number of crystal layers is increased; we observed for instance with surface elastic waves that the LR band gap was apparently not as efficient as the Bragg band gap (Figure 10.16). In the case of Bragg band gaps, the theoretical shape of the transmission as a function of frequency and of the number of layers can be obtained quite confidently, as follows.

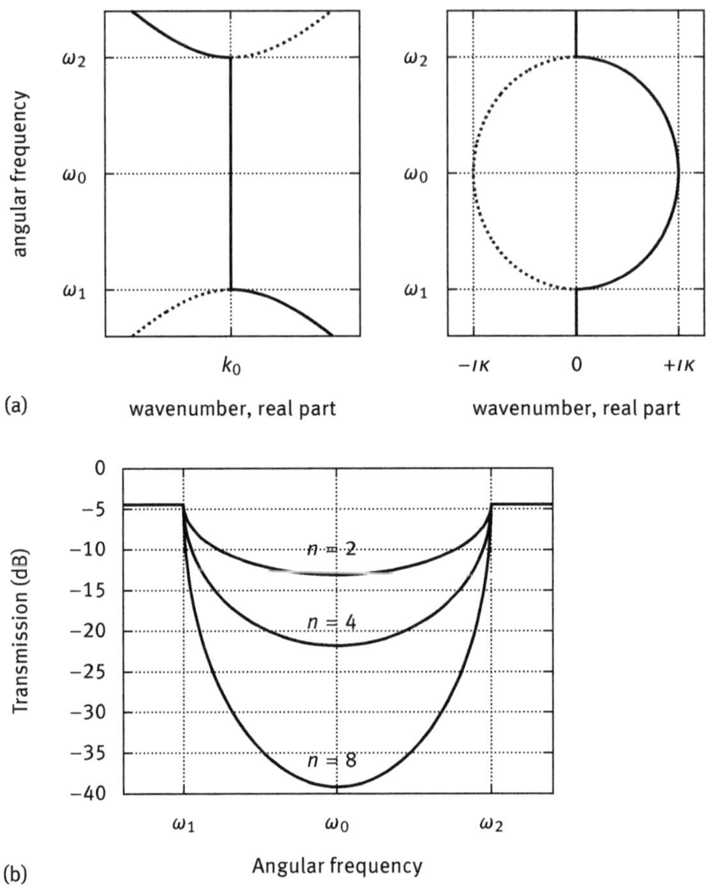

Fig. 11.1: Simple model for the transmission coefficient within a Bragg band gap. (a) The complex band structure as described by (11.2) describes two counter-propagating Bloch waves forming a frequency band gap $[\omega_1, \omega_2]$. The Bloch wavenumber is $\pm\imath\kappa$ at the center of the band gap. (b) The transmission (11.3) is depicted as a function of the frequency and of the number of crystal layers, for $\kappa a = 0.4$ and $T_0 = 0.6$.

Bragg mirror efficiency. Based on the various explicit complex band structure expressions we outlined in Chapter 2, we can write an approximate model for the complex band structure around a Bragg band gap as

$$\frac{4}{B^2}(\omega - \omega_1)(\omega - \omega_2) = \frac{1}{\kappa^2}(k - k_0)^2. \tag{11.1}$$

In this implicit expression, ω_1 and ω_2 are the entrance and exit band gap frequencies, $B = \omega_2 - \omega_1$ is the gap width, and k_0 is the high-symmetry point wavenumber (for instance $k_0 = \frac{\pi}{a}$ at the X point of the first Brillouin zone for the square lattice). κ is the extremal value of the imaginary part of the wavenumber, obtained in the center of the band gap, for frequency $\omega_0 = (\omega_1 + \omega_2)/2$. Inside the band gap, for $\omega_1 < \omega < \omega_2$, the complex band structure represented in Figure 11.1 (a) then has two branches with equation

$$k(\omega) = k_0 \pm \jmath \frac{2\kappa}{B} \sqrt{|(\omega - \omega_1)(\omega - \omega_2)|}. \tag{11.2}$$

This relation is of course only an approximation and models only the least evanescent Bloch wave. But after a few crystal layers, transmission will be dominated by this least evanescent Bloch wave. We consider n layers of a crystal with lattice constant a. Assuming that a wave is incident from the outside of the crystal, it will convert to the least evanescent Bloch wave with efficiency T_0 at the entrance interface. The same transmission coefficient will also apply at the exit of the crystal if the exit medium is the same as the incident medium, by reciprocity. In a first approximation, we can then write the intensity transmission as

$$\begin{aligned} T(\omega) &= T_0^2 \exp\left(-2n|\operatorname{Im}(k(\omega))|a\right) \\ &= T_0^2 \exp\left(-n\frac{4\kappa a}{B} \sqrt{|(\omega - \omega_1)(\omega - \omega_2)|}\right). \end{aligned} \tag{11.3}$$

Expressed in decibel units, the transmission is

$$10\log_{10}(T(\omega)) = 20\log_{10}(T_0) - 10\log_{10}(e)\left(n\frac{4\kappa a}{B} \sqrt{|(\omega - \omega_1)(\omega - \omega_2)|}\right). \tag{11.4}$$

The logarithm of the transmission decreases linearly with the number of crystal layers. The minimum transmission is obtained at the center of the band gap, where

$$\begin{aligned} 10\log_{10}(T(\omega_0)) &= 20\log_{10}(T_0) - 20\log_{10}(e)n\kappa a \\ &\approx 20\log_{10}(T_0) - 8.69 n\kappa a. \end{aligned} \tag{11.5}$$

An example of this function is plotted in Figure 11.1 (b). The value of the imaginary part of the wavenumber times the lattice constant is thus a direct measure of the attenuation per layer that can be obtained with a sonic or phononic crystal. For a nominal value $\kappa a = 1$ (which is hardly achieved in practice even for a large Bragg band gap, see the example complex band structures in Chapter 9), one could expect approximately 8.7 dB attenuation per crystal layer.

Fresnel coefficients for a sonic band gap mirror – Let us specify the evaluation of reflection and transmission coefficients for a sonic crystal mirror. The method we will follow is related to the Fresnel coefficients derivation in Chapter 3. We consider a finite-thickness sonic crystal. The thickness is $L = na$ with n the number of layers and a the lattice constant in the direction of interest. We assume that a monochromatic plane wave is incident normally from the surrounding medium, as depicted in Figure 11.2. The waves in the incident region are the superposition of the incident wave and of a reflected wave as

$$p(t, x) = \left(a_i e^{-\imath\beta x} + a_r e^{+\imath\beta x}\right) e^{\imath\omega t}, \tag{11.6}$$

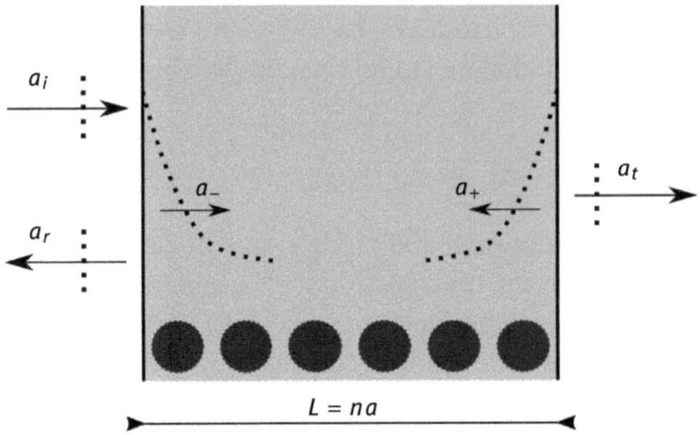

Fig. 11.2: Schematic of reflection on and transmission through a sonic band gap mirror for frequencies within a band gap. The thickness is $L = na$ with n the number of layers and a the lattice constant in the relevant direction. The amplitudes a_i and a_r of the incident and of the reflected plane waves are defined with respect to the entrance interface. The amplitude a_t of the transmitted plane waves is defined with respect to the exit interface. The amplitudes of evanescent Bloch waves are defined with respect to the entrance (exit, respectively) interface for propagation to the right (left, resp.).

where a_i and a_r are the incident and reflected amplitudes, respectively, and $\beta = \omega/c$ with c the celerity in the incident medium. Within the sonic crystal and for frequencies with a band gap, we assume the wavefield is the superposition of only the evanescent Bloch waves with the least imaginary part, or

$$p(t, x) = \left(a_- e^{-\alpha x} + a_+ e^{+\alpha(x-L)}\right) e^{\imath\omega t}, \tag{11.7}$$

where a_- and a_+ are the amplitudes of right-decreasing and left-decreasing evanescent Bloch waves, respectively, and $\alpha = |\operatorname{Im}(k)|$. Note that the left-decreasing evanescent Bloch wave is defined with respect to the exit interface. In the exit medium, the wave field is simply

$$p(t, x) = a_t e^{-\imath\beta x} e^{\imath\omega t}, \tag{11.8}$$

with a_t the amplitude of the transmitted wave. Both pressure and the normal velocity are continuous at both entrance and exit interfaces. At the entrance interface, we thus have

$$\begin{cases} a_i + a_r = a_- + a_+ e^{-\alpha L}, \\ -\dfrac{\imath\beta}{Z_1} a_i + \dfrac{\imath\beta}{Z_1} a_r = -\dfrac{\alpha}{Z_2} a_- + \dfrac{\alpha}{Z_2} a_+ e^{-\alpha L}, \end{cases} \tag{11.9}$$

where Z_1 is the acoustic impedance of the (homogeneous) incident medium and Z_2 is the effective acoustic impedance for the evanescent Bloch wave, defined as the ratio of pressure to normal velocity at the unit cell boundary. We transform this system of equations to matrix form as

$$\begin{pmatrix} 1 & 1 \\ -1 & 1 \end{pmatrix} \begin{pmatrix} a_i \\ a_r \end{pmatrix} = \begin{pmatrix} 1 & 1 \\ \imath\xi & -\imath\xi \end{pmatrix} \begin{pmatrix} a_- \\ e^{-\alpha L} a_+ \end{pmatrix}, \tag{11.10}$$

with $\xi = \frac{\alpha}{\beta}\frac{Z_1}{Z_2}$. This last matrix equation is easily transformed to

$$\begin{pmatrix} a_i \\ a_r \end{pmatrix} = \frac{1}{2}\begin{pmatrix} \gamma & \gamma^* \\ \gamma^* & \gamma \end{pmatrix}\begin{pmatrix} a_- \\ e^{-\alpha L} a_+ \end{pmatrix}, \tag{11.11}$$

with $\gamma = 1 - \imath\xi$.

Similarly, we have at the exit interface

$$\begin{cases} e^{-\alpha L} a_- + a_+ = a_t, \\ -\frac{\alpha}{Z_2} e^{-\alpha L} a_- + \frac{\alpha}{Z_2} a_+ = -\frac{\imath\beta}{Z_1} a_t, \end{cases} \tag{11.12}$$

which is transformed to

$$\begin{pmatrix} e^{-\alpha L} a_- \\ a_+ \end{pmatrix} = \frac{1}{2}\begin{pmatrix} \delta \\ \delta^* \end{pmatrix} a_t, \tag{11.13}$$

with $\delta = 1 + \imath\xi^{-1}$.

Next we eliminate the evanescent Bloch waves amplitudes to get

$$\begin{pmatrix} a_i \\ a_r \end{pmatrix} = \frac{1}{4}\begin{pmatrix} \gamma\delta e^{\alpha L} + \gamma^* \delta^* e^{-\alpha L} \\ \gamma^* \delta e^{\alpha L} + \gamma\delta^* e^{-\alpha L} \end{pmatrix} a_t. \tag{11.14}$$

From these relation we obtain the transmission and the reflection coefficients for the finite-thickness sonic crystal as

$$t = 2(2\cosh(\alpha L) - \imath(\xi - \xi^{-1})\sinh(\alpha L))^{-1}, \tag{11.15}$$

$$r = \frac{\imath(\xi + \xi^{-1})\sinh(\alpha L)}{2\cosh(\alpha L) - \imath(\xi - \xi^{-1})\sinh(\alpha L)}. \tag{11.16}$$

It is not difficult to see by direct inspection that the energy conservation relation $|r|^2 + |t|^2 = 1$ holds, as one should expect.

As the sonic crystal is made thicker, it is clear that the modulus reflection coefficient should tend to unity, while the transmission should diminish exponentially. The asymptotic limits of the Fresnel coefficients are more precisely

$$r \approx -\frac{\xi + \xi^{-1}}{(\xi^{1/2} + \imath\xi^{-1/2})^2}, \qquad |r| \approx 1, \tag{11.17}$$

$$t \approx \frac{2\imath}{(\xi^{1/2} + \imath\xi^{-1/2})^2} e^{-\alpha L}, \quad |t| \approx \frac{2}{\xi + \xi^{-1}} e^{-\alpha L}. \tag{11.18}$$

In particular, the asymptotic value of the transmission coefficient is an exponential function of the imaginary part of the wavenumber of the least evanescent Bloch wave, and is governed by the value of $\alpha(\omega)a$. For consistency, we see that for the sonic crystal we should consider $T_0 = \frac{2}{\xi+\xi^{-1}}$ in (11.3).

A similar derivation can be attempted for phononic crystals (elastic waves instead of acoustic waves), but the issue of the polarization of elastic waves makes it a bit more difficult; a framework for general Fresnel coefficients is given in Chapter 12. Anyway, it can quite confidently be concluded that the exponential law on transmission applies as well. The reflection coefficient in intensity can then, if needed, be obtained in the absence of loss and of diffraction by $|r|^2 = 1 - |t|^2$.

Sound shields. From the previous models of the transmission and reflection coefficients it is clear that Bragg mirrors composed of sonic or phononic crystals can be quite efficient. If the value of $\alpha(\omega)a$ is large enough, then a small transmission can be achieved with only a limited number of crystal layers. For application as a sound shield, however, the practical usability depends on the space available for implementing the crystal. In high-frequency applications, such as SAW and BAW filters, the elastic wavelength is short (of the order of the micron) and the solution is very compact. Inside water and at MHz frequencies, the acoustic wavelength is of the order of a few millimeters, so a total thickness of a few centimeters for the sonic crystal is realistic. In air and for audible frequencies, the acoustic wavelength is a few tens of centimeters. As a result, sonic crystals based on the Bragg interference effect may be too bulky for real applications. The practical alternative is the use of local resonances, as we discussed in Chapter 10. Because the direct relation between lattice constant and wavelength no longer applies, rather thin sound shields can be obtained by proper design [33, 36]. The difficulty with using resonances, however, is obtaining wide frequency band gaps, or sound attenuation over a wide bandwidth.

Mirrors for surface elastic and Lamb wave devices. Resonators can be built from very good mirrors that are able to trap the energy of waves in a confined space. This principle is central to laser resonators, but also to high-frequency microacoustic devices. Surface acoustic wave (SAW) resonators for instance make use of Bragg reflectors consisting of a periodic array of metal strips on a piezoelectric surface. The combination of the mass-loading brought by the added metal and of electrical regeneration in the electrodes brings in a periodical modulation of the effective impedance seen by surface elastic waves and induces a 1D Bragg band gap. The contrast of this periodic modulation remains rather limited, however, so that the band gap is not very wide (a few percent, typically), on the one hand, and the required number of electrodes to obtain a large reflection coefficient is rather large (hundreds of metal strips are typically required), on the other hand. Still, this technical solution is very robust and is widely implemented in commercial SAW devices.

As an alternative, Wu et al. proposed a design that combines a two-port SAW device – composed of two interdigital transducers separated by some free surface acting as a delay line – and phononic crystal mirrors replacing the reflective gratings, see Figure 11.3 [169]. The distance separating the IDT's and the phononic crystal mirrors is found to be of paramount important, since in order to build the resonance efficiently the waves emitted on one side and reflected from the other side must be recombined so that they interfere constructively. A layered ZnO/Si SAW device and square-lattice phononic crystal mirrors composed of cylindrical holes on silicon were fabricated. With 15 layers ($L = 15a$) of holes, the experimental insertion loss was improved by 7 dB at 212 MHz, the central frequency of operation. This corresponds to a very strong reduction in the size of the reflective mirrors as compared to classical IDT gratings.

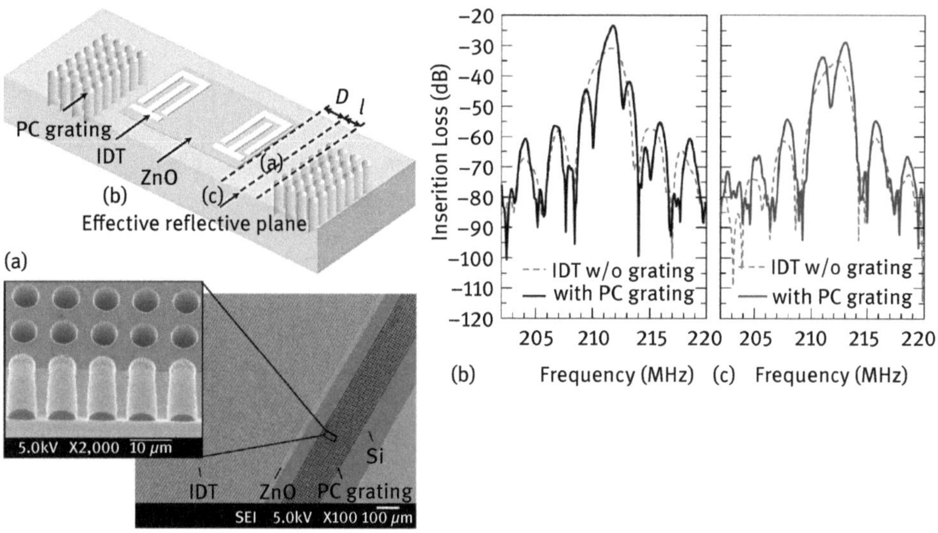

Fig. 11.3: Two-port surface acoustic wave resonator using phononic crystal mirrors. The square-lattice phononic crystal of holes is deeply etched in silicon, with hole diameter $d = 6$ μm and lattice constant $a = 10$ μm. A piezoelectric thin layer of zinc oxide is deposited under the interdigital transducers. Distance $l = 1.32\,\lambda$ μm is estimated as the effective delay distance defining the reference for reflection on the phononic crystal at a frequency of 210 MHz ($\lambda = 23.4$ μm). Distance D is given the values (c) λ and (d) 1.25λ (after Wu et al. [169], copyright 2009 American Institute of Physics).

The same line of thought applies to Lamb wave resonators, as demonstrated by Huang et al. [55]. These authors again considered a two-port ZnO/silicon microacoustic device, but this time using a 15 μm thick silicon membrane instead of a semi-infinite substrate. The Lamb wave resonator was built using 2D phononic crystal mirrors, as shown in Figure 11.4. Again, constructive interference between transduced and reflected waves was obtained by properly selecting the distance between wave sources and the phononic crystal mirrors. The measurements showed that with 15 rows of holes in the phononic crystal mirrors, the quality factor could reach $Q = 2269$ at a resonant frequency within the phononic band gap.

11.3 Defect cavities

It is a common place idea in physics that by introducing a defect inside an otherwise perfectly periodic crystal certain particles or waves can become localized or confined. In the case of sonic and phononic crystals, a complete band gap is generally necessary in order to induce a local confinement of the energy of waves. In the case of an extended defect extending orthogonal to the direction of propagation and forming a line in 2D or a plane in 3D, only a directional band gap is required. Such a structure is

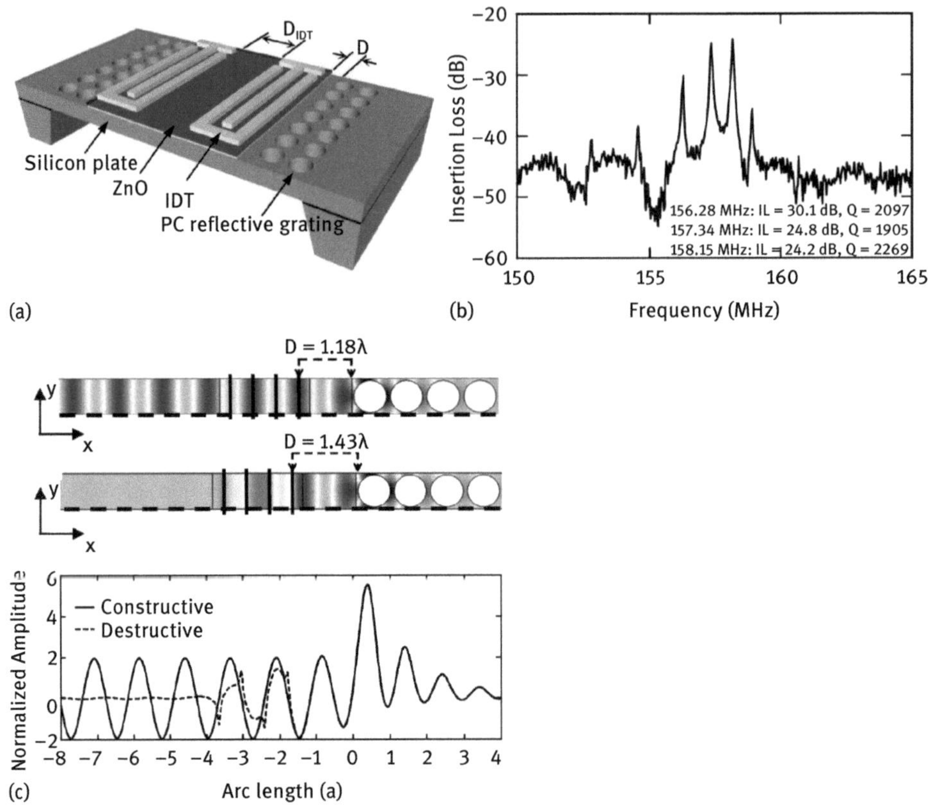

Fig. 11.4: Two-port Lamb wave resonator using phononic crystal mirrors. (a) The square-lattice phononic crystal of holes is etched in silicon, with hole diameter $d = 17.8\,\mu m$ and lattice constant $a = 20\,\mu m$. The thickness of the silicon membrane is $h = 12\,\mu m$. (b) The electrical transmission shows a series of sharp resonances. The frequency distance between resonances is mostly dictated by the distance between IDTs, D_{IDT}. (c) Distance D must be chosen so as to provide constructive interference between directly transduced and reflected Lamb waves. The examples provided show that destructive interference is obtained for $D = 1.43\lambda$, with λ the wavelength of the relevant Lamb wave in silicon, while constructive interference is obtained for $D = 1.18\lambda$. Note the difference $\lambda/4$ between these two values (after Huang et al. [55], copyright 2010 American Institute of Physics).

usually referred to as a Fabry–Perot interferometer. We will start with some examples of this "hyperplane" situation, before moving on to defects truly requiring complete band gaps.

Fabry–Perot interferometers. In optics, a Fabry–Perot interferometer is a thin-layer device comprising a cavity defined by two mirrors with very high reflectivity. Even though the reflection coefficients of the mirrors can approach unity, there is a series of resonant frequencies at which transmission is almost unity. The resonant frequencies are mostly defined by the width of the cavity, L_0. Away from resonance frequencies, the

Fabry–Perot interferometer reflects almost all incident light. In analogy, a Fabry–Perot resonator can be constructed using sonic or phononic crystal mirrors, as depicted in Figure 11.5. The central cavity width must be tuned so that a resonance occurs inside a band gap. Note that only a directional band gap is needed in principle, as long as the resonator is interrogated with plane waves.

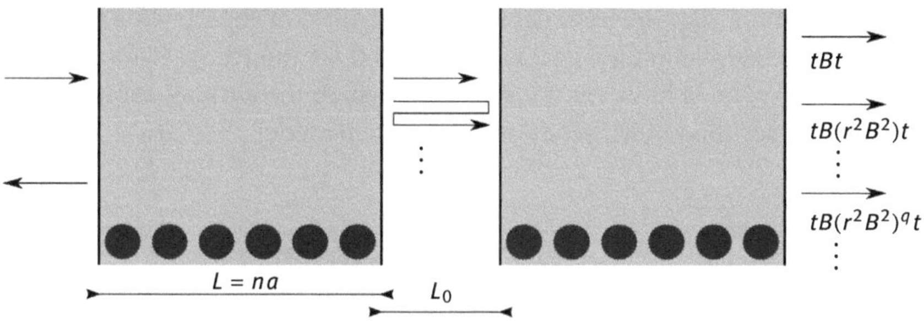

Fig. 11.5: Schematic of a Fabry–Perot resonator defined using phononic crystal mirrors. Transmission of plane acoustic or elastic waves is decomposed in an infinite series of partial waves defined by successive transmissions (t) and reflections (r) at the crystal interfaces. $B = e^{-\imath\beta L_0}$ represents the phase acquired by traversing the central cavity of length L_0 with wavenumber β.

With the notations of Figure 11.5, the transmission coefficient through the resonator can be computed in the following way, based on the previous analysis of the Bragg band gap mirror. First, we notice that there is a partial wave that can travel through the whole structure directly, with amplitude t^2B, where $B = e^{-\imath\beta L_0}$, meaning that it must go twice through the crystal mirror and once through the cavity. t is the transmission through the mirror. The second partial wave to exit will have been reflected twice on the mirrors and will have traveled double the cavity width, thus having amplitude $t^2Br^2B^2$. An infinite number of partial waves will then exit from the resonator, so the total transmission coefficient is given by

$$\begin{aligned} T &= t^2B(1 + r^2B^2 + \cdots + (r^2B^2)^q + \cdots) \\ &= B\frac{t^2}{1 - r^2B^2}. \end{aligned} \tag{11.19}$$

In order to evaluate the infinite series, we have assumed $|r| < 1$. The denominator will be minimum every time $r^2B^2 = |r|^2$, corresponding to the resonance condition

$$\beta L_0 - \operatorname{Arg}(r) = n\pi, \tag{11.20}$$

with n an integer. At resonance, the transmission is $T = Bt^2/|t|^2$, which is a complex number of unit modulus since $|B| = 1$.

Fabry–Perot resonators based on sonic crystals were built by Van Der Biest et al. [154]. These authors considered either 2D or 3D sonic crystals. The 2D square-lattice crystals were composed of steel rods immersed in water, with rod diameter

$d = 0.8$ mm and lattice constant $a = 1.5$ mm. These parameters do not provide a complete band gap but only a directional one, which is enough in the case considered. The 3D close-packed FCC crystals were composed of tungsten carbide spheres in water, with diameter $d = 0.8$ mm. A complete band gap is expected around 1 MHz for this arrangement, as we have shown in Chapter 4. The [111] direction of the crystal is normal to the cavity plane. Two identical sonic crystals were placed around a 7.05 mm thick aluminum slab in either case. The transmission experiments depicted in Figure 11.6 show clear resonance peaks, as expected for the Fabry–Perot interferometer. Unit transmission is however not observed, which the authors explained by incorporating the absorption of water in their theoretical model.

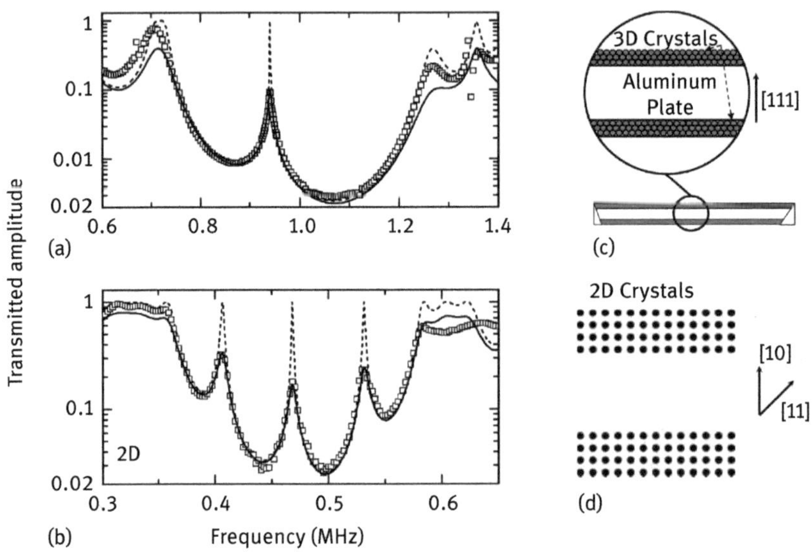

Fig. 11.6: Fabry–Perot sonic crystal slab resonator. Amplitude transmission coefficient through (a) 3D and (b) 2D double phononic crystals. Theoretical predictions with and without absorption (solid and dashed lines, respectively) are compared with the experimental data (symbols). The corresponding crystal structures and crystallographic orientations are shown in (c) and (d) (after Van Der Biest et al. [154]).

As an example of a practical Fabry–Perot resonator constructed using phononic band gap mirrors, let us consider the structure shown in Figure 11.7 [104]. The design basis is a honeycomb lattice of cylindrical holes in a silicon membrane. The corresponding phononic crystal slab has parameters $d = 12.8$ μm (hole diameter), $a = 26$ μm (lattice constant) and $h = 15$ μm (slab thickness). The phononic crystal resonator is defined by removing four rows of holes from the perfect crystal, as the schematic drawing and the SEM image in Figure 11.7 illustrate. The width of the cavity cannot be unambiguously defined, because the defect cut does not follow the unit cell, but the central

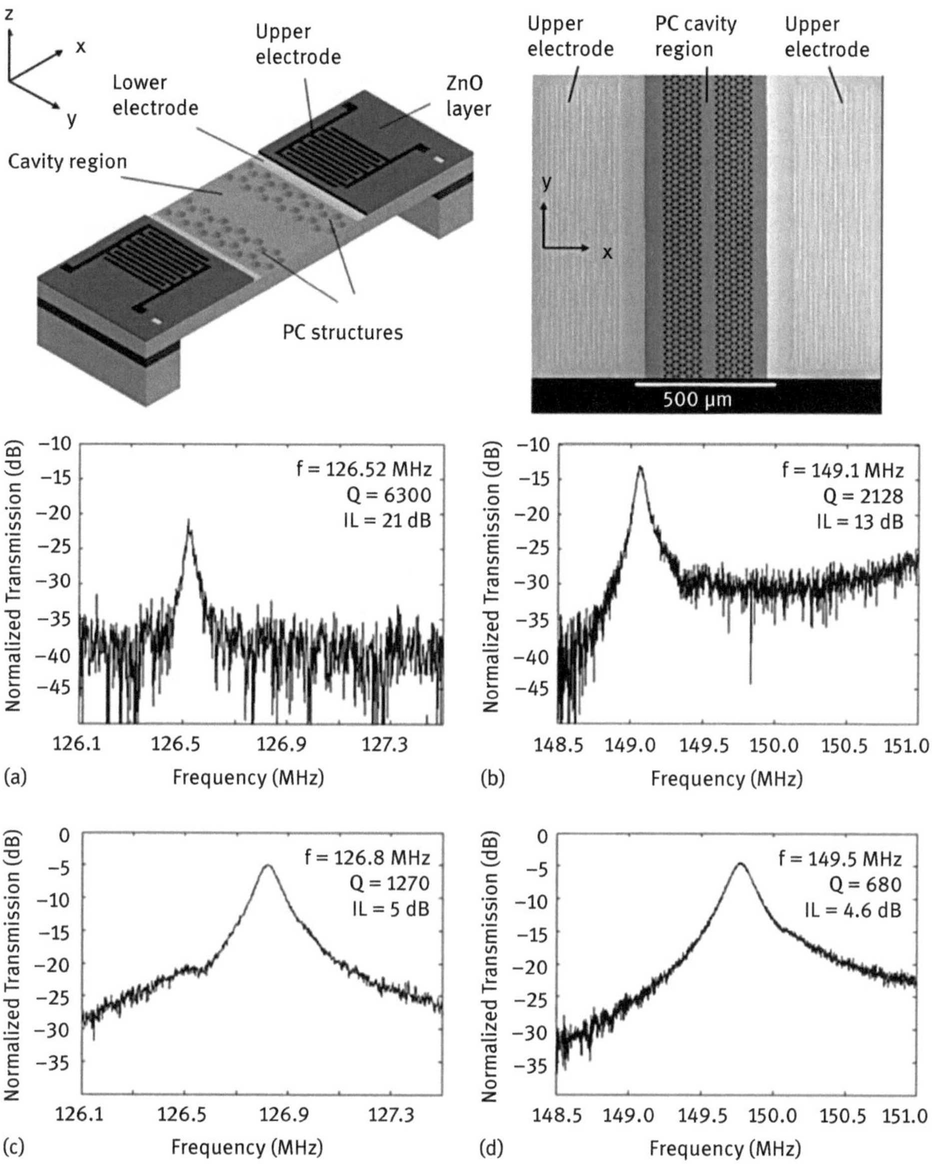

Fig. 11.7: Fabry–Perot phononic crystal slab resonator. The PC slab resonator structure with exciting and receiving transducers on its two sides is depicted schematically. The cavity region is surrounded by four periods of the crystal on each side. The top SEM image of a fabricated PC slab resonator shows the transducer electrodes on each side. The cavity region is surrounded by three periods of the crystal on each side. The normalized transmission through the PC cavity slab structure is plotted for (a) the first and (b) the second mode for the structure with three periods of the crystal on each side of the cavity, and for (c) the first and (d) the second mode for the structure with only two periods of the crystal on each side of the cavity. The peak frequency f, quality factor Q, and insertion loss IL are given in each figure (after Mohammadi et al. [104], copyright 2009 American Institute of Physics).

cavity filled with plain silicon is 45 μm wide, or $L_0 = \sqrt{3}a$. The phononic crystal slab has an experimentally verified complete band gap extending from 115 to 152 MHz. The cavity has two different resonances in this frequency range, at 127 and 150 MHz, approximately. Two different numbers of crystal layers were considered for the phononic band gap mirrors: either 12 rows of holes ($L = 3\sqrt{3}a$) or 8 rows of holes ($L = 2\sqrt{3}a$). It can be seen in Figure 11.7 that the number of crystal layers indeed has a strong effect on the resonance peaks that are obtained experimentally. The quality factor of the resonance is enhanced by the number of crystal layers while the insertion loss is correspondingly degraded. Note that transmission around the second resonance frequency is not symmetrical due to its proximity with the upper band edge: Bloch waves around 150 MHz are only weakly evanescent and become propagating above 152 MHz approximately. The theoretical result that the transmission should be close to unity at resonance is again not confirmed by the measurements in this case. The reason might be connected with different sources of loss within the device. Apart from the obvious possible losses caused by intrinsic attenuation of elastic waves in silicon, which can be enhanced for evanescent Bloch waves in the phononic band gap mirrors, there are possible losses arising from imperfections of the crystal causing in-plane diffusion of waves. In addition, in the special case of elastic waves in plates, there are generally more than one available mode for propagation, causing modal conversion losses at every interface between the homogeneous plate and the phononic crystal slab.

A similar Fabry–Perot phononic crystal resonator was studied by Wang et al. but based on a square-lattice crystal [160]. The silicon phononic crystal slab parameters were $d = 16.36$ μm (hole diameter), $a = 18.18$ μm (lattice constant) and $h = 10$ μm (slab thickness). A central cavity $L_0 = 3a$ wide was surrounded with 4-layer phononic crystal mirrors. The complete band gap was found experimentally to extend from 143 to 186 MHz. A single resonance at 152 MHz with quality factor $Q = 1016$ and insertion loss $IL = 13$ dB was obtained. The same group of authors proposed improvements to the basic Fabry–Perot phononic crystal structure, including adding smaller holes in the central cavity to obtain $Q = 1624$ and insertion loss $IL = 11$ dB [159].

Zero-dimensional defect cavities. At frequencies within a complete band gap, a sonic or phononic crystal has in principle the ability to trap waves inside a defect. Indeed, suppose an excitation is applied inside the defect, for instance by focusing a laser, then the waves outgoing from the source point can only couple to evanescent Bloch waves as they reach the interface between the defect and the perfect crystal. The resulting wavefield can be thought of as a central oscillating field inside the defect dressed with evanescent waves outside of it.

Figure 11.8 shows experimental results obtained with sonic crystals of cylindrical steel rods immersed in water [65]. As we have discussed in Chapter 4, this crystal has a complete sonic band gap extending from 260 to 312 kHz for the geometrical parameters of Figure 4.12. Three different defect configurations are considered in the fig-

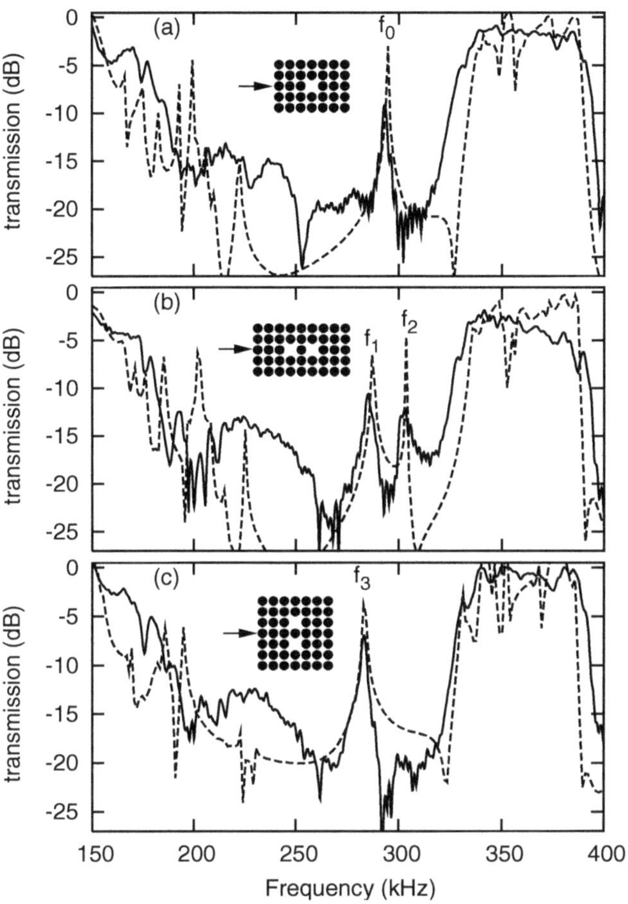

Fig. 11.8: Defect cavities in a square-lattice sonic crystal of steel cylinders in water, with full band gap extending from 260 to 312 kHz. The diameter of the rods is $d = 2.5$ mm and the lattice constant is $a = 3$ mm. Experimental (solid lines) and calculated (dashed lines) transmission power spectra versus frequency for interacting defects. (a) With one cylinder removed from the otherwise perfect crystal, a single defect induces a resonance mode in the stop band with frequency f_0. (b) Two defects in line with the incident wave induce two split resonance modes at frequencies f_1 and f_2 with an intercavity distance equal to one period. (c) Same as (b), but the two defects are aligned perpendicularly to the incident wave (after Khelif et al. [65], copyright 2003 American Physical Society).

ure. First, when a single rod is "removed' from the crystal, one resonant frequency f_0 appears in the transmission spectrum inside the complete band gap. The theoretical transmission, here obtained using a FDTD computation, leads to the expectation that the resonant transmission is not so far from unity. This result of course depends on the vertical size of the supercell used for the computation, but outlines the fact that one should not be surprised to meet a resonant transmission exceeding the mere geometrical filling fraction defined by the ratio of the surface of the defect to the surface of the supercell. This is somehow reminiscent of extraordinary acoustic transmission [25, 37, 91]. The measured transmission through the defective crystal is smaller than the theoretical value. This should not be attributed to acoustic losses, but rather to the fact that there is a single defect in the experiment while the computation assumes a periodically repeated defect. As a further remark, as the square geometry of the defect cavity allows, it is found that the resonance is degenerate, i.e. there are two orthogonal modes with the same eigenfrequency f_0 [65]. One of these modes is deaf, or antisymmetrical with respect to the direction of propagation. The other one is symmetrical and is the one that is excited in the experiment.

When two defects are placed sequentially in the direction of propagation, two resonant transmission peaks are seen to appear in Figure 11.8 (b). Since the two defects are identical, they would have the same resonance frequency if they were uncoupled. Since they are placed only two rows apart, however, they are not isolated by the surrounding crystal and they couple through the evanescent tails of their wavefields. As the double-defect cavity is symmetrical with respect to the vertical axis of the figure, it supports both an antisymmetrical mode with frequency f_1 and a symmetrical mode with frequency f_2. Again, both of these modes are degenerate because of the symmetry of the structure with respect to the propagation direction, and only the horizontally symmetrical modes leave an experimental trace in the transmission. Finally, when the double-defect cavity is placed perpendicularly to the propagation direction, only one transmission peak with frequency f_3 is again observed, corresponding to a mode that is symmetrical with respect to the propagation direction.

As a remark, the modes of zero-dimensional defects are localized spatially around the defect. Their spatial Fourier transform thus spans all of k-space. Now since they are not propagating waves, they cannot have a dispersion in the $(\mathbf{k}, \omega)$ representation. In a phononic band structure, obtained using a supercell computation, they must then appear as horizontal lines. In practice, any deviation from this behavior should be interpreted as the supercell not providing enough isolation between artificially periodized defects, so that the defects are effectively coupled. If this coupling is intentional, conversely, then a tunneling waveguide can be formed according to the principles specified in Section 11.4.4.

11.4 Defect waveguides

While we have considered line defects oriented orthogonally to wave propagation as Fabry–Perot resonators in the previous section, line defects also have the potential to guide waves for frequencies within a complete band gap. This idea has been extensively explored in the literature and many different waveguides have been formed in 2D and 3D crystals, in phononic crystal slabs, and for surface elastic waves on semi-infinite phononic crystals. In this section we illustrate some guidance concepts and some practical issues with defect waveguides in sonic and phononic crystals.

11.4.1 Waveguides in 2D sonic crystals

The most practical system to test the principle of the complete band gap waveguide is probably the two-dimensional sonic crystal of steels rods in air or water. As the crystal can be made much longer than the wavelength along the third dimension, the result is a one-dimensional waveguide that can be easily characterized in transmission with transducers. Because of its simplicity, this system has received quite a lot of attention for its own merits [60, 64–66, 86, 100] and for its combination with grafted resonators [10, 120]. As an example, Khelif et al. considered linear waveguides obtained by removing one or two rows of rods in a square-lattice sonic crystal, as shown in Figure 11.9 [64]. The crystal dimensions are exactly the same as the sonic crystal in Figure 4.12, with a complete sonic band gap from 260 to 320 kHz. When one row of rods is removed along the X direction to form a W1 waveguide – Wn means a linear defect waveguide with n rows removed – the band structure obtained with the supercell shown in the second line of Table 11.2 displays two propagating bands inside the complete band gap. Since there are only evanescent Bloch waves in the surrounding crystal in this frequency range, the Bloch waves are composed of a central pressure distribution in the defect dressed with evanescent tails in the crystal. Their wavenumber is real valued, implying that they are guided and propagating inside the defect region. The phase velocity of the guided Bloch waves is simply given by ω/k, as usual. More strikingly, part of the lower guided band indeed shows a negative group velocity, $\frac{\partial \omega}{\partial k} < 0$.[1] The experiment shows, however, that whenever a propagation band is present, transmission through the W1 waveguide occurs irrespective of the sign of the group velocity. The pressure distribution for the W1 waveguide in Figure 4.12 is obtained at a frequency of 287 kHz, and shows the transmission of a wave incident from the left. When two rows of rods are removed to form waveguide W2, the dispersion of guided Bloch waves is deeply modified as compared to the W1 case. Significantly, a band gap for guided waves appears now in the middle of the complete band gap. Indeed, as the defect crystal can be thought of as a particular one-dimensional sonic crystal for guided waves, it may well be that periodicity induces a band gap (for guided waves) within the band gap (for the perfect crystal). The pressure distribution for the W2 waveguide in Figure 4.12 is obtained at a frequency of 287 kHz again; total reflection is seen to occur at this frequency, with a clear evanescent decay of the Bloch wave amplitude.

---

1 Though we write that the group velocity is negative, it is to be understood that the signs of the phase and the group velocity are opposites. As always with periodicity, negativity or opposite signs result from band foldings with the implication that left- and right-propagating Bloch waves are always simultaneously present.

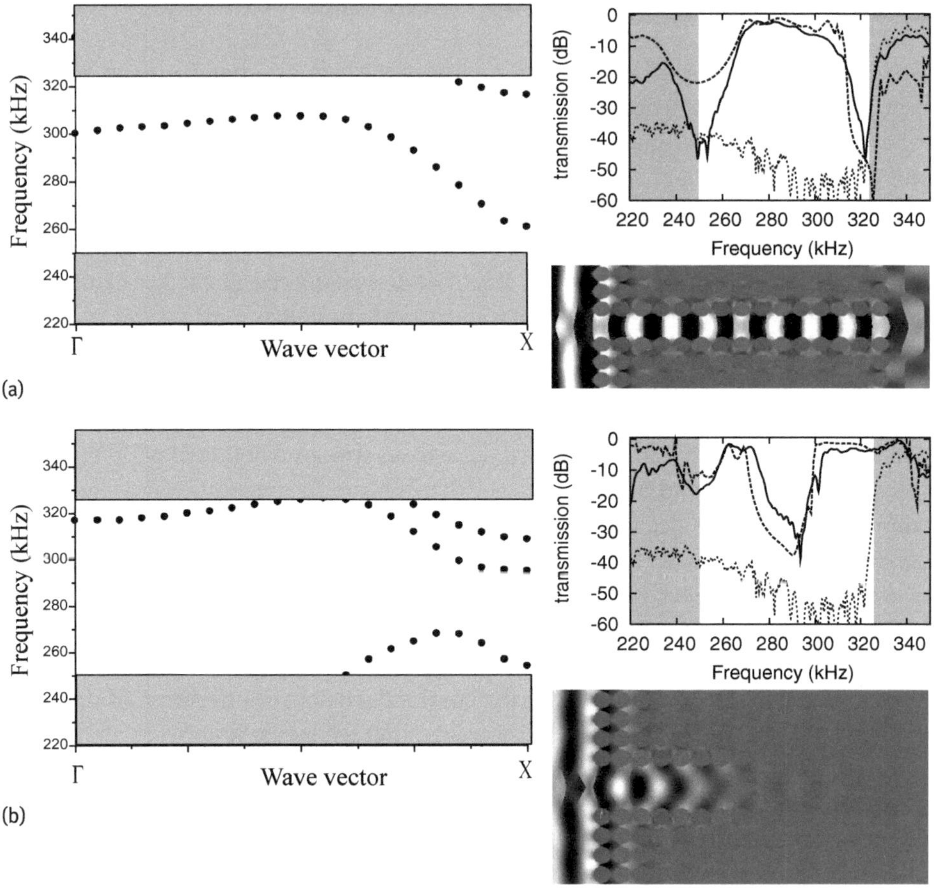

Fig. 11.9: Linear sonic crystal defect waveguides. A line defect is inserted in an otherwise perfectly periodic crystal. Two different waveguide widths are considered: (a) first exactly one row is missing, (b) second exactly two rows are missing. In each case the band structure is shown on the left, and a transmission measurement, its comparison with an FDTD computation, and an example pressure field distribution are shown on the right. The sonic crystal is a square-lattice array of steel rods in water, with $a = 3$ mm and $d = 2.5$ mm (after Khelif et al. [64], copyright 2004 American Institute of Physics).

The principle of the defect wave waveguide defined in a crystal is not limited to straight linear defects of the type Wn. Actually, it is possible to sharply bend the propagation path while still maintaining propagation. As an example, Figure 11.10 shows the spectral transmission of a linear defect that is sharply bent at two different locations. It is apparent by comparison with the W1 case that transmission is still obtained over a large part of the complete band gap, though strong filtering or transmission canceling occurs at certain frequencies. Actually, the sharp bends have resonances whose excitation reshapes the transmitted spectrum.

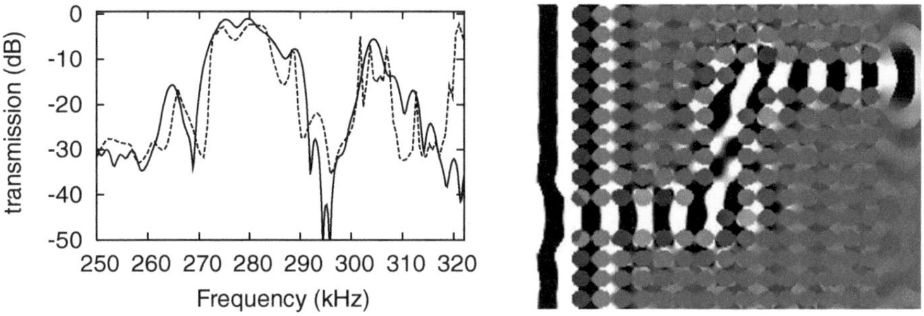

Fig. 11.10: Bent defect waveguide in a sonic crystal of steel rods in water. The sonic crystal is a square-lattice array of steel rods in water, with $a = 3$ mm and $d = 2.5$ mm. The experimental transmission is compared to an FDTD computation and the pressure distribution is shown at a frequency of 275 kHz (after Khelif et al. [64], copyright 2004 American Institute of Physics).

11.4.2 Waveguides in phononic crystal slabs

The phononic crystal slab structure lends itself quite naturally to the definition of phononic waveguides in three dimensions. The two-dimensional periodicity in the plane can trap waves efficiently around a defect, for frequencies within a complete band gap, and the top and bottom surfaces ensure their vertical confinement. As an example, let us consider the experiment conducted by Hsiao et al. [54] shown in Figures 11.11 and 11.12. The experiment is performed with two phononic crystal slab samples, both composed of a square-lattice monolayer of steel beads (diameter $d = 4$ mm) in epoxy. One of the samples is oriented in the ΓX direction while the other one is oriented in the ΓM direction. The beads are close packed ($a = d$) and the slab thickness is also the bead diameter ($h = d$) by construction. Lamb waves were launched in the slab by attaching a transducer to one side. A prism in contact with the surface is used to launch waves with a nonzero wavenumber. The out-of-plane displacements are recorded at any position on the other side with a heterodyne optical interferometer. Figure 11.11 shows the measured displacements as a function of frequency and of the position along the direction of propagation. A complete band gap is found between 250 and 340 kHz. Line scans at different frequencies show that waves are rather well transmitted below the band gap, exponentially decreasing within it, and not very strongly retransmitted above it. The attenuation in the high frequency range may be attributed to propagation losses in epoxy. Figure 11.12 shows the optical measurement of waves propagating inside a line defect waveguide created by omitting a row of beads along the ΓX direction. For a frequency below the band gap, plane wave fronts are observed, accompanying transmission through the crystal and the defect line as well. For a frequency within the band gap, guided waves are observed to be well confined inside the defect line.

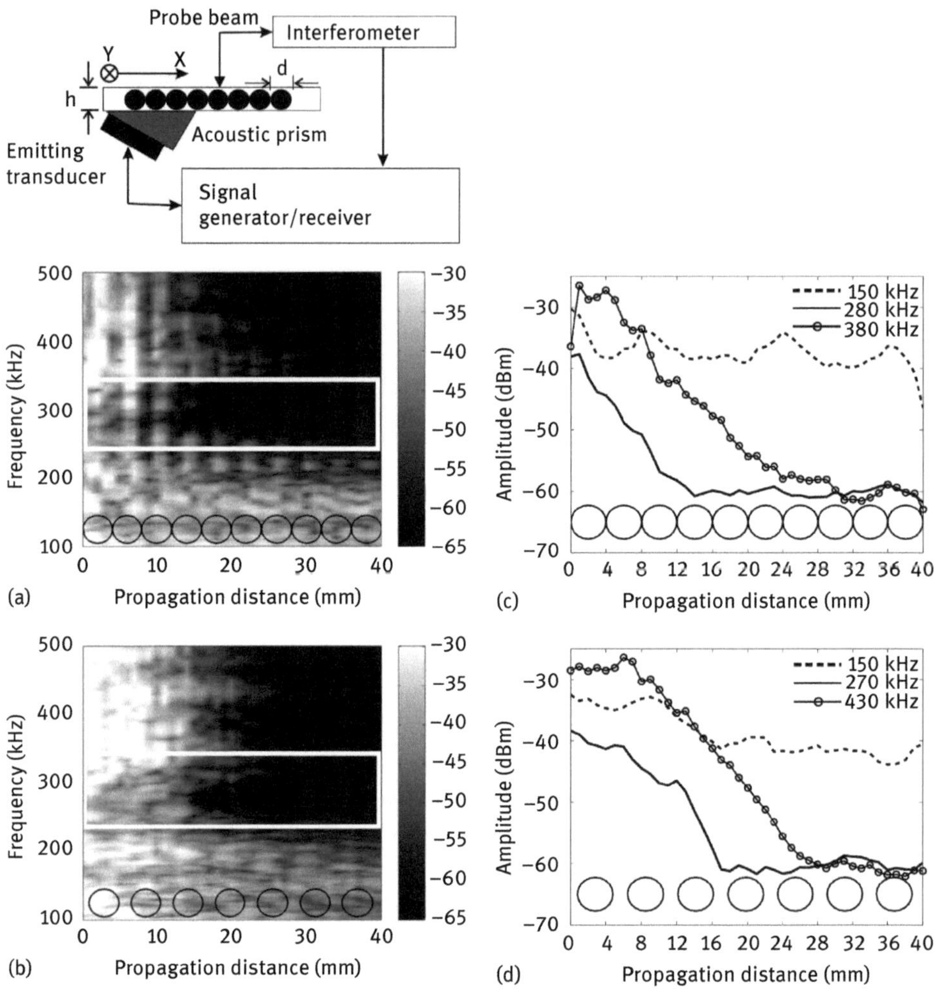

Fig. 11.11: Phononic crystal slab of steel beads in an epoxy matrix. Waves are excited in a phononic crystal slab using an ultrasonic transducer through a prism with an incidence angle of 30°. The phononic crystal slab is formed by a one layer thick array of spherical steel beads with a diameter of 4 mm arranged according to a square lattice in an epoxy matrix. The dependency of the displacement amplitude on frequency and propagation distance is measured for the ΓX and the ΓM oriented phononic crystal slabs. The white lines in the gray maps limit the complete band gap frequency range, while the black circles indicate the position of steel beads. The graphs (c) and (d) are extracted from the data in (a) and (b), respectively, and show line scans at three particular frequencies. The chosen frequencies lie before the complete band gap, inside it, and above it (after Hsiao et al. [54], copyright 2007 American Physical Society).

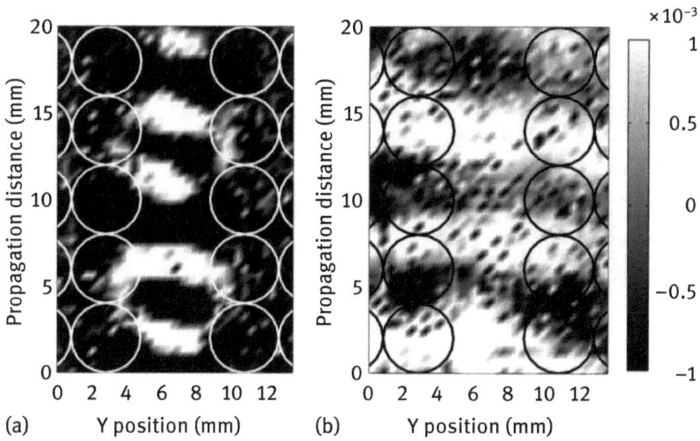

Fig. 11.12: Waveguide in a phononic crystal slab of steel beads in an epoxy matrix. The real part of the displacement amplitude of acoustic waves in the line defect waveguide is displayed at (a) a frequency of 275 kHz within the complete band gap and (b) a frequency of 134 kHz below the complete band gap. The white circles in (a) and the black circles in (b) indicate the positions of the steel beads (after Hsiao et al. [54], copyright 2007 American Physical Society).

The phononic crystal slab waveguide has been the subject of a rather extensive literature, for the most part based on numerical simulations. Besides the solid-solid composition that we just discussed, phononic crystals of holes in a solid slab [69, 156, 157] and phononic crystal of pillars on a membrane were considered [115, 170]. In all these structures, the basic guidance mechanism relied on defects in an otherwise perfect crystal and was only effective for frequencies within a complete band gap.

11.4.3 Waveguides in surface phononic crystals

Waveguides for surface elastic waves on two-dimensional phononic crystals have been investigated theoretically for both arrays of holes [147] and arrays of pillars [4]. Experimental realizations [113, 114] have remained scarce so far, however, because of the technological implications of the fabrication of the structures. The combination of vertical guidance by the surface – as in the widespread surface acoustic wave devices – and of lateral phononic crystal guidance does work very effectively. As an example, Figure 11.13 shows experiments with line defect waveguides in a square-lattice phononic crystal of holes in silicon [114]. Excitation and detection of surface waves is achieved in a pump-probe configuration with a femtosecond laser. All signals are hence recorded in the space and time domain, and a further temporal Fourier transform is used to obtain images of wave propagation at particular frequencies. At a frequency of 322 MHz, transmission through the straight waveguide is observed neatly. When a sharp bend is added, transmission is still ensured, though with some

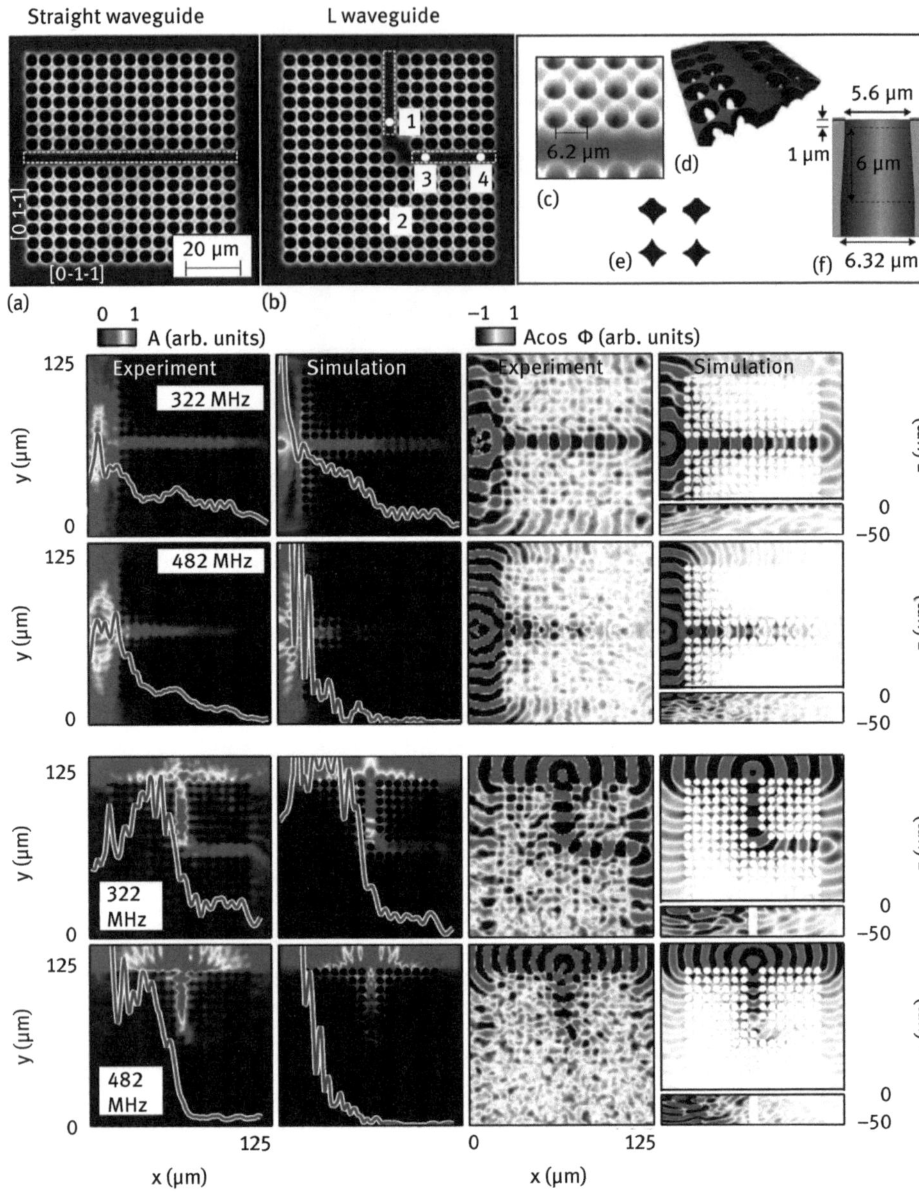
Straight waveguide
L waveguide
[0 1-1]
[0-1-1]
20 µm
1
2
3
4
6.2 µm
5.6 µm
1 µm
6 µm
6.32 µm
(a)
(b)
(c)
(d)
(e)
(f)
0 1
A (arb. units)
−1 1
Acos Φ (arb. units)
Experiment
Simulation
322 MHz
482 MHz
y (µm)
z (µm)
x (µm)
125
0
−50

reflection at the bend. At the time this text was written, much work was still required to obtain reliable surface phononic waveguides and to combine them with efficient transducers. The same observation applies to waveguides in microscale phononic crystal slabs.

11.4.4 Coupled-resonator acoustic waveguides (CRAW)

Guidance in a sonic or phononic crystal is not limited to linear lines of defects. Actually, any chain of defects acting as cavities can become a waveguide, providing the cavities are not too far away from one another and they have the same initial resonance frequency. This is the basic idea of the coupled-resonator acoustic waveguide. To illustrate the concept, consider Figure 11.14 [65]. A sequence of sonic crystal cavities is formed inside a square-lattice sonic crystal. The cavities are separated by only one crystal row, with the consequence that their coupling is rather strong. As a result, the local resonances hybridize to form a pass-band within which waves can be guided.

The pressure field plotted in Figure 11.14 at a frequency in the middle of the pass-band illustrates that the guided mode is apparently composed of an alternation of individual cavity modes with opposite phases. This idea leads to an analytic method to obtain the dispersion relation of the guided waves, based on a classical analog of the tight-binding model of quantum mechanics, as originally proposed by Yariv et al. for coupled-resonator optical waveguides (CROW) [174]. The basic idea is to look for the guided mode as a weighted superposition of resonant modes attached to a cavity site, with the weights depending on the wavenumber, according to

$$u(\mathbf{x}) = \sum_n \exp(-ikx_n) T_n u_0(\mathbf{x}), \qquad (11.21)$$

where $x_n = n\Lambda$ is the coordinate of the n-th resonator and T_n is the translation operator from the origin to location x_n. This expression means that the guided mode is approximated by a superposition of translated cavity modes, each having the same weight

◄ **Fig. 11.13:** Phononic crystal waveguide for surface elastic waves in holey silicon. (a) and (b) are scanning electron microscope (SEM) images of the straight and L-shaped phononic crystal waveguides. Marked regions are for Fourier analysis. (c) is an oblique SEM image. A model of the structure is shown (d), as well as (e) its horizontal cross section at a depth of 0.7 mm, and (f) its vertical cross section. Frequency-filtered images are presented in (g) and (h). (g) is for the straight waveguide at 322 MHz (row 1) and 482 MHz (row 2). Columns 1 and 2 show maps of the experimental and simulated $|\bar{u}(\mathbf{x}, \omega)|$, together with sections (vertical curves) through the center of the waveguide; columns 3 and 4 display $\mathrm{Re}(\bar{u}(\mathbf{x}, \omega))$. Column 4 includes cross sections in a vertical plane running through the waveguide centers. Both the simulation and experimental data are normalized, with identical scales at the two frequencies in each case. (h) gives the same information as (g) for the L waveguide. Cross sections are based on lines or planes though the center of the waveguide (after Otsuka et al. [114]).

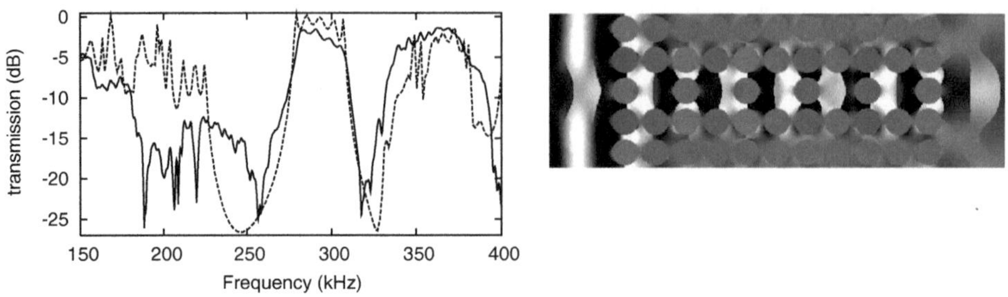

Fig. 11.14: Coupled-resonator acoustic waveguide inside a square-lattice sonic crystal. (a) Experimental (solid line) and calculated (dashed line) transmission power spectrum versus frequency for five consecutive defects. (b) Calculated longitudinal displacement amplitude averaged over one period of oscillation. The two-dimensional crystal is composed of steel rods with $a = 3$ mm and $d = 2.5$ mm. The incident wave is a longitudinal monochromatic plane wave with a 290 kHz frequency. The figure demonstrates the confinement and the guiding of the incident wave along the defect line (after Khelif et al. [65], copyright 2003 American Physical Society).

in modulus but a relative phase kx_n. Replacing this *Ansatz* in the wave equation of the periodic structure, a dispersion relation linking the frequency and the wavenumber of the guided wave can be obtained explicitly. The coefficients appearing in the dispersion relation can be found from overlap integrals involving the cavity mode distribution. Escalante et al. have shown that this direct approach yields a model dispersion relation that only crudely approximates the actual dispersion relation obtained from a supercell computation [34]. Alternatively, they proposed to consider a model where the coupled cavities are described as a linear chain of coupled harmonic oscillators, similar to the description of phonon propagation in a one-dimensional lattice of atoms. The starting expression is

$$-\frac{\mathrm{d}^2 U_n}{\mathrm{d}t^2} = \sum_{m=-\infty}^{\infty} \gamma_m U_{n+m}. \tag{11.22}$$

In this equation, U_n is a scalar variable similar to a displacement from equilibrium position describing the state of the resonator. Coefficient γ_0 equals the square of the resonant angular frequency Ω^2 in the decoupling limit, and the coefficients γ_m for $m \neq 0$ are coupling constants with the same units as γ_0. The underlying assumption is that the response to small perturbations of the equilibrium positions of resonators is linear and translation invariant. Next we consider the discrete Fourier transform (DFT) of the sequence U_n

$$\bar{U}(k) = \sum_n U_n \exp(iknΛ) \tag{11.23}$$

and its inverse

$$U_n = \frac{Λ}{2\pi} \int_{-\pi/Λ}^{\pi/Λ} \mathrm{d}k \bar{U}(k) \exp(-iknΛ). \tag{11.24}$$

In these relations, the wavenumber k is a real variable defined in the first Brillouin zone. Λ is the cavity separation and also the width of the supercell in the direction of propagation. Making use of the orthonormality and completeness of the DFT, the following dispersion relation is obtained upon substituting (11.24) in (11.22)

$$\omega^2 = \sum_{m=-\infty}^{\infty} \gamma_m \exp(-ikm\Lambda). \tag{11.25}$$

For a symmetric CRAW, the dispersion relation simplifies to

$$\omega^2 = \gamma_0 + \sum_{m=1}^{\infty} 2\gamma_m \cos(km\Lambda). \tag{11.26}$$

The dispersion relation involves a Fourier series, thus reflecting the periodic repetition of the cavities. The coupling coefficients γ_m can be fitted from the computed dispersion relation. We can also express directly the angular frequency as a Fourier series

$$\omega = \Gamma_0 + \sum_{m=1}^{\infty} 2\Gamma_m \cos(km\Lambda), \tag{11.27}$$

where the coefficients Γ_m are straightforwardly obtained from the coefficients γ_m. In practice, the Fourier series converges very quickly, in accordance with the intuition that only neighboring cavities have influence on the dispersion of waves guided by resonator coupling. Figure 11.15 compares the computed dispersion relation with the theoretical dispersion relation, in the case of a square-lattice phononic crystal of cylindrical tungsten inclusions in a silicon matrix. An excellent agreement is obtained with

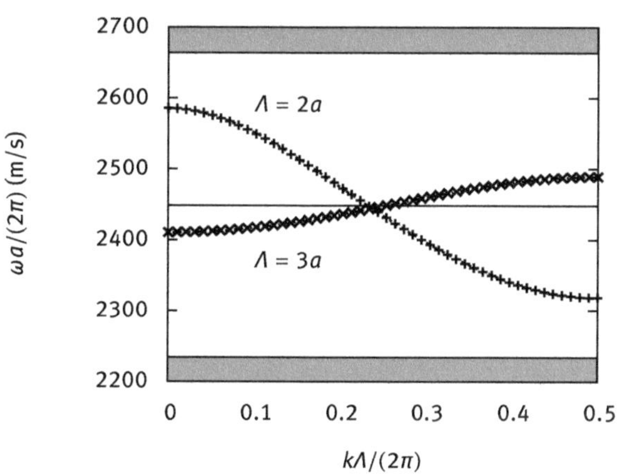

Fig. 11.15: Dispersion relation of a coupled-resonator acoustic waveguide (CRAW). The phononic crystal composed of a 2D square-lattice array of cylindrical tungsten inclusions embedded in a silicon matrix, with a filling fraction of 0.145. CRAWs are formed with single cavities separated by either $\Lambda = 2a$ or $\Lambda = 3a$. In either case, the dispersion relation can be fitted with high accuracy with a Fourier series expansion including only four harmonics (after Escalante et al. [34]).

only the first four Fourier coefficients included, or with couplings included up to the third neighboring resonator. It can be remarked that the property that dispersion relations can be represented by Fourier series expansions is implied by periodicity and linearity alone, independently of the CRAW concept. The same model dispersion relation can thus be considered for any periodic waveguide. The significance of the model in the context of the CRAW, however, is that the Fourier coefficient γ_m can be directly identified as a measure of the coupling of a cavity with its m-th neighbor.

11.4.5 The phononic crystal fiber

All the phononic crystal waveguides we have considered so far employ a two-dimensional periodic structuration to forbid propagation in the plane (x_1, x_2). Assuming that the periodic structure can be made very long, it can also be used to guide waves propagating along the x_3 axis, i.e. along the direction of invariance. This guiding mechanism is directly inspired by the photonic crystal fiber, an optical fiber with a periodic array of holes surrounding a central core [142].

As a first step, let us consider slanted propagation in a phononic crystal. The idea is to look for waves that have a nonzero real-valued k_3 component of the wavevector. To be more explicit, let us reconsider the finite element formulation of the band structure in Section 6.1.4, specifically (6.38)

$$\int_{\Omega} S(\mathbf{v})_I^* c_{IJ} S(\mathbf{u})_J = \omega^2 \int_{\Omega} \mathbf{v}^* \cdot \rho \mathbf{u}.$$

Displacements (u_1, u_2, u_3) depend only on (x_1, x_2), and we let the wavevector be defined in 3D. The strains then simplify to

$$\begin{aligned}
S_1(\mathbf{u}) &= \frac{\partial \tilde{u}_1}{\partial x_1} - \imath k_1 \tilde{u}_1, \\
S_2(\mathbf{u}) &= \frac{\partial \tilde{u}_2}{\partial x_2} - \imath k_2 \tilde{u}_2, \\
S_3(\mathbf{u}) &= -\imath k_3 \tilde{u}_3, \\
S_4(\mathbf{u}) &= \frac{\partial \tilde{u}_3}{\partial x_2} - \imath (k_3 \tilde{u}_2 + k_2 \tilde{u}_3), \\
S_5(\mathbf{u}) &= \frac{\partial \tilde{u}_3}{\partial x_1} - \imath (k_3 \tilde{u}_1 + k_1 \tilde{u}_3), \\
S_6(\mathbf{u}) &= \frac{\partial \tilde{u}_2}{\partial x_1} + \frac{\partial \tilde{u}_1}{\partial x_2} - \imath (k_2 \tilde{u}_1 + k_1 \tilde{u}_2),
\end{aligned}$$

and similar expressions for the test functions $\mathbf{v}$. We can then obtain a series of 2D band structures $\omega(k_1, k_2)$ for any value of k_3 considered as a parameter. If there is a complete band gap, then it applies to slanted wave propagation, because k_3 is nonzero [163].

The next step is to introduce a defect in the phononic crystal that is to serve as a core for waveguiding. Khelif et al. have theoretically considered a square lattice of square cross section tungsten rods in epoxy, and have added a central aluminum nitride (AlN) rod, as depicted in Figure 11.16 [67]. Defect modes appear as flat bands in the phononic band structure in a supercell computation. By examining the distribution of displacements for the defect modes, they are clearly well confined to the central AlN rod and they decay evanescently in the surrounding crystal. They are guided waves propagating along the central rod at the phase velocity ω/k_3.

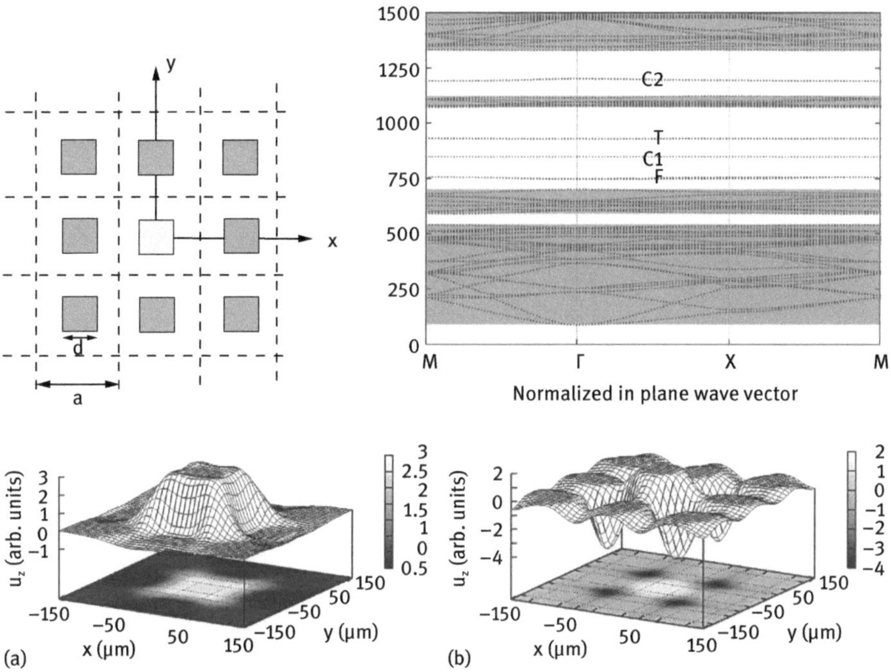

Fig. 11.16: Cross section of a supercell solid-solid phononic crystal, consisting of an AlN rod defect in tungsten/epoxy structure. The lattice constant is $a = 100$ μm and the rod side is $d = 45$ μm. Two-dimensional phononic band structure for $k_3a/(2\pi) = 0.1$, for the tungsten/epoxy/AlN structure. Defect modes appear in two band gap regions. Relative magnitude of out-plane displacements of (a) the compression mode, C1, with $\omega a/(2\pi) = 848.0$ m/s and (b) the compression mode, C2, with $\omega a/(2\pi) = 1198.4$ m/s for the tungsten/epoxy/AlN (after Khelif et al. [67], copyright 2004 American Physical Society).

Gathering together all the band structures for varying k_3, we obtain a dispersion diagram $\omega(k_3)$ giving the dispersion relation of the guided waves. As an example, Figure 11.17 shows such a dispersion diagram for a honeycomb-lattice phononic crystal of holes in silica surrounding a central defect (a missing hole) [81]. It can be noticed that there are no complete band gaps in this dispersion diagram. In the absence of the defect, different regions that are empty of any propagating elastic wave are present in the dispersion relation. When the defect is present, isolated bands corresponding to guided waves appear within them.

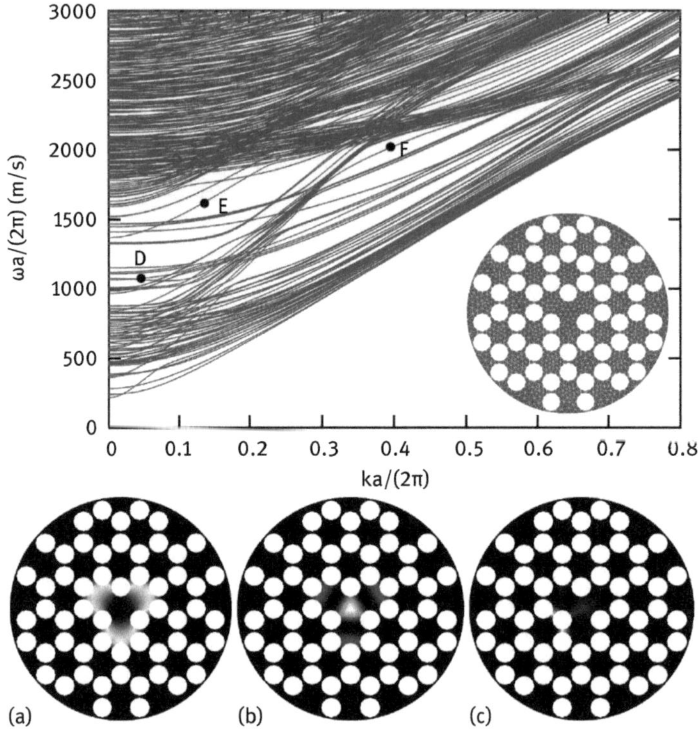

Fig. 11.17: Dispersion diagram for elastic waveguide modes of honeycomb silica photonic crystal fibers with a central defect. The two-dimensional meshes of the cross sections are shown as an inset. The diameter to lattice constant ratio is $d/a = 0.51$. Energy density of guided elastic modes propagating in the core of the photonic crystal fiber by a phononic band gap effect for the points labeled (a) D, (b) E and (c) F in the band structure (after Laude et al. [81], copyright 2005 American Physical Society).

12 Spatial and temporal dispersion

In this chapter, we consider the properties of sonic and phononic crystals from the point of view of their dispersion. In periodic media, temporal and spatial dispersion are intimately connected. We start our discussion with the general dispersion relation for a crystal, in an implicit form, and derive various useful expressions and properties for wave propagation. In particular, the relation between different definitions of group velocity is explained. The concept of the equifrequency surface is next used to discuss various spatial dispersion effects in crystals: positive and negative refraction, collimation, and gradient-index phononic crystals. All these effects are related to the internal details of the crystal, but in practice it is very often necessary to excite wave propagation from the outside. We correspondingly specify a general method to obtain a solution to the reflection/refraction problem at a sonic crystal interface. This methodology is obviously easily extended to phononic crystals. The modal conversion problem at a sonic crystal periodic boundary is then exemplified with the sonic crystal diffraction grating. Finally, the concept of classical tunneling through a finite crystal is illustrated.

12.1 Dispersion relations

We start our discussion of dispersion relations with general considerations for dispersive and anisotropic media.

Implicit dispersion relation as a level set. An implicit dispersion relation has the form

$$\mathcal{D}(\omega, \mathbf{k}) = 0, \tag{12.1}$$

as we already noted in Section 2.1.2. Later in this section we give a general method to obtain an expression of this type for a crystal. Here the dependence on $\mathbf{k}$ actually means dependence on three wavenumbers k_1, k_2, and k_3. Relation (12.1) defines a hypersurface of dimension three in the 4D space $(\omega, \mathbf{k})$ of dispersion relations. In mathematics, this hypersurface is a level set of the real-valued function $\mathcal{D}(\omega, \mathbf{k})$. The first differential of the dispersion relation is

$$\frac{\partial \mathcal{D}}{\partial \omega} \mathrm{d}\omega + \frac{\partial \mathcal{D}}{\partial k_i} \mathrm{d}k_i = 0. \tag{12.2}$$

The group velocity along direction k_i can then be obtained as

$$(v_g)_i = \left.\frac{\mathrm{d}\omega}{\mathrm{d}k_i}\right|_{k_j, j \neq i} = -\frac{\frac{\partial \mathcal{D}}{\partial k_i}}{\frac{\partial \mathcal{D}}{\partial \omega}}. \tag{12.3}$$

This relation is valid for infinitesimal variations of the variables of the dispersion relation and holds exactly on it, i.e. at $(\omega, \mathbf{k})$ points where (12.1) is satisfied. In this definition, each component $(v_g)_i$ of the group velocity vector is the slope of a tangent to the hypersurface defined by (12.1); this tangent lies in a plane containing the frequency axis and direction k_i, as depicted in Figure 12.1 (a). Note that the sign and magnitude of a group velocity component are defined unambiguously by relation (12.3) even if $\mathcal{D}$ is multiplied by an arbitrary nonsingular real function of ω and $\mathbf{k}$.[1]

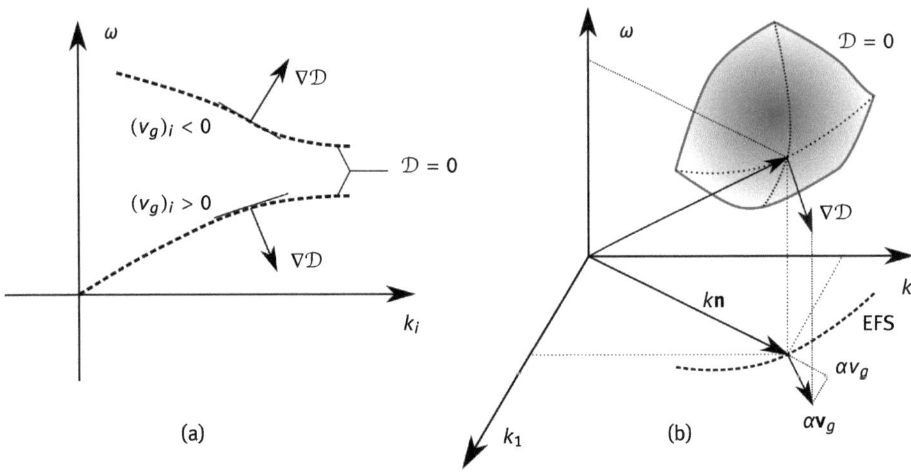

Fig. 12.1: Implicit dispersion relation as a level set. (a) In the (ω, k_i) plane, the implicit dispersion $\mathcal{D}(\omega, \mathbf{k}) = 0$ defines bands. The group velocity is the local slope of the bands. The gradient $\nabla\mathcal{D}$ is orthogonal to the band. (b) In the space $(\omega, \mathbf{k})$, $\mathcal{D}(\omega, \mathbf{k}) = 0$ defines a hypersurface to which the gradient $\nabla\mathcal{D}$ is orthogonal. The group velocity vector is proportional to the projection of the gradient on k-space (the proportionality coefficient $\alpha = -(\frac{\partial\mathcal{D}}{\partial\omega})^{-1}$). The equifrequency surface (EFS) is the cross section of the dispersion relation hyperplane at constant angular frequency.

From the gradient of $\mathcal{D}$ evaluated along the dispersion relation hypersurface, $\nabla\mathcal{D} = (\frac{\partial\mathcal{D}}{\partial\omega}, \frac{\partial\mathcal{D}}{\partial k_i})^T$, the group velocity vector is formed by the scaling operation $\nabla\mathcal{D}/(-\frac{\partial\mathcal{D}}{\partial\omega}) = (-1, (v_g)_i)^T$. It is a mathematical theorem that if function $\mathcal{D}$ is differentiable around a point of its level set, then its gradient is either zero or perpendicular to the level set at that point. Because of its definition from the gradient, the group velocity vector is also orthogonal to the dispersion relation hypersurface. In particular, the group velocity vector is orthogonal to the equifrequency surface (EFS) defined by the cross section of the dispersion relation hypersurface at a constant angular frequency ω.

1 The dispersion relation is unchanged if is multiplied by an arbitrary function $f(\omega, \mathbf{k}) \neq 0$, since $f\mathcal{D} = 0$ is equivalent to $\mathcal{D} = 0$. Furthermore $\nabla(f\mathcal{D}) = f\nabla\mathcal{D} + \mathcal{D}\nabla f = f\nabla\mathcal{D}$ for any point on the hypersurface defined by the dispersion relation. The two gradients are then proportional and point along the same direction.

We can now reinterpret the result in Property 2.5 that $v_g = \mathbf{n} \cdot \mathbf{v}_g$. As depicted in Figure 12.1 (b), let us consider a fixed direction in k-space given by the unit vector $\mathbf{n}$. For point $(\omega, k\mathbf{n})$ in dispersion relation space, $\frac{\partial \mathcal{D}}{\partial k} = n_i \frac{\partial \mathcal{D}}{\partial k_i}$ represents the projection of the gradient with respect to $\mathbf{k}$ on the (ω, k) plane. Dividing by $-\frac{\partial \mathcal{D}}{\partial \omega}$, relation $v_g = \mathbf{n} \cdot \mathbf{v}_g$ follows.

Refraction inside a crystal. Let us illustrate to use of the EFS to graphically obtain the wavevector and group velocity vector upon refraction of a plane harmonic at the interface of a sonic or phononic crystal. The construction is depicted in Figure 12.2 in the case of a 2D square-lattice crystal. The interface between the medium of incidence and the crystal is along x_1. For simplicity we assume that incidence is from a homogeneous and isotropic medium; we also consider only one polarization inside the crystal. The construction can be easily extended to more complex cases.

For the first band, the EFS is centered around the Γ point of the first Brillouin zone. The group velocity vector points outward since $v_g > 0$, according to the previous discussion. In Figure 12.2 (a), we have purposely depicted a slow medium of incidence, so that the range of possible incident wavenumbers extends over a range larger than the first Brillouin zone of the crystal. The wavenumber k_1 is conserved upon refraction. Its value can be tuned by changing the angle of incidence at fixed frequency, through the relation $k_1 = \cos(\theta)\omega/c$. Bloch waves that are excited in the crystal are characterized by a conserved value of k_1, up to a reciprocal lattice vector translation, and a positive sign for $(v_g)_2$. The latter condition is imposed by the requirement that energy propagates in the positive x_2 direction. As θ varies from 0° to 90°, three incidence ranges can be identified. First, a Bloch wave with $(v_g)_1 > 0$ is excited, corresponding to positive refraction (PR). Positive diffraction means that the direction of energy flow in the crystal is on the same side of the normal to the interface as the incident wave. There is then a range of angles of incidence for which no propagating Bloch wave can be excited; there is hence total internal reflection (TIR) at the interface.[2] Next, there is a third range of angles of incidence for which the wavenumber k_1 exceeds the limit of the first Brillouin zone and meets a periodic replica of the EFS. Folding the wavenumber back to the first Brillouin zone, a propagating Bloch wave with $(v_g)_1 < 0$ can be excited, leading to negative refraction (NR). Negative refraction means that the direction of energy flow in the crystal is on the opposite side of the normal to the interface compared to the incident wave.

We still consider in Figure 12.2 (b) the case of the first band, but in a frequency range with a directional band gap for symmetry point X; as a result the EFS is now centered on point M (an actual example of this situation is presented in Figure 12.8).

2 Total internal reflection was introduced in Section 3.2 for an interface separating two homogeneous media with different acoustic velocities. The field in the crystal is here a superposition of evanescent Bloch waves.

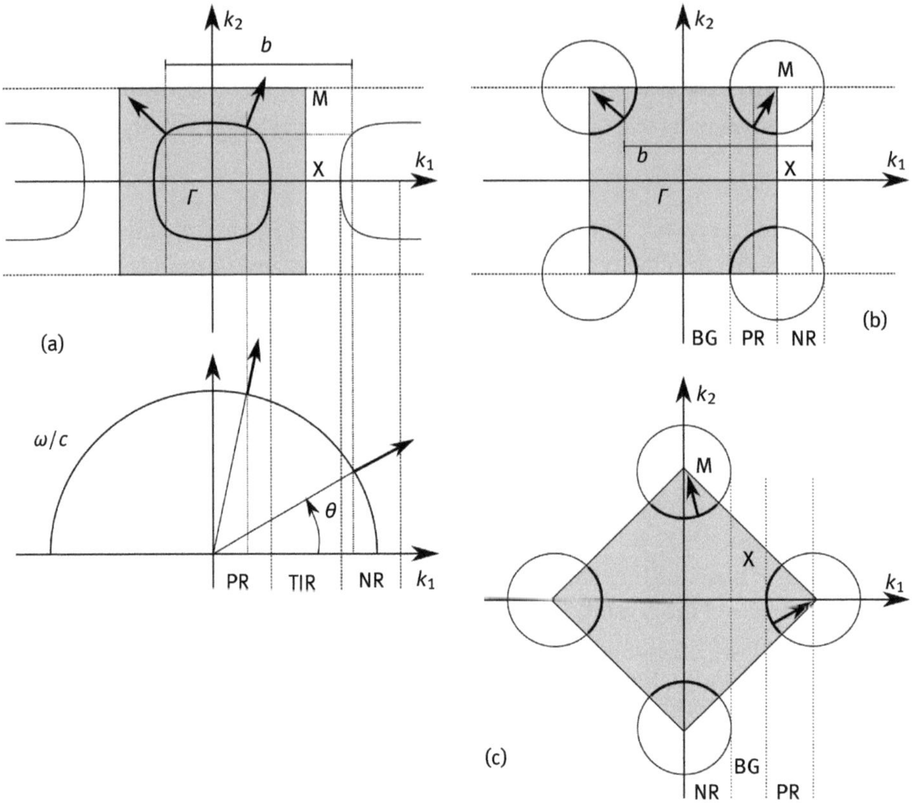

Fig. 12.2: Graphical construction of refraction at an interface separating an isotropic incident medium and an artificial crystal. The case depicted is for a square-lattice crystal. While the dispersion relation in the homogeneous and isotropic medium of incidence can be presented as a circle of radius ω/c in k-space, the equifrequency contours (EFC) of the crystal are periodic with the periodicity of the reciprocal lattice. Only the first band is depicted. (a) For lower frequencies, the EFC is annular and centered on point Γ. The case depicted is such that the interface is aligned with direction x_1. Construction of the refracted Bloch wave follows conservation of the tangential wavenumber and possible folding back to the first Brillouin zone. As the angle of incidence increases, positive refraction (PR), total internal reflection (TIR), and negative refraction (NR) are experienced in sequence. (b), (c) For frequencies such that there is a partial band gap in direction ΓX, the EFC becomes centered on point M. (b) The interface is aligned with direction x_1. As the angle of incidence increases, refraction is to evanescent Bloch waves only (band gap, BG), then positive, then negative. (c) The interface is rotated by 45° from x_1. Refraction is first negative, then to evanescent Bloch waves, and then positive. Note that the group velocity is never negative.

The crystal is oriented so that direction ΓX is along the interface. Following the same procedure as previously, there are again three different incidence ranges as θ varies from 0° to 90°. First the directional band gap induces total reflection for the incident waves. Then until point X there is positive refraction to a propagating Bloch wave. After point X the wavenumber can be folded back to the first Brillouin zone and negative refraction occurs. In Figure 12.2 (c) we have plotted the case of the same crystal and of the same frequency, but with an overall rotation of 45° of the lattice vectors. There is again a succession of three different incidence ranges: first negative refraction, then band gap, then positive refraction.

Further relations for nondispersive media. In the case that a medium of propagation is nondispersive, but possibly anisotropic, there are a few more relations that can be derived for the group velocity vector. The following results apply to bulk wave propagation in homogeneous anisotropic solids, for instance, but not generally to periodic media and artificial crystals. The purpose of presenting them here is to prevent the reader from casually using seemingly classical results in cases where they do not apply.

If propagation is nondispersive, the velocity depends only on the direction of propagation and we can write

$$\omega(\mathbf{k}) = |\mathbf{k}|v(\mathbf{k}), \tag{12.4}$$

defining an explicit representation for the dispersion relation as a cone. We can then obtain by differentiation

$$(v_g)_i = \frac{\partial \omega}{\partial k_i} = \frac{k_i}{|\mathbf{k}|}v + |\mathbf{k}|\frac{\partial v}{\partial k_i} = vn_i + \frac{\partial v}{\partial n_i}. \tag{12.5}$$

With this expression, it is clear that for isotropic propagation the group velocity vector equals the phase velocity vector and that $\frac{\partial v}{\partial n_i}$ measures the deviation from isotropy.

A homogeneous function f of degree q is such that

$$f(\alpha \mathbf{x}) = \alpha^q f(\mathbf{x}), \quad \forall \alpha > 0. \tag{12.6}$$

Euler's homogeneous function theorem states that if f is continuously differentiable, then homogeneity is equivalent to

$$\nabla \cdot f(\mathbf{x}) = qf(\mathbf{x}). \tag{12.7}$$

Let us apply this result to the EFS in (12.4). $\omega(\mathbf{k})$ is a homogeneous function of degree 1 of the wavevector, implying that

$$k_i \frac{\partial \omega}{\partial k_i} = \omega. \tag{12.8}$$

The formula $n_i(v_g)_i = v$ then follows immediately, instead of $n_i(v_g)_i = v_g$. The lesson is that the casual formula $\mathbf{v}_g = \nabla_{\mathbf{k}}(\omega(\mathbf{k}))$ can be misleading in artificial crystals if one does not recognize that ω is not simply a homogeneous function of $\mathbf{k}$.

Implicit dispersion relations for sonic and phononic crystals. The classical band structure of sonic and phononic crystals gives the dispersion relation of propagating

Bloch waves; it can be viewed as a series of cross sections of the dispersion relation hypersurface. We have obtained it in Chapters 4 and 6 from the solution of an eigenvalue problem of the form

$$A(k)\mathbf{x} = \omega^2 B\mathbf{x}, \tag{12.9}$$

where $\mathbf{x}$ is the vector of the unknown coefficients in the expansion of the solution (PWE or FEM, for instance), and A and B are square and symmetric matrices that are formed from material constants and expansion functions. Matrix A explicitly depends on the wavenumber k and on the direction of propagation $\mathbf{n}$. From the eigenvalue problem, a scalar dispersion relation for each eigenvalue ω can be obtained by left-multiplying with the left eigenvector $\mathbf{y}$. The left eigenvector is a right eigenvector of the transposes of matrices A and B. Generally, $\mathbf{x}$ and $\mathbf{y}$ associated to the same eigenvalue are different, unless the matrices are symmetric [152], which is the case here. We thus obtain the dispersion relation in implicit form as [80]

$$\mathcal{D}(\omega, k) = \mathbf{x}^T A(k)\mathbf{x} - \omega^2 \mathbf{x}^T B\mathbf{x} \tag{12.10}$$

where the subscript ^{T} denotes transposition. Note that the eigenvector depends on ω and k. Because it satisfies (12.9) along the dispersion relation, we obtain

$$\frac{\partial \mathcal{D}}{\partial \omega} = -2\omega \mathbf{x}^T B\mathbf{x},$$
$$\frac{\partial \mathcal{D}}{\partial k} = \mathbf{x}^T \frac{\partial A(k)}{\partial k}\mathbf{x},$$

at every point of the dispersion relation. Since we know explicit expressions for all matrices, these formulas allow for a very efficient evaluation of the group velocity at every (k, ω) point of the dispersion relation, given only the knowledge of the eigenvalue and eigenvector. Indeed, the group velocity can be calculated without any approximation along the dispersion relation by the formula

$$v_g = \frac{\mathbf{x}^T \frac{\partial A(k)}{\partial k}\mathbf{x}}{2\omega \mathbf{x}^T B\mathbf{x}}. \tag{12.11}$$

The complex band structure is also found as the solution of a generalized eigenvalue problem of the form

$$C(\omega)\mathbf{x} = kD(\omega)\mathbf{x}, \tag{12.12}$$

as we discussed in Chapter 9. Matrices C and D account for periodicity where required and are nonsymmetric in general. Matrix D explicitly depends on frequency in case of loss. As before, we can obtain an implicit dispersion relation for each particular eigenvalue k, by left-multiply by the left eigenvector $\mathbf{y}$. The left eigenvector satisfies

$$\mathbf{y}^T C(\omega) = k\mathbf{y}^T D(\omega). \tag{12.13}$$

The dispersion relation in implicit form is thus

$$\mathcal{D}(\omega, k) = \mathbf{y}^T C(\omega)\mathbf{x} - k\mathbf{y}^T D(\omega)\mathbf{x}, \tag{12.14}$$

whose first differentials when evaluated on the dispersion relation are

$$\frac{\partial \mathcal{D}}{\partial \omega} = \mathbf{y}^T \frac{\partial C}{\partial \omega}\mathbf{x} - k\mathbf{y}^T \frac{\partial D}{\partial \omega}\mathbf{x},$$
$$\frac{\partial \mathcal{D}}{\partial k} = -\mathbf{y}^T D\mathbf{x}.$$

Again, since we know the explicit expressions for all matrices, these formulas allow for a very efficient evaluation of the group velocity at every (ω, k) point of the dispersion relation, given only the knowledge of the eigenvectors $\mathbf{y}$ and $\mathbf{x}$. The group velocity is explicitly [109]

$$v_g = \frac{\mathbf{y}^T D\mathbf{x}}{\mathbf{y}^T \frac{\partial C}{\partial \omega}\mathbf{x} - k\mathbf{y}^T \frac{\partial D}{\partial \omega}\mathbf{x}}. \tag{12.15}$$

This formula was used to plot the group velocity variation with frequency in Figures 9.11 and 9.12.

12.2 Refractive sonic crystal lenses

Traditional optical lenses are based on refraction at curved interfaces between glass and air. As the refractive index of glass ($\approx$ 1.45 to 1.9) is larger than the refractive index of air ($\approx$ 1), refraction is positive and the angle of refraction is smaller than the angle of incidence. The same principle applies to refractive acoustic lenses that use a shaped piece of material with a longitudinal velocity smaller than the velocity of the host fluid medium. In the low frequency range of the band structure of a sonic crystal, the EFS for the first band initially has the shape of a cone originating from the Γ point. As a consequence, in every direction of propagation there is an effective velocity $v_{\text{eff}}(\mathbf{n})$ that is independent of frequency – at least for low frequencies. This effective velocity is the result of the homogenization of the Bloch wave attached to the first band [21]. For an incident wavelength that is much larger than the lattice constant, and as long as we can disregard any evanescent Bloch wave, the sonic crystal appears as if it were homogeneous [153]. For the hexagonal-lattice sonic crystal of aluminum rods in air shown in Figure 12.3, $v_{\text{eff}}(\mathbf{n}) \approx 270\,\text{m/s}$ for the ΓM direction, resulting in an equivalent acoustic refractive index of 1.3 ± 0.1 with respect to air [21]. The experimental result shows that it is possible to produce a sonic crystal lens that can focus sound waves in air.

The biconvex shape given to the sonic crystal lens in Figure 12.3 is purely heuristic and inspired by the appearance of traditional optical lenses. Håkansson et al. proposed designing this shape by using a numerical optimization procedure [45]; they specifically used genetic optimization, but this choice is not unique. Their design goal was the maximization of the sound enhancement at a given position after the lens and at a given frequency. The design variables are the presence or not of an aluminum rod at every unit cell of a 20×19 array of rods. Optimization time and results obviously

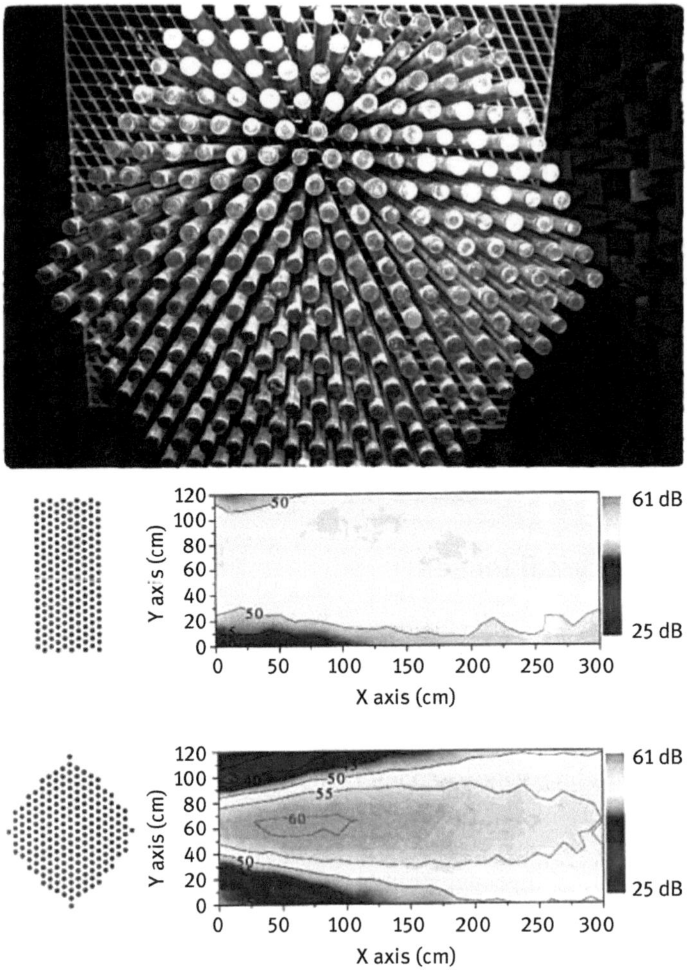

Fig. 12.3: A sonic crystal lens in air. (a) The photograph shows a sonic crystal lens for audible frequencies hanging from the ceiling as seen from below. Part of the frame, the walls and the ceiling of the anechoic chamber can be seen. Sound level maps are measured in air at a frequency of 1 700 Hz (b) after a flat lens and (c) after a biconvex lens. The sketches are drawn to scale: both sonic crystal samples are 1.2 m wide (after Cervera et al. [21], copyright 2002 American Physical Society).

depend on the constraints that are added to reduce the size of the solution space. In Figure 12.4 (a)–(d), the lens was imposed to be symmetric with respect to both the horizontal and the vertical axis, and to have no missing rod inside the lens. The results show the different optimized shapes when the focusing distance varies from 0.65 to 1.25 m. In Figure 12.4 (e)–(g), the constraints are progressively relieved for a focusing distance of 1.05 m. In (e), the lens is permitted to have constrictions, loosening its con-

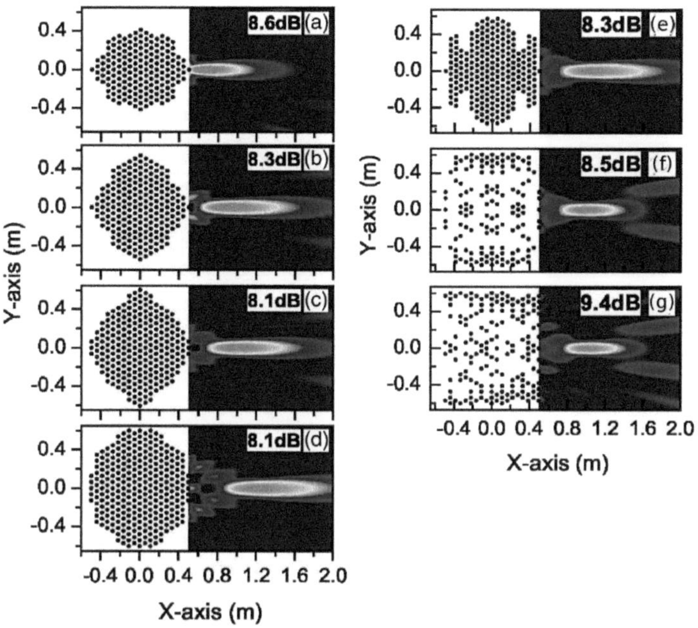

Fig. 12.4: Genetic algorithm optimization of refractive sonic crystal lenses in air. (a)–(d) Four thick sonic crystal lenses are designed for 1 700 Hz and different focal distances: (a) 0.65 m, (b) 0.85 m, (c) 1.05 m and (d) 1.25 m. The maximum amplification in the focus is indicated in each plot. (e)–(g) Three thick SC lenses are designed to focus sound at a distance of 1.05 m with three different symmetry conditions (see text). The symmetry constraint is hardest for (a) and successively lifted for (b) and (c). The maximum amplification in the focus is indicated in each plot (after Håkansson et al. [45], copyright 2004 American Physical Society).

vex character; the enhancement is only marginal compared to (c). In (f), rods are permitted to be removed inside the lens, leading to another small improvement in the pressure level at the focus. Finally, in (g) symmetry with respect to the vertical axis is relaxed, leading to a more significant improvement.

The previous example implies that the convex shape for the lens was not particularly necessary given the optimization goal: maximize the pressure level at a given point in space and at a given frequency. Håkansson et al. proceeded to demonstrate that this goal can be achieved even with flat sonic crystal lenses [46]. The designs in Figure 12.5 are for flat refractive lenses using either five or nine layers of a hexagonal-lattice sonic crystal. In addition to the previous designs, three different diameters for the rods are permitted, either 0 (no rod), 12 mm, or 30 mm. The optimization result is not intuitive and has no clear resemblance to a traditional lens. It has to be pointed out, however, that the flat sonic crystal lens is designed to operate at a single frequency, so it is not guaranteed that it will operate properly at different frequencies and consequently with short acoustic pulses having an inherent wide bandwidth.

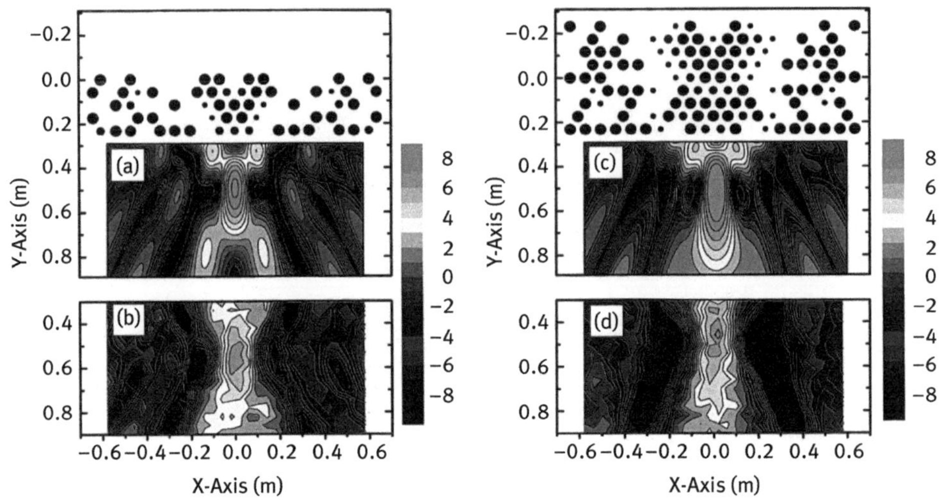

Fig. 12.5: Flat sonic crystal lenses optimized by a genetic algorithm. A hexagonal-lattice sonic crystal of aluminum rods in air is designed as an acoustic lens that is either 5 (left) or 9 (right) layers thick. The pressure enhancement maps obtained numerically (a), (c) are compared with experimental results (b), (d). The lattice constant is $a = 67$ mm and rods with diameters of 12 mm or 30 mm are considered (after Håkansson et al. [46], copyright 2005 American Institute of Physics).

12.3 Negative refraction in sonic crystals

We illustrated in Figure 12.2 that negative refraction can occur at certain frequencies and for certain directions of propagation in sonic crystal. An experimental demonstration was provided by Yang et al. [172] for a 3D FCC sonic crystal of tungsten carbide beads in water. The EFS cross sections in Figure 12.6 are drawn for several frequencies just below 1.6 MHz. For the geometrical conditions considered, the EFS is centered on the Γ point of the first Brillouin zone. As the band involved is a folded band extending above the full band gap, the wavenumber decreases along the EFS as frequency increases; as a result, the group velocity vector points inward to the cross sections of the EFS. For the particular frequency 1.57 MHz, spatial dispersion is quite strong around the ΓL direction: for small angles of incidence the group velocity vector changes rapidly. The negative refraction at the first interface between the incident medium and the sonic crystal is expected to produce a strong focusing of the acoustic beam inside the crystal. At the second interface between the crystal and the substrate, negative refraction again produces a converging beam which can be detected with a hydrophone. The experimental results reported in Figure 12.7 show that a focused beam is clearly observed at a frequency of 1.57 MHz. But at the very close frequency of 1.6 MHz the beam is spatially dispersed and out of focus. This demonstrates the extreme dispersion sensitivity of the sonic crystal.

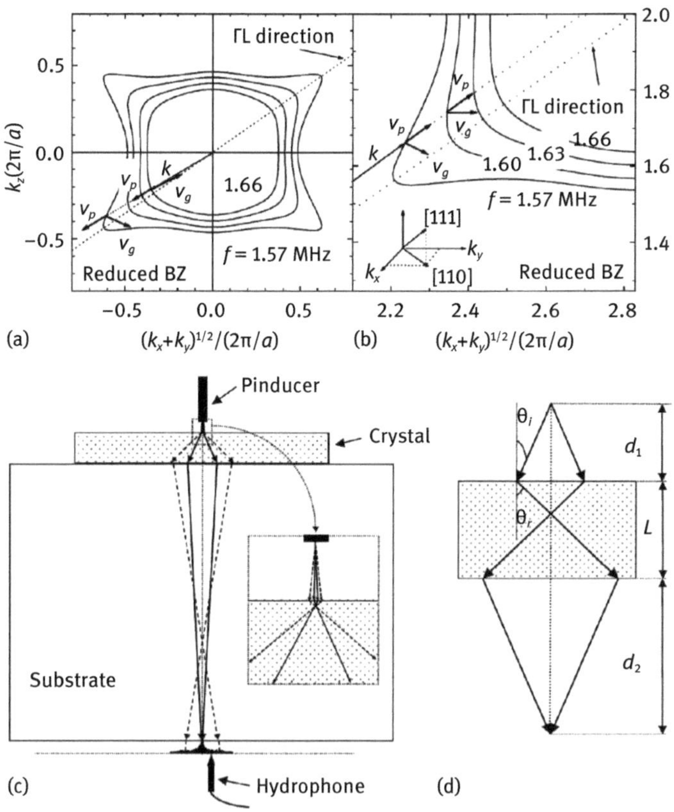

Fig. 12.6: Negative refraction in an FCC sonic crystal of tungsten carbide beads in water. The close-packed crystal is composed of beads with a diameter of 0.8 mm. Cross sections of the equifrequency surfaces at frequencies are drawn near 1.60 MHz in the reduced (a) and extended Brillouin zones (b). The cross sections are shown in a plane containing the [001], [111], and [110] directions, as shown in the inset. As the cross sections are annular around the Γ point and converging to this point as the frequency increases, the group velocity vector always points inward. Negative refraction occurs for directions around the ΓL direction. (c) The experimental set-up is designed to demonstrate negative refraction. The diagram shows rays indicating the predicted directions of the group velocity at 1.6 MHz for angles of incidence of 1.5° and 5°. (d) The focusing condition in a medium with negative refraction is depicted schematically (after Yang et al. [173], copyright 2004 American Physical Society).

As illustrated in Figure 12.2 (b)–(c), the 2D square-lattice sonic crystal potentially shows negative refraction for directions around the ΓM direction for the first band, providing there is a directional band gap in direction ΓX and even in the absence of negative group velocities. This property has been widely exploited in the literature. Zhang et al. discussed the case of steel cylinders in air and of water cylinders in mercury through numerical simulations [176]. They remarked that the fact the EFS is annular around the point M leads to focusing but also to negative refraction around

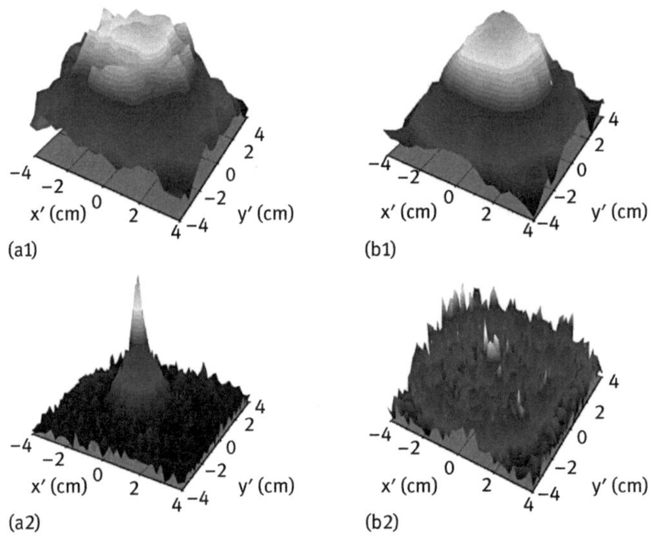

Fig. 12.7: Negative refraction in an FCC sonic crystal of tungsten carbide beads in water. Field patterns are measured at 1.57 MHz without (a1) and with (a2) the crystal in place. The x' and y' axes are in the plane parallel to the sample surface, with x' in the LK direction. Corresponding field patterns for measurements at 1.60 MHz are shown in (b1) and (b2) (after Yang et al. [173], copyright 2004 American Physical Society).

the direction ΓM. As in the example of FCC crystal in Figures 12.6 and 12.7, these properties do not originate from a negative index of refraction as is often sought with metamaterials; they are the consequence of band foldings at the edges of the first Brillouin zone. Feng et al. discussed through numerical simulations how the local shape of the EFS could be shaped around point M by replacing the circular rods with square rods and rotating them in-plane [40]. The tunability range proves to be quite large. He et al. numerically consider a square lattice of rubber-coated tungsten rods in water [47]. As depicted in Figure 12.8, the first band is quite flat just below point M. The cross sections of the equifrequency surface assume a square shape with rounded corners. As a result, refraction is slightly negative in a rather large range of directions of incidence around the ΓM direction. Assuming that a source of pressure waves is placed in water quite close to the entrance surface of the sonic crystal, it is expected that both propagating and evanescent waves emitted by the source can be converted to Bloch waves of the crystal. This canalization mechanism leads to the formation of an image of the source, as shown in Figure 12.9. The image is formed at the exit surface of the sonic crystal and has a width of 0.16λ, with λ the wavelength in the host medium. Capturing incident evanescent waves indeed appears to be the key to obtaining super-resolution [145, 146]. Further numerical simulations show that if the source is placed farther away from the entrance surface of the crystal the image is virtual and forms inside the crystal.

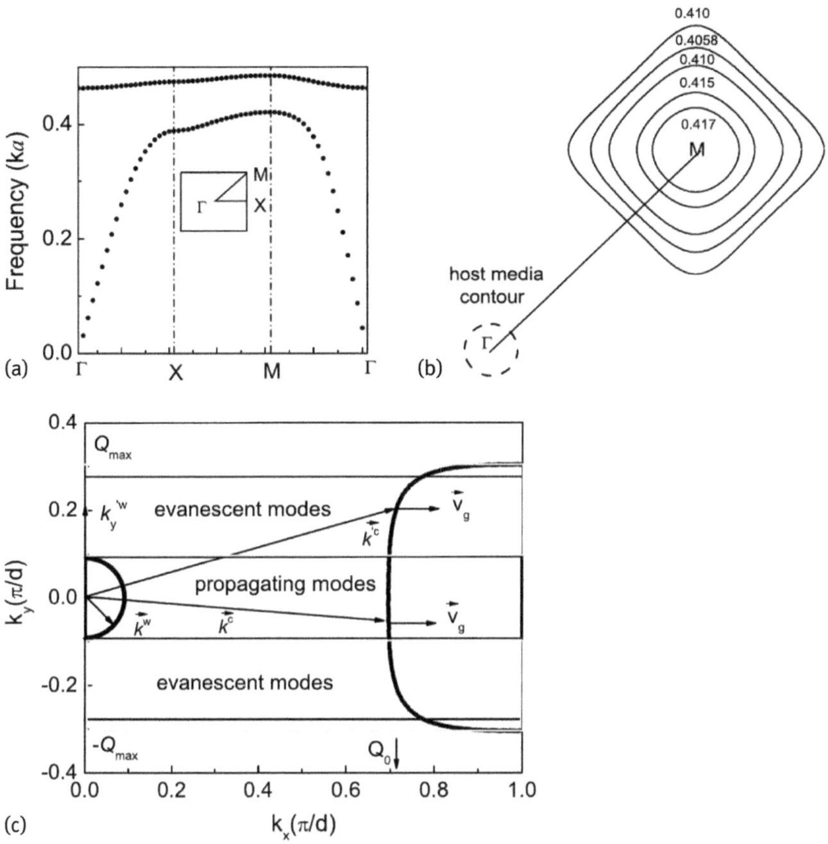

Fig. 12.8: Band structure and equifrequency contours of a square-lattice sonic crystal. (a) The first and second bands are shown for a square array of rubber-coated tungsten rods in water. (b) Equifrequency contours (EFC) are plotted at selected frequencies near the top of the first band, which are closed contours surrounding point M, moving inward to M with increasing frequencies. The small dashed circle is the EFC of water at reduced frequency 0.4058. (c) A schematic diagram of modal conversion at reduced frequency 0.4058 from plane waves in water to Bloch waves of the second band around the ΓM direction is proposed. The governing half of the EFC of water as indicated by the semicircle on the left is smaller than the governing half of the EFC of the sonic crystal on the right. Both propagating modes and evanescent modes in water are able to be canalized by the Bloch modes in the crystal. The flat region of the EFC, for which the group velocity vector remains almost parallel to the ΓM direction, captures both type of incident waves (after He et al. [47], copyright 2008 American Institute of Physics).

12.4 Collimation

The flat equifrequency contours for certain frequencies in connection with the existence of directional band gaps suggest producing collimated beams. Chen et al. presented numerical simulations for a square lattice of rigid solid rods in air [24]. We have

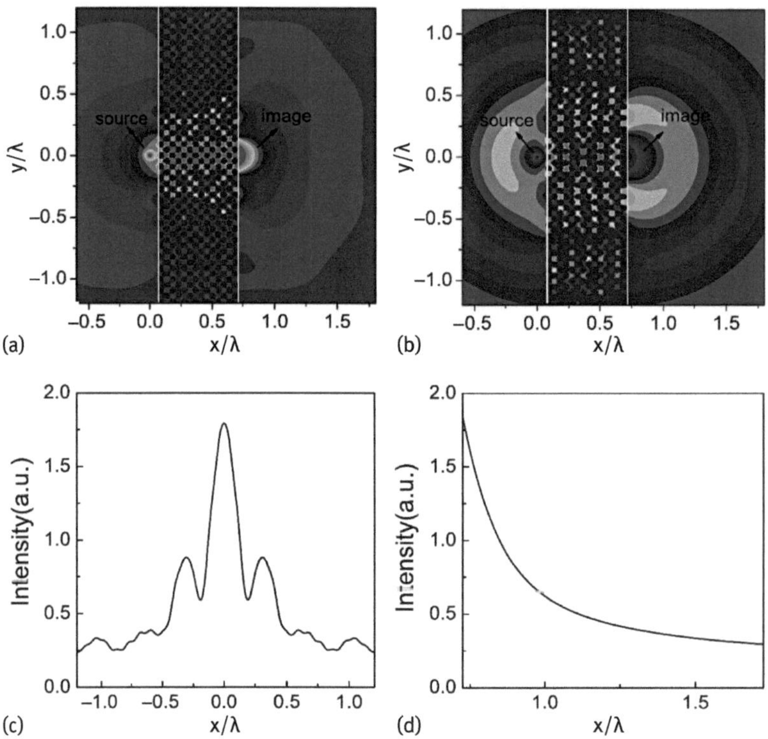

Fig. 12.9: Subwavelength imaging with a flat square-lattice sonic crystal of rubber-coated tungsten rods in water. The reduced frequency is 0.4058. The numerical simulation is performed with a point source placed at a distance of $\sqrt{2}a$ from the left surface of a 14 layer sonic crystal. The intensity (a) and the phase (b) field distributions show that an image of the source is formed on the right surface of the PC. Cross sections show the intensity distribution (c) along the surface and (d) orthogonally to it (after He et al. [47], copyright 2008 American Institute of Physics).

already seen that for the first band the ΓM direction can be used with profit when there is a directional BG for the ΓX direction; these authors also proposed making use of the deaf second band in the ΓM direction to obtain collimated emission in the ΓX direction. In both cases, similar collimation properties and canalization imaging effects are observed in the numerical simulations. Ke et al. proposed a different setting to obtain collimation and more precisely to convert the highly divergent beam emitted by a pinhole transducer into a collimated beam with limited divergence [61]. The system in Figure 12.10 uses an asymmetric Fabry–Perot cavity defined by removing exactly one line of rods from a square-lattice sonic crystal of steel rods in water. The Fabry–Perot cavity supports one resonance inside the full band gap of the sonic crystal. A pinhole transducer is introduced in the cavity. The highly divergent beam emitted by the transducer is trapped in the Fabry–Perot interferometer and exits preferentially from the side with the least number of crystal rows. At the resonant frequency, an exiting col-

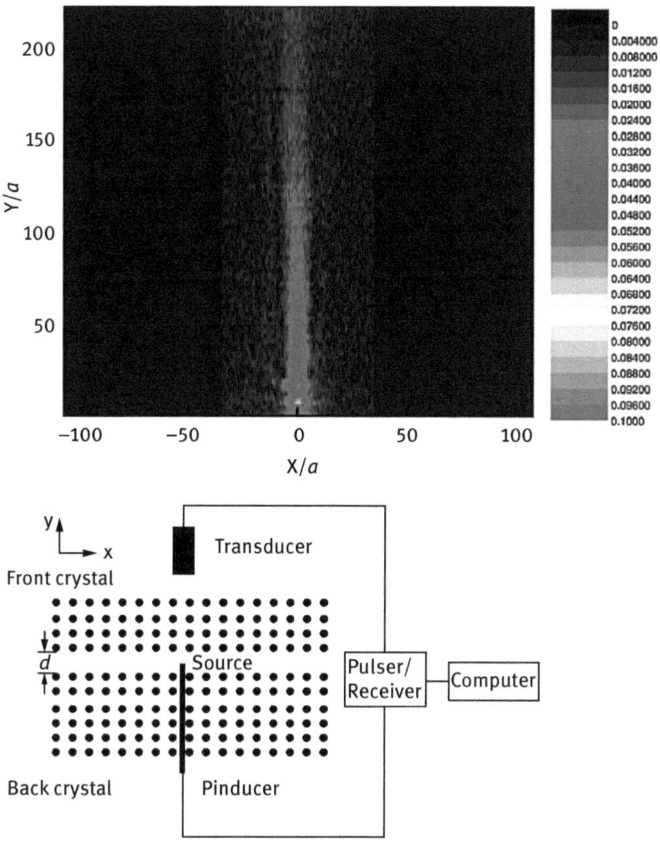

Fig. 12.10: A collimated source of pressure waves in water based on a Fabry–Perot sonic crystal interferometer. The sonic crystal is a square-lattice array of steel rods in water, with diameter $d = 1$ mm and lattice constant $a = 1.3$ mm. A full band gap extends from about 0.57 to 0.72 MHz. The Fabry–Perot cavity consists of four layers for the front crystal and six layers for the back crystal. The separation between the two crystals is exactly one lattice constant. The measured amplitude pressure field distribution at 0.68 MHz after the sonic crystal system reveals the emission of a collimated beam with a divergence less than 6°(after Ke et al. [61], copyright 2006 American Institute of Physics).

limated beam is observed with a divergence slightly less than 6°. All in all, the sonic crystal system transforms the beam from a point source (the pinhole transducer) to an extended plane source. The resolution limit, however, does not reach that of an extended plane transducer that would have the same aperture as the sonic crystal. Indeed, the aperture is about $L = 30\lambda$, with λ the wavelength in water at the operating frequency, so one could expect at best a divergence $\delta\theta \approx \frac{\lambda}{2L} \approx 1°$.

Collimation is not limited to beams propagating in the host medium of a phononic crystal. Actually, since the spatial dispersion of the crystal can be tailored, subdiffractive propagation can be obtained inside the crystal itself [35]. Figure 12.11 shows that

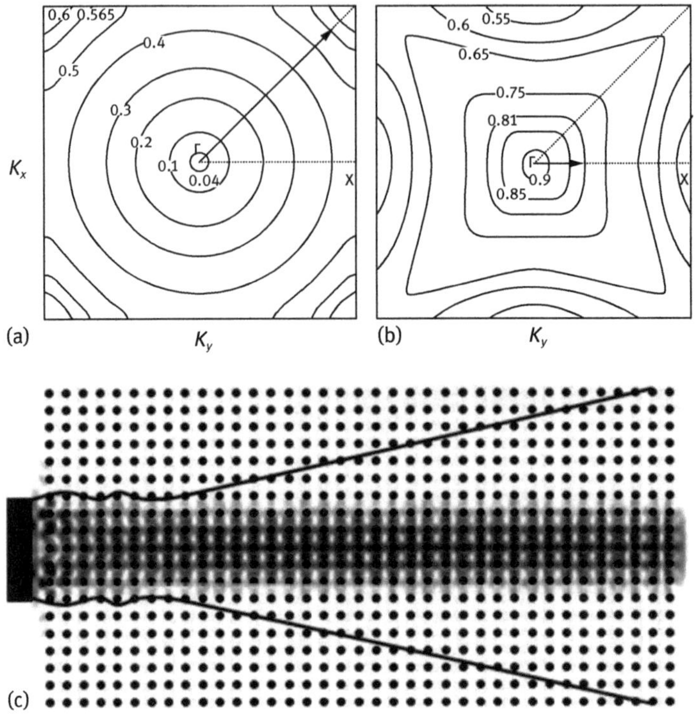

Fig. 12.11: Subdiffractive propagation of ultrasound in a sonic crystal. The square-lattice crystal of steel rods in water is formed for a rod diameter $d = 1.6$ mm and a lattice constant $a = 2.25$ mm. Isofrequency contours are evaluated for the first (a) and second (b) bands, centered at the point Γ. The plot extends over the first Brillouin zone. Numerals denote the reduced frequency $\omega a/(2\pi c)$ with $c = 1483$ m/s the celerity of sound waves in water. The arrows indicate the location of the flat regions, and consequently the direction of wavevectors belonging to a self-collimated beam. Note that the group velocity vector points inward around the Γ point in case (b). The numerical simulation reveals (c) the propagation of an ultrasonic beam in a self-collimated mode in the second band. The lines represent the lobe of the beam propagating in a homogeneous medium. The source frequency is 230 kHz (reduced frequency 0.814). The radius of the plane source is R = 12.5 mm (after Espinosa et al. [35], copyright 2007 American Physical Society).

the equifrequency contours for the second band can be very flat around the Γ point of a square lattice of steel rods in water. For a particular frequency and for propagation in the ΓX direction, the sonic crystal dispersion compensates the natural spreading of the beam that is inherent to isotropic homogeneous media such as water. Espinosa et al. provided an experimental validation of this subdiffractive propagation effect [35].

An obvious drawback of collimation or subdiffractive propagation in sonic crystals is that it can only work in a rather narrow frequency range around the optimal frequency for which the equifrequency contour is the flattest. Coming back to the first band and to the direction ΓM of a square-lattice sonic crystal of steel rods in water, Shi et al. proposed to extend this frequency range by combining two sonic crystals with

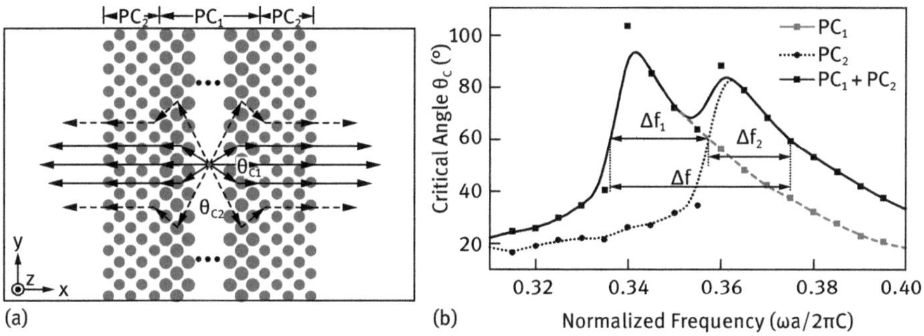

Fig. 12.12: Extension of the frequency range of collimation by using a sequence of sonic crystals. (a) An acoustic collimation lens comprising two sonic crystals (SC1 and SC2) of the same period but different filling ratios is drawn schematically. (b) The dependence of critical angles as a function of reduced frequency for SC1, SC2, and the composite SC1 + SC2 (after Shi et al. [139], copyright 2008 American Institute of Physics).

different filling fractions [139]. As depicted in Figure 12.12, the collimation frequency range is approximately doubled in the case considered. It should be noted, however, that this extension is provided for a punctual source of waves placed in the middle of the central sonic crystal. Indeed, as depicted in Figure 12.12 (a), the frequencies in the collimation range of SC1 remain approximately collimated when they propagate in SC2, while the frequencies in the collimation range of SC2 first undergo negative refraction in SC1 before being collimated in SC2. This principle would not apply to sub-diffractive propagation inside two sonic crystals arranged according to the scheme of Figure 12.11.

12.5 Gradient-index phononic crystals

The idea of combining different sonic or phononic crystals with variable properties in order to obtain given propagation characteristics is natural and is reminiscent of the field of acoustic metamaterials. Lin et al. for instance proposed to define a gradient-index (GRIN) phononic crystal so as to impose the equivalent of a graded optical index of refraction [88]. Indeed, in optics the index of refraction is a measure of the slowness of light in a material as compared to a vacuum. Since light rays tend to bend toward the regions of space with the highest index of refraction, engineering the distribution of the index leads to the formation of continuous waveguides. The hyperbolic secant law is especially known to result in the focusing of an incident plane wave. The design in Figure 12.13 is based on a square-lattice phononic crystal of heavy metal rods in an epoxy matrix. For simplicity, the analysis is limited to out-of-plane shear waves only, which are uncoupled with in-plane elastic waves in 2D phononic crystals composed of isotropic constituents. Wave propagation is considered in the first band, originat-

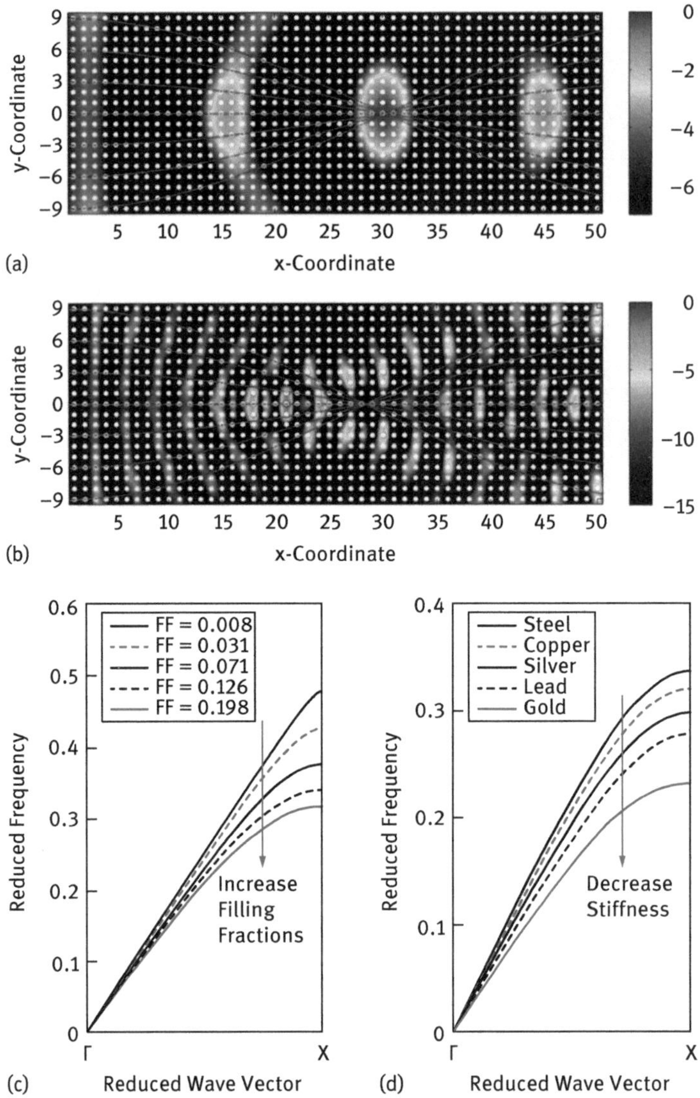

Fig. 12.13: A gradient-index phononic crystal. The square-lattice phononic crystal is composed of metal rods in an epoxy matrix. Only pure shear elastic waves polarized along the rod axis are considered. Elastic wave propagation inside the gradient-index (GRIN) phononic crystal is simulated numerically with FDTD at a reduced frequency of (a) 0.05 and (b) 0.20. The amplitudes of displacement fields are normalized and presented in decibel scale. Reduced frequencies are normalized with respect to the transverse velocity in epoxy. The refractive index profile along the direction transverse to the propagation axis follows a hyperbolic secant law and enables redirection of incident beams inside the medium. The GRIN phononic crystal can be realized by (c) adjusting the filling fraction or (d) changing the elastic properties of cylinders along the transverse direction. The band structures are shown along the ΓX direction only. The group velocity decreases when the filling fraction increases or when the cylinder stiffness decreases (after Lin et al. [88], copyright 2009 American Physical Society).

ing from the Γ point, and is limited to the low frequency range where the phononic crystal can be homogenized. In order to work exactly as planned, it would be necessary that the homogenized phononic crystal be perfectly isotropic. This condition is only fulfilled approximately in the cases considered, when the variable effective shear velocity is provided by either locally tuning the filling fraction or by varying the material composing the rods in every unit cell. The numerical simulations in Figure 12.13 indicate that the focal length achieved by the GRIN phononic crystal is frequency dependent, i.e. the focal length decreases continuously as frequency increases.

Lin and Huang further developed their idea to numerically demonstrate the equivalent of the mirage effect in phononic crystals [87]. The optical mirage effect is an atmospheric phenomenon that results from a gradient in the index of refraction as a function of altitude, causing light rays to bend along their propagation, thus forming a false image of a distant object. The acoustic mirage effect is also observed in seawater, as gradients in sound velocity can result from different variations of temperature, pressure, or salinity. The general rule is that rays bend toward regions with lower velocity. Lin and Huang again considered a square-lattice phononic crystal of steel rods in epoxy with a variable filling fraction. This time, however, the second pure shear band is considered around the Γ point. For all filling fractions in Figure 12.14 refraction is negative. The acoustic mirage effect manifests as a continuous bending of the direction of propagation of wave energy.

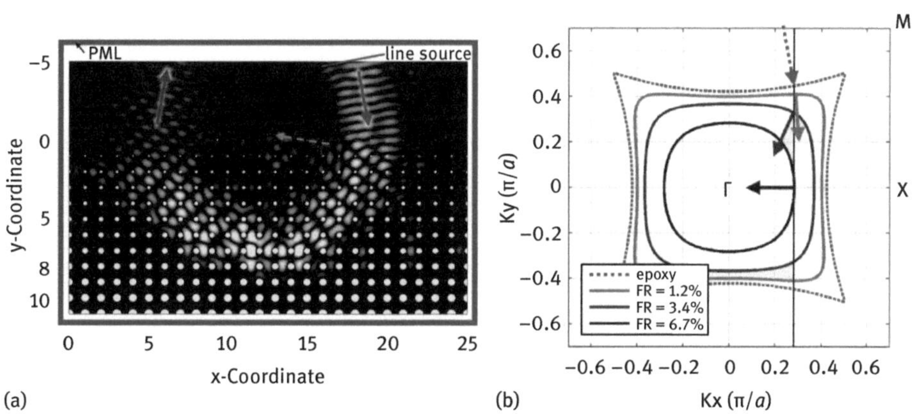

Fig. 12.14: Acoustic mirage effect in a gradient-index phononic crystal. The square-lattice phononic crystal is composed of steel rods in an epoxy matrix. (a) The numerical simulation demonstrates the acoustic mirage effect inside a GRIN phononic crystal illuminated with a pure shear acoustic beam with a width of $4a$, an incident angle of 10° and an operating reduced frequency of 0.79. The light and dark regions correspond to the strong and weak amplitudes of the displacement field, respectively. (b) Equifrequency contours are shown for the filling fractions 1.2 %, 3.4 %, and 6.7 % at a reduced frequency of 0.79. The dotted arrow represents the incident direction. The solid line represents the construction line plotted for an incident angle of 10°. The solid arrows represent the direction of the group velocity vector (after Lin et al. [87], copyright 2009 American Institute of Physics).

Experimental demonstrations of GRIN phononic crystals are not easily performed with 2D crystals, as it is difficult to image the propagation of elastic waves inside them. Zhao et al. provided such a demonstration for flexural waves (the A_0 Lamb wave) in a phononic crystal slab of holes in silicon [178, 179]. Their analysis includes the cases of square-lattice and of rectangular-lattice phononic crystals. Indeed, with the rectangular lattice it is possible in principle to adjust the anisotropy in the direction of propagation while the transverse direction is used to encode the gradient in phase velocity, here tuned by changing the diameter of the holes. An experimental result is shown in Figure 12.15 for the square lattice. Experimental results for the rectangular lattice are also provided by the authors but show only a small improvement compared to the square lattice.

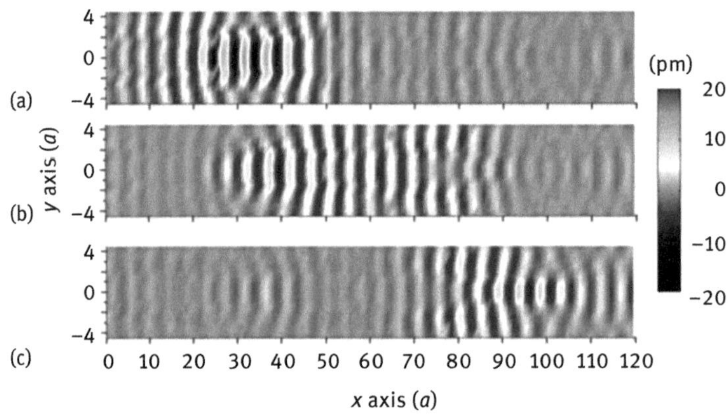

Fig. 12.15: A gradient-index phononic crystal slab of holes in silicon. The amplitude of the out-of-plane displacement in the GRIN square-lattice phononic crystal slab is measured at three different times and demonstrates focusing of the A_0 Lamb wave at different positions along the crystal. The parameters of the phononic crystal slab are $a = 100$ μm and $h = 110$ μm. The diameter-to-lattice constant ratio for the holes (d/a) varies between 0.4 and 0.8 (after Zhao et al. [178]).

12.6 Negative refraction in phononic crystals

Let us now consider negative refraction in phononic crystals. For sonic crystals, propagation in the matrix is isotropic and can be treated with scalar pressure waves. In contrast, propagation in phononic crystals must be described with vector displacement waves and anisotropy in the solid constituents is not uncommon. It is sometimes stated that negative refraction cannot be found in Nature; a counterexample is easily provided for elastic waves in anisotropic solids [58]. Let us consider for instance in Figure 12.16 the slowness curve for the quasi-shear bulk elastic wave in rutile (TiO_2) that we already presented in Figure 5.5. For the range of angles of incidence depicted in gray, refraction is always negative for an interface that is orthogonal to axis x_1. This

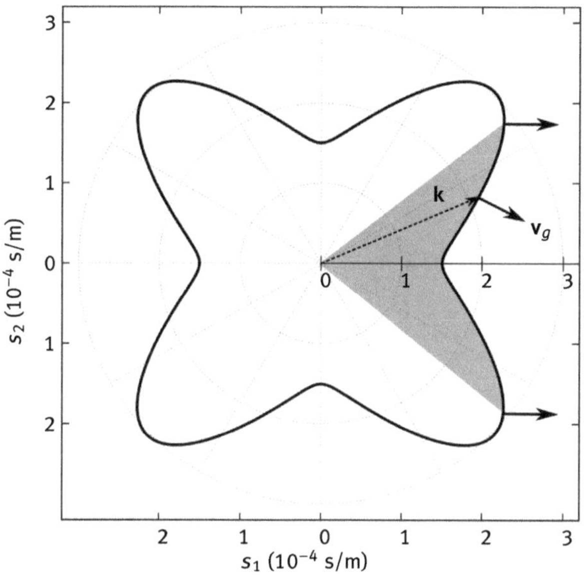

Fig. 12.16: Negative refraction and phonon focusing in rutile (TiO_2). The range of angles for which there is negative refraction with respect to an interface normal to axis x_1 is shown in gray. The construction is based on the slowness curve for quasi-shear elastic waves.

situation leads to phonon focusing in anisotropic solids as can be measured for orthorhombic and tetragonal crystals [144, 165], or in silicon [38].

Morvan et al. provided an experimental demonstration of negative refraction for quasi-shear elastic waves in a phononic crystal [112]. Their experimental arrangement is depicted in Figure 12.17. The 2D square-lattice phononic crystal of holes in aluminum decouples out-of-plane pure shear waves from in-plane waves. The latter subsystem of bands is used. For the chosen geometrical parameters, there is a quasi-shear band that is the sole propagating band in some frequency range. Equifrequency contours in this range are annular around the Γ point in a manner similar to Figure 12.14 (b) and the group velocity is always negative, leading to negative refraction. The experimental demonstration of negative refraction is performed using a prism-shaped phononic crystal and is thus rigorously valid for only one direction, but it potentially extends to all angles of incidence.

Croënne et al. later explored negative refraction for quasi-longitudinal waves in a phononic crystal [28]. The 2D hexagonal-lattice phononic crystal of steel rods in epoxy – of the solid-solid type as compared to the previous solid-hole sample – again decouples out-of-plane pure shear waves from in-plane waves. A quasi-longitudinal band is the only propagating band in a frequency range lying above the first complete band gap for in-plane waves. Probably due to the hexagonal symmetry compared to the square symmetry, equifrequency contours in this range are mostly circular around the Γ point, implying that propagation of quasi-longitudinal waves in-plane is almost

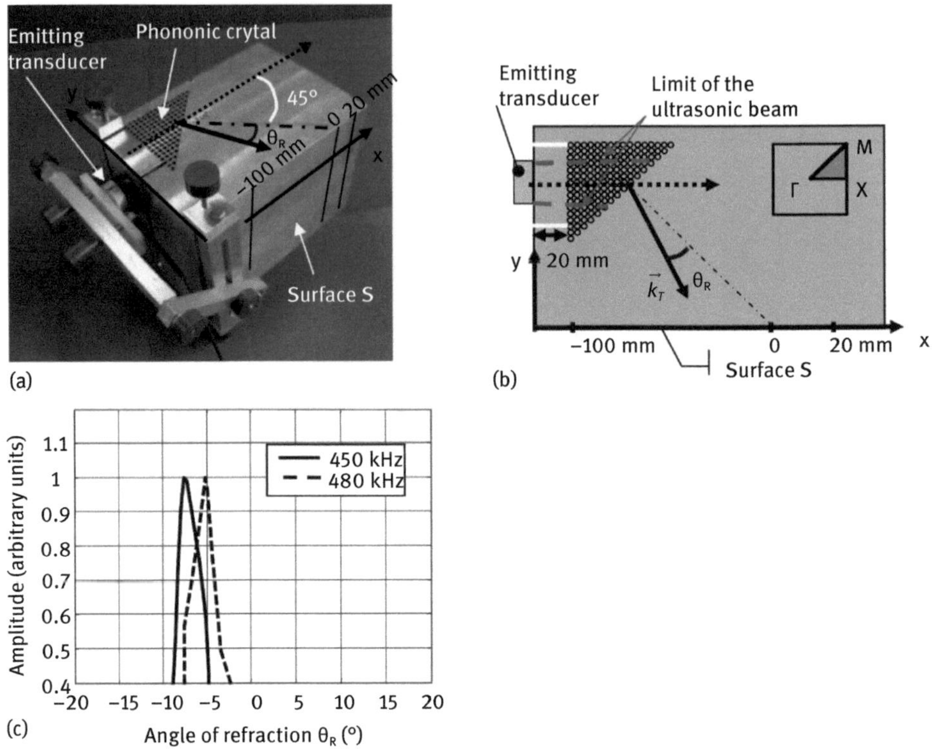

Fig. 12.17: Negative refraction for quasi-shear elastic waves in a 2D phononic crystal of holes in aluminum. The square-lattice crystal has geometrical parameters $a = 3.9$ mm and $d = 3.2$ mm. (a) A photograph of the experimental sample shows the phononic crystal machined in an aluminum block and the transducer of elastic waves placed in mechanical contact with one lateral side. (b) The top view schematic diagram explains the experimental arrangement. (c) The amplitude of the signal received on surface S is plotted in arbitrary units versus the angle of refraction for frequencies 480 and 450 kHz (after Morvan et al. [112], copyright 2010 American Institute of Physics).

isotropic, though still strongly dispersive. The experimental demonstration follows a scheme similar to the one in Figure 12.17, with a prism-shaped phononic crystal.

Negative refraction in a phononic crystal slab of holes in silicon was investigated experimentally by Pierre et al. as shown in Figure 12.18 [122]. With the pump-probe experimental set-up used to generate and detect acoustic phonons, mostly flexural waves (A_0 Lamb waves) are accessed. The experimental configuration again involves a square-lattice phononic crystal shaped in the form of a prism so that negative refraction is observed for a particular direction of propagation.

As a final example of a negative refraction experiment, let us consider the foam-like metallic structure or metal water structure examined by Hladky-Hennion et al. [50]. As depicted in Figure 12.19, this is a honeycomb-lattice phononic crystal with a very special design of the internals of the unit cell. A photograph of the actual sam-

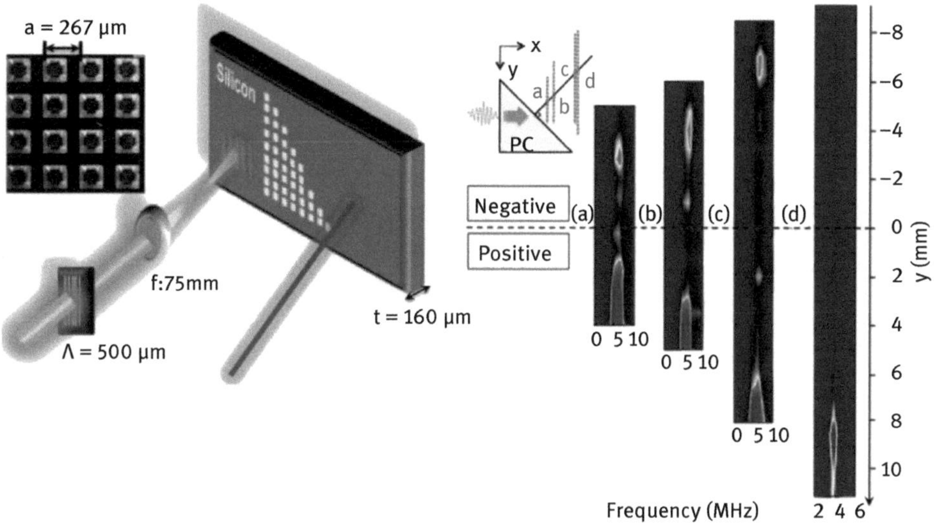

Fig. 12.18: Negative refraction for A_0 flexural Lamb waves in a phononic crystal slab of holes in silicon. Flexural waves are excited and detected in a pump-probe experimental set-up. The square-lattice phononic crystal slab has geometrical parameters $a = 267$ μm, thickness $h/a = 0.6$ and diameter of the circular holes $d/a = 0.71$. The refracted signal is presented in the frequency position domain, and is recorded along four parallel lines. The distances along the normal to the interface are (a) 2.8 mm, (b) 4.2 mm, and (c)–(d) 8.5 mm. The frequency of the acoustic pulse was centered either on (a)–(c) 5.8 MHz or on (d) 4 MHz. The inset shows a scheme of the experiment. The arrow is for the direction of propagation of the Lamb waves in the PC. Two consecutive measurements along the vertical lines are distant by 0.75 mm. (after Pierre et al. [122], copyright 2010 American Institute of Physics).

ple was presented in Figure 6.1 (b). The whole structure is made of aluminum, permeated by air, and mainly consists of masses placed at the intersections of a thin hexagonal web. The lattice constant is $\sqrt{3}a = 11.16$ mm, with a the distance between masses. The dimensions of the thin web and of the masses are chosen so that the effective density, bulk modulus, and shear modulus in the quasi-static regime are very close to those of water.[3] It is actually clear in the band structure of Figure 12.19 that the first quasi-longitudinal band has the same slope at the origin as the water line. The frequency range of operation is chosen to include only an isolated propagating quasi-longitudinal band. The equifrequency contours around the operating frequency 71.2 kHz are almost circular and parallel to the dispersion relation in water. As the group velocity is negative for any point of the equifrequency contours, the all-angle negative diffraction criterion [146] is satisfied: the moduli of the phase velocity in the

3 The quasi-static regime refers to the range of low frequencies where the phononic crystal can be homogenized.

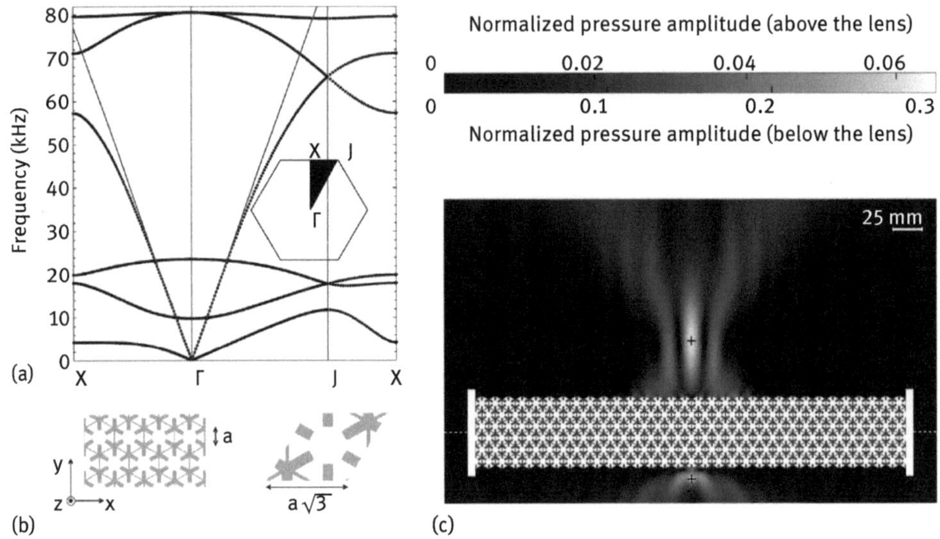

Fig. 12.19: Negative refraction of acoustic waves using a foam-like metallic structure. (a) The foam-like metallic structure follows a honeycomb lattice with additional masses on the corners. The thin solid line is for the water line. The band structure is drawn along the *Γ*-K-M path around the irreducible Brillouin zone (K and M points are denoted J and X on the figure). (b) The unit cell used for FEM calculation is indicated for reference. The gray parts are for aluminum while the white parts are hollow. (c) Normalized (source level equal to 1 arbitrary unit) pressure field maps are obtained for a focusing simulation at a frequency of 71.2 kHz. Aluminum surfaces are shown in gray. The source is located below the slab, and the focusing effect is observed above. Locations of the source and of the focus point are shown as black crosses. For clarity, different gray scales are used below and above the slab (after Hladky–Hennion et al. [50], copyright 2013 American Institute of Physics).

phononic crystal and in the surrounding water match for all angles of incidence. In optical terms, the index of refraction in the phononic crystal equals −1 with respect to water. Despite this excellent phase velocity matching, it should not be concluded that impedance matching is achieved as well. Actually this would be true for low frequencies, but the operating frequency is too far from the quasi-static regime. Figure 12.19 (c) presents a numerical simulation of focusing through the phononic crystal. The image is formed above the flat lens, demonstrating that negative refraction is active for all angles of incidence. The intensity in the image, however, is about 6% of the intensity of the source, as a consequence of poor impedance matching leading to strong Fresnel reflection at the interfaces between the phononic crystal and water.

12.7 Reflection and refraction at a crystal boundary

In this section we discuss modal conversion at the boundary between a finite-size crystal and a host homogeneous medium. Incidence is from the host medium but the derivation can be reversed by reciprocity. As we show, generalized Fresnel formulas can be obtained for reflection and transmission, providing the generation of orders of diffraction in the host medium is taken into account. The analysis follows the derivation in reference [82] and is limited to scalar waves (sonic crystal) and two-dimensional space for simplicity. Extension to the anisotropic case of phononic crystals is straightforward. The geometry of the problem was already considered in Figure 4.19. The crystal boundary is along axis x_1. It supports a periodic corrugation and thus acts as a diffraction grating with period $a_\parallel$, which can be different from the lattice constant of the crystal depending on the orientation of the boundaries of the crystal. The basic idea of the calculation is to use modal expansions inside the crystal and inside the host medium that are matched at the boundary through continuity conditions. This idea makes sense providing the functional bases used for expansion are complete. For this reason, the basis of evanescent Bloch waves described in Chapter 9 is employed for the crystal.

In the host medium, we consider an incident plane wave with angular frequency ω and wavevector $k_0(\cos\theta, \sin\theta)^T$,

$$p_{0i}(\mathbf{r}, t) = \exp(\imath(\omega t - k_0\cos(\theta)x_1 - k_0\sin(\theta)x_2)). \tag{12.16}$$

The distribution of the particle velocity component that is normal to the boundary is

$$v_{02i}(\mathbf{r}, t) = \frac{1}{Z_0} p_{0i}(\mathbf{r}, t), \tag{12.17}$$

with the directional impedance $Z_0 = Z/\cos\theta$ and $Z = \rho c$ the usual impedance. The reason for defining a directional impedance will be clearer later.

Upon reflection on the periodic boundary, multiple reflected diffraction orders are created and have the form of propagating or evanescent plane waves. The m-th reflected diffraction order is

$$p_{mr}(\mathbf{r}, t) = \exp(\imath(\omega t - k_{m1}x_1 - k_{m2}x_2)), \tag{12.18}$$

with $k_{m1} = k_0\cos\theta + 2\pi m/a_\parallel$, and

$$k_{m2} = -\sqrt{\omega^2/c^2 - k_{m1}^2} \quad \text{or} \quad k_{m2} = \imath\sqrt{|\omega^2/c^2 - k_{m1}^2|}$$

for propagating and evanescent diffracted waves, respectively. The form of k_{m1} results from conservation of the tangential component of the wavevector up to an arbitrary reciprocal lattice translation; the form of k_{m2} then results from the dispersion relation in the host medium (see Figures 9.1 and 9.2 and the discussion around them). Reflected diffraction orders in (12.17) can be cast in the form of Bloch waves by specifying their wavevector as $\mathbf{k}_m = (k_0 \cos(\theta), k_{m2})$ and the periodic wavefield as $\bar{p}_{mr}(\mathbf{r}) = \exp(-\imath 2\pi m x_1/a_\|)$. The distribution of the normal particle velocity of reflected diffraction orders further satisfies

$$v_{m2r} = \frac{k_{m2}}{\rho\omega} p_{mr} = \frac{-1}{Z_m} p_{mr}, \tag{12.19}$$

where Z_m is the (directional) mechanical impedance of the m-th diffraction order. The minus sign in the definition of the impedance Z_m accounts for propagation toward decreasing x_2.

In order to make reciprocity apparent in the final expressions, we also consider all possible incident diffraction orders. Incident diffraction orders assume exactly the same form as reflected diffraction orders, but with the reverse sign in the definition of k_{m2}, i.e.

$$k_{m2} = +\sqrt{\omega^2/c^2 - k_{m1}^2} \quad \text{or} \quad k_{m2} = -\imath\sqrt{|\omega^2/c^2 - k_{m1}^2|}$$

for propagating and evanescent diffracted waves, respectively. The distribution of the normal particle velocity of incident diffraction orders satisfies $v_{m2i} = \frac{1}{Z_m} p_{mi}$, with the plus sign accounting for propagation to increasing x_2.

In the crystal, a complete set of Bloch waves can be obtained from the complex band structure for the particular choice $\boldsymbol{\alpha} = (0, 1)^T$ and $\mathbf{k}_0 = (k_0 \cos\theta, 0)^T$ in (9.15). The Bloch wave distributions are noted $p_n(\mathbf{r}, t)$ for the pressure and $v_n(\mathbf{r}, t)$ for the normal component of the particle velocity.

Continuity equations for pressure and normal velocity are next expressed along the boundary. The continuity equations have the form

$$\sum_m i_m v_i(\mathbf{r}, t) + \sum_m r_m v_{mr}(\mathbf{r}, t) = \sum_n t_n v_n(\mathbf{r}, t), \tag{12.20}$$

$$\sum_m i_m p_i(\mathbf{r}, t) + \sum_m r_m p_{mr}(\mathbf{r}, t) = \sum_n t_n p_n(\mathbf{r}, t), \tag{12.21}$$

where the i_m are the incident amplitude, the r_m are the reflected amplitudes, and the t_n are the transmitted amplitudes. For the reflection/refraction problem, it is sufficient to select the incident amplitudes as $i_m = 1$ if $m = 0$ and $i_m = 0$ otherwise. We next multiply by

$$\exp(\imath k_0 \cos(\theta) x_1) \exp(\imath 2\pi m x_1/a_\|)$$

and integrate along the boundary, for all values of m. Because of the orthogonality of Fourier exponentials we end up with the matrix relations

$$\frac{1}{Z_m} i_m - \frac{1}{Z_m} r_m = M_v(m, n) t_n, \tag{12.22}$$

$$i_m + r_m = M_p(m, n) t_n. \tag{12.23}$$

The matrices

$$M_v(m,n) = \frac{1}{a_\parallel}\int_\sigma dx_1 \exp(\imath 2\pi m x_1/a_\parallel)\bar{v}_n(\mathbf{r}) \tag{12.24}$$

and

$$M_p(m,n) = \frac{1}{a_\parallel}\int_\sigma dx_1 \exp(\imath 2\pi m x_1/a_\parallel)\bar{p}_n(\mathbf{r}) \tag{12.25}$$

govern the modal conversion process. They can be easily evaluated from knowledge of the Bloch waves of the crystal. Solving the matrix equations (12.22) and (12.23) is straightforward and results in the following generalized version of the Fresnel formulas for refraction and reflection

$$\mathbf{t} = 2(M_p + ZM_v)^{-1}\mathbf{i}, \tag{12.26}$$

$$\mathbf{r} = (M_p - ZM_v)(M_p + ZM_v)^{-1}\mathbf{i}, \tag{12.27}$$

with the diagonal impedance matrix $Z(m,n) = Z_m\delta(m-n)$. In the above equations, vectors $\mathbf{i}$, $\mathbf{r}$ and $\mathbf{t}$ gather the amplitudes of the incident, reflected and transmitted waves, respectively.

The traditional Fresnel formulas of Chapter 3 must obviously be recovered when the phononic crystal is replaced by a homogeneous medium. In this case, there is only one reflected and one transmitted wave, and the matrices above simplify to scalars. Let us consider medium A as the incidence medium and medium B as the transmission medium. We have $Z_0^{-1} = Z_A^{-1}\cos\theta_A$, with $Z_A = \rho_A c_A$ and θ_A the angle of incidence. For the transmitted wave, $M_v/M_p = Z_B^{-1}\cos\theta_B$ or $-\imath|k_y|/(\rho_B\omega)$ if it is propagating or evanescent, respectively. It follows that $r = \frac{Z_B\cos\theta_A - Z_A\cos\theta_B}{Z_B\cos\theta_A + Z_A\cos\theta_B}$ if the transmitted wave is propagating and $|r| = 1$ if it is evanescent. This is of course the same result as (3.39).

With the generalized Fresnel formulas, it is easy to show that antisymmetric Bloch waves are not excited in the case of normal incidence on a crystal that is symmetric with respect to the direction of incidence, a result that we used explicitly in the discussion of deaf bands in Chapter 4. To that end, we first rearrange the order of Bloch waves of the PC so as to separate symmetric (S) and antisymmetric (AS) Bloch waves

$$\mathbf{t} = \begin{pmatrix} \mathbf{t}_S \\ \mathbf{t}_{AS} \end{pmatrix}. \tag{12.28}$$

Because of normal incidence, we can also recast the reflected diffraction orders into Bloch waves in the incident medium with $\mathbf{k}_m = (k_0, k_{my})$, with $\bar{p}_{mr}(\mathbf{r}) = \cos(2\pi m x/b)$ for $m \geq 0$ (S) or $\sin(2\pi m x/b)$ for $m > 0$ (AS). To achieve this reordering, it is sufficient to recombine the original diffraction orders m and $-m$. Matrices M_p and M_v then become block diagonal, e.g.

$$M_p = \begin{pmatrix} M_{p,\mathrm{S}} & 0 \\ 0 & M_{p,\mathrm{AS}} \end{pmatrix}. \tag{12.29}$$

Since the incident plane wave is itself symmetric, we conclude that $\mathbf{t}_{AS} = 0$ and $\mathbf{r}_{AS} = 0$, i.e. antisymmetric (deaf) Bloch waves are not transmitted and antisymmetric reflected diffraction orders are not excited.

12.8 Sonic crystal as a diffraction grating

From the results of the previous section, it appears that seen from the outside a finite-size sonic or phononic crystal will cause diffraction of incident plane waves and that the diffraction orders are governed by the equivalent of the grating law of optics (conservation of the tangential component of the wavevector up to an arbitrary reciprocal lattice translation). It can be demonstrated, in particular, that the angles of propagation of diffraction orders in the surrounding medium are governed by the periodic boundaries between the incident medium and the crystal [108]. Diffraction efficiencies, however, are governed by a rich interplay with the Bloch waves of the crystal, as summarized by (12.26) and (12.27). The diffraction efficiency of a sonic crystal grating can further be optimized by combining blazing with a phononic band gap [110]. The principle of blazing is depicted in Figure 12.20 (a). Diffraction efficiency for a particular order of diffraction depends on the internal details of the unit cell of the grating. If there is an internal facet in the unit cell such that the incident plane wave is mirrored on reflection into the order of diffraction, then it is said that the grating is blazed. In the particular case of the figure, the sawtooth profile is such that there are two facets oriented at an angle $+60°$ and $-30°$ with respect to the normal to the boundary. For every frequency and angle of incidence such that an order of diffraction happens to be the mirror image of the incident plane wave, it can be expected that the diffraction efficiency will be reinforced.

The square-lattice sonic crystal of steel rods in water of Figure 12.20 (b) is designed to be blazed in the Littrow configuration for order of diffraction -1. The latter means that the diffracted wave is exactly back propagating in the direction of incidence. With square cross section rods, there are two possible blazing angles, $+60°$ and $-30°$. In the considered frequency range, there are only two propagating orders, 0 and -1; all other orders of diffraction are evanescent. The design of the unit cell is adjusted so that specular reflection (or diffraction to order zero) is minimized. This can be achieved by tuning the filling fraction. It remains that transmitted orders of diffraction 0 and -1 exist. Their diffraction efficiency can be minimized if a complete band gap exists. The numerical simulations in Figure 12.20 (b), (c) illustrate these design principles.

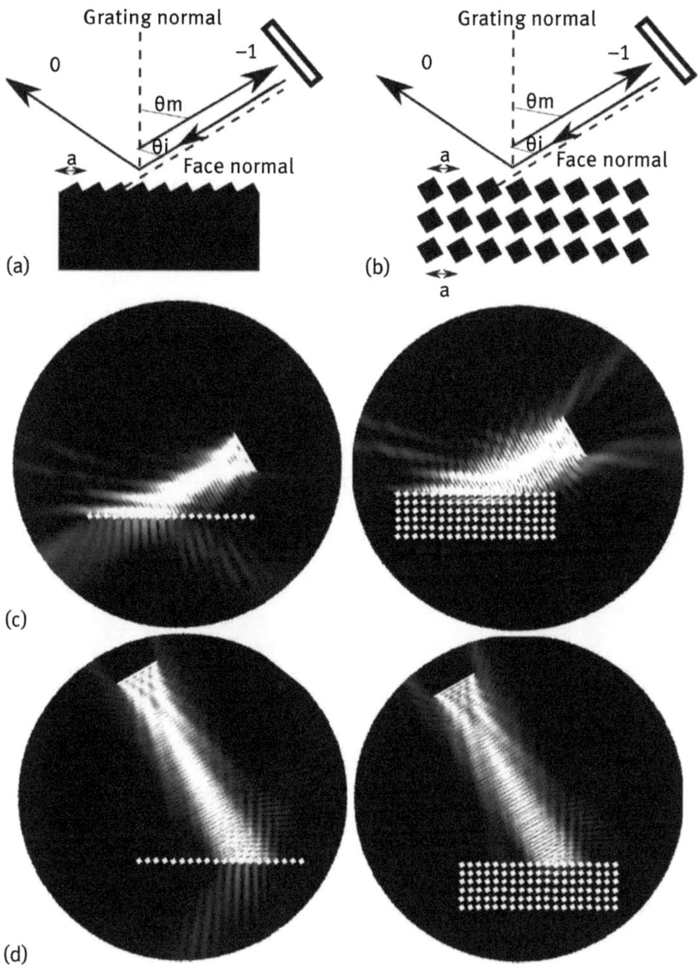

Fig. 12.20: Blazed sonic crystal diffraction grating. (a) Littrow condition for a diffraction grating supporting a surface corrugation with the shape of a sawtooth. An operating frequency can be found so that incidence is perpendicular to the elementary facet and at the same time diffraction order −1 is back propagating. (b) A blazed sonic crystal diffraction grating formed by a square-lattice array of square cross section steel rods in water. The lattice constant $a = 2.6$ mm and the rod width $d = 1.5$ mm. (c) Finite element simulation of intensity distributions for a (left) 1-layer and (right) a 5-layer sonic crystal at frequency 0.329 MHz and incidence angle 60°. (d) Same for frequency 0.569 MHz and incidence angle −30° (after Moiseyenko et al. [110], copyright 2013 American Institute of Physics).

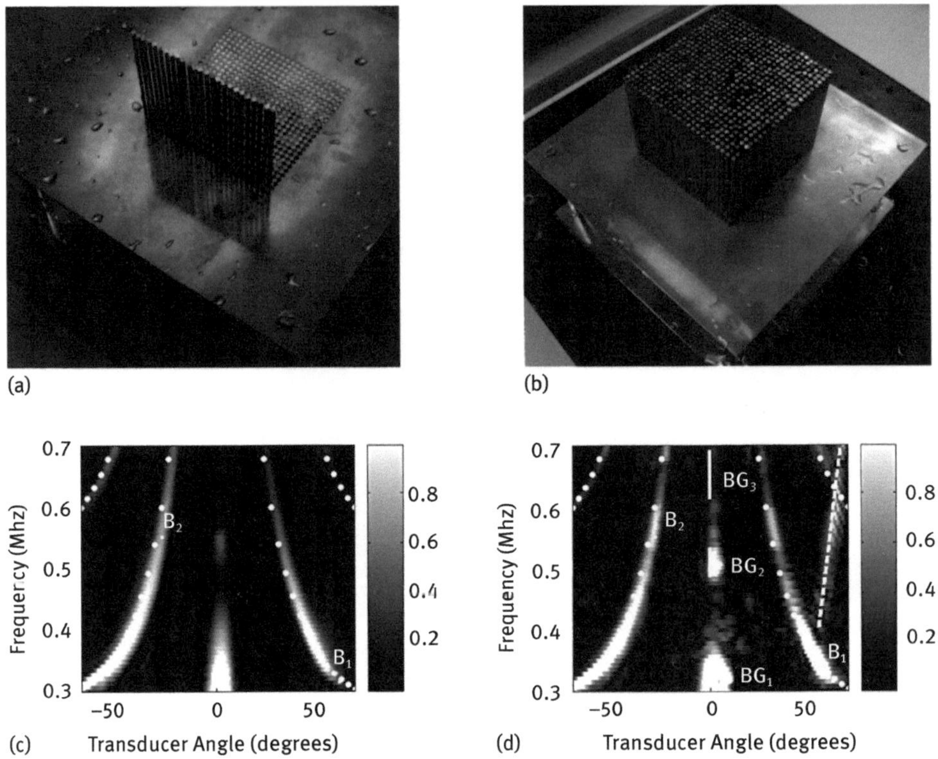

Fig. 12.21: Blazed sonic crystal diffraction grating. The square cross section square-lattice sonic crystal design described in Figure 12.20 is implemented as (a) a 1 × 25 grating sample and (b) a 25 × 25 sonic crystal sample. The samples are immersed in water for measurement. A single transducer is used to generate and detect acoustic waves. The transducer angle thus represents both the angle of incidence and the angle of reception. (c) For the grating sample, the measured intensity as a function of transducer angle and frequency concentrates along lines given by the grating law. (d) For the sonic crystal, the intensity is further enhanced around the blazing angles marked as B1 and B2. For normal incidence, an enhanced specular reflection is observed with band gaps BG1 and BG2 (after Moiseyenko et al. [110], copyright 2013 American Institute of Physics).

Fig. 12.22: Ultrasound tunneling through an FCC sonic crystal of tungsten carbide beads in water. ► The sonic crystal has a complete band gap extending from 0.8 to 1.2 MHz, in correspondence with the band structure in Figure 4.24. The close-packed beads have a diameter $d = 0.8$ mm. (a) A short ultrasonic pulse transmitted through a six-layer sample (right) is compared with the input pulse (left). (b) The Fourier spectra of the two pulses in (a) indicate that most of the signal received originates from the part of the initial pulse spectrum outside the band gap range. (c) The theoretical and experimental frequency dependencies of the amplitude transmission coefficient for a six-layer and a twelve-layer sample are compared. The frequency dependence of the experimental group velocity (actually $L/t_g(\omega)$; see text for a definition) is shown for a five-layer sample. (after Yang et al. [172], copyright 2002 American Physical Society).

Experimental results are shown in Figure 12.21 for both a single layer grating and a 25-layer sonic crystal. There is clear evidence that the grating law applies to determine the direction of waves diffracted in water, irrespective of the internal details of the crystal, as intensity concentrates along lines given by the Littrow condition

$$2\sin\theta = |m|\frac{2\pi c}{\omega a},$$

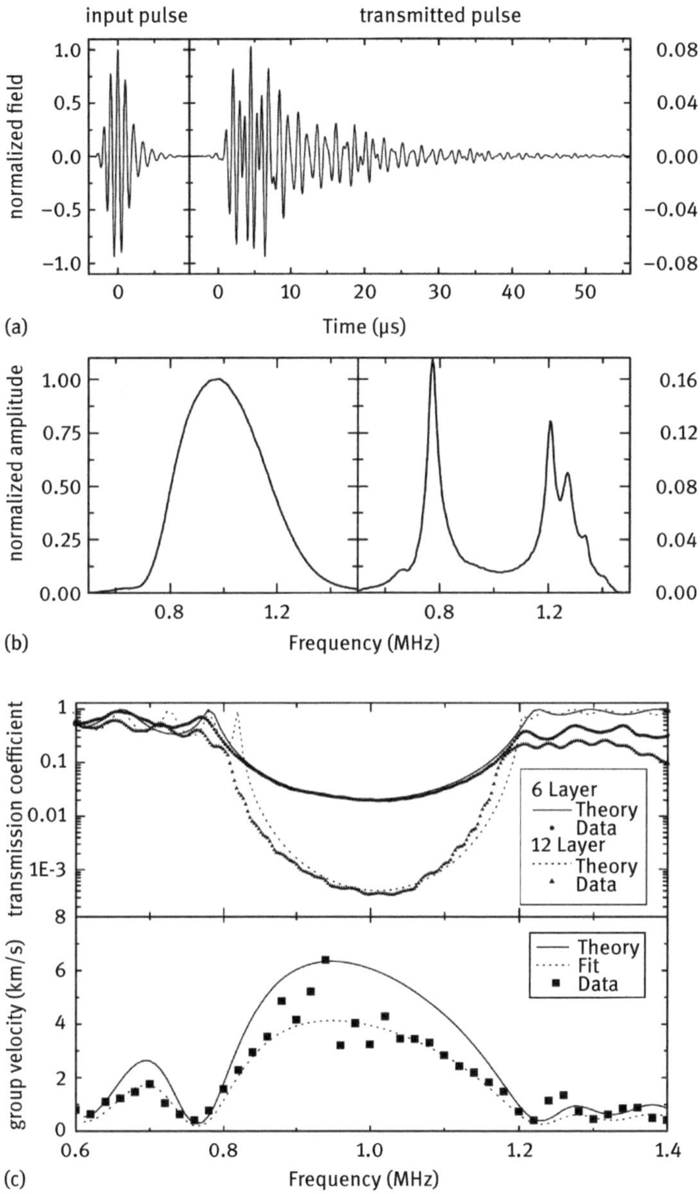

with θ the transducer angle, c the celerity of sound waves in water, and m the order of diffraction. The normal incidence case ($\theta = 0$) shows that enhancement of the redirected energy is obtained within frequency band gaps BG1 and BG2. Experimental data also confirm that the one layer grating is already quite efficient, as the numerical simulations of Figure 12.20 already suggested. Surprisingly, the diffraction efficiency for $m = -1$ approaches 100 % for angles of incidence between 40° and 70°, and also around 50°, even though blazing was not especially expected to operate under such conditions. The diffraction efficiency at blazing points B1 and B2 is around 60 %. It is definitely enhanced by using more layers in the sonic crystal. In the absence of a phononic band gap, at point B2, the diffraction efficiency is experimentally increased to 77 %. Within the complete phononic band gap BG1, at point B1, the diffraction efficiency is increased up to 98 %. In this case, the 25-layer blazed sonic crystal grating achieves an enhancement factor of 1.59 at blazing point B1.

12.9 Temporal dispersion and tunneling

Transmission measurements of waves going through a phononic or sonic crystal have been used several times in this book to show experimental evidence of frequency band gaps. In most reported experiments, only the modulus of the spectrum is shown. There is information attached to the phase on transmission, however, that can additionally be obtained. Following Yang et al. [172], let us write the spectral transmission as

$$t(L, \omega) = A(L, \omega) \exp(\imath\phi(L, \omega)). \tag{12.30}$$

A and ϕ are the modulus and phase of the complex spectral transmission, and L is the length of the crystal. By definition, the derivative of the spectral phase with angular frequency is the group delay

$$t_g(L, \omega) = \frac{\partial \phi}{\partial \omega}. \tag{12.31}$$

For a propagating wave with real wavenumber $k(\omega)$, it is a common place that $\phi(L, \omega) = k(\omega)L$ and then that

$$t_g(L, \omega) = L\frac{\partial k}{\partial \omega} = L(v_g)^{-1}. \tag{12.32}$$

For a propagating Bloch wave, this derivation makes perfect sense. Inside a phononic or sonic band gap, however, there are no propagating Bloch waves and (12.32) should not be employed: the complex-imaginary wavenumber dependence enters the amplitude $A(L, \omega)$ instead of the phase $\phi(L, \omega)$. Equation (12.31) still remains valid and provides a classical equivalent of the tunneling time through a potential barrier in quantum mechanics. This classical tunneling time is actually independent of the thickness of the barrier, L, as this thickness increases, which is known as the Hartman effect [164]. Measurements of the group delay on transmission through one-dimensional photonic crystals have been provided [140, 141] and its exact value for quarterwave dielectric mirrors has been found analytically as a function of the indices of refraction [76]. Yang et al. reported experimentally the measurement of the group delay on transmission through an FCC sonic crystal of tungsten carbide beads in water [172], see Figure 12.22. The experimentally determined value is about 1 μs in their case, independent of the number of crystal layers.

13 Conclusion

Looking back at the contents of this book, we can say that the understanding of the physics of waves in sonic and phononic crystals has reached a form of maturity. The differences and the common properties of acoustic waves in sonic crystals and of elastic waves in phononic crystals are firmly established. Bloch's theorem and the concept of Bloch waves allowed us to reach deep inside the details of wave propagation, incorporating important concepts such as anisotropy in relation to crystal and lattice symmetries, dispersion in relation to periodicity, the formation of evanescent waves whenever periodicity is broken. On the numerical side of things, powerful methods have been developed to explore wave propagation in artificial crystals. We have especially insisted on finite element modeling as a generic approach that is able to cope with all types of waves, materials and geometries, and we hope the reader can find in the previous pages enough indications to conduct numerical experiments.

Among the collective achievements of phononic and sonic crystal researchers, one can cite the understanding and the control of the origins of frequency band gaps, either through Bragg scattering or local resonances of the unit cell; the choice of lattices, of material couples and of internal structure of the primitive cell; the control of spatial dispersion and anisotropy to achieve refractive functions; the use of defects to create various types of resonators, cavities, and waveguides.

There are, however, improvements that remain to be achieved, especially for actual applications of phononic crystals. For medical imaging applications, bandwidth is usually an important requirement, especially if short acoustic pulses are used. In this respect, strongly dispersive periodic structures are in principle more suited to medium to narrow bandwidths, or even monochromatic experiments. In the GHz frequency range, for wireless telecommunications for instance, phononic crystals certainly have the potential to lead to compact filter designs, e.g. to replace existing resonators and Bragg mirrors. One difficulty, however, is technological and resides in the robustness of designs to periodicity imperfections and in the fabrication tolerance of holes, inclusions, or pillars. Any error at fabrication time and because material properties are not known with sufficient accuracy will result in a deviation from the expected response and can hardly be adjusted a posteriori. A related challenge is the need for tunability or reconfigurability of the structures. With this remark, we mean that the functionality of a phononic crystal structure is usually designed by a spatial distribution of materials or a geometrical structure, and is hence fixed by design. Finally, there is a general lack of adequate phonon sources that would match the possibilities of artificial crystals. Indeed, most transducers produce or detect plane or cylindrical waves that do not properly match the Bloch waves of crystals.

Adopting a broader perspective, phononic crystals are one part of a wider emerging force in applied physics: phononics. The word "phononics" is built on the model of electronics and photonics that have preceded it. Phononics can be understood as the analysis and the manipulation of phonons at the nanoscale, microscale or larger scales. Looking closer, phononics encompasses a wide range of disciplines that are usually labeled in terms of the host medium for wave propagation, including phononic and sonic crystals, acoustic metamaterials, phononic superlattices, and periodic structures in mechanical engineering. Applications range from vibration isolation, through frequency sensing in radio frequency communications, to nanoscale thermal transport control in semiconductors. With the recent advent of metamaterials, novel applications are appearing, such as acoustic cloaking and superlensing [26]. The interaction of light and sound in tiny resonators and cavities, i.e. optomechanics [6] and in simultaneous photonic and phononic crystals [93] is currently attracting a lot of attention.

At the very small scale, nanophononics is exploring the fundamental quantum-limited properties of mechanical objects. One general goal is to bring a nanoscale object to its fundamental quantum state by cooling it, i.e. quenching its elementary thermal vibrations, or phonons. The connection with phononic crystals could be the achievement of molecular phononic crystals whose unit cell is composed of a small number of atoms or molecules. Achieving such a program poses formidable challenges in terms of fabrication, but also for the transition from an atomistic to a mesoscopic description of matter. Molecular dynamical or ab initio calculations have to be combined with the continuous matter description we have relied upon in this book. At about the same or slightly larger scale, thermal phonon control and thermoelectricity could benefit from phononic band gaps or from phonon dispersion engineering [94].

At the opposite end of the scale, the control of long wavelengths by arrays of small resonators is finding potential applications. For instance, thin sound screens for audible sound can be implemented with resonances of small holes or slits in a solid plate [33, 36]. The coupling of acoustic waves with elastic waves is an enabling design feature in such structures. Protection against earthquakes is being proposed using acoustic metamaterial cloaks that could redirect the flow of energy so that it avoids a particular building [18]. Acoustic metamaterials are generally conceived as artificial media whose unit cell is very small compared to the operating wavelength. Introducing local resonances within the unit cell, the range of homogenized effective material properties that can be obtained exceeds that of natural microscopically homogeneous materials. In the vicinity of resonances, negative effective material constants in particular can be obtained. Transformation acoustics can then supply the necessary mathematical tools to convert a chosen wave functionality into a spatial distribution of metamaterial [23].

The interaction of photons and phonons in nanoscale structures that are able to store and guide them simultaneously has and will benefit from advances in phononic crystals. The combination of phononic and photonic crystals is now termed phoxonic crystals [118, 133]. Beyond traditional acousto-optics, which describes the interactions of plane elastic and optical waves, phoxonic crystals can simultaneously confine light and sound in a tiny volume measured by their common wavelength scale. In addition to the bulk photoelastic effect, acousto-optical surface couplings also result from the motion of the boundaries of a waveguide or resonator [121]. There are strong relations with the fields of Brillouin scattering in optical and photonic crystal fibers [29], but also with optomechanical crystals [134].

Bibliography

[1] Y. Achaoui, A. Khelif, S. Benchabane and V. Laude, Polarization state and level repulsion in two-dimensional phononic crystals and waveguides in the presence of material anisotropy, *J. Phys. D: Appl. Phys.* **43** (2010), 185401.

[2] Y. Achaoui, A. Khelif, S. Benchabane, L. Robert and V. Laude, Experimental observation of locally-resonant and Bragg band gaps for surface guided waves in a phononic crystal of pillars, *Phys. Rev. B* **83** (2011), 104201.

[3] Y. Achaoui, V. Laude, S. Benchabane and A. Khelif, Local resonances in phononic crystals and in random arrangements of pillars on a surface, *J. Appl. Phys.* **114** (2013), 104503.

[4] M. Addouche, M. A. Al-Lethawe, A. Elayouch and A. Khelif, Subwavelength waveguiding of surface phonons in pillars-based phononic crystal, *AIP Advances* **4** (2014), 124303.

[5] X. Ao and C. T. Chan, Complex band structures and effective medium descriptions of periodic acoustic composite systems, *Phys. Rev. B* **80** (2009), 235118.

[6] M. Aspelmeyer, T. J. Kippenberg and F. Marquardt, Cavity optomechanics, *Reviews of Modern Physics* **86** (2014), 1391.

[7] M. B. Assouar and M. Oudich, Dispersion curves of surface acoustic waves in a two-dimensional phononic crystal, *Appl. Phys. Lett.* **99** (2011), 123505.

[8] B. A. Auld, *Acoustic Fields and Waves in Solids*, Wiley, New-York, 1973.

[9] I. L. Bajak, A. McNab, J. Richter and C. D. W. Wilkinson, Attenuation of acoustic waves in lithium niobate, *J. Acoust. Soc. Am.* **69** (1981), 689–695.

[10] S. Benchabane, A. Khelif, A. Choujaa, B. Djafari-Rouhani and V. Laude, Interaction of waveguide and localized modes in a phononic crystal, *Electron. Lett.* **71** (2005), 570–575.

[11] S. Benchabane, L. Robert, J.-Y. Rauch, A. Khelif and V. Laude, Evidence for complete surface wave band gaps in a piezoelectric phononic crystal, *Phys. Rev. E* **73** (2006), 065601(R).

[12] S. Benchabane, O. Gaiffe, G. Ulliac, R. Salut, Y. Achaoui and V. Laude, Observation of surface-guided waves in holey hypersonic phononic crystal, *Appl. Phys. Lett.* **98** (2011), 171908.

[13] J.-P. Berenger, A perfectly matched layer for the absorption of electromagnetic waves, *J. Comput. Phys.* **114** (1994), 185.

[14] M. A. Biot, General theorems on the equivalence of group velocity and energy transport, *Phys. Rev.* **105** (1957), 1129.

[15] B. Bonello, C. Charles and F. Ganot, Lamb waves in plates covered by a two-dimensional phononic film, *Appl. Phys. Lett.* **90** (2007), 021909.

[16] M. Born and E. Wolf, *Principles of optics: electromagnetic theory of propagation, interference and diffraction of light*, CUP Archive, 1999.

[17] L. M. Brekhovskikh, *Waves in Layered Media*, Academic, New York, 1980.

[18] S. Brûlé, E. H. Javelaud, S. Enoch and S. Guenneau, Experiments on seismic metamaterials: Molding surface waves, *Phys. Rev. Lett.* **112** (2014), 133901.

[19] T. Brunet, J. Vasseur, B. Bonello, B. Djafari-Rouhani and A.-C. Hladky-Hennion, Lamb waves in phononic crystal slabs with square or rectangular symmetries, *J. Appl. Phys.* **104** (2008), 043506.

[20] D. Caballero, J. Sánchez-Dehesa, C. Rubio, R. Martínez-Sala, J. V. Sánchez-Perez, F. Meseguer and J. Llinares, Large two-dimensional sonic band gaps, *Phys. Rev. E* **60** (1999), R6316–R6319.

[21] F. Cervera, L. Sanchis, J. V. Sánchez-Pérez, R. Martínez-Sala, C. Rubio and F. Meseguer, Refractive acoustic devices for airborne sound, *Phys. Rev. Lett.* **88** (2002), 023902.

[22] C. Charles, B. Bonello and F. Ganot, Propagation of guided elastic waves in 2D phononic crystals, *Ultrason.* **44** (2006), e1209.

[23] H. Chen and C. T. Chan, Acoustic cloaking and transformation acoustics, *J. Phys. D: Appl. Phys.* **43** (2010), 113001.

[24] L.-S. Chen, C.-H. Kuo and Z. Ye, Acoustic imaging and collimating by slabs of sonic crystals made from arrays of rigid cylinders in air, *Appl. Phys. Lett.* **85** (2004), 1072–1074.

[25] J. Christensen, L. Martin-Moreno and F. J. Garcia-Vidal, Theory of resonant acoustic transmission through subwavelength apertures, *Phys. Rev. Lett.* **101** (2008), 014301.

[26] R. V. Craster and S. Guenneau, *Acoustic Metamaterials: Negative Refraction, Imaging, Lensing and Cloaking*, 166, Springer Science & Business Media, 2012.

[27] C. Croënne, E. J. S. Lee, H. Hu and J. H. Page, Band gaps in phononic crystals: Generation mechanisms and interaction effects, *AIP Advances* **1** (2011), 041401.

[28] C. Croënne, E. D. Manga, B. Morvan, A. Tinel, B. Dubus, J. Vasseur and A.-C. Hladky-Hennion, Negative refraction of longitudinal waves in a two-dimensional solid-solid phononic crystal, *Phys. Rev. B* **83** (2011), 054301.

[29] P. Dainese, P. S. Russell, N. Joly, J. C. Knight, G. S. Wiederhecker, H. L. Fragnito, V. Laude and A. Khelif, Stimulated Brillouin scattering from multi-GHz-guided acoustic phonons in nanostructured photonic crystal fibres, *Nat. Phys.* **2** (2006), 388–392.

[30] M. B. Dühring, V. Laude and A. Khelif, Energy storage and dispersion of surface acoustic waves trapped in a periodic array of mechanical resonators, *J. Appl. Phys.* **105** (2009), 093504.

[31] E. H. El Boudouti, B. Djafari-Rouhani, E. M. Khourdifi and L. Dobrzynski, Surface and interface elastic waves in supperlattices: transverse localized and resonant modes, *Phys. Rev. B* **48** (1993), 10987–10997.

[32] E. H. El Boudouti, B. Djafari-Rouhani and A. Nougaoui, Surface and interface elastic waves in supperlattices: sagittal localized and resonant modes, *Phys. Rev. B* **51** (1995), 13801–13804.

[33] A. Elayouch, M. Addouche, E. Herth and A. Khelif, Experimental evidence of ultrasonic opacity using the coupling of resonant cavities in a phononic membrane, *Appl. Phys. Lett.* **103** (2013), 083504.

[34] J. M. Escalante, A. Martínez and V. Laude, Dispersion relation of coupled-resonator acoustic waveguides formed by defect cavities in a phononic crystal, *J. Phys. D: Appl. Phys.* **46** (2013), 475301.

[35] V. Espinosa, V. J. Sánchez-Morcillo, K. Staliunas, I. Pérez-Arjona and J. Redondo, Subdiffractive propagation of ultrasound in sonic crystals, *Phys. Rev. B* **76** (2007), 140302.

[36] H. Estrada, J. M. Bravo and F. Meseguer, High sound screening in low impedance slit arrays, *New J. Phys.* **13** (2011), 043009.

[37] H. Estrada, P. Candelas, A. Uris, F. Belmar, F. J. Garcia de Abajo and F. Meseguer, Extraordinary sound screening in perforated plates, *Phys. Rev. Lett.* **101** (2008), 084302.

[38] A. G. Every, W. Sachse, K. Y. Kim and M. O. Thompson, Phonon focusing and mode-conversion effects in silicon at ultrasonic frequencies, *Phys. Rev. Lett.* **65** (1990), 1446.

[39] N. Fang, D. Xi, J. Xu, M. Ambati, W. Srituravanich, C. Sun and X. Zhang, Ultrasonic metamaterials with nagative modulus, *Nat. Mater.* **5** (2006), 452.

[40] L. Feng, X.-P. Liu, M.-H. Lu, Y.-B. Chen, Y.-F. Chen, Y.-W. Mao, J. Zi, Y.-Y. Zhu, S.-N. Zhu and N.-B. Ming, Refraction control of acoustic waves in a square-rod-constructed tunable sonic crystal, *Phys. Rev. B* **73** (2006), 193101.

[41] C. Geuzaine and J.-F. Remacle, Gmsh: a three-dimensional finite element mesh generator with built-in pre- and post-processing facilities, *International Journal for Numerical Methods in Engineering* **79** (2009), 1309–1331.

[42] C. Goffaux, J. Sánchez-Dehesa, A. Levy Yeyati, A. Khelif, P. Lambin, J. O. Vasseur and B. Djafari-Rouhani, Evidence of Fano-Like Interference Phenomena in Locally Resonant Materials, *Phys. Rev. Lett.* **88** (2002), 225502.

[43] C. Goffaux, J. Sánchez-Dehesa and P. Lambin, Comparison of the sound attenuation efficiency of locally resonant materials and elastic band-gap structures, *Phys. Rev. B* **70** (2004), 184302.
[44] M. Gorisse, S. Benchabane, G. Teissier, C. Billard, A. Reinhardt, V. Laude, E. Defaÿ and M. Aïd, Observation of band gaps in the gigahertz range and deaf bands in a hypersonic aluminum nitride phononic crystal slab, *Appl. Phys. Lett.* **98** (2011), 234103.
[45] A. Håkansson, J. Sánchez-Dehesa and L. Sanchis, Acoustic lens design by genetic algorithms, *Phys. Rev. B* **70** (2004), 214302.
[46] A. Håkansson, F. Cervera and J. Sánchez-Dehesa, Sound focusing by flat acoustic lenses without negative refraction, *Appl. Phys. Lett.* **86** (2005), 054102.
[47] Z. He, F. Cai, Y. Ding and Z. Liu, Subwavelength imaging of acoustic waves by a canalization mechanism in a two-dimensional phononic crystal, *Appl. Phys. Lett.* **93** (2008), 233503.
[48] F. Hecht, New development in freefem++, *Journal of Numerical Mathematics* **20** (2012), 159–344.
[49] B. G. Helme and P. J. King, The phonon viscosity tensor of Si, Ge, GaAs, and InSb, *Physica Status Solidi (A)* **45** (1978), K33–K37.
[50] A.-C. Hladky-Hennion, J. O. Vasseur, G. Haw, C. Croënne, L. Haumesser and A. N. Norris, Negative refraction of acoustic waves using a foam-like metallic structure, *Appl. Phys. Lett.* **102** (2013), 144103.
[51] K. M. Ho, C. K. Cheng, Z. Yang, X. X. Zhang and P. Sheng, Broadband locally resonant sonic shields, *Appl. Phys. Lett.* **83** (2003), 5566–5568.
[52] Z. Hou and B. M. Assouar, Modeling of Lamb wave propagation in plate with two-dimensional phononic crystal layer coated on uniform substrate using plane-wave-expansion method, *Phys. Lett. A* **372** (2008), 2091.
[53] F.-L. Hsiao, A. Khelif, H. Moubchir, A. Choujaa, C.-C. Chen and V. Laude, Complete band gaps and deaf bands of triangular and honeycomb water-steel phononic crystals, *J. Appl. Phys.* **101** (2007), 044903.
[54] F.-L. Hsiao, A. Khelif, H. Moubchir, A. Choujaa, C.-C. Chen and V. Laude, Waveguiding inside the complete band gap of a phononic crystal slab, *Phys. Rev. E* **76** (2007), 056601.
[55] C.-Y. Huang, J.-H. Sun and T.-T. Wu, A two-port ZnO/silicon Lamb wave resonator using phononic crystals, *Appl. Phys. Lett.* **97** (2010), 031913.
[56] M. I. Hussein, Theory of damped Bloch waves in elastic media, *Phys. Rev. B* **80** (2009), 212301.
[57] IEEE Ultrasonics, Ferroelectrics, and Frequency Control Society, *Publication and proposed revision of ANSI/IEEE standard 176-1987 "ANSI/IEEE standard on piezoelectricity"*, IEEE Trans. Ultrason., Ferroelec., Freq. Control, vol. 43, no. 5, p. 717, 1996.
[58] K. Imamura and S. Tamura, Negative refraction of phonons and acoustic lensing effect of a crystalline slab, *Phys. Rev. B* **70** (2004), 174308.
[59] K. A. Ingebrigtsen, Surface Waves in Piezoelectrics, *J. Appl. Phys.* **40** (1969), 2681–2687.
[60] M. Kafesaki, M. M. Sigalas and N. Garcia, Frequency modulation in the transmittivity of wave guides in elastic-wave band-gap material, *Phys. Rev. Lett.* **85** (2000), 4044–4047.
[61] M. Ke, Z. Liu, P. Pang, W. Wang, Z. Cheng, J. Shi, X. Zhao and W. Wen, Highly directional acoustic wave radiation based on asymmetrical two-dimensional phononic crystal resonant cavity, *Appl. Phys. Lett.* **88** (2006), 263505.
[62] A. Khelif, Y. Achaoui, S. Benchabane, V. Laude and B. Aoubiza, Locally resonant surface acoustic wave band gaps in a two-dimensional phononic crystal of pillars on a surface, *Phys. Rev. B* **81** (2010), 214303.
[63] A. Khelif, B. Aoubiza, S. Mohammadi, A. Adibi and V. Laude, Complete band gaps in two-dimensional phononic crystal slabs, *Phys. Rev. E* **74** (2006), 046610.

[64] A. Khelif, A. Choujaa, S. Benchabane, B. Djafari-Rouhani and V. Laude, Guiding and bending of acoustic waves in highly confined phononic crystal waveguides, *Appl. Phys. Lett.* **84** (2004), 4400–4002.

[65] A. Khelif, A. Choujaa, B. Djafari-Rouhani, M. Wilm, S. Ballandras and V. Laude, Trapping and guiding of acoustic waves by defect modes in a full-band-gap ultrasonic crystal, *Phys. Rev. B* **68** (2003), 214301.

[66] A. Khelif, B. Djafari-Rouhani, J.-O. Vasseur and P. A. Deymier, Transmission and dispersion relations of perfect and defect-containing waveguide structures in phononic band gap material, *Phys. Rev. B* **68** (2003), 024302.

[67] A. Khelif, M. Wilm, V. Laude, S. Ballandras and B. Djafari-Rouhani, Guided elastic waves along a rod-defect of a two-dimensional phononic crystal, *Phys. Rev. E* **69** (2004), 067601.

[68] A. Khelif, F.-L. Hsiao, A. Choujaa, S. Benchabane and V. Laude, Octave omnidirectional band gap in a three-dimensional phononic crystal, *IEEE Trans. Ultrason., Ferroelec., Freq. Control* **57** (2010), 1621.

[69] A. Khelif, S. Mohammadi, A. A. Eftekhar, A. Adibi and B. Aoubiza, Acoustic confinement and waveguiding with a line-defect structure in phononic crystal slabs, *J. Appl. Phys.* **108** (2010), 084515.

[70] K. Kokkonen, M. Kaivola, S. Benchabane, A. Khelif and V. Laude, Scattering of surface acoustic waves by a phononic crystal revealed by heterodyne interferometry, *Appl. Phys. Lett.* **91** (2007), 083517.

[71] N.-K. Kuo, C. Zuo and G. Piazza, Microscale inverse acoustic band gap structure in aluminum nitride, *Appl. Phys. Lett.* **95** (2009), 093501.

[72] M. S. Kushwaha and B. Djafari-Rouhani, Giant sonic stop bands in two-dimensional periodic system of fluids, *J. Appl. Phys.* **84** (1998), 4677–4683.

[73] M. S. Kushwaha, B. Djafari-Rouhani and L. Dobrzynski, Sound isolation from cubic arrays of air bubbles in water, *Phys. Lett. A* **248** (1998), 252–256.

[74] M. S. Kushwaha, P. Halevi, L. Dobrzynski and B. Djafari-Rouhani, Acoustic band structure of periodic elastic composites, *Phys. Rev. Lett.* **71** (1993), 2022–2025.

[75] J. Lamb and J. Richter, Anisotropic acoustic attenuation with new measurements for quartz at room temperatures, *Proceedings of the Royal Society of London. Series A. Mathematical and Physical Sciences* **293** (1966), 479–492.

[76] V. Laude and P. Tournois, Superluminal asymptotic tunneling times through one-dimensional photonic bandgaps in quarter-wave-stack dielectric mirrors, *J. Opt. Soc. Am. B* **16** (1999), 194–198.

[77] V. Laude, M. Wilm, S. Benchabane and A. Khelif, Full band gap for surface acoustic waves in a piezoelectric phononic crystal, *Phys. Rev. E* **71** (2005), 036607.

[78] V. Laude, Y. Achaoui, S. Benchabane and A. Khelif, Evanescent Bloch waves and the complex band structure of phononic crystals, *Phys. Rev. B* **80** (2009), 092301.

[79] V. Laude, B. M. Assouar and Z. Hou, Computation of plate wave dispersion diagrams and surface wave velocities without explicit boundary conditions, *IEEE Trans. Ultrason., Ferroelec., Freq. Control* **57** (2010), 1649–1654.

[80] V. Laude, J. M. Escalante and A. Martínez, Effect of loss on the dispersion relation of photonic and phononic crystals, *Phys. Rev. B* **88** (2013), 224302.

[81] V. Laude, A. Khelif, S. Benchabane, M. Wilm, T. Sylvestre, B. Kibler, A. Mussot, J. M. Dudley and H. Maillotte, Phononic band-gap guidance of acoustic modes in photonic crystal fibers, *Phys. Rev. B* **71** (2005), 045107.

[82] V. Laude, R. P. Moiseyenko, S. Benchabane and N. F. Declercq, Bloch wave deafness and modal conversion at a phononic crystal boundary, *AIP Advances* **1** (2011), 041402.

[83] F. Lemoult, M. Fink and G. Lerosey, Acoustic resonators for far-field control of sound on a subwavelength scale, *Phys. Rev. Lett.* **107** (2011), 064301.
[84] V. Leroy, A. Bretagne, M. Fink, H. Willaime, P. Tabeling and A. Tourin, Design and characterization of bubble phononic crystals, *Appl. Phys. Lett.* **95** (2009), 171904.
[85] L. Li, Use of Fourier series in the analysis of discontinuous periodic structures, *J. Opt. Soc. Am. A* **13** (1996), 1870–1876.
[86] X. Li and Z. Liu, Bending and branching of acoustic waves in two-dimensional phononic crystals with linear defects, *Phys. Lett. A* **338** (2005), 413–419.
[87] S.-C. S. Lin and T. J. Huang, Acoustic mirage in two-dimensional gradient-index phononic crystals, *J. Appl. Phys.* **106** (2009), 053529.
[88] S.-C. S. Lin, T. J. Huang, J.-H. Sun and T.-T. Wu, Gradient-index phononic crystals, *Phys. Rev. B* **79** (2009), 094302.
[89] Z. Liu, X. Zhang, Y. Mao, Y. Y. Zhu, Z. Yang, C. T. Chan and P. Sheng, Locally Resonant Sonic Materials, *Science* **289** (2000), 1734.
[90] B. Lombard and J. Piraux, Numerical treatment of two-dimensional interfaces for acoustic and elastic waves, *J. Comput. Phys.* **195** (2004), 90–116.
[91] M.-H. Lu, X.-K. Liu, L. Feng, J. Li, C.-P. Huang, Y.-F. Chen, Y.-Y. Zhu, S.-N. Zhu and N.-B. Ming, Extraordinary acoustic transmission through a 1D grating with very narrow apertures, *Phys. Rev. Lett.* **99** (2007), 174301.
[92] R. Lucklum and J. Li, Phononic crystals for liquid sensor applications, *Meas. Sci. and Technol.* **20** (2009), 124014.
[93] M. Maldovan and E. L. Thomas, Simultaneous localization of photons and phonons in two-dimensional periodic structures, *Appl. Phys. Lett.* **88** (2006), 251907.
[94] M. Maldovan, Sound and heat revolutions in phononics, *Nature* **503** (2013), 209–217.
[95] B. Manzanares-Martínez and F. Ramos-Mendieta, Surface elastic waves in solid composites of two-dimensional periodicity, *Phys. Rev. B* **68** (2003), 134303.
[96] R. Martínez-Sala, J. Sancho, J. V. Sanchez, V. Gomez, J. Llinares and F. Meseguer, Sound attenuation by sculpture, *Nature* **378** (1995), 241.
[97] B. Merheb, P. A. Deymier, K. Muralidharan, J. Bucay, M. Jain, M. Aloshyna-Lesuffleur, R. W. Greger, S. Mohanty and A. Berker, Viscoelastic effect on acoustic band gaps in polymer-fluid composites, *Modelling and Simulation in Materials Science and Engineering* **17** (2009), 075013.
[98] F. Meseguer, M. Holgado, D. Caballero, N. Benaches, J. Sánchez-Dehesa, C. López and J. Llinares, Rayleigh-wave attenuation by a semi-infinite two-dimensional elastic-band-gap crystal, *Phys. Rev. B* **59** (1999), 12169.
[99] A. E. Miroshnichenko, S. Flach and Y. S. Kivshar, Fano resonances in nanoscale structures, *Rev. Modern Phys.* **82** (2010), 2257.
[100] T. Miyashita, Full band gaps of sonic crystals made of acrylic cylinders in air – Numerical and experimental investigations, *Jpn. J. Appl. Phys.* **41** (2002), 3170–3175.
[101] T. Miyashita, Sonic crystals and sonic wave-guides, *Measurement Science and Technology* **16** (2005), R47–R63.
[102] T. Miyashita and C. Inoue, Numerical investigations of transmission and waveguide properties of sonic crystals by finite-difference time-domain method, *Jpn. J. Appl. Phys.* **40** (2001), 3488.
[103] P. Moczo, J. O. A. Robertsson and L. Eisner, The finite-difference time-domain method for modeling of seismic wave propagation, *Advances in Geophysics* **48** (2007), 421–516.
[104] S. Mohammadi, A. A. Eftekhar, W. Hunt and A. Adibi, High-Q micromechanical resonators in a two-dimensional phononic crystal slab, *Appl. Phys. Lett.* **94** (2009), 051906.

[105] S. Mohammadi, A. A. Eftekhar, A. Khelif and A. Adibi, Simultaneous two-dimensional phononic and photonic band gaps in opto-mechanical crystal slabs, *Opt. Express* **18** (2010), 9164–9172.

[106] S. Mohammadi, A. A. Eftekhar, A. Khelif, W. Hunt and A. Adibi, Evidence of large high frequency complete phononic band gaps in silicon phononic crystal plates., *Appl. Phys. Lett.* **92** (2008), 221905.

[107] S. Mohammadi, A. A. Eftekhar, A. Khelif, H. Moubchir, R. Westafer, W. Hunt and A. Adibi, Complete phononic bandgaps and bandgap maps in two-dimensional silicon phononic crystal plates., *Electron. Lett.* **43** (2007), 898.

[108] R. P. Moiseyenko, S. Herbison, N. F. Declercq and V. Laude, Phononic crystal diffraction gratings, *J. Appl. Phys.* **111** (2012), 034907.

[109] R. P. Moiseyenko and V. Laude, Material loss influence on the complex band structure and group velocity in phononic crystals, *Phys. Rev. B* **83** (2011), 064301.

[110] R. P. Moiseyenko, J. Liu, N. F. Declercq and V. Laude, Blazed phononic crystal grating, *Appl. Phys. Lett.* **102** (2013), 034108.

[111] R. P. Moiseyenko, N. F. Declercq and V. Laude, Guided wave propagation along the surface of a one-dimensional solid–fluid phononic crystal, *J. Phys. D: Appl. Phys.* **46** (2013), 365305.

[112] B. Morvan, A. Tinel, A.-C. Hladky-Hennion, J.-O. Vasseur and B. Dubus, Experimental demonstration of the negative refraction of a transverse elastic wave in a two-dimensional solid phononic crystal, *Appl. Phys. Lett.* **96** (2010), 101905.

[113] R. H. Olsson III and I. El-Kady, Microfabricated phononic crystal devices and applications, *Measurement Science and Technology* **20** (2009), 012002.

[114] P. H. Otsuka, K. Nanri, O. Matsuda, M. Tomoda, D. M. Profunser, I. A. Veres, S. Danworaphong, A. Khelif, S. Benchabane, V. Laude and O. B. Wright, Broadband evolution of phononic-crystal-waveguide eigenstates in real-and k-spaces, *Scientific reports* **3** (2013).

[115] M. Oudich, M. B. Assouar and Z. Hou, Propagation of acoustic waves and waveguiding in a two-dimensional locally resonant phononic crystal plate, *Appl. Phys. Lett.* **97** (2010), 193503.

[116] M. Oudich, M. Senesi, M. B. Assouar, M. Ruzenne, J.-H. Sun, B. Vincent, Z. Hou and T.-T. Wu, Experimental evidence of locally resonant sonic band gap in two-dimensional phononic stubbed plates, *Phys. Rev. B* **84** (2011), 165136.

[117] R. Peach, On the Existence of Surface Acoustic Waves on Piezoelectric Substrates, *IEEE Trans. Ultrason., Ferroelec., Freq. Control* **48** (2001), 1308–1320.

[118] Y. Pennec, B. Djafari-Rouhani, E. H. El Boudouti, C. Li, Y. El Hassouani, J.-O. Vasseur, N. Papanikolaou, S. Benchabane, V. Laude and A. Martínez, Simultaneous existence of phononic and photonic band gaps in periodic crystal slabs, *Opt. Express* **18** (2010), 14301–14310.

[119] Y. Pennec, B. Djafari-Rouhani, H. Larabi, J. O. Vasseur and A. C. Hladky-Hennion, Low-frequency gaps in a phononic crystal constituted of cylindrical dots deposited on a thin homogeneous plate, *Phys. Rev. B* **78** (2008), 104105.

[120] Y. Pennec, B. Djafari-Rouhani, J. O. Vasseur, H. Larabi, A. Khelif, A. Choujaa, S. Benchabane and V. Laude, Acoustic channel drop tunneling in a phononic crystal, *Appl. Phys. Lett.* **87** (2005), 261912.

[121] Y. Pennec, V. Laude, N. Papanikolaou, B. Djafari-Rouhani, M. Oudich, S. El Jallal, J. C. Beugnot, J. M. Escalante and A. Martínez, Modeling light-sound interaction in nanoscale cavities and waveguides, *Nanophotonics* **3** (2014), 413–440.

[122] J. Pierre, O. Boyko, L. Belliard, J. O. Vasseur and B. Bonello, Negative refraction of zero order flexural Lamb waves through a two-dimensional phononic crystal, *Appl. Phys. Lett.* **97** (2010), 121919.

[123] D. M. Profunser, E. Muramoto, O. Matsuda, O. B. Wright and U. Lang, Dynamic visualization of surface acoustic waves on a two-dimensional phononic crystal, *Phys. Rev. B* **80** (2009), 014301.

[124] I. E. Psarobas, Viscoelastic response of sonic band-gap materials, *Phys. Rev. B* **64** (2001), 012303.

[125] I. E. Psarobas, A. Modinos, R. Sainidou and N. Stefanou, Acoustic properties of colloidal crystals, *Phys. Rev. B* **65** (2002), 064307.

[126] I. E. Psarobas, N. Stefanou and A. Modinos, Phononic crystals with planar defects, *Phys. Rev. B* **62** (2000), 5536–5540.

[127] I. E. Psarobas, N. Stefanou and A. Modinos, Scattering of elastic waves by periodic arrays of spherical bodies, *Phys. Rev. B* **62** (2000), 278–291.

[128] C. M. Reinke, M. F. Su, R. H. Olsson III and I. El-Kady, Realization of optimal bandgaps in solid-solid, solid-air, and hybrid solid-air-solid phononic crystal slabs, *Appl. Phys. Lett.* **98** (2011), 061912.

[129] V. Romero-García, J. V. Sánchez-Pérez, S. Castiñeira-Ibáñez and L. M. Garcia-Raffi, Evidences of evanescent Bloch waves in phononic crystals, *Appl. Phys. Lett.* **96** (2010), 124102.

[130] V. Romero-García, J. V. Sánchez-Pérez and L. M. Garcia-Raffi, Propagating and evanescent properties of double-point defects in sonic crystals, *New J. Phys.* **12** (2010), 083024.

[131] D. Royer and E. Dieulesaint, *Elastic waves in solids*, Wiley, New York, 1999.

[132] C. Rubio, D. Caballero, J. V. Sánchez-Pérez, R. Martínez-Sala, J. Sánchez-Dehesa, F. Meseguer and F. Cervera, The existence of full gaps and deaf bands in two-dimensional sonic crystals, *J. of Lightwave Technology* **17** (1999), 2202–2207.

[133] S. Sadat-Saleh, S. Benchabane, F. I. Baida, M. P. Bernal and V. Laude, Tailoring simultaneous photonic and phononic band gaps, *J. Appl. Phys.* **106** (2009), 074912.

[134] A. H. Safavi-Naeini, J. T. Hill, S. Meenehan, J. Chan, S. Gröblacher and O. Painter, Two-Dimensional Phononic-Photonic Band Gap Optomechanical Crystal Cavity, *Phys. Rev. Lett.* **112** (2014), 153603.

[135] R. Sainidou, N. Stefanou and A. Modinos, Formation of absolute frequency gaps in three-dimensional solid phononic crystals, *Phys. Rev. B* **66** (2002), 212301.

[136] R. Sainidou, N. Stefanou, I. E. Psarobas and A. Modinos, A layer-multiple-scattering method for phononic crystals and heterostructures of such, *Computer Physics Communications* **166** (2005), 197–240.

[137] J. V. Sánchez-Pérez, D. Caballero, R. Martínez-Sala, C. Rubio, J. Sánchez-Dehesa, F. Meseguer, J. Llinares and F. Galves, Sound attenuation by a two-dimensional array of rigid cylinders, *Phys. Rev. Lett.* **80** (1998), 5325–5228.

[138] L. Sanchis, F. Cervera, J. Sánchez-Dehesa, J. V. Sánchez-Pérez, C. Rubio and R. Martínez-Sala, Reflectance properties of two-dimensional sonic band-gap crystals, *J. Acoust. Soc. Am.* **109** (2001), 2598.

[139] J. Shi, S.-C. S. Lin and T. J. Huang, Wide-band acoustic collimating by phononic crystal composites, *Appl. Phys. Lett.* **92** (2008), 111901.

[140] C. Spielmann, R. Szipöcs, A. Stingl and F. Krausz, Tunneling of Optical Pulses through Photonic Band Gaps, *Phys. Rev. Lett.* **73** (1994), 2308–2311.

[141] A. M. Steinberg, P. G. Kwiat and R. Y. Chiao, Measurement of the single-photon tunneling time, *Phys. Rev. Lett.* **71** (1993), 708.

[142] P. St.J. Russell, Photonic crystal fibers, *Science* **299** (2003), 358–362.

[143] M. F. Su, R. H. Olsson III, Z. C. Leseman and I. El-Kady, Realization of a phononic crystal operating at gigahertz frequencies, *Appl. Phys. Lett.* **96** (2010), 053111.

[144] Y. Sugawara, O. B. Wright, O. Matsuda, M. Takigahira, Y. Tanaka, S. Tamura and V. E. Gusev, Watching ripples on crystals, *Phys. Rev. Lett.* **88** (2002), 185504.

[145] A. Sukhovich, B. Merheb, K. Muralidharan, J.-O. Vasseur, Y. Pennec, P. A. Deymier and J. H. Page, Experimental and theoretical evidence for subwavelength imaging in phononic crystals, *Phys. Rev. Lett.* **102** (2009), 154301.
[146] A. Sukhovich, L. Jing and J. H. Page, Negative refraction and focusing of ultrasound in two-dimensional phononic crystals, *Phys. Rev. B* **77** (2008), 014301.
[147] J.-H. Sun and T.-T. Wu, Propagation of surface acoustic waves through sharply bent two-dimensional phononic crystal waveguides using a finite-difference time-domain method, *Phys. Rev. B* **74** (2006), 174305.
[148] J.-H. Sun and T.-T. Wu, Propagation of acoustic waves in phononic-crystal plates and waveguides using a finite-difference time-domain method, *Phys. Rev. B* **76** (2007), 104304.
[149] Y. Tanaka and S. I. Tamura, Surface acoustic waves in two-dimensional periodic elastic structures, *Phys. Rev. B* **58** (1998), 7958.
[150] Y. Tanaka and S. I. Tamura, Acoustic stop bands of surface and bulk modes in two-dimensional phononic lattices consisting of aluminum and a polymer, *Phys. Rev. B* **60** (1999), 13294.
[151] Y. Tanaka, Y. Tomoyasu and S.-i. Tamura, Band structure of acoustic waves in phononic lattices: Two-dimensional composites with large acoustic mismatch, *Phys. Rev. B* **62** (2000), 7387.
[152] R. N. Thurston, Direct computation of the group velocity, *IEEE Trans. Sonics and Ultrason.* **24** (1977), 109.
[153] D. Torrent, A. Håkansson, F. Cervera and J. Sánchez-Dehesa, Homogenization of two-dimensional clusters of rigid rods in air, *Phys. Rev. Lett.* **96** (2006), 204302.
[154] F. Van Der Biest, A. Sukhovich, A. Tourin, J. H. Page, B. A. Van Tiggelen, Z. Liu and M. Fink, Resonant tunneling of acoustic waves through a double barrier consisting of two phononic crystals, *Europhys. Lett.* **71** (2005), 63.
[155] J. O. Vasseur, P. A. Deymier, B. Chenni, B. Djafari-Rouhani, L. Dobrzynski and D. Prevost, Experimental and theoretical evidence for the existence of absolute acoustic band gaps in two-dimensional solid phononic crystals, *Phys. Rev. Lett.* **86** (2001), 3012–3015.
[156] J. O. Vasseur, P. A. Deymier, B. Djafari-Rouhani, Y. Pennec and A.-C. Hladky-Hennion, Absolute forbidden bands and waveguiding in two-dimensional phononic crystal plates, *Phys. Rev. B* **77** (2008), 085415.
[157] J. O. Vasseur, A.-C. Hladky-Hennion, B. Djafari-Rouhani, F. Duval, B. Dubus, Y. Pennec and P. A. Deymier, Waveguiding in two-dimensional piezoelectric phononic crystal plates, *J. Appl. Phys.* **101** (2007), 114904.
[158] G. Wang, X. Wen, J. Wen, L. Shao and Y. Liu, Two dimensional locally resonant phononic crystals with binary structures, *Phys. Rev. Lett.* **93** (2004), 154302.
[159] N. Wang, F.-L. Hsiao, J. M. Tsai, M. Palaniapan, D.-L. Kwong and C. Lee, Numerical and experimental study on silicon microresonators based on phononic crystal slabs with reduced central-hole radii, *Journal of Micromechanics and Microengineering* **23** (2013), 065030.
[160] N. Wang, J. M. Tsai, F.-L. Hsiao, B. W. Soon, D.-L. Kwong, M. Palaniapan and C. Lee, Experimental Investigation of a cavity-mode resonator using a micromachined two-dimensional silicon phononic crystal in a square lattice, *IEEE Electron Device Letters* **32** (2011), 821–823.
[161] Y.-F. Wang, V. Laude and Y.-S. Wang, Coupling of evanescent and propagating guided modes in locally resonant phononic crystals, *J. Phys. D: Appl. Phys.* **47** (2014), 475502.
[162] M. Wilm, S. Ballandras, V. Laude and T. Pastureaud, A full 3-D plane-wave-expansion model for piezocomposite structures, *J. Acoust. Soc. Am.* **112** (2002), 943–952.
[163] M. Wilm, A. Khelif, S. Ballandras, V. Laude and B. Djafari-Rouhani, Out-of-plane propagation of elastic waves in two-dimensional phononic band-gap materials, *Phys. Rev. E* **67** (2003), 065602.

[164] H. G. Winful, Tunneling time, the Hartman effect, and superluminality: A proposed resolution of an old paradox, *Physics Reports* **436** (2006), 1–69.

[165] C. G. Winternheimer and A. K. McCurdy, Phonon focusing and phonon conduction in orthorhombic and tetragonal crystals in the boundary-scattering regime. II, *Phys. Rev. B* **18** (1978), 6576.

[166] T.-T. Wu, Z.-G. Huang and S. Lin, Surface and bulk acoustic waves in two-dimensional phononic crystal consisting of materials with general anisotropy, *Phys. Rev. B* **69** (2004), 094301.

[167] T.-T. Wu, Z.-G. Huang, T.-C. Tsai and T.-C. Wu, Evidence of complete band gap and resonances in a plate with periodic stubbed surface, *Appl. Phys. Lett.* **93** (2008), 111902.

[168] T.-T. Wu, L.-C. Wu and Z.-G. Huang, Frequency band-gap measurement of two-dimensional air/silicon phononic crystals using layered slanted finger interdigital transducers, *J. Appl. Phys.* **97** (2005), 094916.

[169] T.-T. Wu, W.-S. Wang, J.-H. Sun, J.-C. Hsu and Y.-Y. Chen, Utilization of phononic-crystal reflective gratings in a layered surface acoustic wave device, *Appl. Phys. Lett.* **94** (2009), 101913.

[170] T.-C. Wu, T.-T. Wu and J.-C. Hsu, Waveguiding and frequency selection of Lamb waves in a plate with a periodic stubbed surface, *Phys. Rev. B* **79** (2009), 104306.

[171] Z.-Z. Yan and Y.-S. Wang, Wavelet-based method for calculating elastic band gaps of two-dimensional phononic crystals, *Phys. Rev. B* **74** (2006), 224303.

[172] S. Yang, J. H. Page, Z. Liu, M. L. Cowan, C. T. Chan and P. Sheng, Ultrasound Tunneling through 3D Phononic Crystals, *Phys. Rev. Lett.* **88** (2002), 104301.

[173] S. Yang, J. H. Page, Z. Liu, M. L. Cowan, C. T. Chan and P. Sheng, Focusing of sound in a 3D phononic crystal, *Phys. Rev. Lett.* **93** (2004), 024301.

[174] A. Yariv, Y. Xu, R. K. Lee and A. Scherer, Coupled-resonator optical waveguide: a proposal and analysis, *Opt. Lett.* **24** (1999), 711.

[175] D. Yudistira, Y. Pennec, B. Djafari Rouhani, S. Dupont and V. Laude, Non-radiative complete surface acoustic wave bandgap for finite-depth holey phononic crystal in lithium niobate, *Appl. Phys. Lett.* **100** (2012), 061912.

[176] X. Zhang and Z. Liu, Negative refraction of acoustic waves in two-dimensional phononic crystals, *Appl. Phys. Lett.* **85** (2004), 341–343.

[177] X. Zhang, T. Jackson, E. Lafond, P. Deymier and J. O. Vasseur, Evidence of surface acoustic wave band gaps in the phononic crystals created on thin plates, *Appl. Phys. Lett.* **88** (2006), 041911.

[178] J. Zhao, B. Bonello, R. Marchal and O. Boyko, Beam path and focusing of flexural Lamb waves within phononic crystal-based acoustic lenses, *New J. Phys.* **16** (2014), 063031.

[179] J. Zhao, R. Marchal, B. Bonello and O. Boyko, Efficient focalization of antisymmetric Lamb waves in gradient-index phononic crystal plates, *Appl. Phys. Lett.* **101** (2012), 261905.

Index

De Gruyter Studies in Mathematical Physics

Volume 27
Ivan A. Lukovsky, Peter V. Malyshev
Nonlinear Dynamics, 2015
ISBN 978-3-11-031655-1, e-ISBN 978-3-11-031657-5 (PDF),
e-ISBN (EPUB) 978-3-11-038973-9, Set-ISBN 978-3-11-031658-2

Volume 26
Vincent Laude
Phononic Crystals, 2015
ISBN 978-3-11-030265-3, e-ISBN 978-3-11-030266-0 (PDF),
e-ISBN (EPUB) 978-3-11-038791-9, Set-ISBN 978-3-11-030267-7

Volume 25
Sergey D. Algazin, Igor A. Kijko
Aeroelastic Vibrations and Stability of Plates and Shells, 2015
ISBN 978-3-11-033836-2, e-ISBN 978-3-11-033837-9 (PDF),
e-ISBN (EPUB) 978-3-11-038945-6, Set-ISBN 978-3-11-040491-3

Volume 24
Igor Olegovich Cherednikov, Tom Mertens, Frederik F. van der Veken
Wilson Lines in Quantum Field Theory, 2014
ISBN 978-3-11-030910-2, e-ISBN 978-3-11-030921-8 (PDF),
e-ISBN (EPUB) 978-3-11-038293-8, Set-ISBN 978-3-11-030922-5

Volume 23
Boris A. Arbuzov
Non-perturbative Effective Interactions in the Standard Model, 2014
ISBN 978-3-11-030292-9, e-ISBN 978-3-11-030521-0 (PDF),
e-ISBN (EPUB) 978-3-11-038805-3, Set-ISBN 978-3-11-030522-7

Volume 22
Sergei M. Kopeikin (Ed.)
Frontiers in Relativistic Celestial mechanics, Volume 2
Applications and Experiments, 2014
ISBN 978-3-11-034545-2, e-ISBN 978-3-11-034566-7 (PDF),
e-ISBN (EPUB) 978-3-11-037953-2, Set-ISBN 978-3-11-034567-4

Bei Fragen zur Produktsicherheit wenden Sie sich bitte an:
If you have any questions regarding product safety,
please contact:

Walter de Gruyter GmbH
Genthiner Straße 13
10785 Berlin
productsafety@degruyterbrill.com